THE PAPERS OF THOMAS A. EDISON

FINANCIAL CONTRIBUTORS

Public Foundations

National Science Foundation
National Endowment for the Humanities

Private Foundations

The Alfred P. Sloan Foundation
Charles Edison Fund
The Hyde and Watson Foundation
Geraldine R. Dodge Foundation

Private Corporations and Individuals

Alabama Power Company
Amerada Hess Corporation
AT&T
Atlantic Electric
Association of Edison
  Illuminating Companies, Inc.
Battelle Memorial Institute
The Boston Edison Foundation
Cabot Corporation Foundation,
  Inc.
Carolina Power & Light
  Company
Consolidated Edison Company
  of New York, Inc.
Consumers Power Company
Corning Glass Works
  Foundation
Duke Power Company
Exxon Corporation
Florida Power & Light
  Company
General Electric Foundation
Gould Inc. Foundation
Gulf States Utilities Company
Idaho Power Company
International Brotherhood of
  Electrical Workers
Iowa Power and Light
  Company
Mr. and Mrs. Stanley H. Katz

Matsushita Electric Industrial
  Co., Ltd.
McGraw-Edison Company
Middle South Services, Inc.
Minnesota Power
New Jersey Bell
New York State Electric & Gas
  Corporation
North American Philips
  Corporation
Philadelphia Electric Company
Philips International B.V.
Public Service Electric and Gas
  Company
RCA Corporation
Robert Bosch GmbH
Rochester Gas and Electric
  Corporation
San Diego Gas & Electric
Savannah Electric and Power
  Company
Schering-Plough Foundation
Texas Utilities Company
Thomas & Betts Corporation
Thomson Grand Public
Transamerica Delaval Inc.
Westinghouse Educational
  Foundation
Wisconsin Public Service
  Corporation

THE PAPERS OF THOMAS A. EDISON

Volume 1

Thomas A. Edison at the age of sixteen.

Volume 1

The Papers of Thomas A. Edison

THE MAKING OF AN INVENTOR
February 1847–June 1873

EDITORS
Reese V. Jenkins
Leonard S. Reich
Paul B. Israel
Toby Appel
Andrew J. Butrica

Robert A. Rosenberg
Keith A. Nier
Melodie Andrews
Thomas E. Jeffrey

SPONSORS
Rutgers, The State University of New Jersey
National Park Service, Edison National Historic Site
New Jersey Historical Commission
Smithsonian Institution

THE JOHNS HOPKINS UNIVERSITY PRESS
BALTIMORE AND LONDON

Thomas A. Edison Papers, a project of Rutgers, The State University, is endorsed by the National Historical Publications and Records Commission, which has also provided generous financial assistance in the publication of this volume.

The Johns Hopkins University Press
701 West 40th Street   Baltimore, Maryland 21211
The Johns Hopkins Press Ltd., London

The paper used in this publication meets the minimum requirements of the American National Standard for Information Sciences—Permanence of Paper for Printed Library Materials, ANSI Z39.48-1984.

Library of Congress Cataloging-in-Publication Data
Edison, Thomas A. (Thomas Alva), 1847–1931.
  [Works. 1989]
  The papers of Thomas A. Edison / editors, Reese V. Jenkins . . . [et al.].
    p.   cm.
  Bibliography: p.
  Includes index.
  ISBN 0-8018-3100-8
  1. Edison, Thomas A. (Thomas Alva), 1847–1931.   2. Edison,
Thomas A. (Thomas Alva), 1847–1931—Archives.   3. Inventors—
United States—Biography.   I. Jenkins, Reese.   II. Title.
TK140.E3A2   1989
600—dc19   88-9017
                                                    CIP

Edison signature on case used with permission of McGraw-Edison Company.

Contents

Calendar of Documents

Preface

Publication of *The Papers of Thomas A. Edison* represents the effort of a team of historical editors to bring to the public and the scholarly community an intimate view of the personal, entrepreneurial, and creative activities of America's greatest inventor. Out of a documentary record comprising over three and a half million pages, the edition will provide the pieces to the puzzle that is Edison's place in history. These documents reveal high moments of inventive genius and times of intricate business transaction, encounters with a host of powerful leaders in their esteemed places and with day laborers in dirty machine shops, and the consequences of Edison's many successes and his equally many failures. The neglect of generations of scholars has left a mist of popular myth surrounding the historical Edison. This edition provides for the first time the resources that can significantly enhance public knowledge and encourage scholarly research into the technical creativity, innovation, and entrepreneurship that propelled the United States into a position of industrial leadership. Such research is vital for understanding American history and for evaluating critical technical and economic issues today and in the future.

In a sixty-year inventive career from the end of the Civil War to the onset of the Great Depression, Edison obtained 1,093 patents from the United States Patent Office, by far the largest number ever issued to a single individual. At the dawning of the era of mass production, he and his "invention factory" set an enduring record for output of patents. To this day, people immediately associate his name with the electric light—an artifact that has come to symbolize creative insight—and his association with the phonograph and motion picture is nearly as strong.

Less well known is the diversity of Edison's other significant achievements. He contributed notably to telegraph, telephone, mimeograph, electric power systems, electric traction, ore concentration, electric batteries, chemical production, cement manufacture, and home appliances. In an era of rapid industrialization and growth in scale of enterprise, Edison enlisted a group of able technical associates and demonstrated to financiers and corporate executives the efficacy of team industrial research. He pursued many of his inventions from conception through development and into the marketplace, establishing large companies at home and abroad for their exploitation. He was a storyteller out of the American Midwest, and the press exploited this eager raconteur for an adoring audience that accepted him as a hero of mythic proportions.

To historians, social scientists, scientists, engineers, and business people, Edison's cascade of technical achivements raises critical questions. Was he a genius? How could any one person achieve so much? How original were his ideas? Did he steal ideas from his laboratory associates and others? Was he just lucky—"born at the right time"—a time when many technical and business opportunities were uniquely ripe for the plucking? Do his designs reveal his own distinctive technical marks? Do they reflect an indigenous style of his era? What influence did his European associates have on his work in the laboratory? How effective was he as a business person? Are there lessons that educators and policymakers can learn that would help foster future Edisons?

The Edison Papers project at Rutgers University was established to help answer such questions through publication of selective microfilm and book editions. The indexed microfilm edition, for specialist researchers, will eventually include about four hundred thousand documents, offering access to a substantial body of the most important documentary evidence while freeing the researcher from the morass of peripheral material that has impeded scholars in the past.

The book edition is intended for a broader audience, including not only specialist researchers but others who seek documentary evidence: general historians preparing textbooks and related materials; scientists, engineers, and business persons seeking to understand the details of Edison's technologies and entrepreneurship; social scientists and administrators concerned with the context and processes of creative activity; and students of all fields. To meet the requirements of such an audience, the book edition reflects the diversity and evolution of Edison's expansive career in an annotated selection of documents that includes the inventor's personal and business correspondence, patent records, business contracts, and related materials. However, Edison's fundamental historical significance lies in his technical creativity, and this edition necessarily focuses upon those achievements and their context—experimental notes and drawings, laboratory and production models, and other technical material.

Such a focus presents at once wonderful opportunities and special burdens: translating the visual and technical languages involved, comprehending the problems and issues Edison was addressing, and understanding the work of his peers in America and Europe. For the editors, issues of interpretation and annotation also are a particular concern. The canons of traditional documentary editing discourage editorial interpretation. Although this poses little problem for editors of a comprehensive documentary edition in a field already well researched, it presents a quandary for editors of a highly selective edition of documents in a field as little explored as the history of technology. Selection itself presupposes interpretation. Moreover, Edison's work in recondite, abandoned technologies, as well as the paucity of contextual scholarship, necessitate further interpretation. This explanatory material appears in chapter introductions, headnotes, endnotes, and illustrations, all of which are intended to help the reader understand the documents' content and context.

The edition is designed so that readers may select the level at which they wish to use it. Chapter introductions and headnotes allow nontechnical readers to follow at a general level Edison's activities and to select the documents they wish to pursue in detail. Endnotes to both headnotes and documents identify persons and issues, elucidate technical points, and direct the reader to related material. The initial statement in an explanatory endnote is general; subsequent information is more detailed and technical. The editors hope that everyone interested in Thomas Edison will find illumination in this record of his life, work, and world.

The Thomas A. Edison Papers project has benefited from substantial assistance, ranging from the sponsorship and financial support of major institutions to research assistance and documentary information from scholars, archivists, collectors, administrators, and students. From its inception in the mid-1970s, the project has been a team effort. Leaders from the New Jersey Historical Commission, including John T. Cunningham, Bernard Bush, Paul A. Stellhorn, Richard Waldron, and Ronald Grele, initiated discussions with Arthur Reed Abel, William Binnewies, Lynn Wightman, and Elizabeth Albro of the National Park Service's Edison National Historic Site. The two groups soon attracted the participation of Brooke Hindle and Bernard Finn of the Smithsonian Institution. With financial aid from the Edison Electric Institute, they then engaged James Brittain, a distinguished historian of technology at the Georgia Institute of Technology, to prepare a feasibility study. After considering Brittain's recommendations, the representatives of the National Park Service, the New Jersey Historical Commission, and the Smithsonian Institution invited Rutgers, The State University of New Jersey to join in cosponsoring the Thomas A. Edison Papers project. President Edward Bloustein, Paul Pearson, Richard P. McCormick, James Kirby Martin, and Tilden Edelstein of Rutgers responded with enthusiasm, and in 1978 the project began its work with headquarters at Rutgers University in New Brunswick and a second office at the Edison National Historic Site in West Orange.

Since the early days of the project, a number of people and organizations have assisted with obtaining financial support. In the private sector, William C. Hittinger of RCA Corporation chaired the Edison Corporate Associates and recruited distinguished executives as members of this fund-raising group. His outstanding leadership elicited similar efforts from Morris Tanenbaum of AT&T, Arthur M. Bueche and Roland W. Schmitt of the General Electric Company, Robert I. Smith and Harold W. Sonn of Public Service Electric and Gas Company, Paul J. Christiansen of the Charles Edison Fund, Cees Bruynes of North American Philips Corporation, Paul Lego and Philip F. Dietz of Westinghouse Electric Corporation, and President Bloustein of Rutgers University. John Venable of the Charles Edison Fund, Samuel Convissor of RCA Corporatoin, Edmund Tucker of the General Electric Foundation, George Wise of the General Electric Company, and Robert T. Cavanagh of North American Philips Corporation also devoted special efforts to fund raising. Bruce Newman, Ronald Miller, Denise Lyons, and Robert Lusardi of the Rutgers University Foundation have served as genuine partners in the financial program of the Edison Papers.

Private and public foundations have awarded grants that are vital to the project. Two long-term grants from the Alfred P. Sloan Foundation and awards from the Geraldine R. Dodge Foundation and the Hyde and Watson Foundation are notable. The preparation of this volume was made possible in part by grants from the Program for Editions of the National Endowment for the Humanities, an independent agency, and by grants from the Program in History and Philosophy of Science and the Engineering Division of the National Science Foundation (SES-8109665, SES-8213220, DIR-8521524). Ronald Overmann and Nam Suh of the National Science Foundation, and Katherine Fuller and George Farr of the National Endowment for the Humanities, have provided valuable counsel. Any opinions, findings, conclusions, or recommendations expressed in this publication, however, are those of the editors and do not necessarily reflect the views of the National Science Foundation or any other financial contributor.

The editors of this project have benefited from the advice and assistance of many historical editors. Special appreciation is due Edward C. Carter II and Darwin H. Stapleton of the Papers of Benjamin Henry Latrobe; Nathan Reingold, Arthur P. Molella, Marc Rothenberg, and Kathleen Waldenfels Dorman of the Papers of Joseph Henry; Arthur Link and David Hirst of the Woodrow Wilson Papers; Louis Galambos of the Dwight D. Eisenhower Papers; and Carl Prince of the William Livingston Papers. Frank Burke, Rogers Bruns, Mary Giunta, Richard Sheldon, and George Vogt of the National Historical Publications and Records Commission have encouraged and guided the editors since the project's inception.

Many other scholars have given advice, shared information, discussed issues, or provided special assistance. The editors are especially grateful to Brooke Hindle, Bernard Finn, James Brittain, Thomas P. Hughes, Melvin Kranzberg, Hugh Aitken, Merritt Roe Smith, Alfred D. Chandler, Jr., Robert Schofield, Edwin T. Layton, Robert Friedel, Michael Mahoney, Brian Bowers, and Kenneth Rowe.

The project is indebted to many institutions and their staffs for the provision of documents, photographs, photocopies, and research assistance. Staff members of Greenfield Village and the Henry Ford Museum have extended themselves, especially John Bowditch, Tim Binkley, Steven K. Hamp, Jeanine Head, Mary Beth Kreiner, Larry McCans, William S. Pretzer, and Cynthia Read-Miller. Similar help came from Joyce Bedi of the IEEE History Center, Bernard Finn of the Smithsonian Institution, Florence Bartechevski and Robert W. Lovett of the Baker Library at Harvard University, Kenneth A. Lohf of the Butler Library at Columbia University, Millie Daghli of the Corporate Research Archives at AT&T, Mrs. E. D. P. Symons of the Institute of Electrical Engineers in London, Margaret Jasko of Western Union Telegraph Company, Charles F. Cummings of the Newark Public Library, Steven Catlett of the American Philosophical Society, Don C. Skemer of the New Jersey Historical Commission, Deborah S. Gardner of the New York Stock Exchange Archives, and Mrs. C. I. Constantinides of the Post Office Archives in London. The project is also indebted to several persons in the Federal Archives and Record Service: Marjorie Ciarlante and John Butler in Washington, D.C.; James K. Owens in Waltham, Mass.; and Joel Buchwald and Donna Cabarle in Bayonne, N.J. Others who have provided resources or special assistance include Charles Hummel, Patty Dahm Pascoe, Charles F. Kempf, Dorothy C. Serrell, Frank Andrews, Theodore Edison, and John Edison Sloane.

Many people associated with the National Park Service have contributed significantly to this project. Especially notable are Herbert S. Cables, Jr., and Edward Dallop of the Boston Regional Office and Fahy Whittaker, Mary Bowling, Edward Pershey, Roy W. Weaver, Lynn Wightman, William Binnewies, Ray Kremer, Arthur Reed Abel, and Elizabeth Albro of the Edison National Historic Site.

The editors are especially appreciative of friends at Rutgers University. Encouragement and assistance have come from colleagues in the History Department, especially from Richard P. McCormick, James W. Reed, Susan R. Schrepfer, Philip Pauly, Gerald N. Grob, Seth Scheiner, Michael P. Adas, John R. Gillis, and Richard L. McCormick. Notable administrative support has come from Edward Bloustein, T. Alexander Pond, Tilden G. Edelstein, Nathaniel Pallone, Jose Steinboch, David A. Cayer, Kenneth Wheeler, Paul L. Leath, Jean Parrish, Peter D. Klein, John Salapatas, Robert F. Pack, David R. Scott, Andrew B. Rudczynski, Albert Hanna, David A. Rumbo, Donna Estler, Muriel Wilson, Helen Gertler, Joseph Harrigan, and Ruth Scott. Major assistance has come from the staff of the Rutgers University Libraries, especially Ruth Simmons, Francis Johns, Ronald L. Becker, and Clark L. Beck, Jr.

Several staff members, interns, and students not named on the title page contributed in various ways to this volume. These include Susan Schultz, W. Bernard Carlson, Andre Millard, Mary Ann Hellrigel, Douglas G. Tarr, David W. Hutchings, Ann Nigro, David Fowler, Harry Goldsmith, Jacquelyn Miller, Matthew McCright, Dana Norvila, David Rhees, and Alain Canuel. Significant photographic work was done by James Anness, Patricia Matteson, Richard W. Mitchell, Rudy Ruzicska, and William Winter.

Any book owes a debt to the press that publishes it. In the case of the Edison Papers, the complexity of the material magnifies the debt manyfold. The staff of the Johns Hopkins University Press has worked heroically throughout the production of this volume, and the editors particularly wish to thank Penny Moudrianakis and Barbara Lamb for their painstaking editorial guidance; Martha Farlow and Jim Johnston for the design and production of a very difficult volume; Henry Y. K. Tom, George F. Thompson, and Denise Dankert for their counsel; and, finally, Director Jack Goellner for his unstinting support.

The editors conclude with a tribute to two very special people: our assistant to the director, Helen Endick, and our secretary, Grace Kurkowski. The talent, devotion, and good humor they bring to their administrative, secretarial, and editorial tasks are truly extraordinary and vital to the project.

Introduction

Why should humanists, scientists, engineers, policymakers,
students, and lay people want to examine or study Edison's
personal and business correspondence, legal agreements, laboratory notebook entries, and patent materials? Since he has
been the subject of dozens of biographies, what more need be
said about this inventive genius? Every American school child
knows about him. He is among the small pantheon of national
heroes that includes Washington, Jefferson, and Lincoln.
While generations of historians have studied the writings of
such august political figures and written profusely about them,
only a small number of scholars have examined the creative
thinking of one of the most heralded inventors in history[1] or
the activities of less prominent technical people who nevertheless are regarded as central to the emergence of the United
States as a world industrial leader in the twentieth century.
Few have passed through the doors of the Edison archives;
fewer have probed its rich resources. Consequently, most
people continue to harbor the simplistic views of Edison and
invention that are perpetuated in popular biographies.[2] Yet
this is a time when an informed understanding of technical
creativity and innovation is vital for technical and business
leaders, policymakers, and the electorate, whose decisions
will shape the world of the twenty-first century.

Unfortunately, misleading views of Edison and of technical
creativity persist. How many think of the young Tom Edison
as an addled youngster and a country bumpkin? How common
is the view of him as an unsophisticated businessman who
went to the big city not knowing how to cash a check or drive
a good bargain with the sharks of Wall Street; as the lone in-

ventor or wizard who single-handedly made miraculous inventions; as an uneducated tinkerer who used "cut-and-try" methods, mindlessly tested everything, bumbled onto technical breakthroughs, and, by a streak of luck, transformed the world; and as the white-haired, grandfatherly old gent who was at once miraculously prolific, personally affable, and a symbol of benign technology? How ingrained is the view of nineteenth-century invention as the product of the inspired insight of the isolated tinkerer?

In this first volume of Edison's papers, his own early documents challenge traditional interpretations of his epochal position in world history and suggest new ways of understanding technical creativity and innovation. These documents address such important issues as (1) the character and extent of Edison's education; (2) his relationship to the national and international technical and scientific communities; (3) his entrepreneurial instincts and business acumen; (4) the source of the problems he and his contemporaries sought to solve; (5) his inventive methods and style; (6) the role of his early technical and business experience in giving direction to his inventive career; and (7) the relationship of his work to that of other inventors. Let us briefly consider what the documents of this volume suggest about these issues.

Edison, like most of his American contemporaries, received limited formal education, but through his reading and experience he acquired a broad and detailed understanding of the society in which he lived. He attended two schools, where he learned the rudiments of reading, writing, and science. He also received intellectual stimulation from the instruction given him by his mother, a former teacher, and from the ideas and small home library of his politically radical father. By age fifteen he was frequenting the library at the Detroit Young Men's Society, and writing and printing his own newspaper, the *Weekly Herald.* As an itinerant telegrapher in his late teens, he systematically read newspapers, frequented public libraries, developed a beautiful handwriting, and gained a reputation as an excellent press-wire receiver for the Western Union Telegraph Company. These achievements not only attest to his coding and verbal skills and discount claims that he was dyslexic but also confirm his recollections that he commanded the details of the daily news of politics and business.

Significantly, the hearing-impaired young inventor learned to exploit the work of varied technical communities. He knew the machine shops, read the telegraph journals and manuals, and used libraries to find scientific and technical ideas and data. Before he was twenty, he knew of the technical works of Michael Faraday and Robert Sabine as well as the *Proceedings* of the Royal Society of London. His 1868 correspondence with John Van Duzer reveals his detailed knowledge of European telegraph design. Far from being a lone inventor, this shrewd young man joined an existing community of telegraph inventors and learned to exploit publications of the national and international technological and scientific communities. Early in his career he demonstrated that technological creativity and innovation are social and intellectual processes.

Edison received his technical apprenticeship in telegraph offices and machine shops in the Midwest and Boston. During his midteens he worked at the telegraph table in a jewelry store in his Michigan hometown. Like his contemporaries, he manipulated the common components of telegraph apparatus and clockwork: ratchet wheels, escapements, adjusting springs, and gearwork. In contrast to clocks, however, telegraph apparatus exploited a new element, the electromagnet. Armed with this new switching and power source, Edison and his contemporaries explored a wide variety of new applications to electrical communications. In the office he often found components of different "generations" of telegraph systems juxtaposed, and he debated the advantages of different systems with colleagues. He also extended his circle of acquaintances to others of the operator "fraternity" in distant telegraph offices. During his late teens in Louisville and Cincinnati, he not only mastered press-wire telegraphy but investigated the creative technical achievements of others. Working at first from standard telegraph manuals and periodicals, he varied the standard designs for relay, repeater, and duplex circuits. After arriving in Boston at age twenty-one, he not only worked as a telegraph operator but also published articles in the nation's leading telegraph journal. In his off-hours he pursued his own projects at the machine shop of Charles Williams, Jr., an inventors' womb that also nurtured the embryonic ideas of Moses Farmer and, later, Alexander Graham Bell.

Edison, like other successful American inventors, learned to work in close association with technical and business people. Perhaps his hearing deficiency encouraged him to depend upon others; and perhaps his experience in machine shops taught him to delegate the making of models to expert precision machinists. From the time of his arrival in Boston in 1868, he joined in partnership with others. He obtained financial support from businessmen such as E. Baker Welch and technical aid from colleagues such as Frank Hanaford and George Anders. He developed a mentor relationship with the two successive editors of the *Telegrapher,* Frank Pope and James Ashley. In 1869 he joined in partnership with them in New York. A year later they were chagrined to find that, after their help and encouragement, he had so quickly collaborated with others in the telegraph industry—for instance, with Marshall Lefferts of the Gold and Stock Telegraph Company. Edison's relationship with Lefferts fostered his growing financial and legal sophistication. Increasingly he exhibited an entrepreneurial drive, entering into a complex array of agreements and enterprises that would eventually lead him into a series of lawsuits. By 1873 he had attracted the interest of key leaders in the telegraph industry as they grew increasingly sensitive to the significance of invention and technical innovation as a business strategy.

The leaders of the telegraph industry were keenly aware of the needs of their markets and operations. Accordingly, they interacted with Edison and other telegraph inventors, identifying what improvements in telegraphy could reduce costs and might lead to patents that would either deter competitors or serve as a means of entry into a market dominated by others. At first these business leaders set the technical and economic agenda for Edison and other inventors. Increasingly, however, Edison learned the subtleties of the economics of business and developed technical goals and solutions that reflected the influence of the market.

For example, during the early 1870s improvements that would increase message density on a single line were of general interest to the leaders of the American telegraph industry. However, because the leaders of each sector of the industry perceived the critical cost factors differently, they pursued different technical approaches. With a network of lines across the nation, Western Union bore the substantial capital cost of building and maintaining telegraph wires. Thus, in 1872 it enthusiastically adopted Stearns's duplex system with its potential for capital savings. In turn, Edison renewed his developmental work with duplex, diplex, and other forms of multiple telegraphy. He also worked on automatic telegraphy—another approach to increasing message density and one that was attractive to former Associated Press manager Daniel

Craig, a Western Union rival. Edison responded to perceptions of the importance of labor costs in the transcription of long press messages and sought to introduce the labor-saving roman-letter automatic telegraph. Indeed, the business and social environment broadly defined the technical problems or agenda and contributed to his solutions; however, the economics of the marketplace did not alone determine the character of his inventions. As indicated below, Edison's personal style, the traditions of the technical fields within which he worked, and other cultural factors also shaped his inventions.

Another issue that has attracted the attention of popular writers for many years is Edison's "method." Traditionally, biographers have described his method as "cut-and-try." His search for lamp filaments in the early 1880s and his search for a substitute for rubber in the late 1920s epitomized this characterization. Unfortunately, it implied that he did not employ principles or theory but instead "mindlessly" tested every conceivable thing until he found something that would work. A number of the documents in this first volume indicate that such a characterization of Edison's approach to invention is very misleading.

Edison consistently consulted the technical and scientific literature and let ideas guide his work. Although he did not have formal training in science or mathematics and likely was not conversant with all of the latest physical theories, he kept alert to new developments in electrical science and did operate from technical principles if not from scientific theory. For example, from the early days of his career he employed the polarized relay in various circuit designs, deliberately exploiting its ability to respond to the alternating direction of a current. As early as April 1873 he used concepts of a "balanced line" and a "center of resistance" in a line—a method of dealing with what we today call the resistance, capacitance, and inductance in a line. On the basis of these principles, he offered a number of circuit designs for automatic telegraphy on cables or long land lines. In other contexts he even referred to atoms and waves. Clearly, Edison was not a theoretical scientist, but neither was he a tinkering mechanic, mindlessly stumbling upon workable ideas. The documents of Edison's early career suggest that he was a highly inquisitive individual who used many sources to acquire information, ideas, and principles pertinent to the topics of his investigations.

Although some of Edison's early work seems to reflect the cut-and-try method, closer examination reveals a more con-

ceptual approach, one that could be called "theme and variations." In his rough sketches in 1867, for example, he took
basic designs and made modifications. The laboratory notebooks dealing with both printing and automatic telegraphy in
1871 and 1872 also reflect this approach. Early in 1872 he
devoted an entire notebook to drawings of 100 escapement
mechanisms, thereby preparing for himself a thesaurus of
known ways of controlling motion. When he renewed his work
with duplex telegraphy in late 1872 and early 1873, he again
sketched a large number of design possibilities that were variations on basic principles. In a similar manner he repeatedly
sketched sets of circuits for automatic telegraphy. This approach may reflect in part the influence of the United States
patent system. Edison knew well that it was a good strategy to
cover all possible approaches to a particular problem in order
to gain the broadest claim in a single patent or to obtain many
different patents in order to cover a new technical field. This
motivation likely reinforced his natural propensity for "technological play"—problem-solving in which, for his own in-
trinsic satisfaction, he sought all conceivable solutions.

Edison's telegraph designs reflect the centrality of trans-
mission-reception configurations in the foundations of his
creative work. Some scholars have previously emphasized his
"systems" or holistic approach to design.[3] When Edison focused on a single component of telegraph design, he generally
placed such work in the context of the entire circuit. For example, duplex and diplex telegraphy—transmitting and receiving two messages simultaneously over one wire—involved
circuits with components that interacted with each other. Design and modification of such circuits required viewing the
circuit as a whole. Other telegraph circuits required matching
the transmitter with the receiver in the context of the whole
circuit. When Daniel Craig first engaged Edison to work on
automatic telegraphy, Craig asked him to improve the design
of the perforator for Little's system; Edison, however, soon
moved beyond the perforator, conceptualizing the problem as
one that involved all the components in automatic transmission and receiving. Indeed, his early experience with telegraph transmitters and receivers likely fostered in him an approach that continued into his later work with the telephone,
phonograph, electric light, electric power systems, ore milling,
and motion pictures. For example, when Edison later began
working on incandescent lighting, he first focused on the lamp
but gradually extended his view to the other interacting com-

ponents, treating the system's dynamo as a "transmitter," the distribution system as the "line," and the lamp as a "receiver." As others have noted, he also worked on electric lighting by drawing analogies with gas lighting.

Edison's notes, sketches, and artifacts also suggest the centrality of visual-spatial thinking to his technical creativity, and they stimulate questions regarding the possible role of tactile, kinesthetic, and quantitative thinking in his designs.[4] The documents concerned with mechanical and electromechanical designs, such as those in printing and automatic telegraphy, constitute the first record of his visual conceptualizations. At times, his mechanical constructs remind us of the contraptions in Rube Goldberg cartoons, but with the important difference that Edison ultimately sought to avoid mechanical awkwardness and to make his many inventions commercially practicable. Although his sketches do not represent machines in perfect three-dimensional, orthogonal, fixed-point perspective, they do indicate that he effectively conceived of mechanical operations in three dimensions. The early record of his visual thinking contained in the drawings and artifacts in this volume also suggests that he developed a repertoire of stock solutions and visual forms from his telegraph experience. He repeatedly included the dual solenoid, the ratchet wheel escapement, the rotating drum and stylus, flowing tape and stylus, and the polarized relay as central components in his creative designs. Sometimes these forms dominated the composition of his designs, suggesting that analogy played an important role in his creative thinking.

These "methods" marked Edison's inventive style. He remained concerned with communications technologies during most of the 1870s, returned to them in the middle 1880s, and continued working on them into the twentieth century, pursuing new designs for the telephone, railway telegraphy, the improved phonograph, and motion pictures. The cylinder-stylus form evident in his facsimile telegraph designs of 1868, for example, recurred in automatic telegraphy in the early 1870s, in the electromotograph in the mid-1870s, in the original phonograph in 1877, and in his earliest conception of the motion picture in 1888. Before filing his first patent application for the phonograph, he sketched on a single sheet of paper three different versions of the instrument: cylinder, disc, and tape. His first motion picture drawing in October 1888 was based by analogy on the cylinder and stylus form of his cylinder phonograph and also included the ratchet wheel and solenoids. Later, he added a polarized relay and moved to flowing tape (film), complete with perforations to keep it in register. The latter, as he noted at the time, was analogous to the Wheatstone automatic telegraph. These examples suggest that Edison's early work in telegraphy provided direction and intellectual resources for his subsequent career.

The variety of technical fields within which Edison worked during his lifetime is extraordinary. This diversity even characterized his early telegraphic pursuits. He did not confine himself to one area of telegraphy, such as printing, automatic, or multiple telegraphy, but instead worked across the field, frequently on different areas at the same time. His early documents suggest cross-fertilization of ideas and techniques among the different areas of telegraphy. For example, he employed multiple signaling techniques, including use of the polarized relay, in both printing and multiple telegraphy. Another example is his introduction of condensers for automatic telegraphy for long land lines in 1873. This use of condensers may have derived from his awareness of Stearns's successful 1872 introduction of condensers in duplex circuits. Besides the potential for cross-fertilization of ideas among fields, Edison's diverse work, including business as well as technical design and production, likely prevented him from focusing on just one problem and acquiring a mind-set.

Edison's early experience in business also shaped his later career. In the telegraph industry he worked increasingly with captains of industry, gaining an enviable reputation as a telegraph inventor. His invention of the phonograph in 1877 attracted world-wide acclaim and simply reinforced the high regard the leaders of the telegraph industry had for his creative abilities. Consequently, when he began work on electric lighting in 1878, his old financial friends from the telegraph industry introduced him to their friends in the railroad and banking communities and helped him acquire financial support for his efforts in fields beyond telegraphy.

Edison's early pattern of working with others in the technical world also continued into his later career and became largely institutionalized. In his early shops in Newark, he established experimental facilities where small teams of men assisted him with instrument manufacture and inventive design. This network of manufacturing facilities provided the human, technical, and material resources for his creative work and was the embryo of his later separate laboratories in Newark, Menlo Park, and West Orange.

The documents in this volume are, to borrow a phrase from his close associate Francis Upton, "a record of how inventions are made." Indeed, the reader can see the earliest evidence of Edison's thinking, his false starts, and the subsequent steps and stages in the creative process. Moreover, the documents reveal Edison's collaboration with his associates, his complex legal and financial affairs, and his dependence upon the persons and publications of the larger technical communities of telegraph inventors and scientists. They also hint that there were many other creative people who worked on similar problems and often were more successful than Edison with specific inventions. These rivals of Edison in the race to the patent office have too often disappeared from history because of scholarly neglect and the long shadow of the diverse, prolific, and highly visible Edison. Divorced from its context, Edison's work has traditionally been regarded as wondrous or miraculous. When embedded in its historical setting, his overall achievement remains remarkable but eligible for rational understanding.

Some of Edison's laboratory notebook entries and business documents may appear complicated, detailed, and even forbidding, but, like his letters to Frank Hanaford or those from Daniel Craig, they too reflect his exuberant personality and multifaceted daily life. They unveil the very human inventor and deliberate entrepreneur. Like the America from which he came, he was rough-hewn, energetic, and aggressive. As we enter into his world through his notebooks, we find him exclaiming, "oh god"; "Invented for myself exclusively, and not for any small brained capitalist"; and "My Wife Popsy-Wopsy Can't Invent." In the process, we discover the real Thomas Edison: the exceptional technical thinker, the shrewd if not always successful entrepreneurial strategist, and the man, with all his enthusiasm, persistence, overconfidence, and "warts."

1. Thomas Parke Hughes's work dominates scholarly studies of Edison and invention. See *Thomas Edison, Professional Inventor* (London: HMSO, 1976); *Networks of Power: Electrification in Western Society, 1880–1930* (Baltimore: Johns Hopkins University Press, 1983), esp. chaps. 2 and 3; "Edison's Method," in *Technology at the Turning Point*, ed. William B. Pickett (San Francisco: San Francisco Press, 1977), 5–22; and "The Electrification of America: The System Builders," *Technology and Culture* 20 (1979): 124–61. Hughes's student, Christopher S. Derganc, wrote an important article on Edison's use of science, "Thomas Edison and his Electric Lighting System," *IEEE Spectrum* 16 (Feb. 1979): 50ff.

The most detailed study of Edison's work to date is that of Robert Friedel, Paul Israel, and Bernard S. Finn, *Edison's Electric Light: Biogra-*

phy of an Invention (New Brunswick, N.J.: Rutgers University Press, 1986). See also Bernard S. Finn and Robert Friedel, *Edison: Lighting a Revolution* (Washington, D.C.: Smithsonian Institution Press, 1979).

2. The best biography in print is Matthew Josephson's *Edison* (New York: McGraw-Hill, 1959). The best treatment of Edison's technical work is the 1910 biography by Frank L. Dyer and Thomas C. Martin, *Edison: His Life and Inventions*, 2 vols. (New York: Harper & Bros., 1910). Francis Jehl, an old Edison associate, provided a rambling account of the early electric lighting days at Menlo Park and in New York in *Menlo Park Reminiscences*, 3 vols. (Dearborn, Mich.: Edison Institute, 1937–41). Two other biographies in print are Robert Conot's *Streak of Luck* (New York: Seaview Books, 1979) and Ronald W. Clark's *Edison: The Man Who Made the Future* (New York: Putnam, 1977).

3. Hughes, *Networks of Power*, chap. 2.

4. The role of visual-spatial thinking in technological design has been especially emphasized by Eugene S. Ferguson in "The Mind's Eye: Nonverbal Thought in Technology," *Science* 197 (1977): 827–36; and by Brooke Hindle in *Emulation and Invention* (New York: New York University Press, 1981) and "Spatial Thinking in the Bridge Era: John Augustus Roebling versus John Adolphus Etzler," *Annals of the New York Academy of Sciences* 424 (1984): 131–47.

Chronology of Thomas A. Edison

1847–1873

1847	
11 February	Is born in Milan, Ohio.
1847–54	Lives in Milan.
1854–63	Lives in Port Huron, Mich.
1859–60	
Winter	Starts selling newspapers and candy on the trains of the Grand Trunk Railroad.
1862	
Spring	Publishes and prints on the train his own newspaper, the *Weekly Herald.*
Fall	Studies telegraphy with James Mackenzie, station agent at Mount Clemens, Mich.
1862–63	
Winter	Begins work as a telegraph operator in Micah Walker's book and jewelry store in Port Huron.
1863	
Late Spring–Summer	Starts job as a telegrapher for the Grand Trunk Railroad at Stratford Junction, Ont.
1863–64	Returns briefly to Port Huron.
	Works the night shift as a railroad telegrapher near Adrian, Mich., where he meets Ezra Gilliland for the first time.
	Is employed for two months as a railroad telegrapher in Fort Wayne, Ind.
1864–65	
Fall–Winter	Works in the Indianapolis, Ind., office of the Western Union Telegraph Co. and experiments on improvements in telegraph repeaters.
1865	
Spring–Fall	Works in the Cincinnati, Ohio, office of Western Union and experiments on self-adjusting relays.

17 September	Becomes a founding member of the Cincinnati District of the National Telegraphic Union.
September	Is promoted to telegraph operator first class.
	Begins designing devices for multiple telegraphy.
1865–66	
Fall–Spring	Becomes the regular press-wire operator in the Memphis, Tenn., office of the South-Western Telegraph Co.
	Conducts repeater experiments.
1866	
Spring	Enters Western Union's Louisville, Ky., office as a press-wire operator.
4 June	Transfers his membership in the National Telegraphic Union to the Louisville District.
1 August	Leaves for New Orleans, La., planning to embark for Brazil.
Fall	Returns to the Western Union office in Louisville after a short stay in Port Huron.
1867	
Summer	Returns to the Western Union office in Cincinnati.
October	Returns to Port Huron.
1868	
March–April	Begins work as an operator at the main Western Union office in Boston, Mass.
11 April	Publishes in the *Telegrapher* the first of several articles on his telegraph inventions and on the Boston telegraph community.
11 July	Makes the first of several agreements with E. Baker Welch, a Boston businessman who helps finance his early inventive work.
28 July	Signs a caveat for a fire alarm telegraph and assigns the invention to Welch.
13 October	Signs a patent application for an electric vote recorder, which later issues as his first patent.
1869	
21 January	Sells rights in his first successful printing telegraph, the Boston instrument, to Boston businessmen Joel Hills and William Plummer.
30 January	Announces his resignation from his job with Western Union in order to devote himself full time to inventing and to pursuing various telegraph enterprises.
Winter–Spring	Joins Frank Hanaford in establishing a business to produce and sell private-line telegraphs at 9 Wilson Lane in Boston.
13 April	Tries and fails to make his new double transmitter work between Rochester and New York City.
April–May	Moves to New York City.

22 June	Is issued his first telegraph patent (for the Boston instrument).
c. 1 August	Replaces Franklin Pope as superintendent of Samuel Laws's Gold and Stock Reporting Telegraph Co. in New York City and makes improvements on Laws's stock printer.
12 September	Moves to Elizabeth, N.J., and boards with Pope's mother.
2 October	Joins his partners Pope and James Ashley in advertising their newly formed Pope, Edison & Co. as a firm of electrical engineers and telegraph contractors.
Fall	Operates a small shop in the electrical instrument factory of Leverett Bradley in Jersey City, N.J.

1870

10 February	Signs two contracts with George Field and Elisha Andrews of the Gold and Stock Telegraph Co. that provide funds for developing inventions and establishing a shop.
c. 15 February	Joins William Unger in establishing his first major shop, the Newark Telegraph Works.
18 April	Joins Pope and Ashley in assigning to the Gold and Stock Telegraph Co. their rights to printing telegraph patents.
May	Engages Lemuel Serrell as patent attorney.
1 July	Joins Pope, Ashley, Marshall Clifford Lefferts, and William Allen in establishing the American Printing Telegraph Co., an enterprise for providing private-line telegraphs.
3 August	Signs an agreement with Daniel Craig to invent an improved perforator for automatic telegraphy.
1 October	Signs an agreement with George Harrington making them partners in the American Telegraph Works and providing Edison with funds for automatic telegraph experiments.
19 October	Negotiates with Marshall Lefferts to sell his newly designed universal private-line printer to Gold and Stock.
c. 26 October	Charles Batchelor begins employment at the American Telegraph Works.
28 November	The Automatic Telegraph Co. is incorporated and Harrington is named president.
1 December	Pope, Edison & Co. announces its dissolution.

1871

Winter–Spring	Designs perforators, transmitters, ink recorders, and typewriters for automatic telegraphy.
4 April	Gives Harrington power of attorney for disposition of Edison's share in all inventions relating to automatic telegraphy.
9 April	Edison's mother, Nancy, dies in Port Huron.
April–May	Moves the Newark Telegraph Works from Railroad Ave. to Ward St. and changes the company's name to Edison and Unger.

April–May	Manufactures his cotton instrument, developed for Gold and Stock under his contract with Field and Andrews.
26 May	Sells the rights to his existing and future printing telegraph patents to Gold and Stock and becomes the company's consulting electrician.
28–29 July	Begins series of four notebooks to record his inventive work on printing, automatic, and other forms of telegraphy.
August	Begins manufacturing his universal stock printer for Gold and Stock.
October	Employs Mary Stilwell for his News Reporting Telegraph Co., which sought to provide general and commercial news in Newark.
22 November	Purchases his first house, located on Wright St. in Newark.
28 November	Buys stock in the street railway of his brother, William Pitt Edison, in Port Huron.
25 December	Marries Mary Stilwell.

1872

15–17 January	Designs a district telegraph that he assigns to the American District Telegraph Co.
27 January	Transforms his universal private-line printer into an electric typewriter for automatic telegraphy.
January–February	Fills laboratory notebooks with variations for his universal stock printer and his universal private-line printer.
5 February	Becomes a partner in J. T. Murray and Co., which later becomes Edison and Murray.
May	Delivers first models of improved universal private-line printer to Gold and Stock.
May–June	Supplies his universal stock printer to the Exchange Telegraph Co. of London.
3 July	Agrees to purchase Unger's share in Edison and Unger, thereby dissolving their partnership.
5 November	Makes an agreement with Josiah Reiff to provide Edison with an annual salary while he works on automatic telegraph improvements.
14 December	The Automatic Telegraph Co. opens for business using Edison's automatic telegraph improvements.
Fall	Conducts tests of his automatic telegraph system on the lines of the Automatic Telelgraph Co. and the Southern and Atlantic Telegraph Co.

1873

c. 10 February	Meets with William Orton, president of Western Union, and makes a verbal agreement to develop duplex telegraphy.
18 February	Edison's first daughter, Marion Estelle ("Dot"), is born in Newark.

31 March	Agrees to develop a roman letter automatic telegraph for Harrington and Reiff.
9–22 April	Prepares ten patent applications on duplex telegraphy.
23 April	Leaves for England.
23–27 May	Conducts tests of his automatic telegraph system for the British Post Office.
c. 1–15 June	Conducts tests of his automatic telegraph on a cable stored at the Greenwich works of the Telegraph Construction and Maintenance Co.
25 June	Arrives back in the United States.

Thomas A. Edison
and His Papers

The extensive collection of papers preserved in the archive at the Edison National Historic Site—close to three and a half million pages in all—is the product of Thomas Alva Edison's sixty-year career as inventor, manufacturer, and businessman. For years the sheer size and organizational complexity of the archive deterred researchers from delving extensively into its wealth of documentary resources. The recent publication of a selective microfilm edition of the papers (*Thomas A. Edison Papers: A Selective Microfilm Edition* [*TAEM*]), however, makes these historically significant papers readily available to scholars and other researchers. Because the arrangement of the documents on the microfilm parallels the organizational structure of the archive itself, it is helpful to understand how the records of Edison's laboratories and companies were generated during his own lifetime and how the archivists entrusted with their guardianship have subsequently treated them.

Although Edison began his career as a professional inventor in 1869, he did not attempt to keep systematic records of his experiments until 1871. A pocket notebook from October 1870, in which he entered specifications and drawings for printing telegraph apparatus, contains the concluding note: "all new inventions I will here after keep a full record." This determination to keep a full record of inventive activity in anticipation of patent and contractual litigation soon led to the creation of more than a dozen technical notebooks of varying sizes and shapes, many of which contained beautiful drawings. Frequently, the entries in these notebooks were dated, signed, and witnessed. Edison and his associates also used numerous loose pieces of paper to record ideas in the midst of research. When Edison moved from Newark to Menlo Park in 1876, many of the unbound notes and drawings were lost. Perhaps as a result of the difficulty experienced in keeping these pieces of paper together, many of them were subsequently pasted into scrapbooks. Edison did not, however, adopt a systematic method for doing this. Many unbound notes and drawings were not put into the scrapbooks, and related materials within the books were often widely separated.

Edison and his associates used ledger volumes, pocket notebooks, and unbound scraps of paper on an irregular basis throughout the inventor's career. In 1877, however, Edison instituted a more regular practice for notekeeping that, with some refinements, continued throughout his life. At first he used 9″ × 11″ softcover tablets, whose sheets could be torn from the top edge. Only a few of these notebooks were retained in their original condition. The majority were taken apart and, together with material on other loose pieces of paper, organized according to the specific invention to which they related. By the fall of 1878, the number of notes and drawings from Edison's work on the electric light had grown so large that he adopted a standard-size hardbound notebook that would remain intact as a permanent record. Like the tablets, the notebooks were placed around the laboratory, and often the work of more than one experimenter was recorded in them. The first of these notebooks dates from November 1878. Edison and his associates continued to use such notebooks for recording experimental work at Menlo Park and at Edison's later laboratories. The archive at the Edison National Historic Site (ENHS) has over 3,000 such notebooks, each measuring 6″ × 9″ and containing approximately 280 pages.

Not all of Edison's notebooks are experimental in nature. Some of the earliest books contain accounting records, and sometimes experimental and financial material are indiscriminately mixed together in the same book. At his West Orange laboratory, Edison used some of the standard-size books, as well as numerous pocket notebooks, to record ideas about the operations of his factories and businesses. In addition, there are several hundred notebooks containing evaluations of phonograph recordings and tests of storage batteries.

Although Edison seems to have retained most of his technical notes and drawings, some of them were lost or misplaced after their introduction as exhibits in civil court and patent interference cases. Not infrequently, the facsimiles and transcripts found in the printed record of these cases constitute the only surviving record of these documents. This is well illustrated in the case of the many drawings entered into evidence in the telephone interference proceedings in 1880.

The large collection of patent records at the ENHS constitutes another important source for understanding Edison's activities. In the early 1870s, Edison began to retain correspondence from his patent attorneys and from the U.S. Patent Office. In addition, rough drafts and finished copies of his patent applications were occasionally entered into the notebooks. However, Edison did not initially keep comprehensive files of his applications or of the correspondence between his attorneys and the Patent Office examiners. Thus, the major body of documentation on Edison's early patents is in the Patent Office files, located in Record Group 241 at the National Archives and Records Service. Around the turn of the century, Edison established his own patent department, and its extensive collection of patent folio files is preserved in the archive at the ENHS.

Additional important information about Edison's technical work is recorded in his scrapbooks. Edison and his associates prepared scrapbooks from an early date and included in them both unbound technical notes and articles clipped from newspapers and journals. As in the case of the notebooks, Edison initially employed several different types of books and only later instituted more systematic record-keeping procedures. In 1878 or 1879, William Carman and Francis Upton began using a standard-size scrapbook to keep a record of the growing number of articles about Edison and the technologies in which he was involved. Approximately 150 of these Menlo Park scrapbooks are still extant, and there are indications that the series may once have comprised over 200 books. The first 57 scrapbooks are numbered and indexed, and they deal primarily with technical subjects. The other volumes are largely concerned with scientific matters. After the move to West Orange in 1887, Edison and his associates continued to keep scrapbooks, but most of these later books were either photo albums or collections of published interviews with Edison.

In addition to generating an enormous quantity of technical materials, Edison also maintained extensive correspondence files. Prior to 1878 he did not have a secretary, although various individuals working at the laboratory and shops had some responsibility for filing correspondence. The fame that invention of the phonograph brought to Edison in 1878 resulted in a substantial increase in the volume of his mail. Edison was deluged with unsolicited fan mail and letters seeking advice or requesting information about his inventions. As a result, he hired Stockton L. Griffin, an old friend from his days as a telegrapher, as his secretary. Griffin developed a system of filing correspondence by subject or, in the case of frequent correspondents, by the name of the individual. This system continued, with some modifications, to the end of Edison's life.

Edison did not preserve copies of his outgoing letters until 1875, when he began using letterpress copies to record outgoing correspondence. The earliest letterpress copybooks relate primarily to routine business matters. However, the researcher may discern the nature of the replies to many other incoming letters from the notes written in their margins by Edison and his secretaries. After 1881 Edison became more systematic in preserving copies of outgoing correspondence, but he and his secretaries continued the practice of writing comments in the margins of incoming letters.

The letterbooks provided a less than ideal means of recording outgoing correspondence. Letterpress copies were made by wetting the leaf of the copying paper and then placing the original letter on the leaf and blotting it. This process often produced bad copies as a result of bleedthrough, faint ink, or smearing or spreading of the ink. Moreover, the thin tissue pages were easily susceptible to wrinkling, fraying, and tearing. These problems were more common in the early letterbooks, but they can be found in the later ones as well.

Edison was continually establishing companies to manufacture and market his inventions, and the majority of the documents at the ENHS deal with his business operations. In addition to the correspondence, Edison and his associates generated a substantial number of legal records such as agreements, patent assignments, powers of attorney, stock certificates, mortgages, deeds, and insurance policies. Edison's companies also kept extensive financial records, including bound account books, draft account sheets and trial balances, bills and receipts, payroll and time records, and lists of tools and supplies. At Edison's earliest manufacturing companies in Newark, different types of accounts were often mixed in the same ledger, and some books even contained entries for several different companies. Later, at Menlo Park and West Orange, Edison employed professional bookkeepers, who systematized the accounting records.

By the twentieth century, Edison had built in New Jersey a

small industrial empire that culminated in 1911 in the organization of Thomas A. Edison, Inc. As a result of the increasing number, scale, and bureaucratization of his businesses, the amount of documentation for Edison's twentieth-century activities is much greater than that for the nineteenth century, and the character of the twentieth-century material is more heavily oriented toward business records. The laboratory continued to play an important role in Edison's career, but it became merely one part of a large business organization. Much of its work was directly related to improvements for the various manufacturing companies. Because of the increasing size and organizational complexity of the Edison companies, many new types of business records were generated. Included among these were (1) interoffice correspondence and memoranda; (2) legal files maintained by key Edison associates such as Harry F. Miller and Richard W. Kellow and by the company's legal department; (3) advertising materials for the phonograph, storage battery, and other Edison products; (4) folios for patents assigned to the various Edison companies; (5) manufacturing records such as inspection reports, blueprints, and test runs; and (6) labor records for a vastly expanded work force.

Even before his move to West Orange in 1887, Edison had taken steps to preserve the records of his earlier activities in Newark, Menlo Park, and New York. Sometime before 1887, the correspondence, financial papers, and other unbound laboratory and business records were gathered into small packages, labeled, and placed in eighteen large wooden packing cases. These were subsequently moved to West Orange and placed in Edison's library in the main laboratory building. In 1912 Edison constructed a fireproof concrete vault near the laboratory for the safekeeping of his early notebooks, important legal papers, and other valuable documents.

Edison and his secretaries did not adopt a systematic plan for storing and preserving the ever-increasing quantity of correspondence, business records, and other papers dating from the period after 1887. Some of the documents for the early West Orange years were later placed in the wooden packing cases with the pre–West Orange material, while others were either left on shelves and floors in the galleries of the library or packed into crates and boxes to be stored in the library's numerous cabinets. As the volume of material increased, other rooms on the second and third floors of the laboratory were pressed into service as storage areas. In 1918 a Vault

Service Department was created within Thomas A. Edison, Inc., to identify and store obsolete business records generated by the department heads and other company officials.

Not until late in the 1920s did Edison or his associates appreciate the need for a systematic appraisal of a body of records that now totaled several million pages. In the summer of 1928—three years before Edison's death—the company hired Mary Childs Nerney to head a new Historical Research Department, and she prepared a detailed inventory of the large packing cases containing Edison's earliest business records. Although much of this material had originally been labeled and packaged, Nerney found that, over the years, the papers had "been combed many times . . . and many of them thrown back helter skelter." In the process the contents of different packets had been mixed, and many papers were lying loose in the boxes. Before leaving the company in 1930, she organized a few of the most important documents according to subject categories established by Edison's early secretaries and placed them in metal file drawers.

The organization of the Document File, as this collection was called, continued under the leadership of Norman R. Speiden, who served as director of the Historical Research Department and curator of the Edison laboratory from 1935 until 1970. Before the work was interrupted by World War II, the papers for the years through 1882 had been identified, organized by subjects within each year, and indexed on 3″ × 5″ cards. The processing of the documents resumed in 1948 when Kathleen L. Oliver (later McGuirk) became archivist of the collection. By the time of McGuirk's retirement in 1971, much of the correspondence and other unbound material through 1912 had been integrated into the Document File.

At the same time that Speiden and his associates were arranging the unbound documents, they were also collecting, organizing, and indexing the many hundreds of laboratory notebooks scattered throughout the laboratory and factory complex. Each of the standard-size notebooks was assigned a six-digit "N-number," and its contents were summarized on a 3″ × 5″ index card. As voluminous numbers of notebooks, accounting records, legal papers, letterbooks, and other documents were integrated into the collection, the existing buildings could no longer provide adequate storage facilities, and in the spring of 1942 the company completed construction of a spacious, air-conditioned underground vault.

The years following World War II witnessed the culmination of long-standing plans to turn the laboratory into a museum and to make its collections of documents and artifacts more accessible to researchers. In June 1946 the Thomas Alva Edison Foundation was incorporated as a nonprofit organization, and on 11 February 1948—the 101st anniversary of Edison's birth—the Foundation opened the library and laboratory to the public and assumed responsibility for the operation of the museum and the administration of its collections. In September 1955, after twenty months of negotiations with Charles Edison and other officials of Thomas A. Edison, Inc., the National Park Service made a formal offer to take over custodianship of the Edison laboratory complex. President Dwight D. Eisenhower officially designated the laboratory buildings and 1.5 acres of land surrounding them as the Edison Laboratory National Monument on 14 July 1956, and six years later the laboratory and Glenmont, Edison's home in nearby Llewellyn Park, were collectively designated as the Edison National Historic Site.

The contents of the library and the underground vault were officially donated to the National Park Service in late 1957. However, the McGraw-Edison Company, the successor to Thomas A. Edison, Inc., still retained ownership and possession of more than a million pages of business records generated prior to Edison's death in 1931. Over a period of fifteen years, these records were gradually removed from the company's vaults and transferred to the custody of the National Park Service. The transfer continued until 1972, when the company moved from West Orange. In addition to these company records, the ENHS also received donations of important papers from the descendants of Charles Batchelor, Francis R. Upton, and other Edison associates.

These massive acquisitions put a severe strain on the resources of the archival staff and much of this new material remained inaccessible to researchers. Moreover, despite efforts by Nerney and her successors to remove the manuscripts housed in the library and in other rooms within the laboratory, an estimated 200,000 pages of documents remained in these areas, many of them stored in the same boxes and crates into which they had been placed by Edison's secretaries.

In the last two decades the archival staff of the ENHS has made considerable progress in organizing these materials into record groups and making them available to researchers. Since 1979 the staff of the Thomas A. Edison Papers has worked closely with the ENHS archival staff to improve the organization of the collections selected for publication. The current publication of the selective microfilm edition of Edison's papers is making the core of this material (about 10 percent of the total) available to scholars and other researchers. Equally important, introductory targets for each collection and folder in the microfilm edition contain references to related material not selected for inclusion and thus serve as an important entrée into the larger collection. The final part of the microfilm will also include all relevant material from collections outside the ENHS. The extensively annotated book edition represents a further distillation of the archive (about 0.2 percent of the total) and includes material—such as artifacts and newspaper articles—not on the microfilm. References in the book lead researchers into the microfilm, the unfilmed archive, and the outside collections.

Editorial Policy

This book edition of Thomas Edison's papers provides the scholar and the general reader with definitive transcriptions of documents that unveil the man's personality, technical creativity, and business efforts. Introductory headnotes and annotations establish the context within which these documents were created, including the family and immediate associates who sustained him, the technical and financial communities that supported and challenged him, and the society and culture that stimulated and unleashed his technical and innovative energies. It is recognized that despite Edison's promethean achievements, his inventions, business adventures, and personal relations were not always unblemished successes as measured by the values of his day or ours. In preparing this edition of his papers, therefore, the editors are selecting documents that represent the full range of his activities—his triumphs and his failures.

Selection

This edition will finally include nearly 7,000 documents, a significant portion of which Edison himself created. The editors are selecting from three and a half million pages of extant Edison-related materials: incoming, outgoing, and third-party correspondence; laboratory notes and drawings; patent materials; legal and litigation records; and financial accounts. They are also selecting drawings and artifacts, because these materials provide critical evidence of visual-spatial thinking, an important dimension of creative technical thinking and design. Accordingly, these artifacts and drawings are treated as independent documents, in a manner parallel to the treatment of traditional verbal documents. For the period covered in Volume 1, the editors have included every entry from Edison's Newark laboratory notebooks. Aided by the editors' annotations, the interested reader can reconstruct the notebooks. In cases where the meaning of a notebook entry is dependent on a previous or following entry, the first endnote to the dependent entry directs the reader to the related document.

This first volume covers Edison's boyhood and early career, a period for which there are relatively few extant documents. For this reason, the initial volume contains about 90 percent of the known Edison-related materials (about 50 percent if one includes accounting records). The editors have diligently sought records relating to Edison, his family, and the settings in which he grew up and worked as a young telegraph operator and inventor-entrepreneur. They discovered small caches of previously unknown Edison documents in this country and in England, and also found Edison publications and related notices in obscure telegraph journals. Because of the paucity of extant Edison documents for the period of his childhood and early career, the editors have selected for this volume a small number of third-party materials that illuminate such matters as the nature of the curriculum in a school he attended, public notices of his activities, and published references to his publications or inventions. In addition, the editors have systematically searched numerous retrospective sources and have used for annotation the important information they provide regarding Edison's formative years. One of these sources is the collection of Edison's autobiographical notes that he prepared in 1908 and 1909 for Frank Dyer and Thomas Martin's biography of Edison. These notes are presented at the end of this volume, in Appendix 1. Another source is the extensive collection of Edison's late nineteenth and early twentieth century correspondence that relates to the Edison family in Milan and Port Huron and to his itinerant years. Included among these sources are a significant number of letters by Edison himself, in which he reflected upon people and events pertinent to his early years. This manuscript material and related published works are discussed in Appendix 2.

Thomas Edison obtained 1,093 U.S. patents (and many more from nearly two dozen foreign countries), the largest number ever issued to an individual by the U.S. Patent Office. Clearly, these documents could not be reproduced in their totality in this edition. However, a complete set of Edison's U.S. patents is included on reels 1 and 2 of the microfilm edition of the Edison Papers. The issued patents that Edison executed during the period covered by this volume are listed in Appendix 4. Other patent application materials are cited, where pertinent, in the editors' annotations to the documents.

The editors have not provided a comprehensive calendar of Edison documents. The vastness of the archive made preparation of such an aid unfeasible. The editors have, however, included in their annotations references to relevant documents in the microfilm edition. These references are indicated by the acronym *TAEM*. Thus, the book edition may serve as an entrée into the microfilm edition, just as the latter may serve as an entrée into the archives.

Organization

The documents in this edition are organized in chronological order. The editors chose chronological rather than subject or document-type organization in order to maximize the historical understanding of Edison's work and to eliminate the need for multiple publication of the same document. Edison generally pursued several projects at one time, and often they were technologically interrelated. With chronological organization the juxtaposition of documents from different projects reveals the interconnectedness of the technical work and the significance of its historical context. A subject approach would have isolated Edison's developmental thinking and in some cases would have placed later labels on work originally conceived quite differently. Organization by document type—that is, personal, business, or technical materials—would have ripped activities from their historical fabric and would have isolated related ideas and events. The chronological approach presents Edison's work in the richness of its personal, technical, financial, and social interrelationships.

Accordingly, the documents are grouped chronologically within chapters that are themselves chronologically ordered. Organizing documents in chronological order is basically straightforward; however, arbitrary rules apply when elements of the date are missing or when two or more documents of the same date are selected. If the date of a document does not contain the day but is accurate to the month, season, or year, the editors have placed the document at the *end* of that month, season, or year. For example, if a letter is dated "October 1870," it appears immediately *after* all fully dated documents for October 1870. If two or more documents carry the same date and no logical order is dictated by their content, they are ordered by document type:

correspondence
other nontechnical documents
technical material
artifacts

Multiple notebook entries are kept in sequence. Within the category of correspondence, the order is:

Edison outgoing
Edison incoming
third-party correspondence

Two or more Edison incoming or third-party letters are ordered alphabetically by author's last name; two or more Edison outgoing letters are ordered alphabetically by recipient's last name. The date and authorship of an artifact are based upon its *design*, not the date of construction or the builder of the artifact.

Annotation

The editors' goal is to make this both an enduring edition and one that is accessible to a variety of scholars and the general public. Because of the limited scholarly literature and the technical complexities discussed in many of the documents, there are more introductory headnotes than is common in historical documentary editions published in America since 1945. The editors recognize that in a selective edition the process of selection and annotation is fundamentally interpretive; nonetheless they have tried to avoid a level of interpretation in the headnotes that would diminish the useful life of the edition. Accordingly, this first volume begins with an interpretive essay that identifies current historical issues and seeks to relate them to the material in the documents.

The chapter introductions, document headnotes, and endnotes are intentionally less interpretive than the volume introduction. From them the reader can derive a context for the documents, including the identification of individuals, organizations, technical terms, and events not otherwise specified in the documents. Each chapter begins with a brief introduction that highlights Edison's personal, technical, and business activities during the period covered by the chapter. Within chapters there are occasional headnotes to groups of documents. Artifacts and drawings without accompanying text are preceded by headnotes. These introductions and headnotes serve as guides for the general reader. Usually the initial portions of discursive endnotes also deal with general matters; the remaining portions refer to business or technical details that are likely to be of more concern to the specialized reader. In general, the editors have provided more detailed information for technical issues that have received little scholarly attention than for topics that are treated in the secondary literature. The reader who wishes to reconstruct the original documents may do so with the aid of the textnotes.

Form

The editors of this edition have considered the nature of the documents, their potential use, and the transcription principles that are most appropriate for them. The technical and nonverbal characteristics of the documents contribute to their complexity. Their verbal qualities, however, present fewer problems than many historical documents of an earlier era because the literary conventions of Edison's era are still largely understandable to the modern reader. Therefore, in order to provide a readable text, it is not necessary to adopt a "clear-text" approach, a method used by modern literary editors which employs neither critical symbols nor note numbers in the text but buries all of the editorial emendations in notes at the end of the volume.[1] It is also not necessary to adopt the more liberal of the "expanded" methods, which have been popular with recent American historical editors.[2] Such an approach standardizes many elements in transcription and leaves the reader without specific knowledge of many editorial changes. A decade ago historical editors who employed the "expanded" method were criticized in a detailed analysis of their work by G. Thomas Tanselle.[3] Tanselle's critique has encouraged deliberative approaches to the principles for historical documentary editing, and it is in that context that editorial decisions have been made for the Edison Papers.

The editors believe that it is important that the readers of this edition have the evidence of Edison's creative mind at work—that they see the first primitive sketches of a new design; sense the hurried hand that makes false verbal starts, leaves out letters in words, and disregards the conventions of capitalization and punctuation; and note the evidence of Edison's verbal facility as well as his visual-spatial capabilities. Only obvious mechanical errors in typed copies have been silently omitted. By following a conservative expanded approach that does not try to "clean up the text," the editors have sought to strike a balance between the needs of the scholar for details of editorial emendation, the requirements of all users for readability, and the desire of the editors that all readers obtain a feel and flavor for Edison, his associates, and their era.

This edition incorporates the principles of publication style presented in the thirteenth edition of the *Chicago Manual of Style*.[4] Where the range and complexity of documents and annotations presented problems that transcend those treated in the *Chicago Manual*, the editors adapted the basic principles to meet the needs of this edition.

Titles. A title precedes each document. When Edison is the author or recipient of a document, his name is omitted from the title. For example, a letter from Edison to the former head of the Associated Press, Daniel Craig, appears under the title "To Daniel Craig." When Edison is coauthor, corecipient, or coinventor, "And" precedes the name(s) of his associates: "And Frank Hanaford from D. N. Skillings & Co." Third-party correspondence is presented in the following format: "Milton Adams to Frank Hanaford." Notebook entires authored by Thomas Edison are presented as "Notebook Entry."

Place and date lines. Regardless of where the information appears in the original document, the editors have positioned the place and date on the first line of the published document, flush to the right margin. The form, spelling, and punctuation of the place and date are as they appear in the document unless otherwise indicated by a textnote. The use of letterhead stationery is noted in the textnote. If the editors have supplied any element of the place or date from a source other than the document or its immediate context (e.g., the journal or newspaper in which a document appeared), that element is set in brackets. A conjectured date or place appears with a question mark within brackets. Routine inside addresses, return addresses, and dockets have not been transcribed.

Transcription principles. The editors have sought to provide as literal a transcription of the documents as is possible in print; at the same time, they have tried to avoid elaborate editorial apparatus that would intrude unduly upon a smooth reading of the text. They have not corrected errors in capitalization, spelling, punctuation, grammar, and syntax, nor have they silently supplied punctuation or removed repeated words or slips of the pen. In distinguishing between capital and lowercase letters, they have considered the form and size of the letter, the author's style, and usage. When uncertainty has remained, they have followed current convention. Edison often capitalized words at the beginning of a line; the editors have retained such capitalization according to the general capitalization policy. Because of Edison's idiosyncratic use of commas and periods, the editors have distinguished between them on the basis of usage and meaning. Where Edison or others, according to current conventions, omitted punctuation, the editors have inserted extra space in lieu of commas, semicolons, colons, question marks, exclamation points, or periods. Paragraph indentations and the length of dashes and hyphens have been standardized.

When necessary for clarity, contractions and abbreviations have been expanded and the supplied text has been placed in brackets. Cancellations are indicated with overstruck type. Interlineations made at the time the document was created have been brought into the text and identified in the textnotes. Where significant marginalia accompany a document, they appear in the transcription enclosed in angle brackets. Unless otherwise indicated by a textnote, marginalia are in Edison's hand. Other interlineations are identified in the textnotes. Raised letters and numbers have been silently brought into alignment with the rest of the line. Revenue stamps, seals, and similar markings that accompany legal and commercial documents have not been transcribed, but because they may help establish the authenticity of documents, they are identified in the textnotes. Within a document, preprinted text is indicated in a textnote. Underlined manuscript text is underlined; double underlined text is singly underlined and is flagged by a textnote.

Words that serve only to indicate a page turn, such as "over," are identified in the textnotes rather than transcribed in the body of the document. Page numbers generally are not transcribed but are cited in the textnotes to facilitate finding documents in notebooks or large groupings of pages. Page breaks in the original document, and page numbers or other markings made at a later date, such as those of archivists or manuscript dealers, are not identified unless they are of special interest. Edison often used a flourished "and" placed in a circle; this has been transcribed as normal text. In transcribing multiple signatures, such as those appearing in agreements and notebooks, the editors have placed the first signature flush left and the second flush right, two per line, with the last signature flush right. The witness signatures appear flush left on the same line as the last signature unless there are too many of them for all to fit on that line, in which case they appear on the following line. The editors have supplied the term "witness(es)" in brackets when it does not appear in the text of a document.

Drawings and artifacts. Drawings and artifacts have been photographically reproduced. The widely varying size of the drawings makes it impossible to reproduce them all at a standard scale. The editors have attempted to reproduce the drawings within a given document, notebook, or set of identically sized notebooks at a consistent scale.

Endnotes. Substantive and bibliographic endnotes are grouped at the end of the editors' headnotes and after the textnote of each document. In substantive notes, the most general information is provided first and is followed by more details if appropriate. A person or technical term mentioned in the text is fully identified only once within the edition. The reader can locate that identification by looking up the boldface page number under the appropriate name or subject in the index. Bibliographic information in the endnotes appears in a shortened form that employs (1) the author-date system of citation for publications and (2) abbreviations for standard references, journals, and manuscript collections. For full names and titles, the reader is referred to the List of Abbreviations and the Bibliography.

Textnotes. A textnote immediately follows each document and is structured in five parts. The first part describes the document type, using a modified version of archivists' and manuscript dealers' abbreviations (see List of Abbreviations). The second part identifies the archival location of the document, using a Library of Congress repository symbol (see List of Abbreviations), and also indicates the document's location within the repository. If the document has already been published in the microfilm edition of the Edison Papers, the third part then identifies that edition by the acronym *TAEM* and locates the document by reel and frame number. This information is contained within parentheses. The fourth part identifies special characteristics of the docmument such as its having been written on letterhead stationery. The fifth part consists of lettered notes that concisely identify textual problems such as interlineations and missing words. A superscript lowercase letter (*a, b, c,* etc.) is placed at a problem area in the text of the document, and the same letter introduces the note. If a textual problem arises more than once in a document, the same letter is used to mark each occurrence. If a note does not repeat or otherwise indicate words from the text, it refers to only the word or signature in the text to which the superscript letter is attached. If the textual problem involves more than one word, the note indicates all the words and punctuation involved, either by repeating them or by providing an elliptical sequence of words and phrases, each set off by quotation marks.

An example of a textnote is as follows:

ALS, NjWOE, DF (*TAEM* 12:42). Letterhead of the *Telegrapher.*
[a]Interlined above.

This means that the document is in the hand of its author, is a letter, and is signed. It is located at the archives of the Edison National Historic Site in West Orange, New Jersey, in the Document File; and it appears in the *Thomas A. Edison Papers: A Selective Microfilm Edition*, on the twelfth reel, in the forty-second frame. The letter was written on letterhead stationery of the journal, the *Telegrapher.* Within the text of the transcribed letter, one word that the author wrote between the lines of the original has been brought into the line of print and is identified with the superscript lowercase letter *a*.

1. For an excellent discussion of textual methods and the history of their application, see Mary-Jo Kline, *A Guide to Documentary Editing* (Baltimore: Johns Hopkins University Press, 1987), chaps. 1 and 5. A recent example of a modified use of the clear-text method in historical editing is Frederick Burkhardt and Sydney Smith, eds., *The Correspondence of Charles Darwin*, vol. 1 (Cambridge: Cambridge University Press, 1985).
2. Kline, *Guide*, 18–21, 125–29.
3. G. Thomas Tanselle, "The Editing of Historical Documents," *Studies in Bibliography* 31 (1978): 451–506.
4. *Chicago Manual of Style*, 13th ed. (Chicago: University of Chicago Press, 1982).

Editorial Symbols

~~Boston~~   Overstruck letters
   Legible manuscript cancellations; crossed-out or over-written letters are placed before corrections.
[Boston]   Text in brackets
   Material supplied by editors
[Boston?]   Text with a question mark in brackets
   Conjecture
[Boston?][a]   Text with a question mark in brackets followed by a textnote
   Conjecture of illegible text
⟨Boston⟩   Text in angle brackets
   Marginalia; in Edison's hand unless otherwise noted.
[   ]   Empty brackets
   Text missing from damaged manuscript
[---]   Hyphens in brackets
   Conjecture of number of characters in illegible material

Superscript numbers in editors' headnotes and in the documents refer to endnotes, which are grouped at the end of each headnote and after the textnotes of each document.

Superscript lowercase letters in the documents refer to textnotes, which appear at the end of each document.

List of Abbreviations

ABBREVIATIONS USED TO DESCRIBE DOCUMENTS

The following abbreviations describe the basic nature of the documents included in the first volume of *The Papers of Thomas A. Edison:*

AD	Autograph Document
ADf	Autograph Draft
ADS	Autograph Document Signed
AL	Autograph Letter
ALS	Autograph Letter Signed
AX	Autograph Technical Note
AXS	Autograph Technical Note Signed
D	Document
Df	Draft
DS	Document Signed
L	Letter
LS	Letter Signed
M	Model
PD	Printed Document
PDS	Printed Document Signed
PL	Printed Letter
TD	Typed Document
TL	Typed Letter

In these descriptions the following meanings are assumed:

Document Accounts, agreements and contracts, bills and receipts, legal documents, memoranda, patent applications, and published material, but excluding experimental notes, letters, and models

Draft   A preliminary or unfinished version of a document or letter

Technical Note   Technical notes or drawings not included in letters, legal documents, and the like

Letter   Correspondence

Model   An artifact, whether a patent model, production model, or other

The symbols may be followed in parentheses by one of these descriptive terms:

fragment   Incomplete document, the missing part not found by editors

copy   A version of a document made at the time of the creation of the document by the author or other associated party

transcript   A version of a document made at a substantially later date than that of the original, by someone not directly associated with the creation of the document

photographic transcript   A transcript of a document made photographically

historic photograph   A photograph of an artifact no longer extant or no longer in its original form

abstract   A condensation of a document

STANDARD REFERENCES AND JOURNALS

Standard References

BDAC 1928	*Biographical Directory of the American Congress, 1774–1927*
DAB	*Dictionary of American Biography*
DBI	*Dizionario biographico d'egli Italiani*
DNB	*Dictionary of National Biography*
DSB	*Dictionary of Scientific Biography*
Dict. des Bio.	*Dictionnaires des Biographes*
NCAB	*National Cyclopedia of American Biography*
NUC Pre-1956	*National Union Catalog: Pre-1956 Imprints*
OED	*Oxford English Dictionary*

Journals

Annales télég.	*Annales télégraphiques*
Bull. Soc. l'ind. nat.	*Bulletin de la Société d'encouragement pour l'industrie nationale*

Elec. Engr.	*Electrical Engineer*
Elec. W.	*Electrical World*
J. des télég.	*Journal des télégraphes*
J. Frank. Inst.	*Journal of the Franklin Institute*
J. Soc. Teleg. Eng.	*Journal of the Society of Telegraph Engineers*
J. Teleg.	*Journal of the Telegraph*
Phil. Mag.	*Philosophical Magazine*
Proc. Roy. Soc.	*Proceedings of the Royal Society of London*
Sci. Am.	*Scientific American*
Teleg. Age	*Telegraph Age*
Teleg. and Tel.	*Telegraph and Telephone Age*
Telegr.	*The Telegrapher*
Teleg. J. and Elec. Rev.	*Telegraphic Journal and Electrical Review*

ARCHIVES AND REPOSITORIES

In general, repositories are identified according to the Library of Congress system of abbreviations. Parenthetical letters added to Library of Congress abbreviations have been supplied by the editors. Abbreviations contained entirely within parentheses have been created by the editors and appear without parentheses in citations.

DLC	Library of Congress, Washington, D.C.
DNA	National Archives, Washington, D.C.
DSI-NMAH	Archives, National Museum of American History, Smithsonian Institution, Washington, D.C.
DSI-NMAH(E)	Division of Electricity, National Museum of American History, Smithsonian Institution, Washington, D.C.
MdSuFR	Washington National Records Center, Suitland, Md.
MH-BA	Baker Library, Graduate School of Business Administration, Harvard University, Boston, Mass.
MiD-B	Burton Historical Collection, Detroit Public Library, Detroit, Mich.
MiDbEI	Library and Archives, The Henry Ford Museum & Greenfield Village, Dearborn, Mich.

MiDbEI(H)	Henry Ford Museum & Greenfield Village, Dearborn, Mich.
MiHC	Michigan Historical Commission, State Archives Library, Lansing, Mich.
(MiPHStCo)	St. Clair County Building, Port Huron, Mich.
MWalFAR	Federal Archives and Records Center, Waltham, Mass.
NjBaFAR	Federal Archives and Records Center, Bayonne, N.J.
(NjHum)	Private Collection, Charles Hummel, Wayne, N.J.
NjN	Newark Public Library, Newark, N.J.
(NjNECo)	Essex County Hall of Records, Newark, N.J.
NjWOE	Edison National Historic Site, West Orange, N.J.
(NjWU)	Western Union Corporate Records, Western Union Corporation, Upper Saddle River, N.J.
NNAT	Corporate Research Archives, American Telephone and Telegraph Company, New York, N.Y.
NNC	Rare Books and Manuscripts, Butler Library, Columbia University, New York, N.Y.
NNHi	New-York Historical Society, New York, N.Y.
(NNYCo)	New York County Hall of Records, New York, N.Y.
(NNYSE)	Archives, New York Stock Exchange, New York, N.Y.
(OSECo)	Erie County Courthouse, Sandusky, Ohio
(ONHCo)	Huron County Courthouse, Norwalk, Ohio
PPAmP	American Philosophical Society, Philadelphia, Pa.
(PPDR)	Department of Records, Philadelphia, Pa.
(UKLPO)	Archives, Post Office, London, United Kingdom

MANUSCRIPT COLLECTIONS AND COURT CASES

References to documents included in *Thomas A. Edison Papers: A Selective Microfilm Edition* (Frederick, Md.: University Publications of America, 1985–) are followed by parenthetical citation of reel and frame of that work; for example, Cat. 1185:34, Accts. (*TAEM* 22:562). Documents found at NjWOE after contemporaneous material had been microfilmed will be filmed as a supplement of the next part of the microfilm edition. For (*TAEM* Supp. III), see *Guide to Thomas A. Edison Papers, Part III.*

Accts.	Accounts, NjWOE
Anders v. Warner	*Anders v. Warner,* Patent Interference File 5603, RG-241, MdSuFR
APT	American Printing Telegraph Company Minutes, NjWU
ATF	Automatic Telegraphy File, UKLPO
Batchelor	Charles Batchelor Collection, NjWOE
BC	Biographical Collection, NjWOE
DF	Document File, NjWOE
Digest Pat.	Digest of Patent Assignments, RG-241, DNA
EBC	Edison Biographical Collection, NjWOE
Edison Coll.	Edison Manuscript Collection, NNC
Edison v. Lane	*Edison v. Lane v. Gray v. Rose v. Gilliland* and *Edison v. Lane v. Gray v. Edison and Johnson,* Patent Interference Files 8027–28, RG-241, MdSuFR
EP&RI	Edison Papers and Related Items, MiDbEI
G&S Minutes	Gold and Stock Minutes, NjWU
G&S v. Pearce	*Gold and Stock Telegraph Company v. Pearce and Jones,* Equity Case File D-1911, RG-21, NjBaFAR
G&S v. Wiley	*Gold and Stock Telegraph Company v. Wiley,* Equity Case File D-1913, RG-21, NjBaFAR
GF	General File, NjWOE
Harrington v. A&P	*Harrington, Edison, and Reiff v. Atlantic & Pacific, and George Gould et al.,* Equity Case File, 3980, RG-276, and Equity Case File 4940, RG-21, NjBaFAR

Lab.	Laboratory notebooks and scrapbooks, NjWOE
Lbk.	Letterbooks, NjWOE
LBO	William Orton Letterbook, NjWU
Lefferts	Marshall Lefferts Papers, NNHi
Libers Pat.	Libers of Patent Assignments, U.S. Patent Office Records, MdSuFR
LS	Legal Series, NjWOE
Meadowcroft	William H. Meadowcroft Collection, NjWOE
MM Coll.	Miscellaneous Manuscript Collection, PPAmP
Nicholson v. Edison	*Nicholson v. Edison*, Patent Interference Files 8689–90, RG-241, MdSuFR
Pat. App.	Patent Application Files, RG-241, MdSuFR
Pioneers Bio.	Edison Pioneers Biographical File, NjWOE
Port Huron Schools	Annual Reports of the School Inspectors of the City of Port Huron, County of St. Clair, to the County Clerk, Microfilm, MiHC
PS	Patent Series, NjWOE
Quad.	Quadruplex Case, NjWOE (see list below)
RGD	R. G. Dun Collection, MH-BA
Scraps.	Scrapbooks, NjWOE
Sig. Corps	Letters Sent and Letters Received, RG-111, DNA
Welch v. Edison	*Welch v. Edison*, Equity Case File 1995, RG-21, MWalFAR
Wiley v. Field	*Wiley v. Field*, Patent Interference File 9219, RG-241, MdSuFR
WU Coll.	Western Union Collection, DSI-NMAH

Quadruplex Case Citations

Volume 70 (*Atlantic & Pacific v. Prescott & others*)
1 Amended Bill of Complaint
2 Answer of Prescott
3 Answer of Edison
4 Answer of Western Union
5 Answer of Prescott to Amended Complaint
6 Plaintiff's Exhibits 1 through Z.3

THE MAKING OF AN INVENTOR, 1847–1873

*Map of the eastern United
States showing the cities
and towns associated with
Edison's youth and
early career.*

–1– Edison's Boyhood and Itinerant Years

February 1847–March 1868

On the fourth of July 1839 the citizens of Milan, Ohio, jubilantly celebrated the opening of a three-mile canal connecting their small village to the navigable portion of the Huron River. As fireworks crackled and a flurry of cannon bursts reverberated across the waterfront, the people of Milan toasted the economic emancipation of their community as well as the sixty-third anniversary of the nation's independence. In the decade that followed, the canal fulfilled all expectations and Milan, eight miles south of Lake Erie, became one of the busiest harbors on the Great Lakes, an outlet for the region's burgeoning grain crop and other commodities.[1]

Among those flocking to Milan to share in the canal-sparked prosperity was a thirty-five-year-old fugitive Canadian firebrand, Samuel Edison. Samuel's arrival in Milan marked the end of a long, harrowing odyssey. Two years earlier his participation in a failed insurrection against the Canadian provincial government had forced him to flee his homeland, leaving his wife Nancy and four children behind. When Canadian authorities subsequently indicted him for treason, Samuel made his stay in the United States permanent.[2]

As Milan flowered into a thriving shipbuilding and regional manufacturing center, Samuel found a ready market for his carpentry skills. Reunited with his wife and children, he made shingles and speculated in land. In time other relatives also moved to the thriving Ohio port. Samuel soon prospered enough to build a comfortable brick home overlooking the canal, and there, on 11 February 1847, Nancy gave birth to her seventh and last child, Thomas Alva.[3]

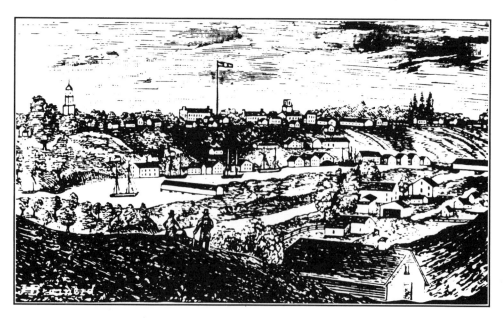

Milan, Ohio, 1846. What may be the Edison family home can be seen at the far left of the picture.

Known in childhood as Al, the youngest Edison was a sickly boy, a source of concern to a mother who had already lost two children in infancy and would suffer the loss of a third before little Al's first birthday. Nancy's three older surviving children, Marion, Pitt, and Tannie, were all teenagers. By the time Al was three, Marion had married and Pitt was often away on business. Only Tannie still lived at home with her brother on a regular basis, although the family occasionally took in boarders.[4]

In later life Edison said little about his early childhood. What he remembered of Milan included a trip to Vienna, Ontario, with his parents, covered wagons bound for the California gold fields, and the drowning of a friend with whom he was swimming in the creek.[5] The documentary record for these years is equally scant. The accounts that have survived are primarily anecdotal and so encrusted with the embellishments of both Edison and his admirers that an objective rendering of them is difficult, though not altogether impossible. Most of the stories of Edison's youth portray a precocious, yet typically American boy whose energy and initiative foreshadowed his success. Samuel remembered his son as mischievous and exasperatingly inquisitive, but not as the child prodigy some biographers have described.[6]

The boom that brought the Edisons to Milan proved short-lived. In the early 1850s the town's economic growth peaked

Edison at the age of ten.

and the railroads took much of the canal's traffic as they by-passed Milan.[7] In the spring of 1854, at the age of seven, Al moved with his family north to the booming lumber town of Port Huron, Michigan, on the Canadian border. Though isolated and raw, Port Huron boasted a population of 3,100, twice that of Milan. Samuel and Pitt joined other aspiring entrepreneurs eager to make their fortunes on the freewheeling northern frontier. Samuel engaged in various enterprises in Port Huron, including lumbering and land speculation, occupations endemic to the area. With the assistance of young Al, he also cultivated a ten-acre farm and sold the produce in town.[8] In his most unusual venture Samuel constructed a tall observation tower near the family home. It provided anyone willing to pay the fee an expansive view of Lake Huron and the surrounding countryside and was, Edison later recalled, a "folly that paid."[9]

The move to Port Huron significantly altered the social circumstances of the Edisons. In Milan, Samuel and Nancy had lived among a host of relatives. In Port Huron they lived more than half a mile outside of town on the St. Clair River and initially had little extended family. Changes also occurred within the immediate family; Tannie and Pitt both married, leaving Al at age nine the only child living at home.

Separated from neighbors as well as most of her family, Nancy devoted much of her time to the education of her son.

Advertisement in the Port Huron Commercial *(11 Aug. 1855) describing the school established by the Reverend George Engle and attended by Edison for a brief period in the 1850s.*

FAMILY SCHOOL!

FOR

Boys and Girls

PORT HURON, MICH.

Rev. G. B. ENGLE. Rector.

Teachers, Mrs. ENGLE.
            Miss D. EDISON, in Music.

ITS NAME indicates its CHARACTER: a family of young persons pursuing a course of thorough instruction, upon christian principles.

The number of pupils is limited to thirteen, besides the Rector's family.

The rules of a well-regulated Christian family are the only restrictions.

TERMS, [payable quarterly in advance.]
Board, fuel, and tuition in English
    branches,--------------------- $120 per annum.
Latin and French—extra--------- 12   do
Music—piano and melodeon------- 40   do
Sacred Music------------------- 12   do

The Term includes the entire year, except the holidays, July and August.

No deduction for absence except when pupils are sick.

Pupils received at any time and charged from the time of entering.

Those who remain during vacation will pay accordingly.

Port Huron, August, 1855.

Nancy Edison.

Al briefly attended a small private school headed by the local Episcopal minister, receiving instruction in piano in addition to traditional academic subjects.[10] For a short time he also attended the union school in Port Huron, where he may have studied the physical sciences.[11] But with the exception of these two brief interludes and some Sunday School attendance, the youngest Edison received his basic education at home.

Nancy, a former school teacher, organized lessons and an extensive reading program. The Edisons possessed a home library and encouraged young Al to master it. Samuel, an outspoken democrat and freethinker, accepted the political radicalism and anticlerical deism of Thomas Paine and shared Paine's writings with his son. The boy also avidly read popular science periodicals and contemporary fiction.

Young Al led a busy life in Port Huron. Inspired by his reading, he conducted chemical experiments and joined his friends in a variety of mechanical pursuits. His boyhood projects reputedly included the construction of a miniature steam-powered saw mill and railroad. Along with his lessons, reading, projects, and chores, he pursued an adolescent romance with a local girl, Carrie Buchanan.[12] Her father was one of the town's lumbermen, and several of her brothers were Al's companions.

In November 1859 the Grand Trunk Railway expanded across the Canadian border and opened a branch line to Detroit from Fort Gratiot, less than half a mile from the Edison

Edison's membership card for the Detroit Young Men's Society provided him access to the Society's extensive library.

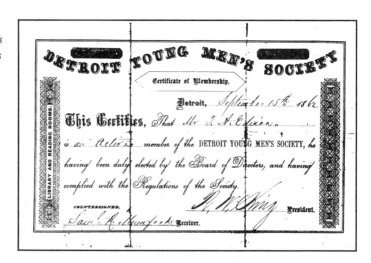

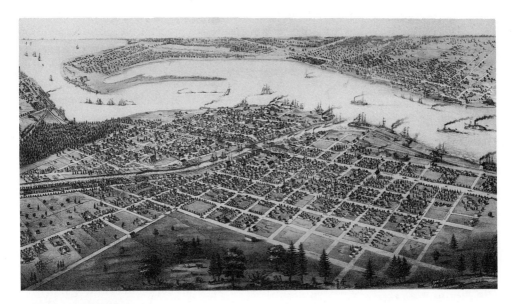

Port Huron, Mich., and environs in 1867. The Edison tower and home are on the left side of the picture at the edge of the forested area, and the Grand Trunk Railroad depot is shown just beyond this. In actuality this area was not so densely forested, and the depot was much farther away.

home and observation tower. Fascinated by the steam locomotives and eager for adventure, twelve-year-old Al soon peddled candy, newspapers, magazines, and dime novels on the sixty-mile run to Detroit. Detroit was ten times the size of Port Huron and offered a wide range of attractions. During his half-day layovers in the city, he frequented the newspaper and telegraph offices as well as the library of the Young Men's Society, where he reportedly read the entire *Penny Cyclopedia* and Thomas Burton's *Anatomy of Melancholy* and struggled with Newton's *Principia*.[13] He also enjoyed reading the popular literature he sold on the train.[14] At this time Edison first noticed a deterioration in his hearing, a loss that grew worse later in his life.

In spite of his hearing difficulties, Edison responded enthusiastically to the new opportunities available to him and soon revealed qualities of entrepreneurship equal to those of his father and brother. In Port Huron he opened two stands and hired other boys to run them. One of the stands sold newspapers and magazines; the other marketed produce that Edison purchased in Detroit or along the line and transported free of charge on the train. For about six months in 1862 the teenager also published his own newspaper, the *Weekly Herald*, which he printed in the baggage car of the train with the help of a friendly conductor. Concerned largely with local news and railroad matters, the paper attracted subscribers along the

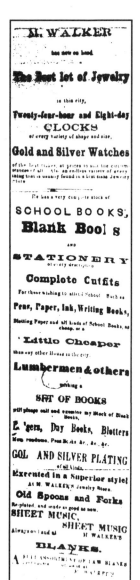

Advertisement in the Port Huron Press *(1 May 1864) describing Micah Walker's store where Edison obtained his first job as a telegrapher.*

Grand Trunk's Michigan line.[15] Edison used some of the profits from his various businesses to establish a small chemistry laboratory in the baggage car.[16] But his experiments on the train and his career as a railway newspaper publisher came to an abrupt end when a bottle of phosphorus fell and set the baggage car on fire. After extinguishing the blaze, the conductor ejected Edison, his laboratory, and his printing press at the next station. Despite this misadventure, the boy retained his job on the railroad.

Edison's work on the Grand Trunk quickened his interest in telegraphy, an essential part of railroad operations. Lingering about the depots, he watched the telegraph operators and eventually persuaded a couple of them to give him some basic instruction. He and a friend erected a half-mile telegraph line between their homes so that they could practice. Like many Americans of the time, young Edison marveled at this new electrical wonder capable of conquering time and space.

In the summer of 1862, during a stop at the Mount Clemens depot, Edison rescued the young son of stationmaster James MacKenzie from the path of a rolling freight car. In gratitude, Mackenzie taught Edison railroad telegraphy.[17] The youth would ride the train to Mount Clemens in the morning, study and practice with MacKenzie until afternoon, and then rejoin the train as it returned to Port Huron.

After several months of study with MacKenzie, the fifteen year old took the position of Western Union telegraph operator for Port Huron. A table in Micah Walker's jewelry store served as Edison's office.[18] Because wire traffic was usually light, he had ample opportunity to observe the other activities carried on in the shop, including fine metalwork, electroplating, and clock and watch repair. The teenager also read the technical books and scientific magazines sold in the store and used the jeweler's tools and equipment to duplicate some of the experiments he read about.[19]

Eager to improve his skills, Edison sometimes remained at the store in the evening to practice receiving wire-service press reports. Because of their considerable length and rapidity of transmission, press reports were excellent training for operators trying to increase their receiving proficiency. After about six months at Walker's store, Edison secured a job as a railroad telegrapher for the Grand Trunk line at Stratford Junction, Ontario—the first in a four-year series of jobs that would take him throughout the Midwest.

Edison's attraction to telegraphy is understandable. Many

In their 1876 engraving "The Progress of the Century," Currier and Ives gave the telegraph a central place among the new inventions for improving communication and transportation.

Americans held electricity and the telegraph in awe and attached considerable prestige to the position of telegraph operator. The electrical telegraph industry was a new enterprise, not much older than Edison himself. The first regular commercial service in America, based on the designs of Samuel Morse, Alfred Vail, and others, began in 1846 and spread with astonishing speed.[20] A telegraph line passed through Milan in the year of Edison's birth, and fourteen years later ribbons of wire stretched across the continent.

In many respects the telegraph molded the world in which Edison matured. American railroads and newspapers quickly adopted this new technology. When Edison started to work on the Grand Trunk, major newspapers already relied heavily on telegraph reports purchased from news services. The telegraph revolutionized the concept of news, giving it an immediacy never possible before. Edison recognized early this potential of telegraphy and exploited it to his economic advantage. On at least one occasion during his stint as a newsboy, he increased his sales of Detroit newspapers at the local stations on his route by having news of a Civil War battle telegraphed ahead of the train.

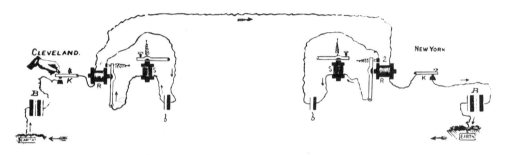

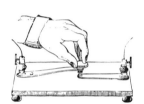

Morse telegraph circuit. At each end: **B** *is a main-line battery,* **K** *a key,* **R** *a relay,* **S** *a sounder in the local circuit, and* **b** *the local battery.*

Simple telegraph key.

The system of telegraphy that Edison learned was relatively simple.[21] The basic components originally included a transmitting key, an electromagnetic relay for receiving the electrical impulses, and a register for recording the coded messages on paper tape. The code consisted of a system of dots and dashes—long and short electrical impulses—that represented letters of the alphabet. Miles of bare iron wire, strung on insulators mounted on wooden poles, connected major urban telegraph offices and small intermediate stations. Batteries of Grove cells[22] at the urban terminals powered the main lines and drove all the stations' relays; Daniell cells[23] in each office powered the local receiving instruments, which responded to the making and breaking of the local circuit by the main-line relays.[24] To send a message, an American operator first "opened" the key, causing a "break" in the circuit. This signaled the operator at the other end of the line to await a message. The sender would then manipulate the key to transmit the electrical impulses.

In the early years of American commercial telegraphy, registers recorded the signals as indentations on a paper tape. In the 1850s sounders replaced most registers. These electromechanical instruments operated on the same principle as the registers but turned the electric impulse into the click of a lever against a sounding piece. Operators "read" the clicks directly and transcribed the message rather than reading it from a tape as it came from a register. The simple sounder increased transmission speeds, reduced errors, and cost less to build and maintain than the Morse register.[25]

In the spring of 1863, as Edison left Port Huron for his new job at Stratford Junction, he joined thousands of other young men hoping to find a rewarding career in telegraphy. The Civil War produced a surge in telegraph use and created an urgent need for skilled operators.[26] Among telegraph companies, Western Union benefited most from the war, receiving generous government subsidies. Moreover, its predominantly east-

Morse telegraph relay (right). Morse telegraph register (below).

Grove batteries (right) were used in main offices, and Daniell batteries (below) were used in local offices.

Sounders such as this one had replaced registers on most Morse telegraph lines by the 1860s.

west lines survived the conflict unscathed. Western Union emerged from the war with 44,000 miles of telegraph lines, more than the combined total of its two major competitors, the American Telegraph Company and the United States Telegraph Company.

Telegraph traffic declined sharply with the end of the war, and many telegraph companies felt the financial pressure. By engaging in price wars, Western Union aggravated the economic difficulties of its competitors. In 1866 Western Union absorbed both of its principal rivals and dominated the industry. But independent regional companies persisted and new companies soon challenged the telegraph giant on its more profitable routes.

Consolidation and turmoil within the industry reduced opportunities and wages for operators, furthering the growth of the National Telegraphic Union (NTU). Founded in 1864 primarily as a benevolent and fraternal society, the union opposed the arbitrary employment and salary policies within Western Union.[27] Membership grew rapidly and the union started a journal, the *Telegrapher*, as a means of communication for the far-flung "telegraphic fraternity."[28] It publicized technical and business developments within the industry, allowed members of the telegraph community to keep track of former colleagues, and represented the interests of operators. In 1867 Western Union began publishing for company employees a similar periodical, the *Journal of the Telegraph*.[29]

Within American society telegraphers belonged to a subculture with its own language and lore, reflected in the operators' jokes and anecdotes about the ignorance and confusion of the uninitiated.[30] In time, public attitudes toward the work and lifestyle of operators developed into stereotypes. One of the most colorful of these images, the "tramp" telegrapher, portrayed a carefree youth who drifted from city to city, always found friends and employment, but never saved his money. Accounts of Edison's itinerant years cast him in this role.

Like Edison, many telegraphers were young, unmarried men. At this time the profession included a few women. They usually worked in rural, one-person offices but were practically excluded from the National Telegraphic Union.[31] Friendships often developed among operators who might never meet but who recognized each other's "touch" on the telegraph key. The NTU and the telegraphic press encouraged this camaraderie, the sense of belonging to a larger technical community that spanned the Atlantic and included ma-

"Is it not a feat sublime?
Intellect hath conquered time."

This banner from the Te-
legrapher, 29 May 1865,
was drawn by the journal's
editor, Franklin Pope, who
later became associated with
Edison in his printing tele-
graph inventions.

chine operators, manufacturers, mechanics, engineers, and inventors.

Within the telegraph industry a hierarchy of skill and status quickly emerged. Operators clearly ranked above clerks, and full-time operators above those who worked part-time. Main-line operators enjoyed more status than branch or circuit operators, as did fast senders and receivers compared to those who worked more slowly. Press-wire operators constituted the telegraphic elite because they received large volumes of material at high speed (very few were senders).[32] Operators usually valued experience more than formal training. Most telegraphers developed their skills in the same way that Edison did, through self-instruction using handbooks and practice keys and through informal apprenticeship with established operators.[33]

By maintaining the telegraph instruments and batteries and sometimes acting as line inspectors, operators learned the basic mechanical and electrical operation of the telegraph system. Ambitious operators like Edison, hoping to advance themselves professionally, enhanced their technical understanding by reading the telegraph literature. A few members of the telegraph community who aspired to higher status turned to invention.[34]

Between Edison's birth in 1847 and his first patent application in 1868, the nation witnessed a sharp rise in inventive and patent activity.[35] Pioneering national corporations such as Western Union increasingly recognized the business potential of technological innovation and created strategies to control and exploit both the inventive process and its products in order to protect and maintain its virtual monopoly of the telegraph industry. Western Union actively sought ways to increase the technical efficiency of its system.

Edison began his career during the early phase of American corporate industrialization, when Americans accepted alle-

A typical small-town telegraph office, showing (right to left on the table) *the key, register, relay, and* (on the floor) *the battery.*

giance to "Yankee ingenuity" as a fundamental national characteristic and regarded machine technology as one of the essential engines of progress. Newspapers and popular magazines lionized inventors and entrepreneurs. As prominent technological innovators like Samuel Morse acquired great wealth and prestige, they became role models for aspiring young operators like Thomas Edison.

Between 1863 and the spring of 1868 Edison worked as an itinerant telegraph operator, an apprenticeship that ushered him into the exciting business of invention. He enjoyed opportunities to observe and experiment with diverse electrical apparatus. Working at night, he took advantage of the lull in wire traffic to conduct telegraph experiments and practice taking press reports. As he refined his receiving skills, he also sought to broaden his understanding of the scientific principles underlying the technology. He read the electrical works of British scientist Michael Faraday and studied a variety of other scientific and telegraph literature.[36]

Many of these publications included visual images that were critical to technical communication. Telegraphy relied on electricity for its motive force, but the design of instruments depended largely upon nineteenth-century mechanical technology. Published drawings and engravings enabled mechanics and potential inventors like Edison to visualize how machines looked and how their parts moved and interacted in space. The circuit drawings included in telegraph studies allowed the reader to follow the flow of electrical current step by step from the ground at the transmitting station, through the sections of the circuit, to the ground at the receiving station. As Edison's technical expertise grew, he started filling notebooks with drawings derived from his readings and from his careful study of the machines he worked with. He also began to reshape these images into new designs of his own.[37]

Edison reputedly developed one of his first designs while working the night shift at Stratford. In order to take short naps on duty, he devised an alarm that woke him in time to send the required half-hourly signals to his supervisor. This clever idea nearly resulted in a train collision and as a consequence abruptly ended the teenager's career with the Grand Trunk.

After a brief return to Port Huron, Edison obtained a job as a night telegraph operator for the Lake Shore and Michigan Southern Railroad at Lenawee Junction, near Adrian, Michigan. He lost that position when, on his supervisor's orders, he interrupted a transmission by the division superintendent. Crossing into Ohio, Edison unsuccessfully sought employment in Toledo before moving on to Fort Wayne, Indiana, where he worked for a short time as a telegraph operator for the Pittsburgh, Fort Wayne, and Chicago Railroad. Then, in the autumn of 1864, he headed for Indianapolis, Indiana.

Indianapolis's position as a major rail shipping center had earned it the name "Railroad City" by the mid-1860s. Western Union ran most of the nation's railroad telegraph lines, and Edison found a job with the company as a railroad telegraph operator at the Union Depot.[38] Because the depot served as a junction for all the rail lines in the area, its telegraph office was busy and important. Contemporaries described the depot as an unsavory place, a haven for pickpockets and other criminals, and a perpetual trade circus where women and children clamorously hawked apples, pies, and other goods at all hours of the day and night.[39]

Assigned to work at a table next to the press-wire operator, Edison continued to practice receiving press reports in between his regular duties. Unable to keep up with the fast pace of the press wire, he arranged two Morse registers so that one recorded the incoming signal and the other played it back slowly enough for him to comprehend.[40] Edison and fellow operator E. L. Parmelee used the device to receive press reports until the newspapers complained about receiving uneven copy. The pair also adjusted the registers to repeat messages from one circuit to another and tested them on a line from Pittsburgh to St. Louis.[41]

By the spring of 1865 Edison had moved to Cincinnati, the commercial hub of the Ohio River valley and one of the nation's largest cities.[42] Sometimes known as "Porkopolis," Cincinnati specialized in hog slaughtering, meat packing, and a host of associated industries such as leather goods, oils, candles, and soaps. Procter and Gamble, one of these ancillary manufacturers, reportedly paid Edison to install a private-line telegraph for it during his stay in the city.[43] The young operator boarded with two actors and telegrapher Ezra Gilliland, an acquaintance from Adrian who worked with him on the night shift at the Western Union office.[44] Gilliland sometimes transmitted plays to Edison as a telegraphic exercise.[45]

On 17 September 1865 Edison joined his fellow telegraphers in establishing a local chapter of the NTU. After the meeting the operators celebrated their accomplishment by getting drunk. Edison, who was not a drinking man, declined to join in the festivities. When most of the telegraphers, including the regular press operator, failed to appear for the evening shift, Edison recorded the incoming press reports himself. His long hours of practice served him well and the copy proved adequate for the newspapers. Office manager J. F. Stevens rewarded Edison for his diligence by promoting him to operator first class.[46]

Shortly after his promotion, Edison moved to Memphis, Tennessee, the most important commercial city on the Mississippi between St. Louis and New Orleans. With the return to civilian government in the aftermath of the Civil War, widespread lawlessness, gambling, and violence descended upon the city. "The whole town," Edison said later, "was only 13 miles from Hell."[47] In the midst of this chaos Edison took a job as a press-wire operator and persevered in his course of self-improvement. He purchased books, including a biography of Thomas Jefferson, and began learning Spanish, French, and Latin.[48] Edison's familiarity with the press and appreciation of its influence increased as he literally surrounded himself with the newspaper business by living in a building that housed a newspaper and a city directory publisher.[49]

During his stay in Memphis, Edison conducted a series of experiments on repeaters (devices that retransmitted messages from one line to another). He also worked on duplex systems (circuits that permitted the sending of two messages on the same wire simultaneously). Edison later claimed that in November 1865 he used a repeater of his own design to bring New York and New Orleans into direct contact for the first time since the end of the war. The local newspaper noted the feat without indicating the person responsible for it.[50] Unfortunately, the young operator's constant experimentation did not endear him to his supervisor and led to his eventual firing.

In the spring of 1866 Edison arrived in Louisville, Kentucky, an Ohio River community that had emerged from the Civil War as a prosperous industrial center. He went to work as a press-wire operator for Western Union but left Louisville with two companions in late July in search of adventure and fortune in Brazil, a country actively seeking American immigrants and specifically advertising for telegraphers.[51] Edison's journey ended abortively in New Orleans, however, where a bloody race riot had occurred just a few days before his arrival. He heard that the steamship scheduled to take him to Brazil had been commandeered under martial law and he went home to Port Huron instead. A month later he returned to Louisville and again took a position with Western Union.

During the next year Edison worked hard to refine his telegraphic skills, concentrating in particular on his handwriting and his store of background information. He experimented with ways of writing more rapidly and attempted to learn shorthand, or phonography, as it was then called. He also tried to keep abreast of the significant political and economic developments of the day. He regularly perused the newspapers for potentially useful information. He learned the names of important politicians, read the proceedings of congressional committees, and kept up with market prices. He befriended the founder and editor of the *Louisville Journal*, George Prentice, who liked to engage the Louisville agent of the Associated Press, George Tyler, in predawn political discussions. Edison often listened in as the two men talked over whiskey and crackers.[52] Consequently, on those frequent occasions when bad weather or poor insulation of the wires caused messages to be garbled or lost entirely, Edison drew upon his own knowledge to furnish the missing text. He later claimed to have supplied as much as one fifth of some transmissions in this way.[53]

On 1 July 1867 Western Union moved its Louisville office from the dilapidated building it had previously occupied into sumptuous new quarters, described by the local newspaper as "a perfect palace." Edison had only a fleeting opportunity to enjoy this luxurious workplace, however. Two months after the move he was fired for spilling sulfuric acid that ate through the carpet and floor into the manager's office below.[54] Undaunted, the young telegrapher returned to Cincinnati and resumed working for Western Union.

Seeking to further his education and improve his technical capabilities, Edison frequented the Mechanic's Library and the Cincinnati Free Library. He read more works on telegraphy and electricity, copied information for future reference, and made sketches in his notebooks.[55] He also met Charles Summers, superintendent of telegraphs for the Indianapolis, Cincinnati, and Lafayette Railroad, who encouraged him to continue his experiments. They worked together on a self-adjusting relay, and Summers provided Edison with scrap apparatus from the railroad's telegraph facilities. Edison rented a room on the top floor of an office building and equipped it with a foot lathe and other tools so that he could alter this material for his experiments. At the urging of George Ellsworth, former telegraph operator for Confederate cavalry captain and raider John Morgan, Edison experimented with a secret signaling system that he and Ellsworth hoped to sell to the government for a considerable sum.[56]

Edison left Cincinnati in October 1867 and returned home to Michigan.[57] His parents had experienced some economic setbacks during their son's wandering years. Forced out of Edison's boyhood home, they now lived in another house at Fort Gratiot, where they again took boarders. The family no longer had the farm, but Sam continued to speculate in land and had become a justice of the peace. Pitt still lived in Port Huron and managed a new street railway that ran from the railroad depot to the middle of town.[58]

During his six-month stay Edison continued to design improved telegraph apparatus. He may have even installed a working duplex arrangement on the Grand Trunk Railway's underriver cable between Fort Gratiot and Port Edward.[59] He also began writing descriptions of his telegraph designs for publication in telegraph trade journals. When some of his work met with an encouraging response from the editor of the

Telegrapher, Edison submitted a description of his duplex telegraph, which the journal agreed to publish.[60]

Edison's ambitions clearly transcended the limited opportunities available to him in Port Huron. The restless twenty-one year old contacted Milton Adams, an acquaintance from his Cincinnati days, about the prospects for employment in Boston, where Adams now worked for the Franklin Telegraph Company. Adams responded that if Edison came quickly, Adams could obtain a job for him in the Western Union office. Using a free pass to ride the Grand Trunk as far as Montreal, Edison apparently arrived in Boston early in March 1868.[61] There he entered a sophisticated technological and business environment that spurred the inventive talents and entrepreneurial proclivities he had honed during his youth and itinerant years in the Midwest.

1. For the opening of the Milan Canal, see Frohman 1976, 35–36. Much of the narrative of Edison's boyhood and adolescent years is based on critical use of Appendix 1 ("Edison's Autobiographical Notes"). Except where specific citations are given in footnotes, other references for this chapter introduction are found in Appendix 2 ("Bibliographic Essay: Edison's Boyhood and Itinerant Years").

2. Samuel Edison was born in 1804 in Digby, Nova Scotia, to Loyalist parents who had left the United States following the War for Independence. In 1811 his family moved to the small town of Vienna, Ontario, located just inland from the northern shore of Lake Erie, about midway between Toronto and Detroit. Nancy's family, the Elliotts, moved to Vienna from New Berlin, in Chenango County, N.Y., where Nancy had been born in 1810. Samuel and Nancy were married in 1828 in Vienna, which was then part of the British colony of Upper Canada. In Vienna, Nancy taught school for a time and Samuel was an innkeeper.

3. Samuel and Nancy had business, church, and social associations with some of Milan's prominent families, and Samuel had enough income to pay court judgments in excess of $100 (OSECo: Deeds 3:96, 6:201–2, 14:335–36; and Common Pleas Execution 3:456, 554). Edison was named Thomas after an uncle; his middle name apparently came from a family friend, Capt. Alva Bradley, who was a master of ships on Lake Erie. Dyer and Martin 1910, 1:21.

4. The Edison children were: Marion (1829–1900), William Pitt (1831–1891), Harriet Ann ("Tannie") (1833–1863), Carlile (1836–1842), Samuel O. (1840–1843), Eliza (1844–1847), and Thomas Alva (1847–1931). Marion married Homer Page on 19 December 1849 and moved to a farm near town.

5. App. 1.D8; Dyer and Martin 1910, 1:17–18; App. 1.D92 and *Norwalk Huron (Ohio) Reflector,* 31 Aug. 1852.

6. For the evolution of Thomas Edison's image as a cultural hero, see Wachhorst 1981; for his father's reminiscences, see Lathrop 1894, 7.

7. Milan's peak year for grain shipments was 1847. In 1846 a railroad started operating from the lake port of Sandusky, by-passing Milan, and

reaching Mansfield, in north central Ohio. The railroad made inland grain collection depots more attractive than Milan. Milan's grain business quickly declined and other enterprises replaced it.

8. Samuel apparently was a grocer for a time. The family resided in a house of size and distinction, owned other property, and employed Michael Oates, a "chore Boy," and Truey, a maid. For Oates see App. 1.A6; for Truey see Doc. 99, n. 4.

9. App. 1.D88.

10. Edison probably attended the Rev. Engle's school for a few months in 1854 or 1855.

11. See Doc. 1.

12. Carrie Buchanan Harrington (c. 1847–1921) wrote to Edison in 1888 asking if she could visit his laboratory. Edison wrote on her letter the following notation to his secretary: "Tate = This is my old flame when a boy   say would be happy to have her call with friends & take lunch at House— I will sign letter." Carrie Buchanan Harrington to TAE, Apr. 1888, GF. See also Edison's 1870 inquiry about her to his parents in Doc. 129.

13. Edison later claimed that he had decided to read through all the books in the library but "gave it up after reading about 10 books that were pretty dry reading" (GF: TAE marginalia, David Heieman to TAE, 3 Aug. 1909; and Edgar Swift to TAE, 29 Nov. 1908). The Detroit Young Men's Society library served as a public library and was later absorbed into the Detroit Public Library after that institution's founding in 1864. Catlin 1923, 305–7, 310–11, 546.

14. Edison recalled a special fondness for the works of Sylvanus Cobb, Jr., particularly *The Gunmaker of Moscow* ("Sunday 1 Mr. Edison's Interview," galley copy in EBC). For other titles that Edison remembered reading and for his dealing in dime novels in general, see "Making Good Friends with Lady Luck," *Literary Digest* 100 (16 Feb. 1929): 54–55.

15. See Doc. 2; and App. 1.A6.

16. The chemistry textbook that Edison was reading at the time was either Carl Fresenius's *Elementary Instruction in Chemical Analysis* or his *System of Instruction in Qualitative Chemical Analysis*, each of which went through numerous editions and printings in the 1840s and 1850s.

17. Railroad telegraphy used standard equipment and Morse code but had a special set of abbreviations.

18. Edison may have begun working at Walker's as early as the end of 1862; the previous operator joined the Union army's telegraph corps.

19. Two craftsmen repaired watches while one worked on clocks and jewelry (Micah Walker to O. M. Carter, 31 Mar. 1908, EBC). Telegraphy did not constitute a large portion of the shop's business and was not mentioned in its advertising, though Walker's store had been the Western Union office for Port Huron since the line's establishment (*Port Huron Press*, 22 Mar. 1865, 3). Edison's duties as an operator required only about a third of his time. Because he did not wait on customers or do other work at the store, he had considerable free time for his reading and experimentation in the basement. Walker's recollections are reported in O. M. Carter to Thomas Martin, 14 Dec. 1907, EBC.

20. European systems were in use before this date, and other systems were later introduced in the United States. For the early history of

American telegraphy, see Reid 1879; Thompson 1947; and Brock 1981, 55–88.

21. Descriptions of the Morse and other telegraph systems are found in Shaffner 1859; Prescott 1863; idem, 1877; Pope 1871; and Maver 1892.

22. See Doc. 336, n. 4.

23. See Doc. 336, n. 3.

24. In American telegraph systems, when messages were not being transmitted the circuits were closed and electric current flowed continuously through the line. In most European telegraph systems the circuits were open when the line was not in use, preventing the flow of current. Consequently, every station required a transmitting battery.

25. Prescott 1863, 93; Fischer and Preece 1877, 336–37.

26. The impact of the Civil War on telegraph operators is discussed in Gabler 1986, 95–97.

27. A successor to the NTU engaged in an unsuccessful strike in 1871. For a social and economic study of telegraph operators, see Gabler 1986, especially chapter 3. Ulriksson 1953 focuses on the history of the operators' unions.

28. Gabler 1986, 88, 110–11, 117–23. Although the *Telegrapher* soon became a separate venture from the union, it generally supported the union and often opposed Western Union's actions and policies.

29. Reid 1879, 666–70.

30. Telegraph lore and anecdotes are found in Huntington 1875; Johnston 1880; and Phillips 1876.

31. Beginning in the late 1860s, Western Union supported a school for female operators at Cooper Union in New York City. The company was undoubtedly interested in using female operators to reduce labor costs, because women were paid three fourths of the salary offered male operators.

32. For the demands on press-wire operators, see "In Memoriam—Night Work," *Telegr.* 2 (1865–66): 98.

33. Gabler 1986, 68–71. There were a number of telegraph schools in operation during this period but many were of dubious quality.

34. Gabler 1986, 103–4; Fischer and Preece 1877, 316–28.

35. The annual number of patent applications grew from about 1,500 in 1847 to over 20,000 in 1868. More than half of this growth occurred after the Civil War. In 1847, when the population of the country was 21 million, there was one patent application for every 14,000 persons. By 1868, with a population of 38 million, there was one application for every 1,900 persons. U.S. Bureau of the Census 1975, 8, 959.

36. See Docs. 7–9 and 46.

37. See headnote, pp. 29–32, and Docs. 10–24.

38. Edison boarded at Macy House, on the corner of Market and Illinois, just four blocks from the Union Depot. *Indianapolis Directory* 1865, 46, 109.

39. *Indianapolis Daily Journal,* 2 Dec. 1864, 2.

40. This combined register is sometimes referred to as Edison's first invention. See Doc. 10 and App. 1.D203.

41. On repeaters see headnote, pp. 30–31. Using Edison's device, Edison and Parmelee recorded press reports on tape, playing them through the second, slower register to transcribe them. When the regu-

lar operator arrived and began receiving the later portion of the report, transcribing the message as it came in, Edison and Parmelee would finish transcribing the beginning of the report at their slower speed. App. 1.D203.

42. Edison claimed that he left Indianapolis because he disliked the city, but one of Edison's biographers states that Edison lost his position because he was an hour and a half late in reporting Lincoln's re-election (App. 1.D205; Simonds 1940, 58). However, Lincoln's re-election occurred in November and Edison continued to work for Western Union until at least January 1865. The election returns were reported late because a storm knocked down many telegraph lines and prevented others from operating. *Indianapolis Daily Journal,* 9 Nov. 1864, 2; and 10 Nov. 1864, 2. For the record of Edison's employment at Indianapolis in November, see Western Union office account, EP&RI.

43. E. L. Bernays to W. Meadowcroft, 27 July 1931, GF; Lief 1958, 37.

44. App. 1.D46; Nat Hyams to TAE, 11 Feb. 1912, GF; Joshua Spencer to TAE, 8 Apr. 1878, DF (*TAEM* 15:478); "Thomas Alvey Edison," *Cincinnati Commercial,* 18 Mar. 1878, 5. The actors were Nat Hyams and Harry Howard. Edison and Hyams were listed at Bevis House at the corner of Court and Walnut Sts. in *Williams' Cincinnati Directory* 1865, 129, 217, 424.

45. Quotations from Shakespeare, particularly the opening line of *Richard III*—"Now is the winter of our discontent made glorious summer by the sun of York"—appear occasionally in Edison's laboratory notebooks.

46. "District Proceedings: Cincinnati District," *Telegr.* 1 (1864–65): 172; App. 1.D209–13. Some of the telegraph operators working with Edison in Cincinnati had applied to the NTU in June 1865 to form a local. Ulriksson 1953, 15–20; "Telegraphic Miscellanea," *Telegr.* 1 (1864–65): 114.

47. TAE to William Dealy, c. Jan. 1909, GF.

48. See Doc. 5. Edison apparently did not pursue the languages for long. An 1865 edition of E. A. Andrews and S. Stoddard's *Grammar of the Latin Language,* with the inscription "Thomas A. Edison, Telegraph operator, Memphis Tenn, Mch 29 1866," is at the Edison National Historic Site. TAE marginalia, Clyde Grissam to TAE, 7 Apr. 1919, GF.

49. Edison lived at 40 North Ct. and boarded next door at Delta House, 42 North Ct. *Memphis Directory* 1866, 110.

50. According to the *Telegrapher,* the *Memphis Bulletin* of 26 November reported the first direct connection of New York and New Orleans. "Telegraphic Feat," *Telegr.* 2 (1865–66): 27.

51. See Doc. 6.

52. App. 1.A20, D37, D100. "The Telegraph and Proper Names," a newspaper article published during Edison's stay in Indianapolis, prescribed that telegraph press operators "should have as good a knowledge of distinguished characters, and of geographical points as an intelligent printer has of orthography, punctuation, and grammar." *Indianapolis Daily Journal,* 26 Jan. 1865, 2.

53. App. 1.D189.

54. "Western Union Telegraphs," *Louisville Daily Journal,* 5 Aug. 1867, 2; App. 1.D199–200.

55. See headnote, pp. 29–32, and associated documents.

56. App. 1.D219–20.

57. Why Edison returned to Port Huron is unknown, but there is reason to believe that he was ill during the winter of 1867–68, which may explain his long stay there. Mrs. M. J. Powers to TAE, 18 Feb. 1880, DF (*TAEM* 53:88).

58. The Port Huron and Gratiot Railway opened for business on 10 October 1866. See also Doc. 4.

59. App. 1.A18; Mrs. M. J. Powers to TAE, 18 Feb. 1880, DF (*TAEM* 53:88).

60. See Docs. 26 and 28.

61. App. 1.A18; Doc 36, n. 1; "The Napoleon of Science," *New York Sun*, 10 Mar. 1878, Cat. 1240, Batchelor (*TAEM* 94:119–20).

–1–

Union School Report

PORT HURON, JAN. 18, 1858.

PORT HURON UNION SCHOOL.[1]

HON. IRA MAYHEW, *Superintendent of Public Instruction:*

DEAR SIR—

Your Circular, asking information in relation to the success and usefulness of the Union School effort in the State, was handed to me by one of our School Board, who wished me to answer it. I will endeavor to do so briefly.

The Port Huron Union School was established in the fall of 1849, and has been in successful operation since that time.

The size of the site is 100 feet square, fronting on a Public Park 200 feet square, and is enclosed with a good post and rail fence. The House is 70 feet in length, 30 in width, costing about $2,500.[2] About $30 worth of apparatus.

No. of Departments, 3. 5 Teachers—4 female and 1 male.[3] Female from $4 to $5 per week; Principal $800 per year; costing in all, for salaries for the year, about $1,600.

Average No. attending School, 200.

We have the following branches taught in School,[4] viz: Reading, Writing, Orthography, Book-Keeping, Prof. Davies' Mathematical Course, English Grammar, Geography, Astronomy, Physiology, Philosophy, and Chemistry.[5] Advantages do result, in our experience, from the co-education of the sexes.

The expenses of the School are met in part by a Rate Bill.[6]

Tuition is less, under the Union School organization, than under the Single District arrangement.

The Union School System is calculated to produce a good influence upon the community. I am well satisfied, from my experience of 4 years under the District organization, and 9 under the Union School, in this place, that it is in every way very desirable, and highly satisfactory to the community.

There are some things I would be pleased to dwell upon more at length, but time will not permit at present. Very respectfully yours,

A. CRAWFORD, *Principal of Union School.*[7]

PD, Mayhew 1858, 473–74.

1. Edison attended this school for an unknown period about 1858. Aside from his limited experience with Engle's school and with Sunday School, this was his only formal education. During a visit to Port Huron in 1896, Edison reminisced publicly about his school days. (N. W. King to TAE, 9 Nov. 1914, and Eugene Fechet to TAE, 13 Feb. 1924, GF; *Norwalk [Ohio] Reflector,* 10 Mar. 1896, copy in EBC, NjWOE.) In the 1850s in Michigan, district schools were ungraded primary schools, and union schools typically had separate grades. Starring and Krauss 1969, chaps. 1–3; Dain 1968.

2. This building, later called the North Union School, served the region north of the Black River in Port Huron. Jenks 1912, 271–73; Endlich 1981, 45–46.

3. "Departments" refer to grade. Including the principal, there were thus two instructors per grade. Starring and Krauss 1969, 36.

4. At this time the school was noted for its attention to science and for its success in providing advanced students with something approaching secondary education even before the high-school department was established in 1859. *Port Huron Commercial,* 25 Aug. 1855 and 6 Oct. 1855; Port Huron Schools 1859, microfilm copy, MLM.

5. "Philosophy" here meant "natural philosophy," approximating the subject today called physics (Nier 1975, chaps. 2, 5). Edison later stated that he first learned physical science from a standard schoolbook of the era, Richard Parker's *School Compendium of Natural and Experimental Philosophy.* The reports for the Port Huron schools in this period list "Parkers" as the natural philosophy text. Textbook list in Port Huron Schools 1859, MLM.

6. Local property taxes helped support the public school.

Frontispiece from Richard Parker's Natural Philosophy, *the textbook that introduced Edison to telegraphy.*

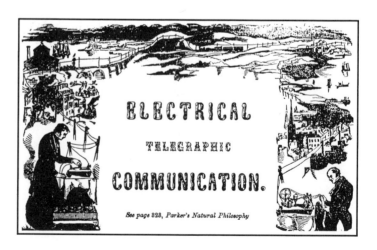

7. Alexander Crawford, an immigrant from Scotland, was then in his last year as head of the Port Huron Union School. His school was known in Port Huron simply as "Crawford's school." Edison recalled that he was highly respected and a strict disciplinarian. *History of St. Clair County* 1883, 566; *Port Huron Commercial,* 7 Apr. 1855, 2; ibid., 18 Aug. 1855, 2; *Norwalk (Ohio) Reflector,* 10 Mar. 1896, copy in EBC.

WEEKLY HERALD   Doc. 2

Edison published the *Weekly Herald* while working as a newsboy on the Grand Trunk Railroad during the winter and spring of 1862.[1] He later claimed 500 subscribers for the single sheet, which contained news and advertising from towns on the line. With some editorial assistance from train conductor Alexander Stevenson, Edison printed the newspaper in the baggage car on a galley proof press bought from J. A. Roys, a Detroit bookseller and stationer.[2] In his autobiographical reminiscences, Edison stated that he bought the type from a junkman, but he later recalled that he purchased "upper and lower case type from William F. Stor[e]y of the Detroit Free Press."[3] When Edison discontinued the paper, he and an apprentice at the *Port Huron Commercial* published an apparently scurrilous and short-lived gossip sheet called *Paul Pry.*[4]

1. Another issue, dated "June," was photographically reproduced in "The Genesis of a Genius," *Magazine of Michigan* 1 (Oct. 1929): 21. Close but not precise copies of that issue are at MiDbEI. Simonds 1940, 322.

2. Alexander Stevenson to TAE, 2 Mar. 1881, DF (*TAEM* 57:76); Jones 1924, 20–33; TAE marginalia, including a drawing of the press, on William Meadowcroft to TAE, n.d., Port Huron Museum of Arts and History.

3. App. 1.A6; TAE marginalia, R. H. Gore to TAE, 12 Oct. 1914, GF. Storey, owner and editor of the *Detroit Free Press,* had left Detroit for Chicago before Edison began his work on the railroad (Walsh 1968, 115). Edison's official biography indicates that he continued printing the *Weekly Herald* at his home in Port Huron after being thrown off the baggage car following a fire caused by one of his chemical experiments. The discovery of 185 pieces of type during an archaeological excavation of the homesite lends credence to this story. Dyer and Martin 1910, 37–38; Stamps and Wright 1982, 35–39.

4. The apprentice, William Wright, later referred to this paper as the *Blowhard* and stated that they surreptitiously printed it on Saturday nights in the *Commercial* office. Wright referred to the paper as "peppery" and "hot stuff." Wright to TAE, 10 Oct. 1878, DF (*TAEM* 16:210); Wright to TAE, 29 June 1900, GF; App. 1.D52; Dyer and Martin 1910, 38–39.

Issue of the
Weekly
Herald[1]

HERALD

THE WEEKLY HERALD.

PUBLISHED BY A. EDISON.

TERMS.

THE WEEKLY Eight Cents Per Month,

LOCAL INTELEGENCE.

Premiums—We believe, that the Grand Trunk Railway, give premiums, every 6 months to their Engineers, who use the least Wood and Oil. running the usual journey. Now we have rode with Mr. E. L. Northrop, one of their Engineers and we do not believe you could fall in with another Engineer, more careful, or attentive to his Engine, being the most steady driver that we have ever rode behind [and we consider ourselves some judge, haveing been Railway riding for over two years constantly.] always kind, and obligeing, and ever at his post. His Engine we understand does not cost one-fourth for repairs what the other Engines do. We would respectfuly recommend him to the kindest consideration of the G. T. R. Offices.

The more to do the more done—We have observed along the line of railway at the different stations where there is only one Porter, such as at Utica, where he is fully engaged, from morning until late at night, that he has everything clean, and in first class order, even the platforms the snow does not lie for a week after it has fallen, but is swept off before it is almost down, at other stations where there is two Porters things are visa a versa.

J. S. P, Hathaway runs a daily Stage from the station, to New Baltimore in connecton with all Passenger Trains.

Profiesser Power has returned to Canada after entertaining delighted audiences at New Baltimore for the past two weeks listening to his comical lectures etc.

Did'nt succeed—A gentleman by the name of Watkins agent for the Hayitan goverment, recently tried to swindle the Grand Trunk Railway company out of sixty seven dollars the price of a valise he claimed to have lost at Sarnis, and he was well nigh succesfull in the undertaking. But by the indominatable perseverance and energy of Mr. W. Smith, detective of the company. The case was cleared up in a very different style. It seams that the would be gentleman while crossing the river on the ferry boat, took the check off of his valise, and carried the valise in his hand, not forgeting to put the check in his pocket, the baggageman missed the baggage after leaving Port Huron, while looking over his book to see if he had every thing with him, but to his great surprise he had lost one piece, he telegraphed back stateing so, but no baggage could be found. It was therefore given into the hands of Mr Smith, to look after, in the meantime Mr Watkins, wrote a letter to Mr Tubman, Agent at Detroit asking to be satisfied for the loss he had sustained in consequence, and refering Mr Tubman to Mr W, A, Howaad Esq, of Detroit, and the Hon. Messrs Brown & Wilson of Toronto, for reference. We hardly know how such men are taken in with such traveling villians, but such is the case, meantime Mr Smith, cleared up the whole mystery by finding the lost valice in his posession, and the Haytian Agent offered to pay ten dollars for the trouble he had put the company to, and have the matter hushed over, Not so, we feel that the villian should have his name posted up in the verious R. R. in the

country, and then he will be able to travel in his true colors.

We have noticed of late, the large quantitys of men, taken by Leftenant Donahue, 14 regt. over the G. T. R. to their rendevous at Ypsalanta, and on inquiring find that he has recruited more men than any other nam in the regiment. If his energy and perseverance in the field when he meets the secesh. is as good as it was in his recruiting on the line of the Grand Trunk R. he will make a mark that the secesh wont soon forget.

Heavy Shipments at Baltimore—We were delayid the other day at New Baltimore Station, waiting for a friend, and while waiting. took upon ourselves to have a peep at things generly: we saw in the freight house of the GTR 400 bbls of flour and 150 hogs, waiting for shipment to Portland

BIRTH

At Detroit Juncton G. T, R, Refreshment Rooms on the 29 inst, the wife of A Little of a daughter.

We expect to enlarge our paper in a few weeks

In a few weeks each subscriber will have his name printed on his paper,

Reason Justice and Equity, never had weight enough on the face of the earth, to govern the councils of men.

NOTICE.

A very large buisness is done at M. V.Millards Waggon and Carriage shop, New Baltimore, Station. All orders promptly attended to. Particular attension paid to repairing.

Port Huron F benary 3rd 1862,

RIDGEWAY STATION,

A daily Stage leaves the above named Station or St Clair, every day, Fare 75 cents,

A Daily stage leaves the above named place for Utica and Romeo, Fare $1.00,

Rate & Burrell, propietors

OPPISITION LINE.

A Daily Stage leaves Ridgeway Station, for Burkes Cor. Armada Cor. and Romeo.

A Daily stage leaves Ridgeway station on arrival of all passenger trains from Detroit. for Memphis R.Quick propietor, tf

UTICA STATION,

A daily Stage leaves the above named Station, on arrival of Accommadation Train from Detroit for Utica, Disco, Washington, and Romeo. S. A. Frink driver. Mr. Frink is the oldest and most careful driver known in the State [Ed.]

Mt CEEMENS,

A daily stage leaves the above named station, for Romeo, on arrival of the morning train from Detroit, our stage arrives at Romeo two hours before any other stage. Hicks & Halsy, prop, tf

THE NEWS,

▰ Cassius M. Clay, will enter the army on his return home.

▰ The thousandth birthday of the Eempire of Russia will be celebrated at Novgorod in august.

▰ "Let me collect myself" as the man said when he was blown up by a powder mill,

GRAND TRUNK RAILROAD

CHANGE OF TIME

Going west.

Express. leaves Port Huron. 7.05 PM
Mixed For Detroit,leaves Pt.Huron at 7,40 A.M

GOING EAST.

Express leaves Detroit, for Toronto, at 6 15 A.M
Mixed, For Pt. Huron, leaves at. 4.00 P.M
Two Freight Trains each way.

C. R, Christie. Supt.

STAGES.

NEW BALTIMORE STATION

A tri-weekly stage leaves the above named Station every day for New Baltimore. Algonac. Swan Creek. and Newport.

S. Graves propietor,

MAIL EXPRESS.

Daily Express leaves New Baltimore Station every morning on arrival of the Train from Detroit. For Baltimore, Algonac. Swan Creek. and Newport. Clark & Bennett. prop.

Pt. HURON STATION,

An Omnibus leaves the station for Pt. Huron, on the arrival of all Trains.

Fare 3 cents, Oley Agent

LOST LOST LOST,

A small parcel of Cloth was lost on the cars The Finder will be liberally rewarded.

MARKETS.

New Baltimore Feb 9th,
Butter at 10 to 12 cts per lb
Eggs, at 12 cts, per dos,
Lard at 7 to 9 cents per lb.
Dressed Hogs, at 3.00 to 3.25 per 100 lbs,
Flour–at 450 to 475 per bbl.
Buckwheat at 1.50 per 100 lbs.
Mutton–at 4 to 5 cts per lb,
Beans–at 1.00 to 1.20 per bush.
Potatoes at 30 to 35 "   " each
Corn at 30 to 35 cts. per bush.
Turkeys–at 50 to 65 cts each.
Chickens at 10 to 12 cts a lb.
Ducks at 30 cents per pair.

ADVERTISEMENTS

RAILROAD EXCHANGE.

At Baltimore Station
The above named Hotel is now open for the reception of Travelers. The Bar will be supplied with the best of Liquors, and every exertion will be made to the comfort of the Guests
S. Davis Propietor,

SPLENDID PORTABLE COPYING PRESSES FOR SALE AT
Mt. CLEMENS ORDERS TAKEN.
BY THE NEWS AGENT ON THE MIXED.

Ridgeway Refreshment Rooms—I would inform my friends that I have opened a refreshment room for the accomadation of the traveling public R. Allen. propietor. tf

TO THE RAILROAD MEN

Railroad Men send in your orders for Butter, Eggs, Lard, Cheese, Turkeys, Chickens, and Geese. W. C. Hulets, New Baltimore Station

---

PD, NjWOE, DF (*TAEM* 12:6). [a]Place and date taken from text; form of date altered.

1. See headnote above.

To Willis Engle

Port [H]uron[a] Mich Aug 10TH [1862][1]

Friend Willie[2]

I have neglected you a long time   Ought to have written long ere this, but my time is all taken up with my business on the cars. you sea that I am on the Grand Trunk road. I dont get home until ten in the evening, and have no time to write except Sundays. there is not [   ][a] news of importance just now: except running fo[r Can?][a]ada because they do not want to be drafted. [They h?][a]ave men stationed at Port-Huron, & at the Fort, to prevent pepole passing over. It is quite ammusing to see some country greeny try to dodge the officers. I think you and me Willie are exempt from "drafting"   are you glad or do you feel very patriotic   They are raising three companys in Port Huron. The Captains are William Sanborn, John Atkinson, Edward Lee.[3]

I see very little of Birt Buchanan, he is working on the river this summer I believe. Bert has got to be a young man and Fred is as tall as his father.[4] Robert Hickling[5] died last week. I suppose you hear from George and Frank often.[6] I have no more time to write you   It is Eleven Oclock P.M. I shall not neglect you so long again. write as soon as you receive this   remember me to your Father's family   I ever remain y[our?][a] friend

Alva Edison.

Edison at the age of fourteen sold newspapers and candy on the trains of the Grand Trunk Railroad.

ALS, MiDbEI, EP&RI. [a]Obscured by damage to paper.

1. This date is based on the commission dates for Sanborn, Atkinson, and Lee. See n. 3.

2. Willis Engle was the son of Rev. George Engle, minister of the Episcopal Church in Port Huron, whose private school Edison briefly attended. U.S. Bureau of the Census 1964, roll 362; Willis Engle to TAE, 24 July 1874, Cat. 1173, Scraps. (*TAEM* 27:412); George Engle to TAE, 13 Aug. 1885, DF (*TAEM* 77:147).

3. William Sanborn, member of a prominent Port Huron family, was commissioned as a major in the 22d Michigan Infantry on 8 August 1862. John Atkinson, who was commissioned as a captain in the 22d Michigan Infantry on 31 July 1862, later became part owner of the *Port Huron Commercial.* Edward Lee was commissioned as a first lieutenant in the 5th Michigan Cavalry on 29 August 1862. *History of St. Clair County* 1883, 362–65, 377–80, 554; Jenks 1912, 280.

4. The Buchanan family lived on the north side of Port Huron near the Edisons. Alexander Buchanan was a lumberman, and his sons, Albert (Birt), Fred, and George, all became raftsmen. By the time this letter was written, George had enlisted. A daughter, Carrie, was Edison's childhood sweetheart (see Chapter 1 introduction, p. 20, n. 12). U.S. Bureau of the Census 1967b, roll 559; *History of St. Clair County* 1883, 458–63, 503–5; Jenks 1912, chap. 23; Crampton 1921, 10–12.

5. Not identified.

6. George and Francis (Frank) Engle were Willie's brothers. Frank stayed in Port Huron, where he worked as a printer, after his family moved away in 1859. U.S. Bureau of the Census 1964, roll 362; idem 1967b, roll 559; Jenks 1912, 319.

–4–

To Family

[Louisville?, Spring?] 186[6?][1a]

Started[b] the store several Weeks   I have growed Considerably = I dont look much like a Boy Now = how is Pitt getting on with street Railway[2]   I will Send you 6 or 7 photographs on the 15th of the month   hows all the folks   did you Receive a Box of Books from Memphis yet   he promised to send them long ageo =

Your Son

ALS (fragment), MiDbEI, EP&RI. Message form of the South-Western Telegraph Co. [a]"186" preprinted. [b]Circled "2" above; probably page number.

1. The comment in the letter regarding receipt of a box of books from Memphis suggests that Edison had already left that town when he wrote this letter.

2. During late 1865 and early 1866, Pitt Edison and a few partners were establishing a horse-drawn street railway from central Port Huron to the Grand Trunk terminal at Fort Gratiot. City authorization was obtained in November, federal approval came in January, the company was formally organized in April, and laying of the rails began in June. Jenks, n.d.; see also Chapter 1 introduction, p. 23, n. 58.

–5–

To Family

[Louisville?, Spring?] 186[6?][1a]

Spanish very good now   before I Come home I will be able to Speak Spanish & Read & write it as fast as any Spaniard   I can also Read French too but Cant Speak it—[2] I want to get Some more Books   not Cost much   wish you would try and get Some of them   Little news boy and give him a letter to that old book man and get them Books right side of the Post Office = wish you would have walker[3] furnish you them regular   take them Every month after this   no more at present

Love to all the folks = it is awful hot here now   people putting Summer Clothes = write soon   have You recd the Second pkg of money   Toney[4] will give you the ten (10)[b] doll[ar]s for the Register[5] he had   get it of him   he says he will sell the other one for 5 dolls   let him sell it   write Soon

A Edison

ALS (fragment), MiDbEI, EP&RI. Message form of the South-Western Telegraph Co. [a]"186" preprinted. [b]Circled.

1. See Doc. 4, n. 1.

2. Edison purchased a Spanish dictionary while in Memphis (Simonds 1940, 60). O. J. McGuire to TAE, 12 Feb. 1912, GF, refers to Edison's studying Spanish in Louisville.

3. Micah Walker.

4. Toney or Tony Bronk, employee at Walker's store. James Wilson to TAE, 14 Feb. 1922, GF.

5. Morse register.

–6–

Notice in the Louisville Daily Journal

Louisville, August 2, 1866

FOR SOUTH AMERICA.—E. A. Edison[1] and J. C. Curran,[2] formerly of the Western Union Telegraph Company, in this city, started yesterday for South America, by way of Nashville and New Orleans. Their exact destination is not known, but they expect to join some one of the various telegraph corps in Brazil or Rio Janeiro. Success to the young adventurers.

PD, *Louisville Daily Journal*, 2 Aug. 1866, 3.

1. That is, T. A. Edison. There is no evidence that an E. A. Edison worked for Western Union in Louisville. See App. 1.D198.

2. On Curran see "District Proceedings," *Telegr.* 2 (1865–66): 15, 140, 172. A list of Louisville employees for November 1866 (EP&RI) does not contain Curran's name, which supports Edison's story about his traveling companions' continuing their journey southward.

MEMORANDUM BOOK Docs. 7–24

The notes and drawings in Docs. 7–24 provide a rare glimpse of Edison's early, self-directed technical education. Selected from a small memorandum book that Edison likely used in the late summer of 1867,[1] these documents suggest his growing mastery of contemporary telegraph technology and related science and reveal his early creative variations on basic telegraph designs. The notebook entries include book lists, a table of electrical conductivity, and more than a dozen sketches of telegraph circuits and components.[2] Probably desiring a record for himself, he copied some of this material from contemporary telegraph manuals and journals found in the Cincinnati Free Library. Lacking the financial resources for extensive acquisition of custom-made apparatus, he employed standard components in his new designs. These early documents are not juvenilia but mark the beginning of an increasingly detailed record of Edison's early technical approaches and ideas, some of which characterize his subsequent inventive endeavors.

Literature and Practice   Docs. 7–10

Documents 7–10 reveal Edison's growing awareness of the informational resources of the technological and scientific communities at home and abroad as well as his design of a practice instrument for apprentice telegraph operators. A table of electrical conductivity and two book lists indicate his knowledge of the *Proceedings* of the Royal Society of London and the scientific works of Michael Faraday and other British scientists. For his table of conductivity, Edison obtained comparative data for silver and copper from the literature and, perhaps on the basis of his own simple calculation, presented copper's conductivity as a percentage of that of silver. His design of a telegraph operator's practice instrument for receiving standard wire messages, storing them on paper tape, and playing them back at slower speed employed apparatus typically found in an American telegraph office of the mid-1860s.

Relays   Docs. 11–13

Many telegraph inventors in the 1860s sought to improve the standard relay, a basic component in telegraph circuitry. Documents 11–13 show Edison's early concern with self-adjusting and polarized relays. A relay was a very sensitive electromagnetic switch that responded to incoming signals from the line and in turn actuated a local circuit that employed a battery and a receiving instrument such as a sounder or a Morse register.[3] Because changing line conditions caused the current to vary widely during long-distance transmission, receiving operators frequently had to adjust the springs on relay armatures in order to receive messages clearly and quickly. Edison, like other inventors, sought to devise self-adjusting relays that would automatically maintain their responsiveness under widely varying conditions. He was also concerned with polarized relays, which used permanently magnetized cores. Although rare in the United States at this time, polarized relays attracted Edison's interest and played an important role in his later telegraph designs.

Repeaters   Docs. 14–22

Another object of interest to Edison was the telegraph repeater, an essential device for long-distance transmission. Prior to the Civil War long-distance telegraphy was not commonplace because, for any given size battery, the strength of the signal was proportional to the length of the line, and increasing battery strength could not generally compensate for distances of more than about 250 miles. When necessary, operators at intermediate telegraph stations on long lines retransmitted messages by hand. The extension of telegraph lines to the Pacific Coast in 1861, the wartime expansion of the nation's telegraph network, and the reopening of southern connections as the war ended focused attention on development of electromechanical repeaters to receive and retransmit telegraph signals at intermediate stations. A repeater used a receiving device to operate an independent transmitting circuit that had its own battery, forwarding incoming signals to the next station. The general introduction of repeaters soon made it economically feasible to transmit messages directly across the continent and to provide press reports to towns on branch circuits.[4]

A number of telegraph inventors, including Edison, sought to improve repeaters. Repeaters worked in only one direction unless a switch was provided to accommodate normal two-way telegraph operation. In this respect there were two types: the button and the automatic repeater. Edison considered both in this notebook. In the button repeater an attendant manually switched the direction of operation whenever appropriate.[5] In the automatic repeater the switching of direction was done electromechanically. Since repeaters were subject to the same line-current variations as relays, inventors also sought to design self-adjusting repeaters.

Multiple Telegraphy   Docs. 23–24

The goal of using a single wire to carry more than one message at a time engaged inventive American and European electricians from the early 1850s. The first such systems were called "double transmission," indicating that they could handle two messages simultaneously.[6] The American inventor Moses Farmer proposed and demonstrated the earliest multiple telegraph system, and many others created workable but limited "duplex" systems. Prior to 1868, however, none had been adopted for regular use on major lines except in the Netherlands, and the feasibility of multiple telegraphy had not generally been recognized in the United States.[7] Therefore, Edison's serious interest in such systems, as indicated in Docs. 23 and 24, was comparatively early for the American context.[8]

Any multiple telegraph design required receiving apparatus

that responded to incoming signals but remained insensitive to outgoing signals from the same station. Most inventors at this time employed a "differential" approach, in which outgoing signals created two opposing and mutually canceling magnetic fields in the local receiver. In contrast, incoming signals created only one magnetic field, thereby activating the receiver.[9] Edison's designs in this notebook followed the differential approach, incorporating a modification of a special type of relay used in some self-adjusting repeaters.[10]

1. A number of the entries could only have been made in 1867; others may have been made earlier. All were written or drawn in the same ink and style. Edison apparently left the book in Port Huron after his return there for the fall and winter of 1867–68. His father used portions of it for personal and financial records between 1869 and 1871.

2. Because many pages of the notebook are loose, blank, or missing, the original temporal and physical order of the entries cannot be determined. The documents presented here are grouped by subject. Those omitted include a page with two less-finished sketches, an incomplete drawing from the front flyleaf, two pages of simple geometric forms, and a few unlabeled calculations.

3. On relays see Pope 1867, 30–31; and Prescott 1877, 439–42.

4. A series of articles about repeaters by Franklin Pope appeared in the *Telegrapher* from 26 December 1864 through 1 November 1865. Much of this material was subsequently included in the 1866 edition of George Prescott's *History, Theory, and Practice of the Electric Telegraph*. In his 1863 edition Prescott had treated long-distance telegraphy as unnecessary and impractical; in his 1866 edition he extolled its ease, thereby indicating the impact of improved repeaters on telegraph practice. Pope 1869, 45–55; Prescott 1863, 142; idem 1866, 459–77.

5. "Button" referred to the knob on the hand-operated switch lever. Close attention was required if such a repeater was to work successfully. "The Button Repeater," *Telegr.* 4 (1867–68): 241.

6. In America the term "double transmission" gave way to "duplex" for the sending of messages in opposite directions and "diplex" for the sending of two messages in the same direction. The terminology of multiple telegraphy was never fully standardized, but the editors have chosen to use these definitions for purposes of clarity.

7. See Pope 1868a; Sabine 1872, 119–29; Prescott 1877, 768–838; Schellen 1888, 779–863; and King 1962b, 307–10. Regular operation of multiple telegraphs in the United States began in March 1868 when Joseph Stearns, president of the Franklin Telegraph Co., introduced a duplex on a line between New York and Boston. It was his own design but was based on the work of the Austrian Wilhelm Gintl, the Hanoverian Carl Frischen, the Prussians Siemens and Halske, and the American Moses Farmer. See Docs. 50 and 283 and the diagram of Stearns's differential duplex (p. 101, n. 3).

8. A common introductory handbook asserted that duplex telegraphy was impossible (Smith 1871, 37), and a survey in 1863 barely mentioned multiple telegraphy, dismissing it as unimportant (Prescott 1863, 142–43). Edison probably knew of a Farmer demonstration on the telegraph

lines of the Indianapolis, Cincinnati, and Lafayette Railroad during the period between Edison's two stays in Cincinnati. During his second stay Edison borrowed instruments from Charles Summers, superintendent of telegraphs for that railroad, and worked with Summers on experiments with relays. App. 1.D218; Reid 1879, 374–75; "Telegraphic Miscellanea," *Telegr.* 2 (1865–66): 16; Pope 1874.

9. Prescott 1877, 793–807; Maver 1892, 169–80. An alternative approach that became important later was the "bridge" method, which employed an arrangement like that in the Wheatstone bridge. See Doc. 285, n. 17; and Doc. 309, n. 2.

10. See Docs. 11 and 17.

–7–

Book List[1]

[Cincinnati?, Summer 1867?]

Further List[2] of Telgh & E[lectrical?] & S[cientific?] Books
Faradays—Researches in Electricity   1 Vol   Cin Free Library—[3]
Electric Telegraph by Dr Lardner[4]   Entirely Re-written by Edward B Bright[5] F.R.A.S.[6]   Secy to the British & Irish Telegraph Co   140 illustrations   8vo—5s—1867—
Handbook of Electricity Magnetism and acoustics by Lardner   Edited & completed by G C Foster[7] B.H.[8]   Prof of Experimental Physics   University College London   New Edition   Small 8vo   400 Illustrations   Price 1867—5s[9]

AX, NjWOE, Lab., PN-69-08-08 (*TAEM* 6:766).

1. See headnote above.

2. It is unclear whether this refers to Doc. 8 (headed "Books") or to a missing entry. However, though loose, this page could not have followed Doc. 8 physically in the original notebook.

3. This entry indicates that by 1867 Edison was acquainted with Michael Faraday's *Experimental Researches in Electricity*. The Cincinnati Public (formerly Free) Library holds an 1867 edition (*NUC Pre-1956* 166:515). It is said that Edison purchased a copy of Faraday's work in Boston in 1868 (Josephson 1959, 60–61; Dyer and Martin 1910, 100–101). A copy he signed and dated 21 August 1873 is located at the Edison National Historic Site.

4. Dionysius Lardner was the first professor of astronomy and natural philosophy at University College in London. A highly successful public lecturer, Lardner was a polymath and devoted most of his career to a prodigious output of encyclopedias and reference libraries. *DNB*, s.v. "Lardner, Dionysius."

5. Edward Bright, one of the organizers of the first Atlantic cable project, was secretary and manager of both the London District Telegraph Co. and the British and Irish Magnetic Telegraph Co. Kieve 1973, 56, 67; Bright 1898, 30–31.

6. Fellow of the Royal Astronomical Society.

7. George Carey Foster did research mainly on electrical topics after

shifting his field from chemistry to physics. He invented the Carey Foster bridge, a modification of the Wheatstone bridge and a common laboratory instrument in the late nineteenth century. He was an original member of both the Society of Telegraph Engineers and the Physical Society of London and was later president of each. Appleyard 1939, 35, 288.

8. Should be "B.A."

9. Some years later Edison obtained copies of other editions of these and other works by Lardner, which are now located at the Edison National Historic Site.

–8–

Book List[1]

[Cincinnati?, Summer 1867?]

—Books—
Culleys—Practical Telegraphy—London[2]
Walkers—Electric Telgh Manipulation[3] "[a]
Dr C. C. Bombaugh—<u>The Gleanings</u> III Edition, Large, pub—Kurtz of Balt[im]o[re] 550 pages—Curiosities Litrature puzzles antiquidated cyphers etc—Refer R. Table Page 333—[4]
Sabine on The Telgh   pub—London 1867 price 6 dollars—Van Nostrand N.Y—[5]

AX, NjWOE, Lab., PN-69-08-08 (*TAEM* 6:779). [a]Ditto mark under "London" in line above.

1. See headnote, pp. 29–32.

2. Richard Culley was a prominent British telegraph specialist. He was chief engineer of the Electric Telegraph Co. and held a similar post in the telegraph department of the British Post Office following the government takeover of domestic telegraph companies in 1870. He published the first edition of *A Handbook of Practical Telegraphy* in 1863 and a second edition in 1867. Edison owned a copy of the 1871 edition (now located at the Edison National Historic Site) and wrote numerous comments and criticisms in the margins. Baker 1976, 94.

3. Charles Walker was a noted British telegraph inventor and a telegraph superintendent and electrician for the South-Eastern Railway. He made one of the earliest significant trials of a marine telegraph cable. His *Electric Telegraph Manipulation* (1850) was published by the Royal Society for the Arts as one of a series of short, elementary handbooks on technical crafts. *DNB,* s.v. "Walker, Charles Vincent."

4. Charles Bombaugh was a medical doctor who wrote several eclectic works. *Gleanings for the Curious from the Harvest Fields of literature, science, and art: A Melange of excerpta curious, humorous, and instructive,* his most widely known book, went through many editions. Page 333 of the third edition (1867) records a day's schedule in the life of Elizabeth Woodville, who became queen of England in 1465. *NCAB* 7:275.

5. Robert Sabine was a British specialist in electricity and the telegraph as well as a consulting engineer, manufacturer, and oceanographic researcher. He was the son-in-law of Sir Charles Wheatstone, scientist

and co-inventor of the first commercially successful electric telegraph in England, and assistant to Sir William Siemens, scientist and industrialist. Sabine was the first treasurer of the Society of Telegraph Engineers (founded in 1872; later renamed the Institution of Electrical Engineers). With J. Latimer Clark he coauthored one of the most widely used electrical handbooks of the nineteenth century, *Electrical Tables and Formulae* (1871). Sabine also wrote *The Electric Telegraph* (1867), which in later editions was entitled *The History and Practice of the Electric Telegraph*. In his laboratory notes Edison cited specific pages of the 1867 version (e.g., NS-undated-005, Lab. [*TAEM* 8:454–55]). *Teleg. J. and Elec. Rev.* 15 (1884): 353.

–9–

Notebook Entry:
Electrical Conductivity[1]

[Cincinnati?, Summer 1867?]

Conductivity Metals[2]

Procdgs Royal Society—[3]

	Conductivity			Condty at 212°
	Silver—32°	Silver 212°		
	= 100	equal 100[4]		
	32° | 212°			
Silver (hard drawn)	100.00 | 71.56	100.00		
Copper "	99.95 | 70.27	98.20		

—No good—

AX, NjWOE, Lab., PN-69-08-08 (*TAEM* 6:776). On the same page there is a pencil sketch of a gate and part of a fence, probably by Samuel Edison.

1. See headnote, pp. 29–32.

2. This table shows relative electrical conductivities for silver and copper, two of the best metallic conductors of electricity. Conductivity is the inverse of electrical resistance; its precise determination was a prominent topic of research and a matter of great technical and financial interest in the second half of the nineteenth century ("Anniversary Meeting," *Proc. Roy. Soc.* 18 [1869–70]: 111–12; Bright 1974, 214-26). Because at ordinary temperatures silver wire has the lowest resistance to a current, it was used as the standard in the comparative data given in this table.

3. The main journal of the Royal Society of London is the *Philosophical Transactions;* in its *Proceedings* the society publishes reports of its meetings, including both business matters and papers read, rather than full texts of lengthy papers. The editors were unable to find the figures in the center column of Edison's table anywhere in the *Proceedings* for the previous decade, but did find those in the left column in A. Matthiessen and C. Vogt, "On the Influence of Temperature on the Electric Conducting Power of Thallium and Iron," (*Proc. Roy. Soc.* 12 [1862–63]: 475). The figures for both columns do appear, however, in Sabine 1867, 266 (see Doc. 8, n. 5). Other reference works give slightly different data.

Sabine mentions Matthiessen as his source but does not cite any particular paper or journal.

4. In this column Edison expressed the conductivity of copper at 212° Fahrenheit as a percentage of that of silver at that temperature. The number he derived for copper is the quotient of the two figures from the second column.

–10–

[Cincinnati?, Summer 1867?][2]

Notebook Entry:
Practice Instrument[1]

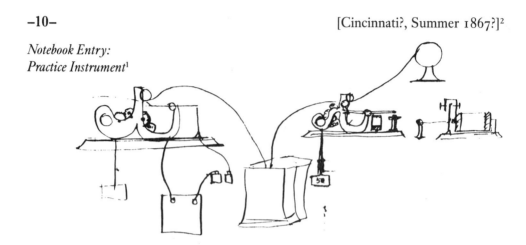

AX, NjWOE, Lab., PN-69-08-08 (*TAEM* 6:769).

1. See headnote, pp. 29–32. This is a sketch of what Edison later called his first invention—a combination of instruments designed to slow the presentation of coded telegraph messages for the apprentice telegraph operator. He used two Morse embossing registers, one to record messages as they came in and the other to reproduce the coded messages on a sounder at a slower speed. While registers were not much used at the time of Edison's experiments, they were still widely available in telegraph offices. Edison later claimed that this invention was in some sense the direct ancestor of the phonograph. App. 1.D203–4; Lathrop 1890, 429; Dyer and Martin 1910, 68.

The registers shown were driven by weights and clockwork, which enabled Edison to adjust the speed at which paper tape ran through them. The device at the far right was the relay that operated the local circuit. Normally it would actuate the sounder (represented here by the square at the lower left with two wires running to it), and the operator would take the message by sound. In this case, however, the relay operated a Morse embossing register (shown here just to the left of the relay, with the roll of paper for it shown above); the embossing register recorded the message on paper tape, which accumulated in the bin. That tape was then drawn from the bin and used to operate a modified register that activated the sounder. Edison described the device further in "A Novel Device," *Operator,* 1 Sept. 1874, 1.

2. Edison devised this practice instrument in Indianapolis in 1865. While he may have made this sketch there or elsewhere prior to 1867, a

common drawing style links it with other items in the notebook. The general form of the registers in the drawing also indicates a date in the mid-1860s. Earlier weight-driven registers had a quite different shape. Later registers were driven by springs and did not have the long, straight, horizontal stylus arm shown in this sketch.

–11–   [Cincinnati?, Summer 1867?]

Notebook Entry: Relay[1]

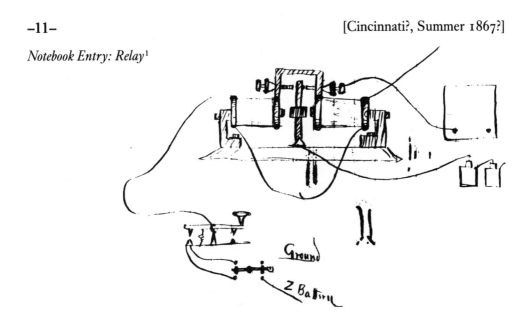

AX, NjWOE, Lab., PN-69-08-08 (*TAEM* 6:773).

1. See headnote, pp. 29–32. In his early duplex telegraph designs, Edison used relays that were of this general form and incorporated an armature pivoted between opposed electromagnets (see Docs. 23 and 24). Several inventors also used such designs in their attempts to devise self-adjusting relays and self-adjusting repeaters (see Doc. 17, n. 1; and Docs. 31 and 40). The intended operation of Edison's design is uncertain. He may have intended that the armature be a permanent magnet, in which case the device could be a modified polarized relay (see Docs. 12 and 21). Otherwise the relay would require parts such as springs, which might be visible in a different view of the instrument and indicate how it worked. The electromagnets (seen in side view) were probably a common U-shaped type with two coils each. The square at the right represents a sounder, and below it are its local batteries.

Notebook Entry:
Relay[1]

AX, NjWOE, Lab., PN-69-08-08 (*TAEM* 6:774).

1. See headnote, pp. 29–32. This drawing shows an early stage of Edison's experimentation with polarized relays. Many of his own designs for printing, automatic, and multiple telegraphy routinely exploited polarized relays that were variations of the original form invented in 1854 by the German electrician Werner Siemens. Apparently Edison was trying here to design a relay without incurring the expense of having one custom-made. At the time, polarized relays were not readily available for purchase, and as late as 1871 the catalog of telegraph supplier L. G. Tillotson of New York did not include them (Smith 1871, app.). The report "Electrical Apparatus at the Paris Exhibition" (*Telegr.* 3 [1866–67]: 214) noted the excellence of Siemens's polarized relays.

Siemens originally devised a polarized relay in order to solve transmission problems caused by self-induction and capacitance in undersea cables. He found a solution in momentarily reversing the current's polarity between signals rather than opening the circuit. The receiving device for these alternating currents closed a local circuit in response to signals of one polarity only. European telegraphers commonly used the Siemens relay on long lines, considering it clearly superior to the Morse relay. It is not certain when knowledge of the polarized relay became common among American telegraphers. Although it was well known by the end of the 1860s, American telegraph circuits did not use it. Edison claimed that he experimented with it in 1865, and he was definitely employing it in 1868 (see Doc. 50). Siemens 1891, 118–23; idem 1968, 130, 291; Brooks 1876, 6; Culley 1871, 216–18; Klein 1979, 159; Maver 1892, 184–87.

Siemens's polarized relay is shown here in its simplest form. The L-shaped bar **NS** is a permanent magnet to which both the pivoted armature **ab** and the electromagnet cores are attached and by which both are magnetized. When no current flows in the coils of the electromagnet, the south-magnetized armature remains stationary, being attracted to whichever north-magnetized pole piece (**n** or **n′**) is closer. The electromagnet is wound so that the coils are oppositely magnetized when the current flows. Current flowing in one direction supplements the permanent north magnetization of one core and neutralizes or reverses that of the other core, thus throwing the armature against the strongly north-magnetized pole piece. When the current reverses, the polarities of the two coils reverse and the armature moves toward the other pole piece.

Werner Siemens's polarized relay (simple form).

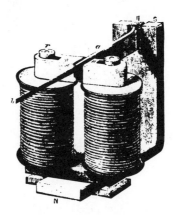

When this device was used as a relay, the local current flowed through the armature **ab**. If the armature was attracted toward **n**, it touched contact point **c** and completed a local circuit. When the armature swung the other way (as in the diagram), it rested against the insulated end of screw **B** and the local circuit was broken.

Werner Siemens's polarized relay in action (top view).

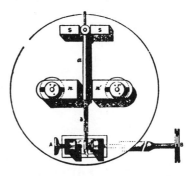

Edison's design uses permanent (horseshoe) magnets to magnetize the electromagnet core and armature so that they repel each other. Winding both electromagnet coils in the same direction would allow a current of the proper polarity to reverse the core's magnetic field, attracting the armature and completing the local circuit with its sounder (represented by the square in the lower center of the drawing) and local battery. A current of the opposite polarity would only more strongly repel the armature.

Notebook Entry:
Relay[1]

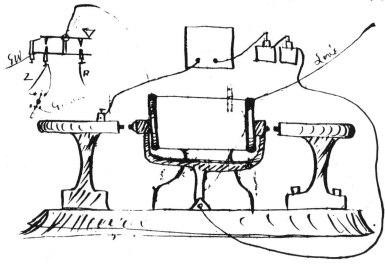

AX, NjWOE, Lab., PN-69-08-08 (*TAEM* 6:773).

1. See headnote, pp. 29–32. This appears to be a polarized relay in which a permanent magnet is in the center of the helix and acts as an armature, rocking left or right on its pivot and thereby making or breaking contact on the left; however, the drawing may be incomplete.

Notebook Entry:
Repeater[1]

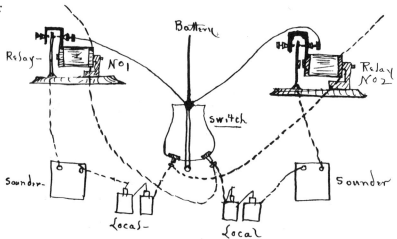

AX, NjWOE, Lab., PN-69-08-08 (*TAEM* 6:767).

1. See headnote, pp. 29–32. This two-sounder button repeater design is a rearrangement of the circuit of the common button (manually

William Maver and Minor Davis's wiring diagram for a button repeater.

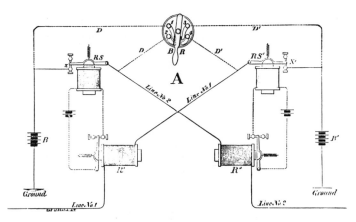

switched) repeater of the era. Edison's design differs from the standard one by using the sounders only to indicate operation; the main relays do the actual repeating. Two problems could arise from it, however. First, general experience showed that using relays as repeaters reduced signal clarity. Second, the repeater operator could not tell by sound when it was necessary to switch the direction of operation, because at such times both sounders but only one relay would react. Therefore, the repeater operator would have to watch the armatures constantly. In the drawing, the relays (shown in side view) are double coils with a U-shaped core; the freestanding vertical piece perpendicular to the core is the relay armature; and each "Local" is a battery for a sounder.

–15–

[Cincinnati?, Summer 1867?]

Notebook Entry:
Repeater[1]

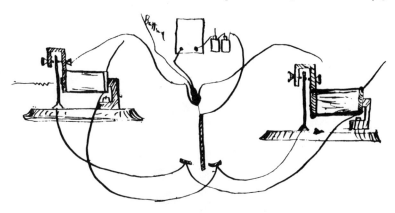

One Sounder Repeater[2]

AX, NjWOE, Lab., PN-69-08-08 (*TAEM* 6:772).

1. See headnote, pp. 29–32.

2. This one-sounder, button repeater design, which is a simpler version of Doc. 14, eliminates the redundancy of the second sounder. Although it still required the operator's constant attention, Franklin

Pope—in an appendix to Pope 1869 (Doc. 71)—presented it as suitable for employment in an emergency.

*Notebook
Entry:
Repeater*[1]

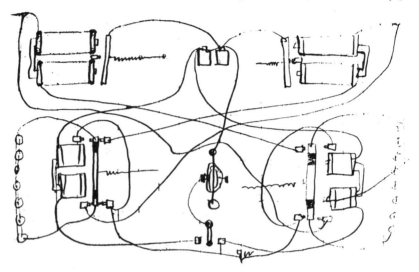

AX, NjWOE, Lab., PN-69-08-08 (*TAEM* 6:775).

1. See headnote, pp. 29–32. In this drawing Edison has copied almost exactly the standard illustration of Charles Bulkley's automatically switched repeater, which was used in 1848 in the first direct telegraph connection between New York and New Orleans. Edison has omitted one wire on each side of his drawing. The electromagnets are shown in top view. On Bulkley's repeater see Prescott 1877, 461–62, 464.

*Charles Bulkley's automatic
repeater.*

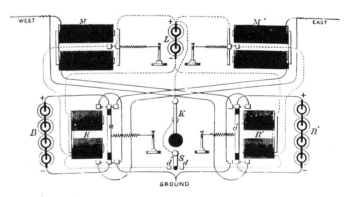

2. The standard illustration of Bulkley's automatic repeater appeared in the *Telegrapher* for 30 January 1865 ("The Telegraphic Repeater," 1 [1864–65]: 141). The style and ink of Edison's drawing are similar to those of other notebook entries that are clearly from 1867.

Notebook Entry:
Repeater[1]

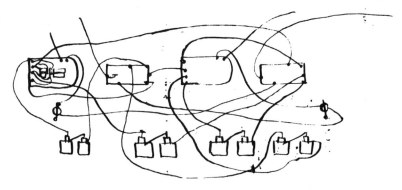

AX, NjWOE, Lab., PN-69-08-08 (*TAEM* 6:772).

1. See headnote, pp. 29–32. This drawing by Edison shows the circuit connections for George Hicks's magnetic self-adjusting repeater, patented in 1862 (U.S. Pat. 34,574). As an operator in Western Union's Cincinnati office in 1865, Edison worked under Hicks, whose automatic repeater remained notable enough in 1867 that the *Telegrapher* reported when one was installed ("Miscellanea: Repeater in Chicago Office," 4 [1867–68]: 16). Edison's drawing shows the office arrangement of the repeater rather than the construction of the individual components. The components are drawn at the top as they would be placed on a table. From left to right they include Hicks's self-adjusting relay, a sounder wired for repeating, a second Hicks relay, and a second sounder. Below these are two switches and eight batteries. Edison's sketch differs from

The patent drawing of George Hicks's self-adjusting repeater shows the sounder on the same board as the rest of the repeater (U.S. Pat. 34,574).

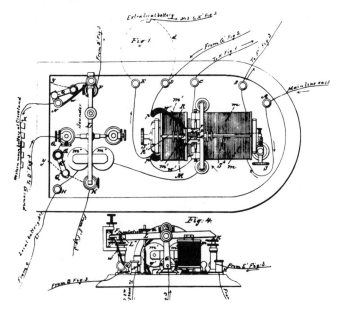

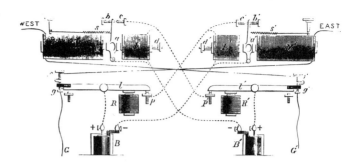

the published patent diagrams by presenting the sounders as separate pieces of equipment. The wiring also differs from the standard diagram of the Hicks repeater; Edison included wires and keys to allow the repeater station to also act as a terminal for each circuit. Pope 1869, 47–51; Prescott 1877, 467–69.

–18–   [Cincinnati?, Summer 1867?]

Notebook Entry:
Repeater¹

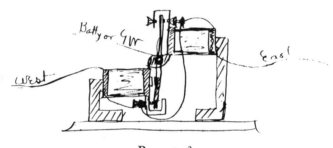

Repeater²

AX, NjWOE, Lab., PN-69-08-08 (*TAEM* 6:769).

1. See headnote, pp. 29–32.
2. This automatic repeater design duplicates almost exactly the drawing in U.S. Patent 13,655, issued to W. A. Peaslee of Indianapolis in 1855. Peaslee's patent was not discussed in the literature of the day and was apparently never used. Edison may have learned of it while in Indianapolis, where he is known to have been interested in repeaters (Cat. 1176:31, Lab. [*TAEM* 6:33]). Edison's drawing differs from Peaslee's patent only in omitting one wire running from the armature pivot to the battery or ground.

Notebook Entry:
Repeater[1]

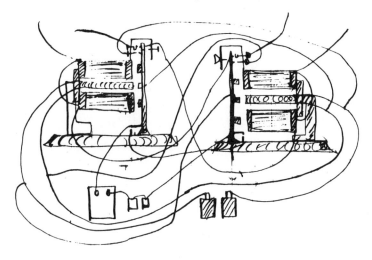

AX, NjWOE, Lab., PN-69-08-08 (*TAEM* 6:770).

1. See headnote, pp. 29–32. This automatic repeater design is quite similar to one patented by James Clark in 1860 (U.S. Pat. 29,247) and almost identical to one patented by Elisha Gray in 1871 (U.S. Pat. 114,938). Edison's drawing shows two electromagnets acting on each armature lever—one in the line circuit and one in a local circuit. The arrangement of the paired magnets assured continuity for either sender. Although the battery connections differ, Edison's design more closely resembles the drawings from Clark's patent than it does the standard illustration of Clark's repeater found in the telegraph literature, which reflected the changes made in actual practice. In practice Clark's repeater was constructed with the two magnets on opposite sides of a centrally pivoted armature (cf. Doc. 20). Edison's design differs from the Clark patent, however, in that the relays rather than the sounders do the actual repeating (cf. Docs. 14 and 15). Consequently only one sounder is necessary for monitoring the repeater's operation. Edison drew the two coils of each relay above and below each other, rather than side by side as usual, with the local electromagnet in between. A sounder, a local battery, and a main-line battery are represented from left to right at the bottom. For Clark's repeater see Prescott 1877, 466–68; for Gray's repeater see Davis and Rae 1877, 15–16 and pl. 9.

Wiring diagram for James
Clark's automatic repeater.

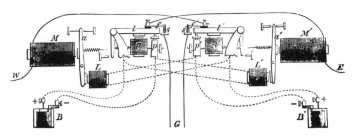

20

Notebook Entry:
Repeater[1]

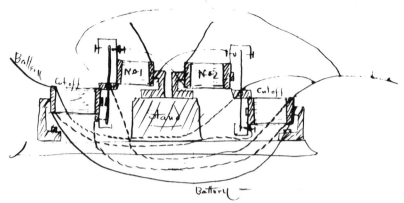

AX, NjWOE, Lab., PN-69-08-08 (*TAEM* 6:768).

    1. See headnote, pp. 29–32. This drawing resembles the standard side-view illustration of James Clark's automatic repeater (see Doc. 19, n. 1). As in his other repeater designs in this notebook, Edison arranged for the relays rather than the sounders to do the actual repeating. This design dispenses with sounders altogether.

21

Notebook Entry:
Repeater[1]

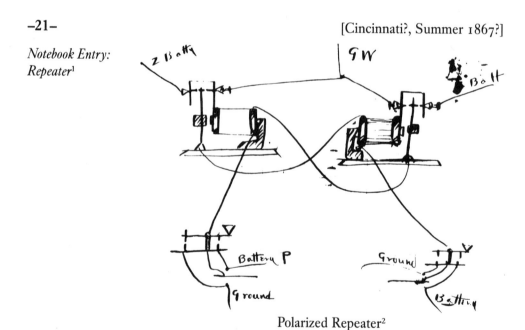

Polarized Repeater[2]

AX, NjWOE, Lab., PN-69-08-08 (*TAEM* 6:767).

1. See headnote, pp. 29–32.

2. This design appears to be an early version of Edison's "combination repeater," which was intended to repeat messages transmitted by Phelps's combination printing telegraph. Edison published a description of his repeater in 1868 in the *Telegrapher* (Doc. 43). This drawing, unlike the published one, includes keys that allow the operator at the relay station to work the line in either direction. Cf. Doc. 22.

–22–

Notebook Entry:
Repeater[1]

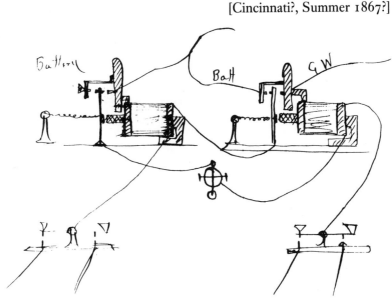

AX, NjWOE, Lab., PN-69-08-08 (*TAEM* 6:768).

1. See headnote, pp. 29–32. This repeater design closely resembles that in Doc. 21, but its intended operation is not known. The drawing represents a side view of the relays. Comparisons with other documents suggest that the vertical pieces that are parallel to the armature levers may represent side views of permanent horseshoe magnets with poles that are close to the projecting poles of the electromagnets (cf. Doc. 12). The projecting, cross-hatched cores of the coils may represent permanent magnets rather than the usual soft iron cores of standard electromagnets, and the small rectangles on the sides of the armatures may represent end views of bar magnets. Cf. Doc. 32.

Notebook Entry:
Multiple
Telegraphy[1]

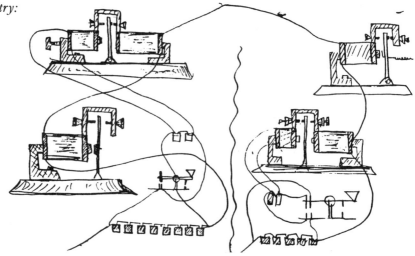

AX, NjWOE, Lab., PN-69-08-08 (*TAEM* 6:770).

1. See headnote, pp. 29–32. This duplex design and that in Doc. 24 are the earliest extant evidence of Edison's work on multiple telegraphy. This arrangement uses a "differential" method, or more specifically a "compensation" method like that of the first successful duplex devised by the Austrian Wilhelm Gintl in 1853 (Schellen 1888, 779–85; Prescott 1877, 769–71). Most differential designs divided a single current rather than providing separate circuits for the opposing electromagnets (see Doc. 28). In Edison's design the opposing magnets are in separate circuits, operated simultaneously by a key. As a message is sent, each magnet of the transmitter's relay negates the effect of the other, but at the receiving end of the line nothing compensates for the effect of the incoming signal. Unlike normal American telegraph practice, the sounder each relay operates (not shown) thus responds to incoming signals but does not respond to outgoing ones. Edison rewired and used relays with opposed electromagnets, like those found in some repeaters (see Doc. 11; and Doc. 17, n. 1). The other relays (shown at the bottom left and top right of the drawing) and the different number of battery cells at each terminal probably represent an attempt (only partially successful) to provide transmitting operators with separate sounders that would respond only to outgoing signals, thus allowing senders to hear better what they were sending. Cf. Doc. 50, n. 3.

2. Edison could have produced this design earlier than 1867 but not later. By early 1868 he had sufficient familiarity with duplex designs that the flaws in this design and that of Doc. 24 would have been obvious. By then he had a workable duplex design (Doc. 28), knew of Stearns's simple device of adding separate local sounders to allow the sender to listen to his work, and had access to the works of Culley and Sabine (see Docs. 8 and 9), both of which clearly explain contemporary duplex telegraphy. Ezra Gilliland recounted in 1878 that Edison had experimented on duplex designs in Cincinnati and had made an experimental

model from discarded repeater parts. *Cincinnati Commercial*, 18 Mar. 1878, Cat. 1240, Scraps. (*TAEM* 94:126); headnote, p. 32, n. 8.

–24–

Notebook Entry: Multiple Telegraphy[1]

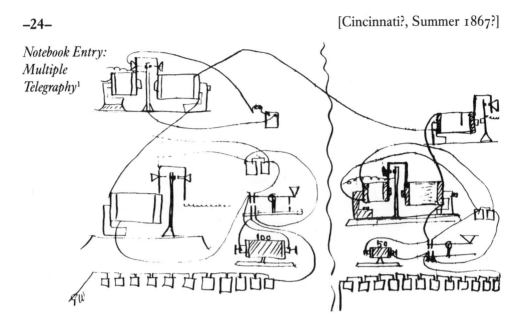

AX, NjWOE, Lab., PN-69-08-08 (*TAEM* 6:771).

1. See headnote, pp. 29–32. This duplex design is clearly related to that in Doc. 23. The two basic differences here are (1) the key puts a resistance into and out of the main-line circuit, and (2) the main batteries are always on the line. Edison seems to have reversed the local circuit (two batteries and the short relay coil) from what is needed; that is, when the sender's key increases the current on the line coil of the main relay, it cuts off the opposing compensating magnet instead of charging it.

–25–

To Samuel and Nancy Edison

Dated, Cincinnati Ohio Sept 30 1867

To[a] Dear Father and Mother

I send you tomorrow five boxes of Things and Two other other articles which please Take Care of   The charges will probably amount to 6 or 7 dollars   I sent you a Valise containing a Lot of clothing Several days ago which I suppose you have received—I am coming

AL (fragment), MiDbEI, EP&RI. Message form of the Western Union Telegraph Co. [a]Preceded by "Received at Cinci 30th"; "Dated," "186", and "To" preprinted.

Editorial Notice in the
Telegrapher[1]

New York, January 11, 1868.

To Correspondents.[2]—H. C. M. Your lines are good and would be published were it not for a pressure of matter of more general telegraphic interest. T. A. E. Port Huron. Good! Come some more.[3]

PD, *Telegr.* 4 (1867–68): 163.

1. The *Telegrapher* began monthly publication in September 1864 as the official journal of the recently formed National Telegraphic Union. By 1868 the journal was issued biweekly and usually comprised eight pages. The *Telegrapher* combined feature articles on telegraph technology or business with many shorter articles and regular columns about members of the "telegraph fraternity." As the chief organ of the union, the journal supported the interests of operators, encouraged them to invent, and decried Western Union's monopoly. Reid 1879, 666–67.

2. This item appeared in the "Miscellanea" section, which contained short notes from the editor.

3. Edison sent to the *Telegrapher* a description of his duplex design (Doc. 28), which was the first of his several contributions to that journal in 1868. Perhaps because of a change in editors in February—James Ashley replaced Franklin Pope—the manuscript was not published until 11 April.

The Nascent Inventor

–2–

March–December 1868

In early 1868 Edison moved from Port Huron to Boston, a city of some 200,000 inhabitants and a bustling center of telegraphy.[1] During the next year the city provided ample resources for Edison's transition from operator to inventor, manufacturer, and entrepreneur, and special opportunities for his participation in a widening circle of technical and business communities. When he arrived in 1868, there were two telegraph companies, each employing a large number of operators, both male and female. The Western Union Telegraph Company had four Boston offices, and the recently founded Franklin Telegraph Company, with lines from Washington to Boston, had three.[2] Boston also boasted two major telegraph manufacturers—Charles Williams, Jr., and Edmands and Hamblet—and a number of minor ones. Several highly acclaimed telegraph inventors lived there, among them Joseph Stearns, president of the Franklin Telegraph Company and inventor of a duplex telegraph system; Moses Farmer, perhaps the most prominent electrical inventor of the day; and George Milliken, manager of Western Union's main office.[3]

George Milliken, manager of Western Union's Boston office and inventor of a much-used repeater and other telegraph apparatus.

Edison's friend from his Cincinnati days, Milton Adams, introduced him to Milliken, who hired Edison because of his experience in handling press reports. Edison worked nights at Western Union's main office principally as a receiver on the "Number 1 wire," a press wire from New York City. He seems to have found the working conditions there more congenial than those in many of his previous jobs and to have found Milliken particularly appreciative of his consuming interest in telegraph invention.[4] Edison visited the city's telegraph and electrical manufacturers and wrote about them for the *Teleg-*

51

rapher. These articles and others about his own inventions brought Edison to the attention of the wider telegraph community.

Little is known of Edison's personal life in this period. He recalled living first in a windowless hall bedroom and then in a series of boarding houses. Much of the time Adams, who was usually broke, lived with him.[5] At work Edison appears to have gotten along well with his fellow operators, on one occasion being placed in charge of collecting a large sum of money for a gift for a departing employee.[6] His co-workers remembered him not only for his ceaseless experimentation and his clever inventions for easing his life in the workplace[7] but also for his conversational ability and good humor. On occasion Edison ignored his own work in order to listen to one operator—a Harvard Law School student—engage others in heated arguments.[8]

All through 1868 Edison sought backers for his experiments and inventions. He found at least three: Dewitt Roberts, a fellow telegrapher; E. Baker Welch, a merchant and director of the Franklin Telegraph Company; and John Lane, a New York banker formerly in the telegraph business. Edison tried his hand in quick succession at nearly every form of telegraph apparatus: a double transmitter, a self-adjusting relay, a method of automatic telegraphy, a stock printer, a fire alarm telegraph, a facsimile telegraph, and an electric vote recorder. These early efforts were largely unsuccessful; either he discovered that his inventions were not new or he was unable to develop them sufficiently to acquire a patent. In October 1868 he executed his first successful patent application—for an electric vote recorder for legislative bodies, a device that no one would use. However, prior to October he had filed a caveat on a fire alarm telegraph and tried to patent several other inventions. Nevertheless, Edison was making progress. By December he was advertising another of his double transmitters for sale and had obtained space in Charles Williams's shop in order to work on his inventions.[9] A month later he resigned his position at Western Union to devote himself to his inventions.

Edison's success depended on more than his technical abilities; it also depended on his proximity to urban centers of telegraphy and finance.[10] He began inventing while working in the larger cities of the Midwest and was to achieve his early successes in Boston, New York, and Newark. These cities offered substantial technical communities located in the large main telegraph offices and in the precision machine shops of telegraph manufacturers. In the machine shops skilled mechanics translated his ideas into actual devices.[11] Edison also needed institutional and financial support. Manufacturers such as Charles Williams, Jr., provided a place to experiment and extended credit for work and materials, while local entrepreneurs and company officials offered more direct financial support. Edison and other inventors attracted local entrepreneurs to the newly emerging urban telegraph inventions for fire and burglar alarms, market reporting and private-line services, and messenger services. These small-scale systems required only limited capital, and Edison obtained such support from entrepreneurs who helped him establish new companies in Boston for providing gold reports and private-line telegraphs.

Renewed competition between Western Union and a growing number of intercity telegraph companies focused attention on increasing the speed of message transmission and the capacity of telegraph lines. Companies started to support two types of inventions that might accomplish these goals. The first type—"minor" improvements—were designed to increase the efficiency and standardization of established systems. Western Union, in particular, focused on "minor" improvements as it consolidated the myriad lines acquired from former competitors and improved the quality of its service to western cities.[12] The second type—new systems such as multiple and automatic telegraphs—were designed to increase either the number of messages that could be transmitted over a single wire or the speed of transmission.[13] Edison received money to develop his double transmitter from a backer who hoped that one of Western Union's competitors would adopt it.

Inventors working on new systems to increase transmission speed and line capacity in long-distance telegraphy usually required a greater knowledge of electricity than did those making improvements in existing components. Electricity provided the motive force for telegraph operations, but precision mechanical design dominated telegraph technology. Accordingly, detailed knowledge of mechanical movements was often more valuable to inventors than advanced knowledge of the science of electricity. Even mechanics who were unschooled in electrical science produced important improvements in electrical devices when the problem was primarily mechanical. Those working on improvements in batteries or chemical recording telegraphs also required some knowledge of chemistry. The elite of telegraph inventors, however, enjoyed proficiency in all facets of telegraph design. These men often worked on a variety of telegraph systems. Edison, for example, made important improvements in printing telegraphs, the mechanical movements of which were a principal element in instrument design. He also invented multiple and, later, automatic telegraphs that required sophisticated electrical techniques or chemical investigations. This combination of chemical, electrical, and mechanical ingenuity gave Edison great versatility and allowed him to do important work that in a few years would be second to none in the field of telegraphy.

1. Edison apparently arrived in Boston in March. App. 1.A18; Doc. 36, n. 1.

2. *Boston Directory* 1868, 720; ibid. 1869, 876.

3. See Docs. 34, 41, and 44.

4. App. 1.A19–23; "The Napoleon of Science," *New York Sun*, 10 Mar. 1878, Cat. 1240, Batchelor (*TAEM* 94:119–20). For descriptions of both the Western Union office at 83 State St. and the conditions of employment there, see "Commendatory," *Telegr.* 5 (1868–69): 70; "A Visit to Boston," ibid., 400–401; "Enlarged and Improved Accommodations," ibid. 6 (1869–70): 122; "The New Western Union Office at Boston," ibid., 151; and Frank Whittlesey to TAE, 3 Aug. 1908, GF.

5. App. 1.A15. The 1868 city directory gives Edison's address as 4 Bulfinch St. (*Boston Directory* 1868, 213); the 1869 directory supplied only his business address, 9 Wilson Lane (ibid. 1869, 216). A fellow operator, George Newton, recalled that he, Edison, and Adams lived together in Exeter Pl. (Newton to TAE, 7 May 1878, DF [*TAEM* 15:615]), and another correspondent remembered that Edison boarded at 44 Cambridge St. (W. E. Sharren to TAE, 31 July 1878, DF [*TAEM* 15:1015]).

6. Corey List, 1 Dec. 1868, DF (*TAEM* 12:12).

7. App. 1A.22; H. S. Martin to TAE, 18 May 1888, GF; Charles Sherman to TAE, 18 Sept. 1916, GF.

8. On the student—Patrick Burns—see *Telegr.* 6 (1869–70): 192, 303, 304; and TAE marginalia, Charles Sherman to TAE, 8 Sept. 1916, GF. W. E. Sharren, a student at M.I.T. who boarded in the same house as Edison, recalled many conversations in which Edison did most of the talking (Sharren to TAE, 31 July 1878, DF [*TAEM* 15:1015]). For recollections of Edison at the Western Union office, see Phillips 1897, 178–80; Charles Sherman to TAE, 18 Sept. 1916, GF; Dewitt Roberts to TAE, 20 Jan. 1911, GF; and "The Napoleon of Science," *New York Sun*, 10 Mar. 1878, Cat. 1240, Batchelor (*TAEM* 94:119–20).

9. App. 1.A25.

10. The majority of telegraph inventors lived in small towns, but the careers of leading telegraph inventors reflected the importance of major urban centers in cultivating and motivating technical creativity. Paul Israel has completed a dissertation, "The Telegraph Industry and the Changing Context of Invention in Ninenteenth-Century America," in

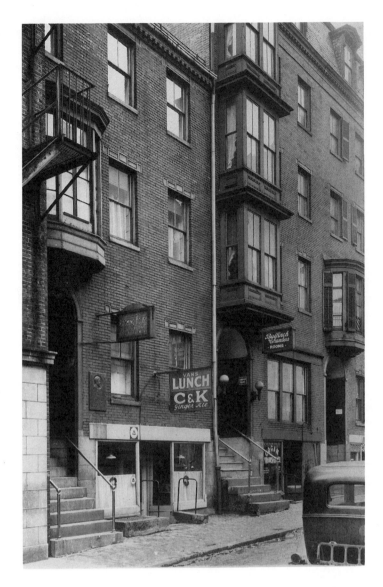

the History Department of Rutgers University. According to his study of residences listed on telegraph patents issued between 1866 and 1886, nearly 80 percent of inventors who had five or more patents resided in or within commuting distance of New York City, the center of the American telegraph industry. Other prominent telegraph inventors lived and worked in Boston, Philadelphia, and Chicago.

11. Israel 1984, 7–8; Jenkins and Israel 1984, 74–75.

12. In 1870 Western Union established an electrician's office, which was responsible for standardizing the company's technical operations and for reporting on new inventions.

13. Israel 1986 (see abstract, *History of Science in America, News and Views* 4 [Nov./Dec. 1986]: 7).

–27–

Editorial Notice in the
Telegrapher[1]

New York, April 4, 1868.

T. A. EDISON, formerly of the W.U. Co.'s office,[2] has accepted a position, and is now with the same Company's office, Boston, Mass.

PD, *Telegr.* 4 (1867–68): 258.

1. This notice is from the "Personal" column, where news of operators regularly appeared.

2. Two weeks later the *Telegrapher* noted that "T. A. EDISON, whose appointment to Western Union Company's Boston Office, was noticed week before last was from same company's Cincinnati office" (4 [1867–68]: 275).

–28–

Article in the
Telegrapher

New York, April 11, 1868.[1]

(Written for the Telegrapher.)[2]

EDISON'S DOUBLE TRANSMITTER.[3]

By means of this ingenious arrangement, two communications may be transmitted in opposite directions at the same time on a single wire. This result is accomplished by the use of rheostats, and the neutralization of the effect of the current from the transmitting station, upon the receiving instrument at the same station.

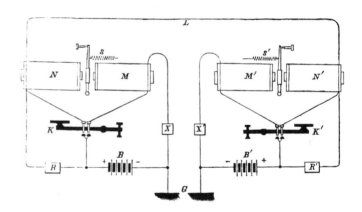

In the diagram two stations are represented, with the necessary connections for working in this manner. M N and M′ N′ are fine wire helices of the usual construction, placed opposite to each other. K and K′ are the transmitting keys, arranged to close two circuits at the same time, as shown in the diagram. R R′ and X X′ are adjustable rheostats or resistances. S and S′ are retracting springs, whose tension is adjusted according to the strength of the main-line current.

The arrangement of the wires may be readily seen by reference to the diagram. The rheostats R R', when both in circuit, should offer such a resistance that the main-line current will not be of sufficient strength to overcome the tension of the springs and work the instruments. The resistance at X is made equal to that of the main line L, added to that of the helix N' and the resistance R.' Similarly X' is made equal to L × N × R.[4]

By inspection of the diagram, it will be seen that when the instruments are at rest there will be a constant current over the line,[5] passing through battery B, rheostat R, helix N, line L, helix N', rheostat R' and battery B'; but, as above stated, owing to the resistance of the rheostats R R', it will be insufficient to affect the instruments.

Now, if the key at K be closed, the rheostat at R is cut out, and the current on the main line is increased to, say 50, passing through the helix N. At the same time, a current of equal strength passes through the other helix M and the rheostat X; the resistance of this circuit being equal to that *via* L and R', therefore the effect upon the armature of the relay M N at the transmitting station will be null.

Suppose the tension of the springs s and s' to equal 20, the armature of the relay M' N' will be drawn towards N' with a force of 50, less the tension (20) of the spring s', there being no current through M'.

If now the key at K' be also closed, the main-line current is increased to 100 by the cutting out of the second rheostat R'; but the effect of the additional current of 50 is neutralized at M' N', as in the former case. The current through N will now be 100, while that through M remains at 50; therefore the armature will be attracted towards N with a force of 30 (100 less the attraction of M, which is 50, and the tension of the spring s, which is 20), and will remain attracted as long as the key at K' remains closed.

If the key K be now opened, the current in the helix N is reduced to 50, and that in M to nothing, and the difference being the same as before, the relay remains closed. But the current at N' also being reduced to 50, while that in M' remains at 50, as before, the two attractions neutralize each other, and the spring s' draws back the armature.

Thus, it will be seen that the writing from the key K will only affect the relay M' N', and *vice versa*.

Local circuits can be attached to the keys for convenience in writing, or a key and sounder may be placed in a local circuit, and the lever of the sounder made use of to work the main circuit instead of the key, as shown in the diagram.

A repeater may also be arranged to work on this system, if required.

The inventor of this arrangement is Mr. THOMAS A. EDISON, of the Western Union Telegraph Office, Boston, Mass.

PD, *Telegr.* 4 (1867–68): 265.

1. Edison had sent a manuscript to the *Telegrapher* in January (see Doc. 26). In the journal's 18 April issue, editor James Ashley wrote, "The drawing and description of Mr. EDISON's instrument was forwarded to us nearly three months since, but was unfortunately mislaid, which prevented it from appearing in our columns at an earlier date." "Miscellanea: Correction," *Telegr.* 4 (1867–68): 275.

2. The authorship is attributed to Edison on the basis of the information given in note 1. The editor usually wrote lead articles such as this, but occasionally other contributors submitted articles. James Ashley probably edited Edison's manuscript. Edison wrote seven articles or letters for the *Telegrapher* in 1868 (Docs. 28, 30, 32, 34, 40, 41, and 44).

3. Edison claimed that he devised this duplex in Cincinnati in 1865. He said that he thought at the time his invention was original but later discovered it was not. Edison's duplex differed from others of the period in its use of closed circuits and the addition of a pair of rheostats (**R R′**). Testimony for Edison, 22–23, *Nicholson v. Edison;* see also Doc. 23.

4. This should read "L + N + R." That correction was made in "Miscellanea: Correction," *Telegr.* 4 (1867–68): 275.

5. The polarity signs for battery **B** are reversed in the diagram.

–29–

Editorial Notice in the Journal of the Telegraph[1]

New York, April 15, 1868.

Double Transmitter.[a]

Mr. Thomas A. Edison, of the Western Union Telegraph Office, Boston, has invented a mode of transmission both ways over a single wire at the same time, which is interesting, simple, and ingenious. Double transmission is not new and has been used for many years in Germany, but Mr. Edison has simplified the process by which it is effected. We will refer to it again.[2]

PD, *J. Teleg.* 1 (15 Apr. 1868): 4. [a]Followed by centered horizontal rule.

1. The *Journal of the Telegraph*, begun in 1867 and published biweekly in New York, was the house organ of Western Union. Edited by James Reid, the journal contained company announcements, business information on other companies, technical articles, and personal news of managers and operators.

2. No further description of Edison's instrument appeared.

[Boston,][a] April 25, 1868.

The Induction Relay.

To the Editor of the Telegrapher.

IN the *Journal of the Telegraph* for April 15th, an "Induction Relay" is described, which is claimed to be self-adjusting.[1] Some years ago, I experimented with an arrangement similar to this, using an Induction Coil;[2] the primary coil being of fine, instead of large wire, as is generally used, was connected with the line, and the secondary coil with a Siemen's "polarized relay."[3]

It worked well where there was no escape;[4] but when placed on a line where the escape was considerable, the increase and decrease in the strength of the magnetism in the iron bar of the induction coil, caused by the variability of the escape current, constantly induced currents of different polarities in the wire passing through the polarized relay.

To make my meaning clearer, suppose that the escape current be represented by 15, and the current from the distant battery by 5, then the magnetism in the iron core of the induction coil would be equal to 20. Now, by taking off the battery at the distant end, it is decreased to 15, and this decrease in the strength of the magnetism induces a current through the helix of the polarized relay of sufficient strength to work the armature.

Now, while the battery from the distant end is off, suppose that the escape increased to 20, this increase would induce a current of different polarity in the helix of the polarized relay from that induced by decreasing the strength of the magnetism, and this increase of the escape current would act precisely in the same manner as if the magnetism was increased by putting on the distant battery.

T. A. E.[5]

PD, *Telegr.* 4 (1867–68): 282. [a]Place not that of publication.

1. Elisha Gray, "The Induction Relay," *J. Teleg.* 1 (15 Apr. 1868): 1. During Edison's second stay in Cincinnati in 1867, he and Charles Summers, superintendent of telegraphs for the Indianapolis, Cincinnati, and Lafayette Railroad, worked on a self-adjusting relay based on an induction coil (App. 1.D219). On self-adjusting relays see headnote, p. 30; "A Self-Adjusting Telegraph Instrument Wanted," *Telegr.* 3 (1866–67): 120; and Pope 1867.

2. Induction coils consist of two separate, concentric coils—primary and secondary. Any change in the current of the primary coil induces a current in the secondary. Until the 1870s the coil was most often used to produce a high-voltage spark for medical treatment, mining, or scientific experiments. Shiers 1971.

3. See Doc. 12, n. 1.

4. Leakage due to imperfect insulation. Most of the escape occurred at the points where the wire came in contact with support structures such as telegraph poles. Knight 1876–77, s.v. "Escape"; "Self-Adjusting Relays," *Telegr.* 4 (1867–68): 397.

5. Authors of articles or letters in the telegraph journals usually signed their articles with initials or a pseudonym. Edison signed his contributions either "T. A. E." or simply "E." See Doc. 34, n. 10.

–31–

Patent Assignment to Dewitt Roberts[1]

Dated Boston April 28. 1868.

I do hereby assign forever to Dewitt C. Roberts[2] of Boston Mass. one third interest in my "Stockbroker Printing Instrument"[3] provided that the said Dewitt C. Roberts shall furnish or cause to be furnished sufficient money to patent and manufacture one or more of the said Stockbroker Printing Instruments.[a]

Witnessed by.   O. J. Waddell.[4]   T. A. Edison.[b]

D (transcript), MdSuFR, Libers Pat. P-10:322. [a]"Recorded June 25th 1868. I.B.P." written in left margin. [b]Representation of 5¢ Internal Revenue stamp canceled with "T. A. E. April 29, 1868"; "100" written at bottom.

1. This is Edison's earliest recorded patent assignment. Typically, an inventor signed and sent the original document of assignment to the U.S. Patent Office in Washington, D.C. There it was copied into the Libers of Patent Assignments. A patent office clerk entered a summary of the assignment into the Digest of Patent Assignments under the name of the inventor and returned the original assignment to the assignee. Patent attorneys Crosby, Halstead, and Gould of Boston handled Edison's assignment. Apparently the Patent Office rejected Edison's application as no patent was issued. Digest Pat. E-2:78; *Boston Directory* 1868, 344, 464.

2. Dewitt C. Roberts (n.d.), a telegraph operator for Western Union in Boston since at least 1865, was an active member of the Boston district of the National Telegraphic Union. Edison assigned Roberts a half interest in his electric vote recorder (Doc. 43) in October 1868. In November 1868 Roberts entered the flour business in upstate New York. *Boston Directory* 1868, 502; "Boston District," *Telegr.* 1 (1864–65): 94; "Personal," ibid. 5 (1868–69): 95; Dewitt Roberts to TAE, 28 May 1877, DF (*TAEM* 14:63); Dewitt Roberts to TAE, 16 July 1896 and 20 Jan. 1911, GF.

3. Stockbrokers' printing instruments were telegraph receivers. They were leased to brokers and supplied current prices of gold and stocks from a central office. See Doc. 51; Doc. 91, n. 4; and Chapter 4 introduction.

4. Orin Waddell was Edison's fellow telegrapher at the main office of Western Union in Boston. He boarded at the same location as Edison, 4 Bulfinch St. *Boston Directory* 1868, 590; Taltavall 1893, 347.

New York, May 9, 1868.

EDISON'S COMBINATION REPEATER.

In repeating the rapid vibrations of the combination printing instrument[1] from one circuit to another, it is of the utmost importance that there should be the least possible loss of time in the transmission of each signal. The repeaters in general use on the Morse lines have been unable to repeat these vibrations with the accuracy, in respect to time, that is necessary for the successful working of the instrument, owing to the use of repeating sounders and spring points, which occasion a considerable loss of time between the closing of the relay and sounder, respectively.

The annexed diagram represents a repeater on a new principle, which is so arranged as to be free from the objections above mentioned, and which, although especially intended for lines on which the combination printer is used, can be arranged to work on the ordinary closed circuit of the Morse lines with equal facility.

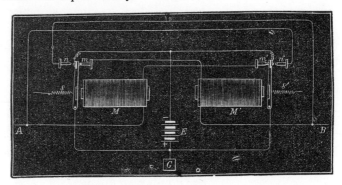

In the diagram A and B represent the two main lines, which, for the sake of convenience in description, may be designated as the eastern and western circuits. These wires are connected through the helices of their respective relays, magnets M and M' to the contact of the armature levers of the opposite relays, as shown at *m* and *m.'* There is, also, an insulated point upon the back of each armature lever, which is connected with the main battery E, the other pole of which is to the ground G. The screw points n,' n are connected, respectively, with the main line wires A and B before entering the relays.

The magnets M, M' are of peculiar construction. The cores are of magnetized steel instead of soft iron, or they may be made of soft iron and kept constantly polarized by contact with a large permanent magnet. Thus it will be seen that when no current is passing the armature will be strongly attracted towards m by the permanent magnetism of the cores. The retracting spring s s' are adjusted to a tension just sufficient to allow the armature to be drawn forward towards the cores of the magnet when the main circuits are open.

If we suppose the holding force of the permanent magnetism in the relay cores to equal, say 20, and the tension of the spiral springs to be 18, the armatures will be retained in contact with the points m m' with a force of 2, and both lines will be connected through the relays direct to the ground, as shown in the figure.

Now, if a current is sent from the distant station over the line A, leaving a force, say of 5, it will pass through the relay M and the point m' direct to the ground at G. The relay being so wound that the magnetism induced by its coils is in opposition to the permanent magnetism existing in the cores, the force of the latter will be reduced from 20 to 15, and the spring having a tension of 18 would draw back the armature lever and close the circuit at n, placing the main battery E in connection with the line B, and thus repeating the signal over that circuit. The other magnet being cut out from the main line, and the current sent around it will not be in any manner affected. If a current be sent in the opposite direction over the line B the reverse action takes place.

This repeater, as will be seen, is very simple in its construction and arrangement, and the inventor states that the principle on which the relays is constructed renders them less liable to be thrown out of adjustment than the ordinary kind. It can be adjusted to work with a current so feeble that its action would not be perceptible upon a Morse relay.

Parties desiring further information upon the subject may address the inventor, Mr. F. A. EDISON, at 83 State street, Boston, Mass.[2]

PD, *Telegr.* 4 (1867–68): 298.

1. In 1859 George Phelps patented the combination printing telegraph (U.S. Pat. 26,003), so named because it combined features of the earlier printing telegraphs of David Hughes and Royal House. This instrument superseded the older printing telegraphs and was quickly placed on several major lines of the American Telegraph Co. The operator of Phelps's instrument transmitted characters by depressing keys on a piano-like keyboard connected to a revolving cylinder that was synchronized with a continuously revolving typewheel in the receiving instrument. The typewheel printed messages in roman characters on a thin strip of paper. Edison here describes a repeater for use with such a

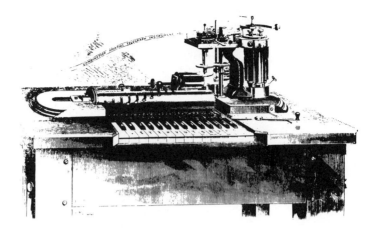

George Phelps's "combination" printing telegraph, which was used primarily between major cities on lines that did not require repeaters.

system. He did not receive a patent on this invention. Prescott 1863, 144–55; idem 1877, 642–47.

2. As in Doc. 28, authorship of the article is attributed to Edison, but with the recognition that James Ashley probably edited the manuscript. The address given for Edison was that of the main Western Union office in Boston.

–33–

Milton Adams Article in the Journal of the Telegraph

BOSTON, May 25[a] [1868].
Automatic Telegraphing.[b]
BY M. F. ADAMS.[1]

In a late number of the TELEGRAPHIC JOURNAL, I noticed a long and interesting article on improvements in Automatic Telegraphing,[2] the innovation of M.M. Chaudassaignes and Lambrigat, two French telegraphers. The system is similar to that of Mr. Alexander Bain (i.e.) the decomposition of salts upon chemically prepared paper, but has more of a resemblance to the system of Mr. Hummiston a New York gentlemen, who in 1855 tried his apparatus between that city and Boston.[3]

In this arrangement a long band of paper was punched out in Morse signals by means of a very complicated and ingenious machine, operated by a key board similar to that of a combination printing instrument.[4]

This paper was passed between two metallic rollers moved by clock work, one of the rollers being in connection with a battery. The line wire was connected with a spring which pressed against the paper and was thus prevented from coming in contact with the roller beneath, but when a fissure in the paper occurred the spring came in contact with the roller and placed the battery upon the line which produced a blue mark upon the chemically prepared paper at the other terminal of a certain length, according to the length of the fissure in the transmitting paper. Owing to a defect in the machinery this invention did not prove as successful as expected.

The difference between this arrangement and that of the Messrs. Chaudassaignes and Lambrigat is that they transmit the signals by means of a metallic band prepared with insulating ink which opens and closes the main line circuit and produces Morse characters upon chemically prepared paper at the distant station, but with the exception of their method of preparing the transmitting paper it will be seen that these two systems are identical.

The trouble of preparing the paper for transmitting signals and the translation of these signals has been the great drawback to the usefulness of this mode of Telegraphing, and I am of the opinion that these methods will never come into genuine use, at least not in America.

Mr. T. A. Edison of Boston, has recently invented a very practical automatic apparatus which is both ingenious and curious; the arrangement is very simple and not so liable to get out of order as the methods described above. The arrangement is as follows: A transmitting Morse Register[5] is placed at one end of the line, and a receiving register at the other, the first register has no battery attached, but is merely used for the purpose of running the paper through. The front standard of this register is connected with a battery, and the armature with the line, the transmitting paper is prepared by operators upon registers placed in local circuits, and when thus prepared, they are cut off in suitable lengths and run through the transmitting instrument at a very rapid rate. This paper in passing between the rollers of the transmitting register by its thickness prevents the armature from touching the standard, but when an indenture in the paper occurs, the pen of the register passes into the groove which allows the armature to come into contact with the standard and places the battery upon the main line, reproducing the Morse signals at the other terminal; this paper is passed through as rapidly as it can be prepared by two operators upon registers as local circuits.

The indented paper at the other end of the line, is then cut off in suitable lengths and passed at a slow pace through two re-writing registers similar to the Morse line transmitting registers, with the exception that the armature and standards are connected with a sounder and local battery, and the writing is received by sound. It will be seen that the indentures in the paper depress the armature and act as a key upon a main or local circuit: thus nearly eighty words per minute can be sent and received upon sounders by two operators at each end of the wire, which is, I think a more convenient mode of transmitting and translating than of the Messrs. Chaudassaigne and Lambrigat.

PD, *J. Teleg.* 1 (1 June 1868): 3. ᵃPlace, month, and day not those of publication. ᵇFollowed by horizontal line.

Milton Adams, one of Edison's fellow operators at Cincinnati in 1865 and at Boston in 1868.

1. Milton F. Adams (1844–1910), described by a contemporary as "a typical bohemian," served in the U.S. Military Telegraph during the Civil War and worked as a telegrapher in a number of cities, including Pittsburgh, Cincinnati, New York, Boston, and San Francisco. Edison met him in Cincinnati in 1865 and through him obtained a job in Boston in 1868. Soon afterwards Adams was laid off and moved into Edison's meager lodgings. In 1869 Adams left for San Francisco and began an odyssey that took him to Latin America, South Africa, Europe, and Great Britain. In 1883 Edison hired him to sell electric lighting systems in South America. "Career of the Late Milton F. Adams," *Teleg. and Tel.*, 16 Nov. 1910, 755–56; Taltavall 1893, 221; App. 1.A15–19; "District Proceedings," *Telegr.* 2 (1865–66): 15.

2. "Improvements in Automatic Telegraphy," *J. Teleg.* 1 (15 Apr. 1868): 1. Adams was referring to the *Journal of the Telegraph* by the name of its predecessor, the *Telegraphic Journal,* edited by Jerry Borst and privately published from March to November 1867. *J. Teleg.* 1 (2 Dec. 1867): 4.

3. As the pressure of increased traffic mounted, automatic telegraphy became desirable as a means to increase the speed of transmission and decrease the need to "break" because of faulty manual sending. Most automatic telegraphs were variations on the pioneering developments by Scotsman Alexander Bain during the 1840s. With Bain's automatic, an

Alexander Bain's automatic chemical recording telegraph, first introduced into the United States in 1848.

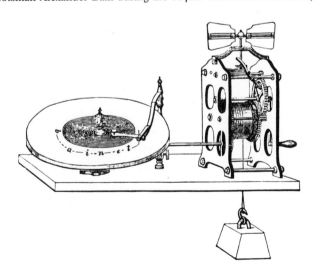

operator perforated messages on a paper strip, the coded holes representing letters. As the paper strip ran through the transmitting device, an electrical contact passed over the paper and closed a circuit at each perforation. At the other end of the line the signals electrically decomposed chemicals soaked into a paper strip or disk. The preparation of the perforated paper strip before transmission meant that an operator's skill did not limit the speed of transmission. Sabine 1872, 199–201; *DNB*, s.v. "Bain, Alexander"; Prescott 1870, 292.

The system of automatic telegraphy developed during the 1860s by Paul Chauvassaignes and Jacques Lambrigot of the French Telegraph Administration recorded dots and dashes with insulating ink on a strip of conductive metallic paper for transmission and used a Bain-like receiver. This system achieved some success in 1867–68 on a line between Paris and Lyons. Chauvassaignes and Lambrigot patented their telegraph in the United States in 1868 (U.S. Pat. 80,452). "Le rapide télégraphe electro-chimique de MM. Chauvassaignes et Lambrigot," *J. des télég.* 3 (Nov. 1868): 6–8.

John Humaston of Connecticut patented an apparatus for perforating paper strips for transmission in automatic telegraphy in 1857 (U.S. Pat. 18,149), but the device was too slow and complicated to be commercially valuable. Prescott 1877, 701; idem 1870, 292. See also Chapter 6 introduction.

4. George Phelps's combination printing telegraph.

5. Morse registers were normally used only for receiving messages. However, Edison had previously used one as a transmitter (see Doc. 10). C. Westbrook of Harrisburg, Pa., filed a caveat for the use of a Morse register as a transmitter before Adams's article appeared, and received U.S. Patent 88,248. App. 1.D201; C. Westbrook to the Editor of the *Telegrapher*, *Telegr.* 6 (1869–70): 26; ibid., 42; Prescott 1877, 740.

–34–

To the Editor of the Telegrapher

BOSTON, Mass., *June* 2ª [1868].

TO THE EDITOR OF THE TELEGRAPHER.

A TELEGRAPH line has just been built over the Dover and Winnepisseogee branch of the Boston and Maine Railroad, from Dover to Alton Bay, a distance of 30 miles.[1]

The wire used upon this line was manufactured by the "American Compound Wire Company,"[2] of this city, and is composed of a steel core and a copper covering, No. 14 in size, and having a conducting power equal to that of No. 8 iron wire.[3]

There seems to be several important advantages in the use of this wire, and one in particular, which is, that on account of the great strength of the steel core, a reduction in the number of posts can be made, and a consequent diminution in the escape of the current.

In a recent experiment upon this line it was found that the

use of a single carbon element[4] was sufficient to work six relays, each having a resistance of 12 miles, No. 8 iron wire.

The city government have adopted this wire on the Roxbury extension of the fire-alarm,[5] and it is also to be used by the Boston Commercial Telegraph, recently organized.[6]

The building occupied by Mr. CHARLES WILLIAMS, Jr.,[7] Telegraphic Instrument maker of this city, was destroyed by fire on Thursday last. Mr. WILLIAMS had a large stock of electrical apparatus, nearly all of which was badly damaged. His loss is estimated at about $2,000. Insurance $5,000.

The Laboratory of MOSES G. FARMER,[8] the well known Electrician, which contained an immense quantity of valuable electrical and experimental apparatus, was also badly damaged. Loss, several thousand dollars. His insurance was light. The same building was also occupied by W. H. REMINGTON, manufacturer of FARMER'S Thermo-Electric batteries,[9] of which a large number were damaged. Loss probably $2,500. No insurance.

E.[10]

PD, *Telegr.* 4 (1867–68): 334. [a]Place, month, and day not those of publication.

1. Dover and Alton Bay are towns in southern New Hampshire.

2. See Doc. 44.

3. The gauge refers to wire size according to the Brown and Sharp system. The most commonly used gauge in the United States at this time was no. 8. The diameter of this gauge was 128.490 mils (about 3.25 mm), while that of no. 14 was 64.084 mils (about 1.6 mm). Before agreement on an international standard unit of electrical resistance (the ohm), electricians customarily made measurements in terms of the known resistance of standard wire for a given number of miles. Maver 1892, 511–12, 514, 515.

4. A "single carbon element" refers to a single zinc-carbon cell.

5. The first fire alarm telegraph was installed for the city of Boston in 1852 (see n. 8). Call boxes were located throughout the city and connected by wires to a central office in City Hall. Boston annexed the town of Roxbury in January 1868. Maver 1892, 437–39; *Boston Directory* 1868, 4, 770.

6. The Boston Commercial Telegraph Co. does not appear in the 1868 or 1869 city directories; it may have been the company established to exploit Edison's "stock printing instrument" (see Doc. 31).

7. Charles Williams, Jr. (1830–1908), electrical manufacturer, had a shop on the third floor and in the attic of 109 Court St., Boston. Inventors like Edison, Bell, and Moses Farmer had custom work done in Williams's shop. Williams also rented work space to electrical inventors; Bell and Thomas Watson did much of their work on the development of the telephone on Williams's premises. A native of Claremont, N.H., Williams began in the early 1840s as a partner in Hinds and Williams, a

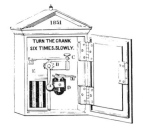

The Channing-Farmer fire alarm call box, first used in Boston in 1852.

Charles Williams Jr.'s telegraph instrument manufacturing firm, located at 109 Court St. in Boston.

Boston manufacturer of electrical equipment and supplies, and continued alone after Hinds retired. In 1868 he employed about ten men and primarily manufactured telegraphic equipment. Moses Farmer worked in a corner of the shop, and Edison had acquired space for his experiments by the end of 1868. "Death of the Oldest Telephone Manufacturer (Charles Williams, Jr.)—1908," Box 1071, AT&T; Watson 1926, 30–34, 47–48, 52–54; Bruce 1973, 92.

8. Moses Gerrish Farmer (1820–1893) was one of the period's most prolific electrical inventors. He and William Channing developed a fire alarm telegraph system that was installed in Boston in 1852. He also invented a duplex telegraph, a multiplex telegraph, a method of aluminum electrometallurgy, an electric clock, a battery-powered electric light with a platinum filament, a self-exciting dynamo, and compound wire (with George Milliken). In 1852 Farmer built for the U.S. Coast Survey a telegraph line that connected Harvard University's Cambridge observatory with the Boston office of the Magnetic Telegraph Co. Reid 1879, 370–76; *DAB*, s.v. "Farmer, Moses Gerrish."

9. W. H. Remington manufactured electrical apparatus. A "thermo–electric battery," or thermopile, was a battery of thermocouples that converted heat directly into electricity. In Farmer's thermopile, jets of burning coal gas heated a pile of alternating sections of antimony and copper. Remington advertisement, *Telegr.* 4 (1867–68): 189; "Miscellanea: A Novelty in Telegraphing," ibid., 195; "The American Compound Wire Company," ibid. 5 (1868–69): 167; *Boston Directory* 1869, 514; Knight 1876–77, s.v. "Thermoelectric Battery."

10. This document and others published in the *Telegrapher* in 1868 and 1869 under the initial "E." are attributed to Edison (see Docs. 41, 44, and 81). The three written in 1868 are datelined Boston or concern topics that center in Boston—Edison's location in 1868. They reflect Edison's interests in electrical manufacturing in Boston and mention people he knew, such as George Milliken (his employer) and Charles Williams, Jr. The document from 1869 is from the fall, when Edison was in New York. The topics of that letter have nothing to do with Boston,

but they reflect Edison's humor and his interest in induction coils. No articles or letters signed "E." appeared in the *Telegrapher* before or after these four.

–35–

Patent Assignment to John Lane

New York, June 22, 1868[a]

These presents entered into by and between Thomas A. Edison of the City of Boston, State of Massachusetts, of the first part, and John W. Lane,[1] of the City, County and State of New York, of the second part, on this twenty second day of June AD 1868 at said City of New York, witnesseth as follows:—

1st.   Said first party has invented a new and useful machine for Telegraphic printing by electro-magnetism, between distant points, and simultaneously between intermediate points, in Roman letters, embracing all letters and figures of The Roman Alphabet, which he calls "Edison's Improved Automatic Printing Telegraph," and upon and for which he proposes to obtain "Letters Patent" under the laws of the United States, and foreign governments embracing a combination of two electro-magnets in one circuit, to be acted upon seperately or together, by currents of different intensities passing in one circuit, and simultaneously moving a ribbon of paper in synchronism with the revolving Type wheel, by means of a peculiarly arranged ratchet wheel and a combination therewith, of a polarized-magnet in the same circuit, by which the circuit can be shortened, or restored at pleasure, so as to include or exclude the magnet so as above named, combined to imprint upon the moving paper aforesaid.[2]

2nd   Said Edison agrees to convey by any needful additional instrument in writing, and hereby conveys absolutely, for a valuable consideration, named in another writing,[3] signed by the said Lane, and bearing even date with these presents, and to the heirs and assignees of said Lane, one undivided half part of said invention, and of all letters patent that shall be obtained therefor, either within or without the limits of the United States of America and in reissue, renewal, or extension of the same, and of and for any improvement of said invention that I, the said Edison shall hereafter make, or discover, to have and hold, or dispose of the same, at his and their pleasure.

3d—And in consideration of the aforesaid additional Instrument, and of the payments and promises, and understandings of the said Lane therein contained, the said Edison

March–December 1868

69

hereby constitutes and appoints the said Lane, sole and exclusive attorney of himself, The said Edison, for the sale of said invention and of any and all letters patent that shall be hereafter obtained therefor, or any improvement thereon, within or without the limits of the United States, or in renewal or extension or reissue of such patents, with like full and ample powers, to sell, negotiate and by lawful deeds or assignments to convey all and parts at the discretion of said Lane of said Edison's rights property and interests therein conjunctively with the rights interests and property of the said Lane therein and not otherwise said Lane accounting to the said Edison his heirs and legal representatives for one (1) moiety of the proceeds of all and every such sale as is fully provided and agreed in and by the aforesaid additional instrument bearing even date herewith and signed by said parties hereby ratifying and confirming whatever my said attorney shall do or perform in the premises as herein contemplated & hereby making said power irrevocable by me during the existance of any Letters Patent in force herein contemplated.

In testimony whereof said parties each for himself hereunto subscribes his name and affixes his Seal on the day and year first mentioned above.[b]
In the Presence of   G. A. Arnoux[c]   Thomas A. Edison[d]

D (transcript), MdSuFR, Libers Pat. M-10:429. [a]Place and date taken from text; form of date altered. [b]Representation of Internal Revenue stamp in margin. [c]Followed by copy of notarization. [d]Followed by representation of seal.

1. John W. Lane (n.d.), telegraph entrepreneur of Portland, Me., built an independent telegraph line from Boston to Springfield, Mass., that merged with the Franklin Telegraph Co. in November 1865. Under Lane's presidency, the Franklin Company's Boston–New York line was completed and opened for business in January 1866. Joseph Stearns replaced Lane as president in March 1867, and in November of that year Lane sold out. Reid 1879, 590–92.

2. Many of the features mentioned in this patent assignment appeared in Edison's Boston instrument (Doc. 51); he did not receive a patent for the invention assigned to Lane.

3. Not found.

–36–

Receipt for E. Baker Welch

Boston, July 11th, 1868.

$5.50

Received of E. B. Welch[1] five 50/100 dollars in consideration of which I hereby agree to assign to said Welch one-half undivided interest in polarized and vibrating double transmitters for use on telegraph lines of companies, private firms, or individuals.[2]

T. A. Edison.

TD (transcript), MWalFAR, *Welch v. Edison.* Welch submitted typed copies of many documents; though sworn to be accurate, there are indications that they contain errors.

1. Ebenezer Baker Welch (1821?–?1902), a commission merchant in Boston, had become one of nine directors of the Franklin Telegraph Co. on 26 November 1867. He claimed to have met Edison in Boston "on or about 26 March 1868." In the course of Edison's year in Boston, Welch provided funds for him to experiment with fire alarm telegraphs, dial telegraph instruments, and a new form of double transmitter. None of these was patented. Testimony for Anders, p. 9, *Anders v. Warner;* Massachusetts, Middlesex County 1912, 910; *Boston Directory* 1868, 602; Reid 1879, 592; Plaintiff's Answer to Defendant's Interrogatories, 18 Dec. 1889, *Welch v. Edison.*

2. This $5.50 was only the first installment of funds provided by Welch for this invention. Edison signed a more formal agreement with Welch on 7 April 1869 (Doc. 61).

[Boston,]^a July 27— 1868

–37–

Caveat: Fire Alarm[1]

Caveat

To the Commissioner of Patents.

The Petition of Thomas A. Edison, of Boston in the County of Suffolk, and State of Massachusetts Respectfully represents: That he has made certain improvements in Indicating Fire alarm telegraph and that he is now engaged in making experiments for the purpose of perfecting the same preparatory to his applying for Letters Patent therefor. He therefore prays that the subjoined description of his invention may be filed as a Caveat in the confidential archives of the Patent Office, agreeably to the provisions of the Act of Congress, in that case made and provided, he having paid ten dollars into the Treasury of the United States, and complied with the other provisions of the said Act.

And he hereby authorizes Joseph H. Adams,[2] of Boston, Mass. to act as Attorney in the matter of the said Caveat, and to receive the certificate of deposit for the same.

Thomas A. Edison

The nature of my invention consists in so arranging the apparatus of a Telegraphic Alarm, that the number of the box and its location will be shown upon an Indicator in printed characters. An arm is connected with the Indicater in such a manner that when an alarm is given the striking arm or hammer is released and rings the bell for several minutes for the purpose of calling attention to the locality of the fire, as shown upon the Indicater.

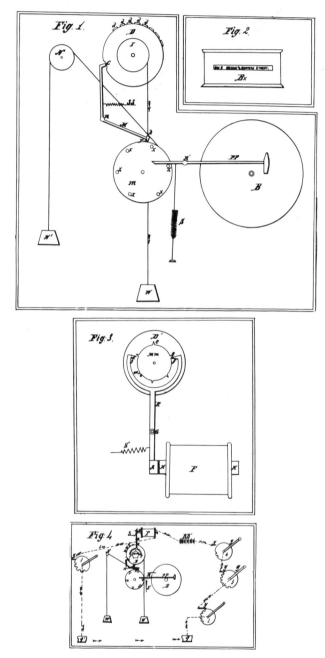

Witnesses.[b]   J. H. Adams[c]   Inventor.   Thomas A Edison[c]

In the accompanying sketch, D, in fig. 1, represents an end view of the indicating cylinder. ~~To~~ Upon the cylinder are ~~attached~~ printed slips d, d, &c. one of which is shown at B,x, in fig. 2, l,[3] in fig. 1, is a drum upon which is wound a cord having a weight W, attached to its end— D', in fig. 3, represents the opposite end of the indicating cylinder having an escapement m,m, attached which latter is controlled by means of the electro-magnet F, arm E, and pallets f, f',—the arm E, being pivoted at G', and made adjustable with the spiral spring S'—

c in fig. 1, on the cylinder D is an indentation into which a projection on the upper portion of the arm or lever M, is drawn by the action of the spring S,S. The arm or lever M is pivoted at n, and is provided at its lower end with a catch h, which engages with a projection r on the wheel m, and serves to detain the latter— On the side of the wheel m, are projections or teeth x, x, &c. which serve to raise the bell striking arm or hammer PP, and allow the same to fall back upon the succeeding tooth.

A quick motion is given to the arm PP, by means of the spring S.

Motion is imparted to the wheel m, by means of weight W', attached to a cord passing over the pulley N— It is released by turning the indicating cylinder D, so that the lever M, will be thrown out of the indentation c^d ~~in the~~ and the catch h, will release the projection r.

B, fig. 1, is the alarm bell. ~~w~~When the electro-magnet F, fig. 3, is active, the pallet f, is in connection with the tooth e, which prevents the cylinder D', from moving—and when the electro-magnet F, loses its magnetism the arm or lever E will be drawn back by the action of the spring S'— the pallet f, then releases the tooth e, and the opposite pallet catches upon the tooth c,c, and holds it until the magnet becomes active, and so the operation continues—

Upon releasing two other teeth, one of the printed slips on the cylinder will appear at the opening in the box as shown in fig. 2—

Fig. 4, represents an alarm circuit with one alarm and indicating apparatus, four transmitting boxes and a battery B,B'.

The printed strips are placed upon the indicating cylinder, corresponding in number with the teeth on the transmitting boxes, 1, 2, 3, 4—

Box 1—for instance has three teeth, the wheel is made of metal and is connected with one end of the wire and the spring y, with the other— These teeth break the circuit three times, which by the action of the electro-magnet F, allows the escapement to be released six times, and consequently the third printed slip upon the cylinder appears at the opening of the box as shown in fig. 2.

Box no. 2, having four teeth, then the fourth strip will be shown and so on.

Two movements of the escapement m,m, backward and forward are necessary to move the cylinder the space of one strip, i.e. when the arm E falls back it moves the strip half the distance and when it is drawn up again to the magnet the strip is moved the other half—)

g, g', are ground wires, the circuit being shown by the arrows—

Thomas. A. Edison

ADDENDUM[e]

Boston, July 28, 1868[f]

Oath

Boston, County of Suffolk   State of Massachusetts[g]

On this twenty eighth day of July A.D., 1868 before me, the subscriber, a Justice of the Peace, in and for the County & ~~State~~ aforesaid, personally appeared the within-named Thomas A. Edison, and made solemn oath that he verily believes himself to be the original and first inventor of the within-described Indicating Fire Alarm Telegraph and that he does not know or believe that the same was ever before known or used; and that he is citizen of the United States.

Jos. H. Adams   J.P. (L.S.)[4h]

D (transcript) and DS, DSI-NMAH, WU Coll. Page with drawings and accompanying signatures is original. [a]Place taken from text. [b]Drawing has heading "T. A. Edison   Caveat. Filed Aug. 1, 1868"; "Witnesses." and "Inventor." written in unknown hand. [c]Signature. [d]Interlined above. [e]Addendum is a D (transcript). [f]Place and date taken from text; form of date altered. [g]Representation of stamp at right. [h]"Ex. J. A. W." written in left margin; "L.S." circled.

1. Edison's petition and oath, together with the specification for this caveat, provide an example of the full caveat application form.

2. Joseph Adams was a solicitor of patents with offices at 33 School St., Boston. *Boston Directory* 1869, 45.

3. Should be "I."

4. "L.S.," for *locus sigilli*, means "the place of the seal."

38

Boston Massachusetts, July 28th 1868.

In consideration of Twenty dollars to me paid by Ebenezer B. Welch of Cambridge Massachusetts, the receipt whereof is hereby acknowledged, I Thomas A. Edison of Boston Massachusetts, do hereby assign, transfer and convey to said Ebenezer B. Welch one undivided half interest in an "Indicating Fire Alarm Telegraph"[2] invented by me and for which I have this day made application for a Caveat[3] through the Agency of Joseph H. Adams Esq of Boston. And I do hereby assign and convey to said E. B. Welch, the right and title to one undivided half interest in any improvements I may at any time invent or make to the said Indicating Fire Alarm Telegraph.[a]

Witness my hand and seal the day and year before written

Witness: Jos. H. Adams   Thomas A. Edison

D (transcript), MdSuFR, Libers Pat. Q-10:391–92. [a]Representation of canceled 5¢ Internal Revenue stamp in left margin.

1. This assignment was recorded at the Patent Office on 3 August 1868. Edison did not receive a patent on the invention. Digest Pat. E-2:81.

2. By 1868 there were at least six U.S. patents on fire alarm telegraphs. George Newton, Edison's fellow operator at Western Union, later recalled that he, Edison, and telegrapher Patrick Burns had gone to Cambridge to try to interest that city in Edison's fire alarm telegraph but had been rebuffed. The contract was instead given to Gamewell and Co., which had installed a number of other fire alarm systems. The *Telegrapher* called Gamewell's telegraph "doubtless the best and most effective system as yet brought into practical operation in the world." Prescott 1877, 237–45; George Newton to TAE, 7 May 1878, DF (*TAEM* 15:615); "Fire Alarm Telegraph," *Telegr.* 6 (1869–70): 11.

3. See Appendix 4.

39

New York, August 1, 1868.

Chirography.

MR. T. A. EDISON, of the Western Union Boston office, is about the *finest* writer we know of. We have received a specimen of press report sheet written by him as the news came over the wire from New York at the usual speed. The sheet is five inches by eight inches, and there are 647 words upon it.[1] Each letter is separate from the other, which is one of the peculiarities of Mr. EDISON's chirography, and the whole plain as print—with the diamond type so fashionable in Boston.[2]

PD, *Telegr.* 4 (1867–68): 400.

we have been driven to adopt this course, when it is proved that had we uniformly received courteous gentlemanly & humane treatment & been subjected to no tyrannical rules had the old standard of salaries been carefully maintained & some system of promotion established so that the far distant future we might see at least a single ray of hope there would have been no cause for complaint. Though I believe that other heads... (Edison's copy 1868)

1. Edison had made a deliberate effort to perfect the art of taking clear and rapid copy by writing small and disconnected letters. One evening the night manager at the Boston office discovered that none of the press report was usable "for the reason that Edison had copied between fifteen hundred and two thousand words of stock and other market reports in a hand so small that he had only filled a third of a page." App. 1.A20; Phillips 1897, 179; Taltavall 1893, 336.

2. Diamond type is a very small type size (approximately 17 lines to the inch). Knight 1876–77, s.vv. "Diamond," "Type."

–40–

Article in the Telegrapher

New York, August 8, 1868.
Self-Adjusting Relays.[1]

IT is a well-known law of magnetism that there is a limit to the magnetic attraction which can be induced in a soft iron bar, and this law may be taken advantage of in the construction of a self-adjusting relay magnet for Telegraphic purposes, simple both in principle and form.

The diagram will convey a good idea as to the size of the helix and core, when compared with the ordinary form of relay magnets.

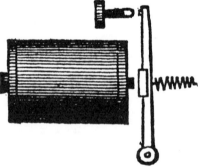

The cores of this magnet are composed of bars of soft iron, one sixteenth of an inch in diameter. It will be seen, by referring to the diagram, that the helix is quite short and thick, 2½ inches by 2 inches, which is another adaptation of a well known law of electro-magnetism, to wit: that short and thick electro-magnets receive and lose their magnetism with more facility than those of greater length, although both may have the same retentive power.

The cores being very small, and wound with a great number of convolutions of fine wire, become magnetized to their fullest extent by a current whose action would scarcely be perceptible upon the ordinary form of magnets.

As before stated, there is a certain limit to the amount of magnetism that can be induced in a soft iron bar, and it will, therefore, be seen that currents of different tensions may pass through the helix, but only a small portion of each will produce an effect upon the core.

To illustrate more clearly, suppose that the smallest force which an ordinary instrument can be worked on be represented by 5, and the strongest by 100, then the force of 5 from the distant battery passes through the helix and thence to the ground (this magnet must necessarily be worked on the open circuit principle).[2] The little cores become magnetized to their highest limit. If the force be now increased to 100, this increase will produce no effect, as the force of 5 had already produced all the power obtainable.

This little core, being always intensely magnetized by the passage of the current through the helix, allows a considerable tension to be placed upon the spiral spring, which tends to make the armature act quick. As no change in the strength of the magnet ever occurs, the armature is drawn up to the core with an uniform force.

T. A. E.

PD, *Telegr.* 4 (1867–68): 405.

1. See Doc. 30, n. 1.
2. For "open circuit," see Chapter 1 introduction.

–41–

Article in the Telegrapher

New York, August 15, 1868.
(Written for THE TELEGRAPHER.)
The Manufacture of Electrical Apparatus in Boston.[1]

A DESCRIPTION of the different establishments devoted to the manufacture of electrical and Telegraphic apparatus in Boston will doubtless prove interesting to many readers of THE TELEGRAPHER, especially as that city has obtained an enviable reputation among Telegraphers and electricians for the superior quality and finish of the work turned out by some of its leading manufacturers. One of the principal firms engaged in this business is that of EDMANDS & HAMBLET,[2] at No. 40

Hanover Street, who are well known to the public and Telegraphic fraternity as the manufacturers of the "Magneto-electric Alphabetical Dial Telegraphs," of which a large number are used upon private lines in different parts of the country.[3] This Telegraph is constructed upon the magneto-electrical principle, dispensing entirely with the voltaic battery. The following is a brief description of this admirable apparatus: The transmittor is contained in a small square box, upon which there is a dial plate, with a circle of thirty equidistant keys or buttons radiating from the same centre. Upon the dial plate are marked the alphabet, three points of punctuation and an asterisk; in an inner circle are the numerals. A pointer in the circle revolves in connection with the handle of the rotating armature, and is stopped at any letter by depressing one of the buttons. Four soft iron cores, with their enveloping helices of fine wire, are fixed upon the poles of a compound permanent magnet, these cores being placed at equal distances from each other in the circumference of a circle. On an axis passing through the centre of this circle, in connection with the handle, revolves a soft iron armature, whose breadth is a little greater than the distance between two adjacent cores. When the armature revolves it approaches one pole as it recedes from the one diagonally opposite, and induces simultaneously in the two coils currents having the same polarity. Immediately under the transmittor is an arm, upon the same axis as the pointer above, whose motion is arrested when a button or key is depressed, and the current which would otherwise pass over the wire is "short-circuited."

The face of the indicator is similar to that of the transmittor, having a small pointer, which is thrown around from letter to letter by a very curious and delicate escapement in connection with a polarized magnet, similar to that invented by SIEMENS, and which is actuated by currents of different polarities, generated by the permanent magnets.[4]

The coils of the indicator and permanent magnets are connected in one common circuit. When the armature of the magnets is turned around by means of the handle, if the pointer is free to move round the dial, a current traverses the line at every letter which the pointer passes, and moves the hand of the indicator correspondingly, but as soon as the carrier-arm on the same axis as the transmitting pointer is stopped, by coming in contact with a depressed key, the currents which would follow are "short-circuited." The pointers of the transmittor and indicator, therefore, stand still upon the same letter until the key is raised and the "short-circuit" removed. Alarm or call bells are also attached in such a manner that when no communication is being sent the indicators are cut out and the call bells put in circuit, and *vice versa*.

In operating this instrument no knowledge of the usual Telegraphic signs or sounds are necessary; the operator simply places his fingers upon the letters of the alphabet which compose the Telegram, and the person receiving simply takes notice of the letters as they are successively pointed out upon the indicator at the other terminus.

Several trials have been made with this instrument over the wires between Boston and New York, to determine their applicability for railroad lines, all of which have proved highly successful. The working of this beautiful instrument, as well as the neatness with which it is constructed, and its advantages over the clumsy apparatus for similar purposes of a foreign manufacture, cannot be too highly spoken of, as it shows that America can successfully compete with Europe in the manufacture of Telegraphic apparatus, even if they are turned out of the shops of a Froment[5] or Siemens-Halske.

This firm also manufacture another piece of curious electrical mechanism, which is called "Hamblet's Electro-magnetic Watch-clock," which is in use in nearly all of the fire alarm offices, hospitals, and prisons in the Union, and in a large number of the principal manufacturing establishments of New England.

It is for recording the rounds of a night watchman every hour, or half hour, which it does upon a paper dial, marked with the hours and subdivisions of time similar to the dial of a common time-piece, and which is made to revolve in such a manner as to receive the impress of a lead pencil bearing thereupon, which, as time passes, makes its mark upon the paper.

The electric current being in the quiescent state of the electrical mechanism, open (*i.e* not actuated at the point of operation), will cause the pencil to make a regular continuous line, which in twelve hours would form a perfect and unbroken circle round the dial.

The instant that the watchman touches a simple piece of mechanism at any point upon his beat—upon which there are several—he causes the circuit to be opened and closed, and the pencil advances a degree towards the centre of the paper dial, leaving its impress as it advances, and then commences its mark on a new concentric parallel, and this action is repeated as often as, and whenever the apparatus is operated upon at different points in the circuit. An angular record is thus produced, which, on comparison of the angles with the marks of subdivision of time, will show not only that watch duty has been done, but will also show the exact time that each point has been visited. When a watchman operates the last point on his round the pencil falls back to its original level, and is ready for the next round.

If the pencil marks are all regular and similar in the different hours, it is proof that the twelve hours' watch duty have been performed; if, on the contrary, there are irregularities in the angles, they will be evidence that something has occurred requiring investigation.

This apparatus may be seem in nearly every Fire Alarm Telegraph office in this country—where, perhaps, many of our readers have observed it in operation.

In this establishment is also made the "Electric Plural Time Dial," an ingenious contrivance, by which the time indicated by one standard regulator clock is shown upon any number of duplicate time dials or electric clocks, situated at any distance from each other, and all connected in one electric circuit. The most curious part of this system is that the duplicate clocks have neither springs, weights, nor trains of wheels, to produce a movement of the index, but contain a simple though curious escapement, operated by an electro-magnet in the regulating circuit; they, therefore, require no winding up or attention. Another curious piece of electrical mechanism is also manufactured by this firm, called the "Electric Pendulum Gauge," for measuring and recording the varying heights, depths, and quantities of gas or water in reservoirs, but is of too complicated a nature for an accurate description without the aid of drawings. This apparatus has, after a series of severe tests, been adopted by the Boston Gas Co.

Electric Wind Indicators, Astronomical Clocks and Apparatus, Chronographs, Printing Telegraph Instruments, Repeaters, Galvanometers, Electrometers, Philosophical Apparatus,[6] Fire Alarms, and every variety of magneto-electric and electro-magnetic mechanism, are also manufactured by this firm—all of which compare favorably with, if they do not excel any similar mechanism of foreign manufacture. Twelve persons are employed here, among whom are several of the best mechanicans in the country. Telegraphers visiting the "Hub"[7] would do well to call at the office of Messrs. EDMANDS & HAMBLET, where all of the apparatus described may be seen in actual operation.

The next on the list is that of CHARLES WILLIAMS, JR. The establishment of Mr. WILLIAMS is located at 109 Court Street, and though but a short time since damaged by fire, is again in full blast. Very little apparatus, except that used for Telegraphic purposes, is manufactured here, and in this particular branch the work is of a most excellent character, consisting of Repeaters, Switch-boards, Relays, Registers, Sounders, Keys, Rheostats, Galvanometers and batteries, all of which are made in large quantities. The most noticeable instrument manufactured here is the well known "Boston Relay," of which an large number are turned out weekly, mostly for use on railroad wires.[8] Ten men are employed here. The office of the well known electrician and Telegraph inventor, MOSES G. FARMER, is also at this establishment.

The next is H. B. & W. O. CHAMBERLAIN,[9] manufacturers, dealers, and importers of Mathematical, Astronomical, Chemical, Electrical and Philosophical Apparatus, at 310 Washington Street. This establishment is probably the largest and best of its kind in the United States. Every conceivable form of experimental apparatus appertaining to the above mentioned sciences can be found here. This firm have recently imported a large number of monster induction coils from the shops of RHUMKOFF,[10] of Paris, one of which is probably the largest in this country.

The next is RITCHIE & SONS,[11] of 149 Tremont Street, manufacturers and importers of Philosophical and Electrical Apparatus, similar to that of the Messrs. CHAMBERLAIN. Mr. RITCHIE is known to the scientific public as the inventor of several important improvements on the original form of the RHUMKOFF, or PAGE[12] Induction Coil, and as the maker of the largest and most powerful induction coil hitherto constructed, now in the possession of M. GASSOIT.[13] A description of this coil may be found in "NOAD's Manual of Electricity," page 326, and in the "Philosophical Magazine," vol. xv, page 466.[14]

The last is THOMAS HALL.[15]

Very little Telegraphic mechanism is manufactured at this establishment at the present time, it being almost exclusively devoted to the manufacture of Electrical Toys and Medical Electrical Machines.

Mr. HALL's shop is situated at No. 19 Bromfield St.

E.[16]

PD, *Telegr.* 4 (1867–68): 413–14.

1. This is the featured article on the first page of the issue.

2. Benjamin Edmands and James Hamblet, Jr., manufacturers of clocks and telegraph apparatus in Boston from 1862 to 1870, patented an electric clock in 1864 (U.S. Pat. 41,217) and a magneto-electric dial telegraph in 1868 (U.S. Pat. 79,741). Edison had instruments made at Edmands and Hamblet's shop, and in 1869 joined in a business venture with one of Hamblet's assistants, George Anders (see Chapter 3 introduction). *Boston Directory* 1868, 213, 278, 1069; Taltavall 1893, 96; *Elec. W.* 35 (1900): 56; App. 1.A26.

3. Private lines were owned or leased by individuals or companies, in contrast to lines owned by the telegraph companies for their regular business. After the Civil War, a market for telegraphy developed among merchants who wanted rapid intracity communication between distant offices or offices and warehouse. Before the development of small, reliable printing telegraphs (see Docs. 97 and 130), such lines used dial telegraphs, on which a pointer indicated transmitted letters. British inventor and scientist Charles Wheatstone patented the first dial telegraph in 1840. Although used widely in Europe, dial telegraphs were not common in the United States. In 1869 Edison invented his own magneto-electric dial telegraph, the magnetograph (see Chapter 3 introduction). *DSB*, s.v. "Wheatstone, Charles"; Prescott 1877, 562–602; App. 1.A26.

4. Werner Siemens received patents in almost all areas of telegraphy. One of his earliest inventions was his 1846 improvement of Wheatstone's dial telegraph. Ten years later he invented a magneto-electric dial telegraph that employed a polarized relay. In 1847 Siemens, his brother Johann, and Johann Halske formed Telegraphenbauanstalt von Siemens & Halske to manufacture telegraphs, electromedical devices, electrical meters, and railway signaling equipment. They later moved into electric power generation and distribution. Siemens's inventive and business activities made him a major force in the worldwide development of electrical industries during the late nineteenth century. Siemens 1968, 130, 160; Weiher and Goetzeler 1977; *DSB*, s.v. "Siemens, Ernst Werner von."

5. Gustave Froment, a graduate of the Ecole Polytechnique, was a prominent Parisian manufacturer of scientific instruments. He manufactured Giovanni Caselli's facsimile telegraph, David Hughes's printing telegraph, and battery-powered electric motors of his own design. *Dict. des Bio.*, s.v. "Gustave Froment"; Laussedat 1865; "Le pouvoir et la science," *Le Cosmos* 10 (1857): 495–97; and "Mort de M. Froment, membre du Conseil de la Société d'encouragement," *Bull. Soc. l'ind. nat.* 64 (1865): 74–80.

6. "Philosophical Apparatus" meant scientific instruments.

7. Boston.

8. Charles Williams's relay is pictured in Pope 1869, opp. p. 31.

9. Henry and Walter Chamberlain. *Boston Directory* 1868, 133, 880.

10. Heinrich Ruhmkorff, the eponymous German maker of induction coils, had in fact improved the device invented by several others, including American physicist and physician Charles Page. Page's long and bitter priority fight over the invention is detailed in Post 1976. See also *DSB*, s.v. "Ruhmkorff, Heinrich Daniel."

11. Ritchie & Sons comprised Edward Ritchie and his sons T. P. and John. In 1857 Ritchie improved the induction coil by changing the method of wrapping wire around the core from overlaid longitudinal spi-

rals to projecting spirals placed side by side. *Boston Directory* 1868, 501; *DSB*, s.v. "Ruhmkorff, Heinrich Daniel."

12. See n. 10.

13. John Gassiot, a wealthy English merchant and physicist, used the induction coil to study the phenomena of striated discharges. *DSB*, s.v. "Gassiot, John Peter."

14. Edison made an error in the first citation. On page 326 of his copy of Noad 1859 there is a reference not to Edward Ritchie but to the English physicist William Ritchie and his torsion galvanometer. The second citation is to John Gassiot, "Description of a Ruhmkorff's induction Apparatus, constructed for John P. Gassiot, V.P.R.S., by Mr. Ritchie, Philosophical Instrument Maker, Boston, U.S."

15. Thomas Hall ran one of the oldest telegraph manufacturing shops in Boston. He began working in 1840 for Daniel Davis, who made the instruments for Samuel Morse's Baltimore-Washington line. When Davis retired in 1849, Hall acquired part of the business and in 1857 took over as sole proprietor. *Boston Directory* 1868, 277, 884; Hall 1874, 7–8.

16. Authorship attributed to Edison. See Doc. 34, n. 10.

–42–

To John Van Duzer[1]

John Van Duzer, who first met Edison in Memphis while working as a military telegrapher, corresponded with him in 1868 regarding Edison's proposed invention of a facsimile telegraph and a method of secret singaling.

Boston ~~Au~~Sept 5—68

Dear Sir

Your favor of the 5th Recd   I have for nearly 3 years been experimenting on a "fac simile" which I intend to use for Transmitting Chinese Characters[2]   It will probably be several months before I will be able to Bring it out, as experiments are rather Costly and there is a scarcity of funds = Have done nothing with it for nearly 3 weeks: being engaged on my automatic Printers, of which four 4 have Already been made by Williams

My system of "fac simile" is entirely different from the systems of Bain Bakewell Casselli and Bronelli as the messages do not have to be prepared before Tramission, and does not require a syncroneous movement[-] as in all others excepting Brouneli who uses 5 wires = [3] I use but one = I hope to attain A speed of from 100 to 125 messages per hour =

Would be happy to hear from you as to what the prospects are; have had ~~Liv~~ some encouragement from Isaac Livermore—Burlingames Bro-in law[4] who is rather incredulous and also from the East India Telegraph Co[5]   73—to you[6]— Respy

T. A. Edison[a]

ALS, PPAmP, MM Coll. [a]"83 state st" written below.

1. John Clark Van Duzer (1827–1898), whom Edison met in Memphis, was a civilian electrician in the U.S. Signal Corps from the summer

of 1868 to September 1869. As a young man he had edited and published newspapers before beginning telegraph work in 1848. During the Civil War he was second assistant superintendent of the U.S. Military Telegraphs and achieved the rank of captain. Around the time of this letter he did some work for the East India Telegraph Co. (see n. 2). Van Duzer correspondence, Sig. Corps; Plum 1882, 1:303–5, 2:293, 340, 342, 348; Taltavall 1893, 305–6; *Boston Transcript,* 7 Mar. 1898, 5.

2. A facsimile telegraph transmitted messages or pictures as written or drawn by the sender (see Doc. 46). Edison's special interest in transmitting Chinese characters may have stemmed from the highly publicized visit of a Chinese delegation to Boston in August 1868. The visit was organized by Anson Burlingame (see n. 4), who had recently obtained a contract from the Chinese government for the East India Telegraph Co., an American concern, to connect several of China's major seaports by cable. Prescott 1877, 741–67; "The East India Telegraph Company," *Telegr.* 4 (1867–68): 392. On the visit of the Chinese delegation, see issues of the *Boston Traveller* from 20 August 1868 to 2 September 1868.

3. The facsimile systems of Alexander Bain, F. C. Bakewell, Giovanni Caselli, and Gaetano Bonelli are described in Doc. 46.

4. An Isaac Livermore served as treasurer of the Michigan Central Railroad Co., with offices at City Exchange in Boston. Anson Burlingame, lawyer, congressman, and diplomat, was minister to China from 1860 to 1867. When he resigned, the Chinese government made him the head of an official delegation to seek diplomatic recognition from the United States and Europe. Burlingame married Jane Cornelia Livermore, whose father was Isaac Livermore of Cambridge, Mass. Whether she had a brother named Isaac is not known. *Boston Directory* 1868, 369, 785; *DAB,* s.v. "Burlingame, Anson."

5. See n. 2.

6. "73" was the commonly used telegraphic abbreviation for "Compliments of sender." Pope 1871a, 16.

ELECTRIC VOTE RECORDER   Doc. 43

Edison invented the electric vote recorder for use by legislative bodies. He may have been spurred by reports in the *Telegrapher* that the Washington, D.C., City Council planned to install an electric vote recorder and that the New York State legislature was considering one as well.[1] In Edison's system, each legislator moved a switch to either a yes or a no position, thus transmitting a signal to a central recorder that listed the names of the members in two columns of metal type headed "Yes" and "No." The recording clerk then placed a sheet of chemically prepared paper over the columns of type and moved a metallic roller over the paper and type. As the current passed through the paper, the chemicals decomposed, leaving

Model of Edison's electric vote recorder, which includes clockwork-driven dials for recording yeas and nays as well as an electrochemical recorder.

the imprint of the name in a manner similar to that of chemical recording automatic telegraphs.[2] Dials on either side of the machine recorded the total number of yeas and nays. Edison was issued a patent—his first—on 1 June 1869, but the vote recorder was never used.[3]

1. "Voting by Machinery," *Telegr.* 4 (1867–68): 155; "Miscellanea: An Electrical Voting Machine," ibid., 304.

2. See Doc. 33.

3. U.S. Pat. 90,646. Edison later claimed that the device failed to gain commercial acceptance because legislatures commonly made political use of the roll-call process and consequently had no interest in instantaneous vote recording. Lathrop 1890, 431.

–43–

Patent Model: Vote Recorder[1]

[Boston, October 13, 1868?[2]]

M, MiDbEI(H), Acc. 29.1980.1294.

1. See headnote above.

2. Edison executed the covering patent application on this date. Pat. App. 90,646.

−44−

Article in the Telegrapher

New York, October 17, 1868.

AMERICAN COMPOUND TELEGRAPH WIRE.[1]

PROBABLY no more important and valuable improvement in the science of Telegraphy has been made for a number of years, and certainly none that will effect a greater stride towards perfection in this science, than the wire manufactured by the American Compound Telegraph Wire Company.[2]

The employment of a steel core with a copper covering was a happy thought on the part of its inventors, Messrs. FARMER and MILLIKEN,[3] of Boston.

The superiority of copper as a conductor, and the great strength of steel, produce, when combined together, as perfect a line wire as could be desired.

The many advantages which will accrue by the adoption of this wire by our Telegraph companies, and the merits which it undoubtedly possesses, deserves the attention of all interested in the progress of the Telegraph.

In this wire the composite parts are steel and copper, the steel forming the core, and serving mainly for strength, while copper is used more especially for its superior conductivity.

The method of manufacture is quite simple—the steel wire, which is first tinned, is covered by being drawn through the plate together with a long thin strip of sheet copper, which in its turn is tinned. The second coating of tin is used for the purpose of preventing moisture from contact with the steel.

The following tables, prepared by Mr. FARMER, from the results of a large number of experiments made during a space of five years, may be relied upon as acurate in every respect:

TABLES.

		Weight per Mile.	Tensile Strength.	Conductivity.	Conductivity compared with.
Table No. 1		375	1,091	1,331	1
Table No. 2	Steel,	56	418	147	....
	Copper,	56	96	1,288	....
	Compound,	112	514	1,435	1.07
Table No. 3	Steel,	187	1,397	490	....
	Copper,	188	325	4,324	....
	Compound,	375	1,722	4,814	3.61
Table No. 4	Steel,	119	889	311	....
	Copper,	119	205	2,737	....
	Compound,	238	1,094	3,048	2.29
Table No. 5	Steel,	52	388	136	....
	Copper,	52	89	1,196	....
	Compound,	104	477	1,196	1
Table No. 6	Steel,	78	583	204	....
	Copper,	297	511	6,831	....
	Compound,	375	1,094	7,035	5.28
Table No. 7	Steel,	357	2,768	935	....
	Copper,	18	31	414	....
	Compound,	375	2,799	1,349	1
Table No. 8	Steel,	136	1,016	356	....
	Copper,	43	74	989	....
	Compound,	179	1,090	1,345	1

Table No. 1 contains the elements for the average of No. 8 galvanized iron wire.[4]

Table No. 2 is the ordinary equivalent of the compound wire to No. 8 galvanized iron wire. The decrease in the tensile strength of the compound wire, when compared with the iron wire, is regained, owing to its lightness, both having the same "relative strength."* Also, a small increase in conductivity.

*Denotes the quotient obtained by dividing the strain which would break the wire by its own weight per mile.

Table No. 3 is the compound wire, having the same weight per mile as the iron wire. In this form it will be noticed that the tensile strength of the wire has increased nearly double, and an increased conducting capacity three times greater than that of the iron wire.

Table No. 4 is the compound wire, with the same strength as the iron wire, but considerably lighter, and with more than double the conducting capacity.

Table No. 5 shows a compound wire of equal conductivity, weighing three times less, and possessing the same relative strength.

Table No. 6 shows a compound wire having the same weight and tensile strength, but with a conductivity five times greater than that of the iron wire.

Table No. 7 shows a compound wire having the same weight and conductivity, but with nearly three times the tensile strength.

Table No. 8 shows a compound wire of the same tensile strength and conductivity, and weighing but 179 pounds to the mile.

It will be seen, by referring to these tables, that the compound wire need have only about one third the weight of the galvanized iron wire, to be relatively stronger, and at the same time possess an equal or greater conducting capacity. It is evident why this should be so, since the best commercial copper possesses more than six times the average conductivity of galvanized iron wire, and the steel wire has nearly three times the tensile strength of galvanized iron wire of the same size.

The relative strength of the best steel wire averages 7.47, that of copper 1.72, while the average strength of galvanized iron wire, as found by testing various samples, is only 2.9.

It is evident that, by varying the proportions of steel and copper in the combination, any desired relative strength can be given between the limits of 1.72 and 7.47, and at the same time, any desired conductivity can be had along with it.

The impossibility of drawing steel into wire containing flaws, which is not the case with iron, prevents the breakages which occur so frequently when iron is used.

The great advantage which this wire has over iron wire is, that its lightness will admit of an average of ten poles to the mile less than would otherwise be necessary; which, according to Mr. FARMER, will effect a decrease of twenty-five per cent. or more in the escape of the current—besides, a reduction in the number of poles will conduce to economy in construction.[5]

Another point in favor of this wire, and there seems to be many, is the imperishable nature of copper, which is the exposed metal, zinc coating of the galvanized iron wire being deteriorated near the sea, from the effects of gases, &c., while copper, under the same condition, is unimpaired.[6]

E.[7]

PD, *Telegr.* 5 (1868–69): 61.

1. The article is featured on the top left of the first page of the issue.

2. American compound wire was invented in 1865–66 by Moses Farmer and George Milliken (U.S. Pats. 47,940 and 59,673). The American Compound Telegraph Wire Co., set up in 1868 to market the invention, had its offices in New York City rather than Boston. Reid 1879, 374; American Compound Telegraph Wire Co. 1873.

3. George F. Milliken (b. 1834), formerly manager of the Boston office of the American Telegraph Co., became manager of the Boston office of Western Union when the two companies merged in 1866. Milliken held several telegraphic patents. He hired Edison to work for Western Union in Boston in 1868. Taltavall 1893, 265–66.

4. On wire gauge no. 8, see Doc. 34, n. 3.

5. Current leaked from the wires where they touched the poles. In 1870 Farmer stated that compound wire required only 15 poles per mile, whereas iron wire required 38 poles per mile. Farmer 1870.

6. Near the sea, airborne salts attacked the wire's zinc coating. The "effects of gases" refers to the sulfur oxides released by the burning of coal and coal gas, which joined with atmospheric moisture to form acids that reacted with zinc. Culley 1871, 138; Maver 1892, 511.

7. Authorship attributed to Edison. See Doc. 34, n. 10.

–45–

Milton Adams Agreement with E. Baker Welch

Boston, Novem 24, 1868.

Agreed between Milton F. Adams & E. B. Welch that said Adams & Thomas A. Edison are to give E. B. Welch their joint note for one half the expenses incurred in making and experimenting with Printing Telegraph Instruments and to assign to said Welch a majority of interest in any Patent they may jointly or separately obtain for any Printing Telegraph Instrument or Transmitter for Telegraph purposes.[1]

Milton F. Adams.

TD (transcript), MWalFAR, *Welch v. Edison*. See Doc. 36 textnote.

1. On 21 July 1868 Adams had executed a caveat and signed an agreement with Welch to give Welch a half interest in an "automatic printing instrument." On 5 December 1868 Adams officially assigned a half interest in the invention to Welch and applied for a patent. Edison's name was crossed out both in the text of the assignment and in the patent application as an assignee. Adams received U.S. Patent 99,047 for an "Automatic Printing-Telegraph Apparatus," a printing telegraph transmitter and receiver intended especially for "brokers, merchants, heads

of departments &c., to correspond directly and privately with their agents in distant parts." Milton F. Adams, Memorandum of an Agreement with E. B. Welch, 21 July 1868, *Welch v. Edison;* Patent Assignment to E. Baker Welch, 5 Dec. 1868, Edison Coll.; Pat. App. 99,047.

–46–

To John Van Duzer

Boston Sunday, Dec. 6 1867[1868][1]

Capt Van—

According to promise I herewith send a general description of the "Fac Simile"

In Casseli's apparatus, There is a synchroneous principle, which is obtained by the vibration of a pendulum. one pendulum [is] so connected by mechanical devices that it passes over tin-foil written upon by an insulating ink, while the other pendulum, similiarly connected, passes over chemically prepared paper, both in unison = when the transmitting pendulum passes along the tinfoil the battery is "short circuited" and consequently no chemical change at the distant end takes place, but when the tramitting pendulum passes of a line of the insulating ink, the "short circuit" is removed an[d] chemical decomposition takes place at the other end— The objection to this system, is that it is not practical enough, on account of the impossibility of producing two pendulums that will vibrate in unision for any length of time. another objection, is that it is very complicated.[2]

Bronelli uses from 4 to 6 wires = and sets up the dispatches in Type— This system is dead, I believe =[3]

Bakewell and Bain use clockwork which is open to the same objection as Casseli[4]

Chavaussaignes and Lambrigot only transmit conventional signs—[5] These are The only system That amount to anything.

Now with these systems all before me, I conceived that an apparatus for tramitting writing, "fac simile," to be of any practical value, must not depend upon synchronious movements, 2nd that it must be <u>practical</u> 3d simple, and 4th Rapid, 5th self acting, (ie) No operators & 6th No prepartion of messages[a]   You will see further on that the apparatus at one end is in reality only a part of the apparatus at the other end (ie) the transmittor and Receiver is but one machine, one part depending on the other just as the flywheel on a train of wheels depends on the drum, one cannot move without the other and if one move fast, the other must also = Now if I only use on[e] machine, there is no synchronism for to produce synchronism there must be two

The manner of tramitting messages without preparation, I think is the most novel part = I have several different methods but the one I describe I think is the best =

Graphite is a conductor, the purest graphite is used in Fabers No 1 pencil. Take one of these pencils and Make a mark, now take, say an intensity battery[6] of ~~say~~ 12 cups, connect with a relay on a short Line, then connect each end or the extremities with two small thin plates of platinum seperated from each of by mica, draw it across the pencil mark and the relay will close. Thus you see ~~than~~ Lead pencil marks may be made to close an electric [-]circuit   Metallic Ink could also be used.

~~If~~When messages are written with a pencil, the receiving clerk does not have to prepare it for transmission but has merely to shove it in the hopper of the apparatus and its transmitted directly from the pencil marks

I suppose you understand the polarization of relays but for fear that you do not, Ill describe one

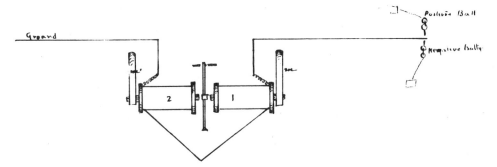

1 & 2 a[re] common magnets, the e~~I~~ron cores of which are permanently magnetized by the steel horse shoe magnets m' and m.   one iron core has North magnetism imparted to it, while the other has South. Now if a positive current be sent through both spools, (which are connected up in the same circuit) it neutralizes the induced magnetism in one spool say No 1 and adds to the other   then the armature will be attracted to magnet No 2, and when the a negative current is used it will act vice versa etc   when there is no current in the spools the armature will remain at either side, both forces being equal which they always are = when properly adjusted

This Little instrument might be called the piston rod of the "fac simile" as you will see = I will show these two instruments in one circuit working the apparatus. The drawing, is not anything Like what I intend the apparatus to be but one which you will be more apt to the get the idea from—[b]

I suppose you are aware of The fact that if a bar of soft iron half way or less into a helix of wire, placed in circuit that it will be drawn in the full length of the helice. Now suppose we take two helices of wire, without Cores—but having a brass tube fitted within. and suppose you take an iron bar a little smaller than the barass tube. place one end of the bar into one helix and the other end of the bar in the other helix—Now if a current is sent through one helix it will be drawn into this helix—and if this helix is discharged an the Current placed into the other helix it will be drawn from the other into the one having the current. Prof Pages engines were constructed this way and they are known to have been the most powerful ever constructed[7] The strength is gained by the peculiar manner of ~~o~~partially overcoming the Law of the increase and decrease of magnetism as the square of the distance. By these hollow helices several inches of play can be given to an armature, which [-] could not be[c] done by the ordinary core magnet and armature unless very power~~o~~ful

M′ S′ & M S as thus constructed— G′ and G figs 1 and 2 ~~is an~~ are[d] armatures connected to the sliding Cores P′ and P. The armature G′   Fig 2   is 2 long pieces of brass ~~s~~insulated from each other by hard rubber   ~~one~~ on the end of this armature is a Little wheel composed of two pieces of platinum insulated from~~e~~ each other by a very thin piece of mica. The thickness of this wheel is about the 100th part of an inch — each platinum portion of the platinum wheel is connected to either of the strips of brass forming the armature   Thus

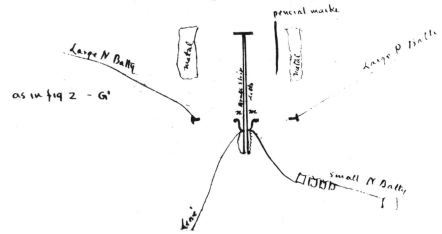

Transmitting armature

The Line com~~inge~~s in and connect~~ings~~ with the brass strip n and one of the platinums of the wheel, and the other strip and platinum with the small battery (decomposing battery) = When the platinum wheel passes over a pencil mark the graphite being a conductor, connects the two sheets of Platinum composing the wheel, together which places the small battery on the Line, and decomposition takes place on the paper at the other Terminal. You will Notice on the transmitting apparatus two strips of metal on each side of the armature. This is used for connecting the platinums of the wheel together and closing the circuit after passing the paper[e] so that the Large batteries can be placed upon the Line to effect a simultaneous change of the armatures of both polarized relays =

Now I will try and Show how the two armatures move exactly together =

Suppose that both the armatures G' and G had just left the points L and T. The the armatures of the polarized relays would be at the points n and K' and conqsequently the [b]Locals mm and BB would actuate the helices S & M' drowing both armatures over = Now the moment the platinum wheel touches one of the [-]metalic plates placed on both sides the continuity of the Line would be complete (the metalic plate connecting the two platinum parts of the wheel together) and then the main Line would be placed simultaneously on the Large positive battery. Negative at Fig 2 and Postive at Fig one   Now this battery passing through the polarized relay being contrary to ~~ma~~permanent magnetism there, Both armatures would fluy[8] to points n' and K together thus taking the Local[d] battery BB & mm of[f] of the magnets S & M' and placing battys BB' & mm' in helices M & S' which would throw the armatures G' & G over to the points L' & L and placing an negative-positive batty on the line which in its turn would throw the polarized armature to the other points, and so on to eternity. These large batteres are merely used to throw the armatures from one point to the other. It is like two men on each side of a swinging effigy   one knocks it one way and the other knocks it backs = The writing is done by a smaller battery when neither armature is touching its points = You will easily see that these armatures must go together, for if some cause ~~should~~ one armature should Li~~k~~g the 1000th of a second, the continuity of the Line would be interrupted and the one would have to wait ~~t~~and both start of[f] together = a correction at every vibration = [9]

There is another magnet ~~so~~ arranged in the Local circuits (~~but~~ which I have not show—) and also[d] connected with a release escapement so that for every movement of the armature the paper of both machines is thrown ahead one Line ~~a~~which Line will average the breadth of a hair

I have only shown the generalities for fear of complicating it so you would not understand.

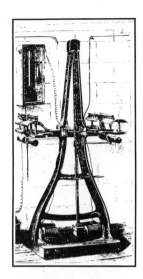

Italian Giovanni Caselli's autographic telegraph, which was known as the "pantelegraph."

I have calculated that the speed would be about 125 messages per hour—and the cost of each set complete about $200. To perfect it will consume abut one year and from 500 to $800 dollars in money as I am quite certain, that[d] to get it perfect 5 or more[f] machines would have to be constructed in different sha[–]pes before the perfect one could be reached besides a large amount of experimenting independent of the machine itself= With sufficient Tools and supplies I could do it nearly all myself=

I am willing to try it one year or even two years providing, some one, furnishes the necessary funds

It can be done ~~and it will be done~~ and if I don't do it somebody will

Will call in tomorrow and see Mr Smith[10]

Write me what[g] you think= Will send you description of that signal Corps appartus as soon as I get time=[11] 73[12]

Yours Respy

Thomas A Edison[h]

ALS, PPAmP, MM Coll. [a]Interlined below. [b]"over" written at bottom right for page turn. [c]"could not be" interlined above. [d]Interlined above. [e]"after passing the paper" interlined above. [f]"or more" interlined above. [g]Written over "=". [h]"83 State St Boston Mass" written below.

1. Although Edison dated this letter 1867, its content logically follows that of Doc. 42. The editors have therefore chosen to place the letter in the 1868 chronology. Moreover, 6 December 1868 was a Sunday.

2. Giovanni Caselli, professor of physics at the University of Florence, invented the "Pantelegraph" in 1856. It was installed on a line between Paris and Marseilles in 1863 and extended to Lyons and Le Havre in 1865. The highest speed attained by Caselli's system was the transmission in three and a half minutes of a message on a surface measuring 30 square centimeters. Caselli received U.S. patents for the invention in 1858 and 1863 (U.S. Pats. 20,698 and 37,563). Sabine 1867, 204–6; Prescott 1877, 744–56; Saint-Edme 1869, 502–4; *DBI*, s.v. "Caselli, Giovanni"; *Annales télég.*, 2d ser., 8 (1865): 365.

3. Gaetano Bonelli, head of the Piedmont and Sardinian telegraph service, invented an instrument he called the typo-telegraph ("tipo-telegrafo") because it used messages set in type. The type passed under a comb with five metallic teeth, each connected to a separate line wire.

The transmitter of Gaetano Bonelli's automatic "typo-telegraph," which may have inspired Edison's later design of a roman-letter automatic telegraph (see Doc. 186).

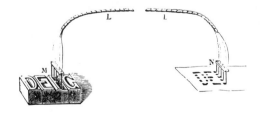

At the receiving end, five metallic styli connected to the line wires passed over a strip of paper soaked with a chemical solution. Clockwork mechanisms advanced the metallic type and the strip of paper to provide synchronous movement between the sending and receiving devices. Although Bonelli's system was a technical success, claiming speeds as high as 2,000 words a minute, it required ten wires and was therefore not considered practical. Bonelli patented his telegraph in the United States in 1863 (U.S. Pat. 37,331). Sabine 1872, 208–12; Prescott 1877, 763–67; *DBI*, s.v. "Bonelli, Gaetano."

4. Alexander Bain modified his automatic telegraph in the 1840s to transmit handwriting. The message was written in insulating ink on a disk of metallic paper and a stylus passed over the revolving disk in a spiral, transmitting except when interrupted by the ink. At the receiving end, a disk of chemically prepared paper revolved by clockwork in synchrony with the first. As a stylus passed over that disk the current decomposed the chemicals, reproducing the message. Sabine 1867, 199–201; Prescott 1877, 690–92.

In 1850 F. C. Bakewell of London improved Bain's system, replacing the disks with rotating cylinders. A metal stylus, moving on a threaded rod, described a continuous spiral line on the surface of the cylinder. Weight-driven clockwork with an electromagnetic governor maintained synchrony. Sabine 1867, 203–4; Prescott 1877, 741–44.

5. See Doc. 33, n. 3.

6. Batteries were commonly described as either intensity or quantity batteries, meaning that their component cells were wired either in series, to produce a high potential (or voltage), or in parallel, to produce a large amount of current. Voltage, amount of current, and resistance are quantitatively related in any circuit in that the voltage equals the product of the current and the resistance ($V = IR$). The German scientist Georg Ohm formulated this relationship in 1827; electrical experts were generally acquainted with Ohm's law by midcentury.

Charles Page's electric motor.

7. Beginning in 1844 Charles Page developed a form of electric motor that he called an "axial engine." Page's motor essentially comprised two solenoids through whose centers ran a common axial rod. As the switching of current alternately activated and deactivated each solenoid, the axial rod moved back and forth between them. Post 1976, 81–82.

8. Fly.

9. Edison's system differs from all previous systems by transmitting not only the message but also the signals for synchronization of the apparatus at each end of the line.

10. Unidentified.

11. See Doc. 47.

12. See Doc. 42, n. 6.

To John Van Duzer

Capt Van—

It will be some time before I will be able to prepare a drawing and description of That "Secret Signalling Apparatus" if you think there is any money to be made out of it, I will have the instruments made.[1] I suppose If the signal corps should buy it, they would not want it Patented.[2]

What it is and what I claim for it is this

1st   The transmission of dot and dash signals over telegraph circuits between terminal stations in such a manner, that the magnets at the intermediate stations will always remain charged at a uniform strength while communications are passing between the terminals.[3]

2nd   The use of only one current (ie) a current having but one polarity, which prevents the signals from being copied off at an intermediate station by means of polarized appartus or by any device known in Telegraphy =

3d   The retransmission of the signals by the receiving apparatus ѳback to the transmitting apparatus over the same circuit and at nearly the same time, ensuring the positive reception of the despatch =

34th =   The instantaneous detection of a person tapping the wire, = simplicity = cheapness = etc

Will assign you ½ if you conclude to take hold of the thing = [4]

Have been testing some Manganese Battys Today[5]—very busy   Respy

Thomas A Edison

ALS, PPAmP, MM Coll. Letterhead of Charles Williams, Jr. [a]"Boston," and "186" preprinted.

1. The idea for a secret signaling apparatus had been suggested to Edison by George Ellsworth, telegrapher for the Confederate cavalry captain and raider John Morgan. While Edison was an operator in Cincinnati, Ellsworth approached him about protecting telegraph lines from being tapped. Edison later regarded one of his unsuccessful designs in this work as the first step toward his later invention of a quadruplex telegraph. Edison claimed to have kept a model of a successful design and to have installed it for a company on a private line sometime while he was at Menlo Park. Testimony and Exhibits on Behalf of T. A. Edison, pp. 3–4, *Nicholson v. Edison;* App. 1.D220; *Telegr.* 3 (1866–67): 255.

2. The issuance of a patent required public disclosure of the details through publication by the U.S. Patent Office.

3. This would prevent operators at intermediate stations from detecting signals. See Doc. 49, n. 2.

4. There is no evidence that Van Duzer financed the development of the invention.

5. Manganese batteries were a recent development, patented in 1867 (U.S. Pat. 64,113), by Georges Leclanché, a chemical engineer employed by a French railroad company. Leclanché's battery had a zinc anode in a sal ammoniac solution separated by an earthenware partition from a carbon cathode resting in a mixture of manganese peroxide and other materials. It produced about 1.5 volts with a moderate current. The French telegraph service adopted this battery, and it was employed in the United States to power police and fire alarms because of its appropriateness for intermittent use. Knight 1876–77, s.v. "Leclanche Battery"; Schallenberg 1978, 348–51; "Constant Galvanic Batteries," *Telegr.* 5 (1868–69): 53; King 1962a, 247; Moise and Daumas 1978, 4:315–16.

–48–

Editorial Notice in the
Telegrapher

New York, December 12, 1868.
Edison's Double Transmitter.

WE would call attention to the advertisement in this paper of "EDISON'S Double Transmitter."[1] We are assured that this is an entirely new and greatly improved instrument, and entirely different from any similar instrument heretofore described in our columns. We shall probably soon be able to publish a full illustrated description of this invention for the information and gratification of the readers of THE TELEGRAPHER.[2]

PD, *Telegr.* 5 (1868–69): 128.

1. Edison's earliest known advertisement for an invention appeared in the *Telegrapher* from 12 December 1868 to 9 January 1869 (5:129, 137, 145, 153, 161) and in the *Journal of the Telegraph* from 15 December 1868 to 15 May 1869 (2:22, 43, 94, 115, 140). Although Edison claimed

several "bona fide offers" (Doc. 68), there is no evidence that he sold any instruments. This is probably not the device described in Doc. 28.

2. No description appeared.

49

To John Van Duzer

Capt Van =

Yours of 14 Recd— The whole apparatus is containded in a box about 1 foot square =[1] The actual[b] Cost will vary from $120 to $150 for a complete set (2) (ie) 60 to 75 dollars each. It will weight about 18 pounds. by turning a switch you can use the apparatus in the ordinary way[c] (ie) suppose the operators were working Morse—one could notify the other to throw his switch over. they would still be working Morse but the circuit would be closed at Intermediate stations. There is no difference in the speed from ordinary apparatus. I use the ordinary batteries Positive at one end & Neg at the other or a single battery at one end. It is very simple.

The Same instruments <u>Cannot</u> be used at way-stations, except by making ~~the~~ it a Terminal by[d] grounding[e] the wire. This is instantly detected by both terminals, and each Terminal may allow him to receiv~~ere~~ or not just as they wish   It is done by Concentric Relays etc =[2]

Suppose I have a set made. Respy

T A Edison

P.S. What length of circuit should I calculate the magnets for   Should I get a set made   E

ALS, PPAmP, MM Coll. Letterhead of Charles Williams, Jr. [a]"Boston," and "186" preprinted. [b]Interlined above. [c]Added in right margin. [d]"it a Terminal by" interlined above. [e]"ing" added below.

1. This was Edison's proposed "secret signaling apparatus." See Doc. 47.

2. Most likely a kind of induction coil arranged as a relay (see Doc. 30, n. 2; cf. also "Duplex No 12" in Doc. 285, and Doc. 310). The details of the design are not known, but Edison's comments indicate that his envisaged system would have maintained a current on the line strong enough to keep all relays closed while using variations in current strength above that level for signaling, rather than opening and closing the circuit as was ordinarily done. Such variations would induce currents in coils wound around the ordinary relay coils, and these induced currents would activate the receiving devices. In addition Edison's system apparently would have arranged the devices for duplex operation, with a repeater returning the signal to the sender as it was received.

50

[Boston?, 1868?][2]

Memorandum: Multiple Telegraphy[1]

Stearnes[3]—When not in use no battery is upon the line.

Edison—Battery always upon the line.

Stearnes—It is an open circuit.

Edison—It is a closed circuit.

Stearnes—Uses the neutralization of currents.

Edison—No neutralization used.

Stearnes uses one battery at each end.

Edison uses two disimilar battery at each end.[4]

Edison—Use rheostal[5] for the purpose of increasing and decreasing the strength of the current.

Stearnes—Use rheostals prevent battery from being eaten up.

Stearnes—Is compelled to equate battery by rheostal buttons.

Edison—No equating principle used.

Stearnes—But one message can be exchanged. (i.e.) One can be sent from B to N.Y. and one from N.Y. to "B", but two cannot be sent from Boston to N.Y.

Edison—B can send two messages at same time to N. York.[6]

Stearnes—No way offices can get a single word.

Edison—Way offices can receive and send as usual.

Stearnes—Breaks must be exchanged between sender and receiver vive voice.

Edison—Breaks come in the usual way.[7]

Stearnes—Is the incomplete parts of two lines.

Edison—Is one line complete, and one uncomplete.[8]

Stearnes—Increase and decrease used for extra wire.

Edison—Reversal of batteries and pulsations upon polarized relays.

Stearnes uses no polarized relay.

Edison uses polarized relay.

Stearnes—The currents can only be half the strength of each battery.

Edison—Full power of battery used.

Stearnes—Has no reversal key.

Edison—Has a reversal key.

In fact there is not the slightest similarity in any form, principle or manner.

Joseph Stearns, president of the Franklin Telegraph Co.

TD (transcript), MWalFAR, *Welch v. Edison.* See Doc. 36 textnote.

1. This list compares Edison's and Joseph Stearns's multiple telegraph systems. No attribution or address is on the document. In his suit against Edison, E. Baker Welch entered it and other communications from Edison as exhibits.

This document illuminates the state of Edison's thinking and work in multiple telegraphy during 1868–69. Regarding Edison's system, the document may be a prospectus or may refer to devices already tried. The description of Edison's system suggests Edison's familiarity with methods and devices used in previous duplex and diplex designs by others.

2. Welch grouped this undated document with others of 1868–69.

Because Edison received money in Boston from Welch to support development of a double transmitter, Welch obtained a half interest in the results (see Docs. 36 and 61). As a director of the Franklin Telegraph Co., Welch knew Stearns's duplex system and would have wanted assurance that Edison's results would not duplicate it.

3. Joseph B. Stearns (1831–1895), Boston inventor, was president of the Franklin Telegraph Co. at this time. He had previously been superintendent of the Boston fire alarm telegraph for twelve years. In 1868 Stearns improved the well-known duplex design of Frischen, Siemens, and Halske (see headnote, pp. 31–32), incorporating local sounders for transmitting operators (U.S. Pats. 78,547 and 78,548), and put his innovation into operation on his company's lines between Boston and New York. In 1872 he invented and patented an improved duplex system that used condensers and thereby rendered duplex telegraphy generally profitable. His breakthrough was the first significant step in dealing with static induction on land telegraph lines. Stearns sold his duplex innovations to Western Union in 1872. *Boston Directory* 1869, 575; Taltavall 1893, 354; "A Telegraphic Novelty," *Telegr.* 4 (1867–68): 236; "Telegraphic Improvements," ibid., 252; *Elec. Engr.* 20 (1895): 37; *Boston Transcript,* 5 July 1895, 7.

Wiring diagram for Stearns's differential duplex.

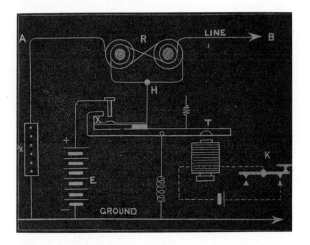

4. Edison and others used such arrangements to vary current strength or to reverse polarity.

5. That is, rheostat, which is an adjustable resistance.

6. That is, diplex.

7. A "break" is defined as the interruption of a sender by a receiver. The "usual way" was for the receiver to open the circuit. On a duplexed line, a pair of operators—a receiver and a sender—worked simultaneously at each end. If a receiver wanted to break, using Stearns's design (or nearly any other duplex), the sender at that end had to notify the receiver at the other end to stop the transmission.

8. When Edison calls Stearns's lines incomplete, he apparently refers to the problem with breaks (see n. 7); using Edison's system, one pair of operators could send and receive in either direction in the normal fashion, while the other pair could apparently transmit in one direction only.

From Operator to Inventor-Entrepreneur

January–June 1869

The first half of 1869 marked a turning point in Edison's career. After working for five years as an itinerant telegraph operator, he left Western Union to "devote his time to bringing out his inventions."[1] During this period he engaged in three new business-technology ventures. One employed his first printing telegraph for stock quotations; the second used his dial telegraph, the magnetograph, for private lines; and the third involved a new double transmitter. This last venture brought him to New York in April, where he eventually settled in order to take advantage of the support available in the nation's telegraph center.[2]

Edison's first important venture as a telegraph inventor involved an improved printing telegraph system that provided stock quotations for bankers and brokers.[3] It marked the beginning of Edison's inventive work in a field that helped him achieve a reputation as one of the premier American telegraph inventors. In contrast to the three-wire printing telegraph marketed by the Gold and Stock Telegraph Company in New York, his new printing telegraph employed only a single wire to transmit information.[4] Although his printer could be used on private lines, Edison designed it primarily for bankers and brokers. He placed all instruments in a single circuit and operated them from a central office by means of a single transmitter. He later sold his first stock ticker invention to Gold and Stock.[5]

The budding inventor also began to develop and exploit his nascent entrepreneurial talents. In January he assigned the patent rights to his printer to two Boston businessmen, Joel Hills and William Plummer, in return for their financial assistance. With their backing, Edison rented two rooms at 9 Wilson Lane, near the Boston Exchange, where he established headquarters for a stock-quotation service. He also attracted to this enterprise Samuel Ropes, Jr., a business promoter; Frank Hanaford, a former telegraph operator in Boston; and Dewitt Roberts, with whom he worked on the vote recorder. Forty years later Edison recalled that he established "a Laboratory over the Gold-room" and "opened a stock quotation circuit with 25 subscribers, the ticker being of my own invention." Roberts later remembered that their first customer was the banking and brokerage house of Kidder, Peabody and Company, then headquartered on State Street in Boston.[6]

At this time Samuel Laws's Gold and Stock Reporting Telegraph Company and the Bankers' and Brokers' Telegraph Company, a New York firm that provided telegraph service in competition with Western Union, were planning to build a line to Boston to establish a stock-quotation service. Laws already used the lines of Bankers' and Brokers' to provide a stock-quotation service to Philadelphia. He abandoned the Boston plan when his Gold and Stock Reporting Telegraph Company merged with the Gold and Stock Telegraph Company in September 1869. Edison may have had some connection with Bankers' and Brokers' at this time—the company operated from 9 Wilson Lane, and Edison used its letterhead stationery.[7]

Edison also used the Wilson Lane address as headquarters for another enterprise. Supported by Boston businessman E. Baker Welch, Edison manufactured and marketed his dial telegraph, the magnetograph. Cheaper and less likely to get out of order than the printing telegraph, the magnetograph provided private-line communication between the head offices of businesses and their factories and warehouses. Jerome Redding later described the instrument as "a small low priced" indicating machine in which the operator spelled out words by turning the hand in the center of the dial to point to successive letters in an outer circle.[8] Edison recalled that the instrument

Edison's signature, from an autograph book containing names from the Boston Western Union office.

Thomas A. Edison

Boston January 11th 1869

"was very simple and practical and any one could make it work after a few minutes explanation."[9] George Anders, a skilled mechanic employed by the telegraph manufacturing firm Edmands and Hamblet, resigned his position in late March to join Edison in the manufacture and sale of magnetographs at 9 Wilson Lane.[10] In May, Edison assigned a patent application for this instrument to himself, Welch, and Anders, but it failed to receive U.S. Patent Office approval.

Edison also continued his interest in multiple telegraphy and assigned to Welch half of his interest in a new double transmitter and in any future improvements on it. Edison's double transmitter seems to have been an unusual combination of duplex and diplex, which he and Welch intended to market to major telegraph companies. They received permission to conduct tests of the instrument on the New York to Rochester line of the Atlantic and Pacific Telegraph Company, which had previously tested Joseph Stearns's duplex. With funds provided by Welch, Edison probably traveled first to New York City to set up his instruments there and then to Rochester, New York, to complete the installation and operate the instruments at that end of the line. Franklin Pope may have been the operator at New York.[11] The instrument failed to perform satisfactorily, however, and Edison returned to New York, where he and Pope conducted further tests that engendered changes in the instruments.[12] While he continued to keep informed of his Boston business ventures, which were not flourishing, Edison also made new contacts in New York that induced him to stay in that city rather than return to Boston.

1. Doc. 55.

2. See Chapter 4 introduction and Doc. 70.

3. In 1868 Edison already knew of opportunities for telegraph entrepreneurs to supply special services to bankers, brokers, and other businesses. Accordingly, he had developed an appropriate instrument for reporting stock prices.

4. Invented by Edward Calahan, the three-wire telegraph required two wires to turn the dual type wheels and one to lift the platen that carried the paper tape. See Doc. 91, n. 4.

5. See Doc. 97.

6. App. 1.A26; Dewitt Roberts to TAE, 28 May 1877, DF (*TAEM* l4:63); Dewitt Roberts to TAE, 16 July 1890 and 20 Jan. 1911, GF. On Kidder, Peabody and Co., see Carosso 1979.

7. New York City 349:974, RGD; Reid 1879, 605; *Boston Directory* 1869, 876. The 1869 directory entry for Edison gives his occupation as "telegrapher" and his address as 9 Wilson Lane (ibid., 216). For Edison's use of Bankers' and Brokers' stationery, see Doc. 60.

8. Jerome Redding to Frank Wardlaw, 1929, Pioneers Bio.

9. App. 1.A26.

10. George Lee Anders (born c. 1836), mechanic and inventor, had been employed since 1866 by Edmands and Hamblet, where he gained experience in the manufacture of dial telegraphs (see Doc. 41, n. 2). Later he recalled that he also joined with Edison "in making instruments or perfecting instruments for stock quotations"—that is, Edison's Boston instrument. Testimony for Anders, pp. 1, 4–6, *Anders v. Warner;* George Anders to TAE, 24 Sept. 1878, DF (*TAEM* 16:121).

11. See, for example, "Thomas Alva Edison," *Operator,* 1 June 1878, 3; Pope to TAE, 27 Dec. 1882, DF (*TAEM* 63:566); and App. 1.D229. It is certain that Pope worked with Edison in trying to improve the instrument (see Doc. 68).

13. In Doc. 68 Edison indicates that he was having his instruments altered, but nothing is known of their original design (although it may be the instrument discussed in Doc. 50) or of the changes he made in New York.

–51–

Agreement with Samuel Ropes, Jr., Joel Hills, and William Plummer

[Boston,] January 21, 1867[1869][1a]

Whereas:[b] Thomas Alfred Edison[2] of Boston has invented an "Improvement in Telegraphing" which he has this day taken measures to secure by letters patent of the United States, And, Whereas—Samuel W. Ropes Jr.[3] has an interest in said invention and is desirous of acquiring an interest in the letters patent which may be obtained therefor— And— Whereas Joel H Hills[4] and William E. Plummer[5] both of Newton have advanced the sum of thirteen hundred dollars to said Ropes expended in perfecting and developing said invention— And have also agreed to pay the expenses of procuring letters patent of the United States for the same. And have also agreed to pay the said Edison the further sum of two hundred and fifty dollars when the practical utility of said invention shall be proved satisfactorily to them.

Now, therefore, be it known: that in consideration of the premises they have severally agreed with each other and among themselves as follows:

That said Edison shall sign papers necessary to obtain said letters patent which shall be issued to and in the names of said Hills and Plummer.[6]

The said Hills and Plummer shall pay the expenses of procuring said letters patent and also pay said Edison two hundred and fifty dollars when the utility of said invention shall be proved as hereinbefore recited. The said Ropes shall use all reasonable diligence to introduce said invention to the public and into general use, but no person or Corporation shall have the right to use said invention without written license signed by said Hills & Plummer   And said Ropes shall be entitled to one third of the profits or net monies accruing from said invention or letters patent. And whenever the said Ropes shall pay to the said Hills & Plummer the sum of thirteen hundred dollars out of his share of said profits or net monies or otherwise repay the amount advanced to him by them as hereinbefore recited, they shall transfer and assign to him one undivided third of said letters patent and the said letters patent shall thereafter be held for their joint and mutual benefit, and each shall receive one third of all profits or net monies accruing therefrom to be from time to time divided between them[7]

In witness whereof the parties have hereunto set their hands and seals this twenty first day of January Eighteen hundred and sixty seven.

| (signed) Thomas A. Edison | Samuel W. Ropes Jr |
| Joel H Hills | William E. Plummer |

D (transcript), MiDbEI, EP&RI. [a]Date taken from text, form altered. [b]"Copy" written in top left margin.

1. Although this document is dated 21 January 1867 in the text and 1 January 1867 in the docket, all other evidence indicates it was signed in January 1869. Neither Edison nor Ropes is known to have been in Boston in 1867, and all other references to Edison's association with Hills and Plummer are from 1869 or 1870. See n. 6.

2. The same error in Edison's name appears in the text of the patent assignment. Libers Pat. I-11:292.

3. Samuel W. Ropes, Jr. (d. 1871), was listed in the Boston city directory only in 1868—as a stock and exchange broker at 81 Washington St., residing in Dover. In addition to partially financing Edison's experiments on stock tickers, he acted as a salesman in Edison's telegraph enterprise. In 1871 Ropes helped introduce Edison's printing telegraphs in Chicago. While working there as an agent for the Gold and Stock Telegraph Co., he died on 4 April 1871. *Boston Directory* 1868, 508; Jerome Redding to Frank Wardlaw, 30 Sept. 1929, Pioneers Bio.; Doc. 128; "Sudden Death of a Telegraph Agent," *Telegr.* 7 (1870–71): 259.

4. Joel H. Hills (d. 1892) of Newton, Mass., and his brother, William Hills, were partners in a profitable Boston flour and storage business, Hills and Brother. R. G. Dun and Co. described them as "smart honest and energetic." *Boston Transcript*, 23 June 1892; *Boston Directory* 1865, 205; ibid. 1869, 311; Mass. 71:125, RGD.

5. William E. Plummer (1844?–1890) of Auburndale, Mass., was a successful hide broker in Boston and an agent for Miller's bark extract. He was active in local Democratic politics and was known to purchase interests in patents. *Boston Transcript*, 2 Aug. 1890; *Boston Directory* 1865, 329; ibid. 1869, 495; Mass. 84:492, 494, 531, RGD.

6. Edison executed the patent application (U.S. Pat. 91,527) and as-

signed his rights to Hills and Plummer on 25 January 1869. Libers Pat. I-11:292.

7. In 1870 Edison repurchased from Hills and Plummer the rights to the patent.

–52–

Joel Hills to William Plummer

[Boston,] Jany 25/69

Dear William,

You furnish Edison with money $41. to pay expenses to N.Y. including opperator—and I will divide the expense with you.[1] Whether I con[c]lude to go any further or not. Yours Truly

Joel H. Hills

ADDENDUM[a]

Boston Jany 25/69

Red. of W. E. Plummer, the sum of Forty nine dollars, on a/c of the business above stated.

$49.[2]   Thomas A. Edison

ALS, MiDbEI, EP&RI. 2¢ Internal Revenue stamp attached and canceled by Plummer. [a]Addendum is a DS; written by Plummer.

1. At this time Samuel Laws and the Bankers' and Brokers' Telegraph Co. planned to construct lines from New York to Boston and to establish a stock-reporting circuit in Boston. See Chapter 3 introduction.

2. The discrepancy between this figure and the $41 may relate to an undated, signed note in Edison's hand:

Extra Expenses in N. York	4.35
Man on Tonight	3.00
	7.35

Recd Payment Thomas Edison
MiDbEI, EP&RI.

–53–

To William Plummer

[Boston, January 25, 1869[1]]

Mr Plummer

Took Model to Stearns[2]   signed and sworn to skeleton of Patent Papers and Signed Patent Ofs Assignment   Can go up There any time and explain the technicalities to the Draftsman.[3]

Edison

ALS, MiDbEI, EP&RI.

1. On this date Edison signed the application for a printing telegraph patent that later issued as U.S. Patent 91,527 (see Doc. 54), and he also

signed the papers assigning rights to it to Plummer and Joel Hills. Pat. App. 91,527; Digest Pat. E-2:101.

2. This was Norman Stearns, of the firm of Teschemacher and Stearns, who were Edison's patent attorneys. The application procedure required that Edison declare under oath his citizenship and his belief that he was the original inventor. *Boston Directory* 1869, 575, 596; Pat. App. 91,527.

3. Typically, a draftsman made the drawings that accompanied a patent application.

BOSTON INSTRUMENT   Doc. 54

The Boston instrument[1] embodied Edison's first patented printing telegraph design.[2] Edison had worked on such devices since April 1868[3] and finally executed a successful patent application on 25 January 1869.

Two classes of printing telegraphs existed at the time. The first class comprised the House,[4] Hughes,[5] and Phelps[6] printers. They were large, fast, complex machines that required external sources of mechanical power and operators at both ends of a circuit. These were employed in America on only a few busy lines. The second class, to which the Boston instrument belonged, included in practice only Edward Calahan's stock ticker.[7] First installed in the New York financial district at the end of 1867, this small machine required only batteries and a sending operator. It had won immediate acceptance and had opened up a new field of telegraphic invention.[8]

A small printing telegraph had to position and ink a typewheel and to press paper against the wheel to print each successive character. Calahan's printer used three wires and a local battery to accomplish these tasks; Edison's design required only one wire and no local battery. Edison used a modified polarized relay as a switch to by-pass either the typewheel advance mechanism or the printing mechanism, allowing each to operate alone.[9] He considered the circuitry "the most perfect device for producing two movements at a distance on one wire by magnetizm as it does not depend upon an even & rapid transmission of waves to effect the result but will act with the slowest as well as the most rapid pulsations."[10] It was a very clever simplification but had a serious weakness: when the switch failed to effect a good contact, both the printing and typewheel magnets acted simultaneously. That proved to be a major problem; several years later Edison wrote, "The permanent magnetism [of the polarized relay] was so feeble that [even with] the most powerful current in the switch magnet the bar would scarcely make contact with the right or left point necessary to shunt the magnets and produce the desired results. Another defect was that the lightning depolarized the switch bars and necessitated remagnetization."[11] Edison and his Boston associates inaugurated a financial-news circuit with this instrument, but by the spring of 1870 the business had failed.[12]

Edison encased his instrument in a wooden box with a glass front, unlike Calahan, who had placed his device under the glass shade later commonly associated with tickers. Edison had some of these instruments built at Charles Williams's shop in Boston and built others with George Anders.[13]

1. Edison called this printing telegraph the "Boston Instrument" in his notes. Cat. 297:48, Lab. (*TAEM* 5:611).

2. U.S. Pat. 91,527.

3. See Docs. 31 and 35.

4. Royal House, a self-educated American inventor, patented a printing telegraph in 1846 (U.S. Pat. 4,464) and improved it through the 1850s, during which time it was used throughout the United States. It employed one wire to indicate the letter to be printed, and it transmitted as many as 40 words a minute. Power for printing came from compressed air that was supplied at the receiving end by a "grinder," a man turning a crank or working a treadle. In the early 1870s steam or electricity began to replace the grinders. *DAB*, s.v. "House, Royal Earl"; Sabine 1872, 190–92; Prescott 1877, 605–9; "The Western Union Telegraph Company's Manufactory," *Telegr.* 8 (1871–72): 105; "Improvements in the New York office of the Western Union Telegraph Company.—A magnificent operating room," ibid., 69.

5. David Hughes, a professor of music, was born in England but spent his youth in the United States. He patented his printing telegraph in 1856 (U.S. Pat. 14,917). A single wire transmitted the signal for printing; weight-driven clockwork powered the printing mechanism. Like the House system, Hughes's printer depended on synchronized revolving typewheels at the sending and receiving stations and an attendant at the receiver. If the transmitter and receiver became unsynchronized, the receiving operator signaled the sender, who then repeatedly transmitted a single character until the receiving machine was reset. George Phelps rendered the Hughes instrument practical. Once improved, it could transmit thirty words a minute (later sixty) and was widely adopted in Europe, where it was used until around 1950. Sabine 1872, 179–90; Prescott 1877, 609–42; Reid 1879, 640–41; Garratt, Goodman, and Russell 1973, 23–24; Preece 1884, 14.

Hughes returned to Europe in 1857 to promote his printer and remained there. In the late 1870s Edison engaged him in a priority dispute over the invention of the microphone. *DNB*, s.v. "Hughes, David Edward"; Marsh 1980.

6. See Doc. 32, n. 1.

7. See Doc. 91, n. 4.

8. Reid 1879, 606–89.

9. Edison and George Anders discussed uses of the polarized relay in printing telegraphs during the winter of 1868–69. In the Boston instrument, Edison transformed the polarized relay into a double-pole switch by replacing what was usually an insulated screw with a second contact point. Preliminary Statement of George Anders, pp. 3–4, and George Anders's testimony, pp. 4–5, *Anders v. Warner.*

10. NS-74-002, Lab. (*TAEM* 7:231–32).

11. NS-74-002, Lab. (*TAEM* 7:233–34). Edison also made this point in the patent specifications for U.S. Patent 96,681.

12. See Doc. 56; "The Printing Telegraph," *Telegr.* 6 (1869–70): 269; "Thomas Alva Edison," *Operator,* 1 June 1878, 3; and Chapter 3 introduction.

13. Jerome Redding to Frank Wardlaw, 1929, Pioneers Bio.; George Anders's testimony, p. 5, *Anders v. Warner.*

–54–

Production Model:
Printing Telegraphy[1]

Boston, [January 25, 1869?][2]

M (historic photograph) (est. 17 cm × 14 cm × 25 cm), NjWOE, Cat. 551:48. This photograph is from an album, assembled around the end of 1872, that contains fifty photographs of printing telegraph instruments. An annotated duplicate is at NMAH-DSI (Cat. 66-42, Box 44, WU Coll.). The artifact in the photograph is marked "Charles Williams & Co. Boston" on the upper left corner of the inner frame.

1. See headnote above.

2. Edison executed the covering patent application on this date (Pat. App. 91,527), but it should be noted that at this time he already possessed the patent model. See Doc. 53.

New York, January 30, 1869.

Mr. T. A. EDISON has resigned his situation in the Western Union office, Boston, Mass., and will devote his time to bringing out his inventions.

PD, *Telegr.* 5 (1868–69): 183.

1. This notice is from the "Personal" column, where news of operators appeared regularly.

–56–

Samuel Ropes, Jr., to Joel Hills

[Boston,] Feb. 1st 1869.[a]

Sir,

With this please find statement of the affairs of the Telegraph concern;[1] please send your check for amount needed to day $(333.84) Three hundred and thirty three & 84/100. we cannot go ahead, until we receive it   Very Respty

S. W. Ropes Jr

P.S. You will also find statement made up to Feb. 13th/69 showing assets on hand.   R

ENCLOSURES

[Boston, February 1, 1869][b]

Amount due Feb. 1st 1869.

2 Machines, Battery, Sounder, Key, Model, Wire, for Offices Labor. Operator in N.Y.	$. 258.01
Self.	25.
Paid out of my own purse as per Memo attached	30.83
	$. 313.84
Due Edison	20.
	$. 333.84[c]

The mony paid T. A. Edison and charged to him, he has paid to men at the W.U. Tel Office for working nights, that he might be on hand mornings.[2] consequently their is due him one weeks wages, $20

[Boston, February 1, 1869][d]

For tools in use	250.00
Planer	350.00
Engine Lathe	400.00
2 Fox Lathes[3] 80.00[e]	160.00
1 No 3 Fox Lathe chocks &c[f]	150.00
	1310.00
Shafting Benches &c Belting	150 —[c]

20 Instruments @ $50.00	1000
Rent for—d[itt]o⁴—per week	~~160~~ 120
Expenses	100. per wk[c]

[Boston, February 1, 1869][g]

Amount needed Saturday February 6th 1869

2 Printing Machines	$ 100.00
5 Battery Cups[5]	14.00
Office wire	2.50
Labor of man	10.00
Edison	20.00
Self	20.00
Telegrams (about)	50.00
Incidental	10.00
Running wires	13.50
Battery fluid	2.00
Total	$242.00.[c]

Amount needed Saturday February 13th 1869.

2 Printing Machines	$100.00
5 Battery jars	14.00
Office wire	2.50
Labor	10.00
Running wires	15.00
Edison	20.00
Self	20.00
Telegrams	50.00
Incidental	15.00
Fluid	3.50
Insulators	10.00
Total	$260.00.[c]

Assets on hand February 13th 1869.[c]

10.—Printing Instruments	$1500.00
65 —Cups Battery	150.00
—Office furniture[6]	125.00
—Sounder and Key	30.00
—Tnansmitter	40.00
—Office wire	25.00
All our lines	175.00
Battery Boxes	15.00
Total Assets	$2060.00[c]

ALS, MiDbEI, EP&RI. ᵃ"9 Wilson's Lane." written above. ᵇThis enclosure is an AD; written by Ropes. ᶜFollowed by centered horizontal line. ᵈThis enclosure is a D; written in an unknown hand. ᵉInterlined above.

1. This concern operated a stock-quotation service that used Edison's Boston instrument. App. 1.A26.

2. On 1 February, Edison received $10.00 from Samuel Ropes. Edison Receipt to Samuel Ropes, 1 Feb. 1869, EP&RI.

3. George H. Fox and Co. manufactured lathes in Boston. Its successor was American Tool and Machine Co. *Boston Directory* 1868, 875.

4. Instruments.

5. "Battery Cups" referred to the unglazed cup-shaped earthenware partitions used in wet cells. The cups permitted the flow of ions between electrodes while separating one electrolyte from the other.

6. On 18 March, Edison signed a receipt for office furniture received from Hills and Plummer and valued at $109.80. The furnishings included two tables, a rotary chair, three armchairs, a parlor stove, and a green woolen carpet. Receipt from Joel Hills and William Plummer, 18 Mar. 1869, EP&RI.

–57–

Receipt from Charles Williams, Jr.

Boston, Mar 12 1869

M T. A. Edison   Bought[1] of Charles Williams, Jr.[a]

30 Glass & Brackets[2]	15	4.50
8 Glass Insulators		.80
1¾ lb B. Vitriol[3]		.37
2 oz Wire		.20
1 Mahogany base		.25
2 castings		.10
9 Block Insulators[4]		2.61
55[b] feet Kerite wire[5]		3.30
2 lb 9 oz Painted wire[6]		3.85
5 " 14 oz Kerite wire Cable		10.00
1¼ mile 22 lb 14 oz Compound wire[7]		87.06
2 Indicating Telgh' Inst[8]	30.00	60.00 $173.04

Re[ceive]d Payment

C. Williams Jr

ADS, NjWOE, DF, 60-002 (*TAEM* 12:34). Billhead of Charles Williams, Jr. a"Boston," "186", "M", and "Bought ... Jr." preprinted. bTraces of Internal Revenue stamp in left margin.

1. These purchases include materials for installing lines for clients. Edison recalled using the roofs of houses to string the wire in the same manner as Western Union. App. 1.A26.

2. Glass insulators with brackets (usually made of wood) were attached directly to telegraph poles or buildings to support the wire with a minimum of current leakage. Pope 1869, 59–60.

3. Blue vitriol (copper sulfate, $CuSO_4$) was used as an electrolyte in many batteries. Pope 1869, 12–14.

4. Block insulators, named for their shape, were made of porcelain. Williams 1876, 4.

5. Kerite-insulated wire, patented in 1866 by Austin Day of New York, was coated with a mixture of tars, various oils, and sulfur. Knight 1876–77, s.vv. "Kerite," "Kerite-wire."

6. Painted wire resisted rust and corrosion, which was especially important if the wire was exposed to large quantities of coal and coal gas fumes. Maver 1912, 511.

7. Regarding compound wire, see Doc. 44.

8. Probably Edison's magnetograph.

–58–

Receipt from Charles Williams, Jr.

Boston Mar 31 1869

$180— Received of T. A. Edison One Hundred and Eighty —   Dollars X[a]100 on account[b]

C. Williams Jr[c]

ADS, NjWOE, DF (*TAEM* 12:35). [a]"X" written above "100". [b]"Boston", "186", "$", "Received of", and "Dollars 100 on account" preprinted. [c]Canceled 2¢ Internal Revenue stamp at left.

–59–

Section from Franklin Pope's Modern Practice of the Electric Telegraph[1]

New York, March[2] 1869

161. EDISON'S BUTTON REPEATER.[3]—This is a very simple and ingenious arrangement of connections for a button repeater, which has been found to work well in practice. It will often be found very convenient in cases where it is required to fit up a repeater in an emergency, with the ordinary instruments used in every office. Fig. 57 is a plan of the apparatus.

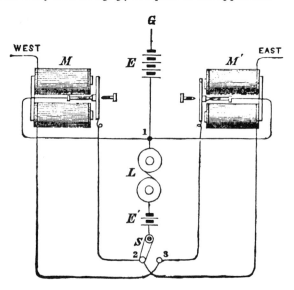

M is the western and M' the eastern relay. E is the main battery, which, with its ground connection G, is common to both lines. E' is the local battery, and L the sounder. S is a common "ground switch," turning on two points, 2 and 3. In the diagram the switch is turned to 2, and the eastern relay, therefore, repeats into the western circuit, while the western relay operates the sounder, the circuit between 1 and 2 through the sounder and local battery being common to both the main and local currents. If the western operator breaks[4] the relay M opens, and consequently the sounder, L, ceases to work. The operator in charge then turns the switch to 3, and the reverse operation takes place; the western relay repeats into the eastern circuit, and the eastern relay operates the sounder. The sounder being of coarse wire, offers but a slight resistance to the passage of the main current.

PD, Pope 1869, 107–8.

1. Franklin Leonard Pope (1840–1895), electrical engineer, inventor, and later patent attorney, held positions as telegraph operator, manager, engineer, and writer before becoming superintendent of Samuel Laws's newly formed Gold and Stock Reporting Telegraph Co. on 11 November 1867. He made several improvements to Laws's indicator for gold quotations, substituting parallel wheels for Laws's overlapping discs. As editor of the *Telegrapher* from 15 August 1867 to 8 February 1868, Pope enlarged the journal and introduced scientific and technical articles (see Doc. 26). The first edition of his *Modern Practice of the Electric Telegraph* appeared in May 1869. It became one of the most popular telegraph manuals of the late nineteenth century. Reid 1879, 565–66, 666–67; *DAB*, s.v. "Pope, Franklin Leonard"; "Pope's Work," *Telegr.* 5 (1868–69): 272.

2. Taken from the preface to Pope's book.

3. See Doc. 15.

4. Opens the circuit.

–60–

To Joel Hills

Boston Mas Apl 2 [1869]

Mr Hills

The Transmittors[1] will be done by Tuesday next—

Thes~~ee~~ instruments which I intend to put in a[re] good ones but are out of order and Look quite dirty   The Bases which they are on have been altered so many times that They are all full of Holes and Look bad   I would advise taking the instruments all apart clean and adjust them and make an alteration in the Paper Cam. And placing Instrument and Transmitter all on one Base Black walnut or Mahogany   The Cost of cleaning altering and fixing up the Instruments in

first class style upon New Bases, with all the Connections
~~w~~Soldered will Cost 23 dollars   and Looks ~~are~~have a good
deal to do in the Success of an instrument Especially on the
first Line   I Advise you having it done quickly
    reply by boy

(Edison)

ALS, MiDbEI, EP&RI. Starts at bottom of back and continues on front
of message form of the Bankers', Brokers' & Commercial Lines of Telegraph.

    1. The mention of a paper cam later in this document suggests that
these are transmitters for printing telegraphs.

–61–

*Patent Assignment to
E. Baker Welch*

Boston,[a] April 7th 1869

    In consideration of various sums of money received by me
at sundry times since July 1st 1868, and of forty dollars received this day of E. Baker Welch, of Cambridge, Massachusetts, I, Thomas A. Edison, of Boston, Massachusetts, do
hereby sell, assign and transfer to said E. Baker Welch one
undivided half interest in a double transmitter which I have
invented and am about to put on the lines of the Atlantic &
Pacific Telephone Company[1]—and I do hereby agree to apply
for a Caveat or patent for the same at such time as said Welch
shall recommend or request me to do,[2] and to assign to him
one half interest in the said patent, or in any rights or privileges accruing under a caveat or patent at any time—[3] I do
also hereby agree and bind myself to sell, assign and convey
to said Welch, for the considerations before mentioned herein one undivided half interest in any improvements which I
may make to the said double transmitter, and one undivided
half[b] interest in any other instrument or any other principle,
method or system, which I may invent, or obtain a Caveat or
patent for, to be used for the transmission of messages ~~or~~on
Telegraph lines both ways simultaneously[4]   signed—

Thomas A Edison[c]

D (transcript), NjWOE, DF (*TAEM* 12:17). In an unknown hand.
[a]Preceded by "(Copy) Original sent to Patent Office January 27th 1875
to be recorded". [b]"undivided half" interlined above. [c]Followed by "The
words 'or patent at' on the first page third line from the bottom [see
location of endnote 3] are interlined in Edison's handwriting with blue
ink—the word at which was in the original being erased also with blue
ink. The figure '7th' in the date also put in by Edison with blue ink.
Boston, January 27th 1875   Having seen the original of this paper
signed by Thomas A. Edison and compared this copy with it, I certify
that this is a true copy.— George L Anders".

1. Edison was about to travel to Rochester, N.Y., to test the device on the lines of the Atlantic and Pacific Telegraph Co. This company, organized in 1865 in the state of New York with a capitalization of $5 million, had built lines from New York to Cleveland via Albany, Rochester, and Buffalo (Reid 1879, 579–80). The company was correctly named "Telegraph" (instead of "Telephone") in the version of the document used in *Welch v. Edison*. The concept of the telephone as a device to transmit complex sounds was already familiar by the time Bell made his invention in 1876. See, for example, "The Telephone," *Telegr.* 5 (1868–69): 309; and *OED*, s.v. "Telephone."

2. Edison filed neither a caveat nor a successful patent application for duplex telegraphy in 1869. See textnote a.

3. See textnote c.

4. On the basis of this statement, Welch later made a claim on Edison's profitable invention of the quadruplex.

–62–

*And Frank Hanaford
Bill from Charles
Williams, Jr.*

Boston, April 10th 1869

M ~~Boston~~ T. A. Edison & Hanaford[1]   Bought of Charles Williams, Jr.[a]

April 5	9 lb 3½ oz G[utta]. Percha[2] wire	2.00	18.44	
6	10 lb Blue Vitriol		2.30	
	1 copper		.75	
	3 P[orous]. cups[3]		.75	
	3 Zincs		1.35	
~~8~~10	2 Indicating Instruments	335.00	70.00	
	2 lb B Vitriol		.50	$ 94.29
Bill rendered Mar 27				185.20
				$279.49
cr[b] By cash as per rcpt' March 31				180.00
				99.49[c]

AD, NjWOE, DF, 69-002 (*TAEM* 12:35). Billhead of Charles Williams, Jr. [a]"Boston," "186", "M", and "Bought . . . Jr." preprinted. [b]Traces of Internal Revenue stamp below. [c]Followed by "Rcd' Payment April 16th C. Williams Jr."

1. Frank A. Hanaford (n.d.), formerly an operator at the Franklin Telegraph Co. in Boston, became Edison's partner at 9 Wilson Lane. For several months after Edison left Boston, Hanaford continued to manage the shop, but by the fall of 1869 he had returned to the Franklin Co. *Boston Directory* 1868, 280; ibid. 1870, 314.

2. Gutta-percha was a common electrical insulation made from the latex extracted from several species of Malaysian trees. Knight 1876–77, s.v. "Gutta-percha."

3. Porous cups, made of unglazed earthenware, separated the copper and zinc electrodes in a Daniell battery. Pope 1869, 13, 92, 94–95.

63

New York, April 17, 1869.
A New Double Transmitter.

ON Tuesday evening last[1] a new double transmitter, on an improved plan, invented by Mr. T. A. EDISON, was tried between New York and Rochester, a distance of over four hundred miles by the route of the wire, and proved to be a complete success. Communications were sent and received at the same time over a single wire without the slightest interference, and the instrument accommodated itself to the disturbance caused by the working of other wires upon the same route as readily as the ordinary apparatus. This invention is materially different from any other of the kind yet brought out, and is much more simple, effective, and easily managed. This is the longest circuit which has ever been practically worked by this system.[2]

PD, *Telegr.* 5 (1868–69): 272.

    1. 13 April.

    2. Edison later called this a "duplex and diplex system" (see Doc. 50). He tested his system from the Rochester, N.Y., office of the Atlantic and Pacific Telegraph Co. This journal notice was in error; Edison readily admitted that the trial failed. App. 1.A29; draft reply to Louis Wiley, in Wiley to TAE, 9 Feb. 1925, GF; Doc. 68.

64

Boston May 1st 1869.
Six months after date, value received I promise to pay to E. B. Welch or order one hundred and twenty eight dollars with interest from date.[1]
Due 1/4 Nov. 1869

Thomas A. Edison.

TD (transcript), MWalFAR, *Welch v. Edison.* See Doc. 36 textnote.

    1. On the same date, Edison signed a second promissory note for the same amount due in twelve months. *Welch v. Edison.*

[Boston?,][1] May 3, 1869[a]
Whereas Thomas A. Edison of Boston, County of Suffolk and State of Mass. has invented an improved "Magnetograph"[2] for which he has made application for Letters Patent of the United States, and has assigned said invention to himself, E. B. Welch of Cambridge Mass. and George L. Anders of Boston aforesaid, according to the Assignment hereto annexed:—

Now therefore it is agreed by and between the said Edison, Welch and Anders that said Invention and any letters patent which may be granted for the same shall be held and owned by them as their joint and mutual property, and shall be worked under for their mutual benefit: that is, that all profits which may accrue from the manufacture and sale or use of said "Magnetographs" either by them together or by either alone or his assigns or representatives shall be for the mutual profit of the parties hereto in the proportion of one third to each:— And neither party shall sell or assign his interest in said invention and said letters patent, without the consent of the others first had and obtained, but the sum realized by either for the sale of his interest shall accrue to the party selling: and either party so disposing of his interest shall do so subject to this agreement.[b]

In testimony whereof We have hereunto set our hands and affixed our seals, this Third (3rd) day of May A.D. 1869.

Thomas. A. Edison[c]
E. B. Welch
Geo. L. Anders[d]

In presence of—Carroll D. Wright

D (transcript), MdSuFR, Libers Pat. V-11:188. [a]Date taken from text, form altered. [b]Representation of 5¢ Internal Revenue stamp in right margin. [c]Representation of seal follows each signature. [d]"Recorded July 14th 1869" written in right margin.

1. The place where this document was written is problematic, for Doc. 68 (written 8 May) implies that Edison had not seen Welch since going to Rochester. The assignment was not registered at the U.S. Patent Office until 14 July 1869. Digest Pat. E-2:116.

2. The magnetograph was an alphabetic dial telegraph for use on private lines. Like James Hamblet's dial telegraph, the magnetograph did not require a battery, and thus avoided the necessity for customers to cope with messy acids. Instead, signals were manually generated by turning an armature in a magnetic field—a magneto—until a pointer reached the proper letter on an alphanumeric dial. At the other end a similar pointer moved to the corresponding letter on another dial. The

receiver easily read the message from the dial, letter by letter. Edison and Anders manufactured magnetographs in the spring of 1869. On 6 July, Anders assigned his share to Samuel Laws, though no patent was issued. In 1870 Franklin Pope offered a pair of "Edison & Anders' Magnetograph[s]" for sale. App. 1.A26; Anders's testimony, p. 6, *Anders v. Warner;* Digest Pat. E-2:116, 126; F. L. Pope advertisement, *Telegr.* 6 (1869–70): 401.

–66–

And Frank Hanaford from D. N. Skillings & Co.[1]

[Boston,] May 5th 1869

Sirs,

We are very much gratified with the Telegraph you have put up for us from our office in Kilby St. to our Yard at E. Cambridge. The simplicity, and certainty with which it works, makes it a very great Convenience in our business and we recommend with pleasure your apparatus to every person in want of Telegraphic facilities   Yours, Truly

D. N. Skillings & Co.[a]

L, NjWOE, DF (*TAEM* 12:19). [a]Company name written in a different hand from body of letter; followed by "Downer. Kerosene. Oil Co. Wm. B Merrill Genl Agent" in a third hand.

1. David N. Skillings & Co. was a prosperous lumber company that had been in business since the 1850s and had a downtown office at 5 Kilby St., Boston (*Boston Directory* 1869, 558; Mass. 71:1a, RGD). For more on the Downer Kerosene Oil Co., which also signed this document, see Doc. 71, n. 3.

–67–

Testimonial from Continental Sugar Refinery

Boston, May 5h 1869[a]

We have now in use between our Office and the Refinery at South Boston, Edison's telegraph with his patent indicators—[1] It works to our satisfaction, and we cheerfully recommend it to those in need of anything of the kind.

Continental Sugar Refy[2] by D. Townsend[3]   Treas

L, NjWOE, DF (*TAEM* 12:20). Letterhead of Continental Sugar Refinery. [a]"Boston," and "186" preprinted.

1. This is not the magnetograph. See Docs. 71 and 73.

2. Continental Sugar Refinery, established about 1865 and owned by Jonathan Cottle, Horatio Harris, and W. A. Kinsman, was described in an 1869 R. G. Dun credit report as "the smartest sugar concern in Boston." The offices were located at 36 Central St., Boston, and the refinery was at the corner of First and Granite Sts. in South Boston. *Boston Directory* 1869, 156; Mass. 77:196, RGD.

3. David Townsend was a commission merchant with offices at 36 Central St., Boston. *Boston Directory* 1869, 606.

To E. Baker Welch

Mr. Welch:

I went to Rochester to put in double transmitters and waited four days to get a chance, but did not get one until late at night, and then the wires worked very poorly on account of bad insulation, and I came to the conclusion not to wait there on expense any longer but return to New York and wait till they trimmed the line which they are doing now. They will get wires all trimmed to Rochester in about three weeks.[2]

When I returned I tested my instruments on the Bankers and Brokers,[3] but they were made for such a long line (A & P)[4] and delicate current that they did not give satisfaction to myself although they worked. What delays me here is awaiting the alteration of my instruments which on account of the piling up of jobs at the instrument makers have been delayed and I will probably have to wait one week longer, and then if everything works as it has got to do for I'll never say "fail" I have the Pacific and Atlantic from Phila. to Pittsburg, from Pittsburg to Cincinnatti, from Cincinnatti to Louisville and Nashville. Two or three sets for the B & B including line to Boston—The Cuba Cable,[5] all bona fide offers. So you see that there is no use letting anything stand in the way of successfully getting this apparatus perfected. Me and Pope[6] have experimented considerably upon it and have several improvements   only one thing stands in the way now and that is Induction, and my alterations will overcome that, so I have concluded to stick here till I am successful.[7]

What is the matter with the shop? If you only knew how many of those instruments[8] could be sold here you would make things howl. Five days ago a man who saw the instrument I brought over (which by the way is in very bad condition) came to me and wanted six of them to put on lines where Chester's Dial[9] was working the lines being 14 miles long and the great objection was the battery. He wanted them immediately and I promised to have them here Saturday. He is very anxious and excited about getting them as Chester's arrangement don't work and it will be to our advantage to have them here quickly or may lose sale to that person although can sell all you send on. Chester Patrick & Co. of Philadelphia wants the exclusive right to sell them in Philadelphia,[10] and as they are a go ahead firm would advise letting them have it. N.Y. appears to be quite different to Boston. People here come and buy without your soliciting. Rochester wants a set. Dyer wants a set. Jacobs here six instruments. Compound Wire Co., set.[11] All quick as possible.

I wrote full explanation to Field[12]—Rec'd letter from Adams—He says:—Big field here for brokers, am awaiting arrival of instruments—etc—[13]

If you cannot sell any instruments there send them all here and I will sell and deposit proceeds with your brother[14] and George[15] can draw on him for money to build more instruments, or will deposit cost with your brother and remit George net profits. $225. per set is what I asked. Respy.

Edison[a]

I have enough money to last ten days.

TL (transcript), MWalFAR, *Welch v. Edison.* See Doc. 36 textnote.
[a]Followed by "Care Pope, Box 6138 N. York."

1. Edison's return address is that of Samuel Laws's Gold and Stock Reporting Telegraph Co. See textnote a.

2. "Trimming" meant cutting tree limbs away from the wires. Maver 1892, 549.

3. Edison did this with fellow telegrapher J. B. Collins. Collins to Stockton Griffin, 27 Dec. 1878, DF (*TAEM* 16:548).

4. Atlantic and Pacific Telegraph Co.

5. At this time there were two cables between the United States and Cuba. The second, from Florida to Havana, had been completed in February 1869. *J. Teleg.* 5 (1869): 77, 92.

6. Franklin Pope.

7. James Ashley, in an editorial in the *Telegrapher* on 16 June (5 [1868–69]: 352), declared that the best double transmitter yet invented was Edison's but noted that it had "not yet been introduced on any line, although it is being tested on the lines of one of our telegraph companies." Edison continued to experiment with the instrument through 1870. Edison's preliminary statements, 27 Apr. 1878 and 31 Mar. 1879, *Nicholson v. Edison.*

8. Magnetographs.

9. The most widely used American-made alphabetic dial telegraph was that of Charles Chester, patented in 1863 (U.S. Pat. 40,324). The firm of Charles and John Chester of New York—"Telegraph Engineers, Manufacturers of Instruments, Batteries and Every Description of Telegraph Supplied"—manufactured the device. They also made Laws's gold indicators. Chester's dial telegraph required a battery to operate, whereas Edison's magnetograph did not. Prescott 1877, 578–79; Charles T. & John N. Chester advertisement, *Telegr.* 5 (1868–69): 300; Reid 1879, 603, 622, 627; *New York Times*, 14 Apr. 1880, 4.

10. Chester, Partrick and Co., owned by Stephen Chester and James Partrick, was founded in 1867 and described in 1869 as a "young but enterprising" telegraph manufacturing company. Partrick had previously served as assistant manager of Western Union's Philadelphia office. Telegraph manufacturers frequently advertised that they were the sole agents for particular inventions. "Miscellanea: A New Firm," *Telegr.* 4 (1867–68): 99; "Chester, Partrick & Co.," ibid. 5 (1868–69): 80, 343.

11. Rochester may refer to the Rochester office of the Atlantic and Pacific Telegraph Co. Joseph Dyer was superintendent of the Bankers'

and Brokers' Telegraph Co. in Philadelphia from 1865 to 1871 (Taltavall 1893, 212). Jacobs is unidentified. Regarding the American Compound Telegraph Wire Co., see Doc. 44.

12. Unidentified.

13. Milton Adams left New York in March 1869 for San Francisco,
where operators' jobs had been advertised. Milton Adams to E. Baker
Welch, 8 Mar. 1869, *Welch v. Edison;* "Personals," *Telegr.* 6 (1869–70):
74.

14. Probably Joseph Welch, a New York lawyer. E. Baker Welch had
two brothers living in the New York area. Wilson 1869, 1161.

15. George Anders.

–69–

To Frank Hanaford

N York = Jaune 10[, 1869]

Friend Frank =

Please write and Let me know how matters stand with
you = What about the rent of the shop, did you pay it etc and
have you got any Show for building any Lines— Has Williams
finished all the instruments and how many have you now altogether— What instrument is O'Doud[1] going to use. Do you
work by the day for him—. Do you know what he gets. Did
he use the wire and is Farrar and Folletts bill paid.[2] Did you
sell him any insulators? Has Hills been round   what did he
say—? What appears to be the matter with the Shop— I have
sold six instruments which were to be delivered Last Monday
and I think I have lost the sale in consequence of the delay =
When will they be done, in your opinion\\\   Does the anti-
qudated, fossil pariential[a] D D Cummings[3] still exist arround
the sacred precincts of that dilapidated, mansion   Does He
still pursue the ignatus fatus[4] of Private Telegraphs, that
Bright beacon light upon the headlands of ~~of~~ eternity where-
on the bright prospects of two sanguine young men were to-
tally wrecked and obliterated and their hopes flown to that far
off land from whence no traveller, Line repairer of Telegraph Operator return. Alas! that the bright vision of untold
wealth should vanish like the fabric of a dream, that two ambitious mortals gifted by the genii of enterprise, and strength,
should drink the bitter dregs of premature failure in their
grand dreams ~~o~~for the advancement of science, etc— — —
Keno[5]   Does Cummings Jr still exercise his matchless skill,
preserverence, and abiquity, which the Lamented Valentine,[6]
was wont to pride himself upon   Has no tidings of the lost
one reached his sorrowings colaborers, or does he still wan-
ders sad and sorrowful among the busy haunts of men bent
upon ~~s~~the self destruction which inevitably awaits those who
pamper to the tastes of "Collateral Brokers"—

Such is the Kingdom of heaven!

Please answer the questions ~~of~~ w in the foreground of this salubrious document and oblige yours Truly—

Thos Edison, Ex-Contractor

only one PS.[b] What is the prospects in the future   E

ALS, NjWOE, DF (*TAEM* 12:21). [a]Interlined above. [b]"only one PS." divided from body of letter by a drawn line.

1. Peter O'Dowd, formerly a telegrapher in Boston, was at this time agent of the Merchants and Manufacturers Telegraph Building Co. He and Hanaford built the lines for the magnetographs and installed the instruments. *Boston Directory* 1865, 311; ibid. 1869, 469; ibid. 1870, 827; Jerome Redding to Frank Wardlaw, 30 Sept. 1929, Pioneers Bio.

2. Farrar, Follett & Co., owned by Daniel Farrar and Dexter Follett, was an importer of metals and manufacturer of wire located at 73 and 75 Blackstone St., Boston. Edison and Hanaford purchased galvanized telegraph wire from this company for $8.74 in May. *Boston Directory* 1868, 222, 931; ibid. 1869, 226, 241; receipted bill of Farrar, Follett & Co., 3 May 1869, 69-002, DF (*TAEM* 12:36).

3. Unidentified. There is a receipt of payment to a J. L. Cummings for $11 dated 12 March 1869. 69-002, DF (*TAEM* 12:34).

4. "Ignis fatuus" means a will-o'-the-wisp or fleeting light.

5. Or "that's life." Keno was a popular game of chance similar to bingo.

6. There is an 1869 receipt of payment to an unidentified Valentine for $.65, signed with an "X." 69-002, DF (*TAEM* 12:36).

-4-

Establishing Connections in New York

July–December 1869

While the first months of 1869 saw Edison drawing upon local Boston associates to aid him in his developing inventor-entrepreneurial activities, the second half of the year witnessed him successfully connecting with senior telegraph men of national reputation in his new business home in New York. In the spring he had arrived in the city of one million people, a center that was at once the focal point of national commerce and finance and also the American capital of telegraph invention and enterprise. As the year went on he grew closer to Franklin Pope, the former editor of the *Telegrapher,* a respected telegraph engineer, and superintendent of Samuel Laws's Gold and Stock Reporting Telegraph Company. When Pope left his position with Laws at the beginning of August, Edison succeeded him. At this time Edison may also have used the company's battery room as a temporary residence.[1]

Edison's growing expertise with printing telegraphs served him well in his inauguration into the New York telegraph world, particularly the area of financial reporting. Samuel Laws had pioneered in this field when he founded the Gold and Stock Reporting Telegraph Company in 1867.[2] The company supplied price quotations for gold from a central transmitter in the "Gold Room" adjacent to the stock exchange to indicators in the nearby offices of bankers and brokers. Laws had invented the indicator and Franklin Pope had improved it. When Edison became superintendent, about 140 indicators were in offices around the city, but the company was not alone in the business.[3] Laws's enterprise was competing with the aggressive Gold and Stock Telegraph Company, which had been organized in August 1867 to exploit Edward

Samuel Laws introduced an electric indicator system for gold prices in 1867.

The dial of Laws's indicator (located at the left end of the balcony) as it appeared in the "Gold Room" of the Gold Exchange Bank and Clearing House in New York City.

The mechanism of Laws's gold indicator as modified by Franklin Pope.

Calahan's new printing telegraph.[4] Laws introduced a printer of his own shortly after Edison's arrival in New York and soon Edison redesigned it.[5] More important to Gold and Stock, Laws also acquired rights to an 1856 patent for a printing telegraph with independent circuits for typewheel and printing mechanism[6]—one of the fundamental claims of Calahan's patent. Unable to circumvent Laws's patent holdings, Gold and Stock acquired the Laws company on 27 August 1869.[7] Edison's patents were included in the sale agreement but his services were not. Calahan retained his position as superintendent of the merged company, and Edison lost his four-week-old job.

Soon Edison joined Pope and James Ashley, editor of the *Telegrapher,* in the new firm Pope, Edison & Company. The new firm advertised as "Electrical Engineers and General Telegraphic Agency." Using a compact printing telegraph that Edison and Pope designed, they also established the Financial and Commercial Telegraph Company, a reporting system to compete with Gold and Stock. During the fall Edison's alliance with Pope extended to his rooming at the Pope family home in Elizabeth, New Jersey.

1. Edison claimed that he slept in the company's battery room. App. 1.A29.

2. Samuel Spahr Laws (1824–1921) had been president of the New York Gold Exchange (*DAB*, s.v. "Laws, Samuel Spahr"). Regarding Laws's gold indicator (U.S. Pats. 72,742 and 75,775) and business, see Prescott 1877, 672–73; Reid 1879, 602–5; and Hotchkiss 1969.

3. See n. 7.

4. Edward A. Calahan (1838–1912) was born in Boston and at the age of twelve began his telegraph career, starting as a messenger. Later he became an operator and manager for the American Telegraph Co. At the time that he invented his printing telegraph, he was a draftsman and assistant to Western Union's engineer, Marshall Lefferts. Obituary, *New York Times*, 13 Sept. 1912, 9; Reid 1879, 606.

Regarding Gold and Stock, see "Certificate of Incorporation of The Gold and Stock Telegraph Company," *G&S v. Pearce;* "The Gold and Stock Telegraph," *Telegr.* 6 (1868–69): 12; "The Gold and Stock Telegraph Company—Its Inception, Development, Business, etc.," *Telegr.* 11 (1875): 127–28; Reid 1879, 602–13; Calahan 1901b; and Hotchkiss 1969, 433–41.

5. Both the new printer design and a new circuit switch that used a polarized relay resulted in Edison patent applications that were executed and assigned to Laws in August. U.S. Pats. 96,567 and 96,681.

6. H. N. Baker's U.S. Pat. 14,759.

7. The terms of the agreement are in G&S Minutes 1867–70, 57–60.

–70–

To Frank Hanaford

New York, July 26 1869[a]

Friend Frank

Your Letter received. I am sorry that Skillings instruments act in that manner. Chas Williams is to blame for it. the last Lot were constructed very badly = Redding[1] is making a Printer and if it works satisfactory it is my intention to order another one and two Transmitters and place them on Skillings Line   It would appear from what has happened already that the Grey-eyed spectre of destiny has Been our gauardian angel, for no matter what I may do I reap nothing but Trouble, and the blues. I had saved [--]θ 160[b] dollars ~~and~~ from experimenting for other parties and with this I came to Boston and paid my debts which took the Entire amount leaving me without a single penny when I arrived in N York. It is all I can do to keep the wolf from the door, and supply Redding with money to make the Printer

It is useless for me to lay around or come to Boston, for I cannot make any money there and we should be farther than ever from the solution of this interesting problem   If I stay here I can eàrn enough money to fix the thing all straight within a reasonable time. I think the whole cause of the frequent disorders of Skillings apparatus is want of battery. I would advise you to suggest to Skillings the purchasing of two or three cups more battery, and have them[-] put on the Line. If you will read the Contract you will notice that it does not mention the purchase of Supplies ~~and it canno~~ neither was it mentioned at the time. Therefore I do not see how Skillings can expect you to buy the Vitrol. Please do all you can till I get the Printers Ready, and put them on the Line. You ought to know me well enough to know that I am neither a dead beat or a selfish person, and that I ~~u~~always do as I agree without some damnable god damnable ill luck prevents it. If I make anything it happens somehow that some dead beat Like Adams eats it all up. However I ll never give up for I may have a streak of Luck before I die   Yours

Edison

PS[c]   I ll try and Send you some money if your hard up. Will $10. help you   all I got—   E

ALS, NjWOE, DF (*TAEM* 12:23). Letterhead of the Gold and Stock Reporting Telegraph. [a]"New York," and "186" preprinted. [b]Interlined above. [c]Postscript written along right margin of last page.

1. Jerome Redding (1840?–1939), instrument maker, had for several years been employed in the telegraph manufacturing firm of Charles Williams, Jr. In 1869 he set up his own shop, Jerome Redding & Co. On 30 July 1869 Redding presented Edison with a bill for $75.97 to cover castings, gear cutting, a base, japanning, and 142¾ hours labor. *Boston Directory* 1867, 422; ibid. 1868, 491; ibid. 1869, 511; Redding to Frank Wardlaw, 1929, Pioneers Bio.; *New York Times*, 25 Nov. 1939, 17; bill enclosed with Redding to TAE, 9 May 1878, DF (*TAEM* 19:511).

–71–

To Frank Hanaford

New York, [July][a] 30— 1869[b]

Hanaford

Please tell Skillings that I am making a Printer for him which will be just what he needs— If Welch wont permit you to use the Magnetographs until I get the Printer done, please have ~~A~~George[1] fix up The Brass Kettles[2] so they will Last till I get the Printers Ready = Also ask skillings to have Little patience = The difference in cost will be 135 dollars,. So Skillings will have to pay $130, being the difference in Cost = I will also give the Continental and Downer[3] Printers at $150, and take their instruments. I am disposed to do all I can even to the extreme Length of my pile = ~~The pri~~ it is not unlikely that the Printers may prove to be worthy of introduction upon other Lines there = please write and Let me know if it is the Lines that give trouble or instruments   Yours

Edison

ALS, NjWOE, DF (*TAEM* 12:25). Letterhead of The Gold and Stock Reporting Telegraph. [a]"July" interlined above in unknown hand. [b]"New York," and "186" preprinted.

1. George Anders.

2. Unidentified.

3. Downer Kerosene Oil Co. was founded by Samuel Downer. With his assistants, Luther Atwood, William Atwood, and Joshua Merrill, Downer introduced on a large scale hydrocarbon fluids for lubricating machinery and for illumination. Edison's telegraph linked the company's downtown office at 108 Water St. with its refinery at 122 First St. *Boston Directory* 1869, 204; *DAB*, s.v. "Downer, Samuel." See also Doc. 66, n. 1.

–72–

Editorial Notice in the Journal of the Telegraph

Franklin Pope joined Edison and James Ashley in 1869 to form Pope, Edison & Co.

New York, August 2, 1869.

FRANK L. POPE has resigned his position as superintendent of the Gold and Stock Reporting Telegraph in this city.

T. A. Edison, of Boston, succeeds Mr. Pope in the above position.

Mr. Pope, like a wise man, thinks health worth more than money, and has started West to breathe Superior air. When he returns he designs paddling his own canoe in some congenial occupation respecting which his shingle will be seen in due season.[1]

Mr. Edison is like his predecessor, a man of genius and skill. Few men are better posted than he.

PD, *J. Teleg.* 2 (1868–69): 197.

1. At this time negotiations were under way for the Gold and Stock Reporting Telegraph Co. to merge with the Gold and Stock Telegraph Co. (see Chapter 4 introduction). By early October, Pope had returned from his travels, the merger had taken place, and he, Ashley, and Edison had formed Pope, Edison & Co. Pope's trip to Chicago and the Lake Superior area is related in "Carpet Baggist in the Northwest," *Telegr.* 6 (1869–70): 25.

–73–

To Frank Hanaford

New York, Aug 3 1869[a]

Friend Hanaford

Did Mr Welch allow you to put Magnetos upon Skillings Line. also what does Skillings Say about Printer. Does he feel Satisfied that we intend to fulfill our Contract   answer with particulars

Edison

ALS, NjWOE, DF (*TAEM* 12:27). Letterhead of the Gold and Stock Reporting Telegraph. [a]"New York," and "186" preprinted.

In the summer of 1869 Edison modified the printing telegraph that Samuel Laws was using on his stock-reporting
lines.[1] In doing so he practically redesigned the entire printer.
Edison reduced its size and the number of moving parts, replaced the inking ribbon with a wheel, eliminated one printing
electromagnet and added one for the unison stop, and altered
the structure of the pawls that turned the typewheel. His patent application for the new design—assigned entirely to
Laws—claimed the pawls and the magnet for the unison stop.
The frame Edison chose looks a great deal like contemporary
sewing machines, although there was nothing in the new mechanical arrangement that required such a frame.

*Samuel Laws's stock
printer, used by his Gold
and Stock Reporting Tele-
graph Co.*

Edison's redesigned printer received a patent (U.S. Pat.
96,567) before Laws's original design (U.S. Pat. 99,273), for
which Laws had applied in January of 1869 and which was in
interference with the reissue application of Edward Calahan's
printer.[2] All of Laws's telegraphic patents went to the Gold
and Stock Telegraph Company with the sale of his company
in September 1869.[3] On 28 December 1869 Gold and Stock
applied for reissue of the patent covering this printer.[4]

1. "Electricity Among the Stock Brokers," *Telegr.* 5 (1868–69): 312;
Reid 1879, 604–5. Reid describes the Laws unison mechanism incorrectly; cf. Laws's patent specification.
2. Reissue 3,810, granted 25 Jan. 1870; Pat. App. 99,273.
3. See Chapter 4 introduction.
4. Reissue 3,870, issued 1 Feb. 1870.

And Samuel Laws
Production Model:
*Printing Telegraphy*¹

M (historic photograph) (est. 15 cm × 5 cm × 13 cm), NjWOE, Cat. 551:14, 36. See Doc. 54 textnote. Except for a relocation of the magnet that activates the unison stop, this machine appears to be essentially identical to the patent specification.

1. See headnote above.
2. Edison executed the covering patent application on this date. Pat. App. 96,567.

New York, Sept. 14, 1869.

Mr. Welch:

Your letter received. Laws seem inclined to give you $500 dollars cash for your third and one third of the profits $250 down. But I am quite sure he will not give you anything for the instruments on hand but will take them and dispose of them and pay you their cost. He thinks of putting the whole thing into some of these private Line men here in N. York.[1] Mr. Laws has sold out his gold Reporting Telegraph to Callahans Co.,[2] and he is to furnish all necessary funds to bring my Double Transmitter out for a portion of my interest.[3] I think everything will come out all right yet.

Edison.[a]

TL (transcript), MWalFAR, *Welch v. Edison.* See Doc. 36 textnote. [a]Followed by "Room 48.   5 & 7 New St."

1. Samuel Laws proposed to purchase Welch's interest in the magnetograph. Laws had already purchased George Anders's one-third interest in the magnetograph on 6 July. The U.S. Patent Office has no record of a reassignment from Welch to Laws. Digest Pat. E-2:126.

2. See Chapter 4 introduction.

3. Nothing further is known about Laws's offer to finance the double transmitter.

FINANCIAL AND COMMERCIAL INSTRUMENT   Doc. 76

On 16 September 1869, Edison and Franklin Pope jointly executed a patent application that specified the design incorporated in Doc. 76. This was the first printing telegraph design Pope and Edison patented together.[1] Late in 1869, Pope, Ashley, and Edison introduced these instruments on the lines of their Financial and Commercial Telegraph Company, an enterprise that provided gold and stock quotations to mercantile and importing companies in lower Manhattan. Manufacture and use of the instruments continued for several years.[2]

The electromechanical heart of the Financial and Commercial instrument consisted of two coordinated but different types of electromagnets in the same circuit. One was a "polarized" electromagnet or modified polarized relay like that employed in Edison's Boston instrument. The cores of this electromagnet projected from one end of a permanent magnet, as in Siemens's original design for the polarized relay, with an armature between them that was attached to the other end of the permanent magnet and was hence of the opposite polarity.[3] The other magnet was an ordinary neutral electromagnet. In his earlier printer, Edison had used the modified polarized relay as a switch to shunt current to either the typewheel or the printing electromagnet. Here he used it solely to advance the typewheel, with every impulse passing through both the printing and typewheel electromagnets.[4] Rapid signals of alternating polarity were transmitted by the breakwheel of a central transmitter that had an indicating dial.[5] At the local receivers in the circuit, these short signals moved the typewheels, whereas longer signals of either polarity activated the printers. To avoid accidental printing, a portion of the current was cut out until the proper letter was reached on the transmitting dial. The diminished current could move the typewheel but not the printing lever. Like the Boston instrument, this machine was designed to operate on one wire without a local battery.[6]

1. U.S. Pat. 102,320; see Doc. 93. George Anders later stated that in 1869 he had worked out the printing and typewheel circuit used in this printer and had described it to Edison. Edison and Jerome Redding also remembered the development of this instrument as having begun in Boston. Anders's testimony, 11 Apr. 1876, pp. 2, 5–6, *Anders v. Warner;* Preliminary Statement of G. L. Anders, 21 Feb. 1876, pp. 3–4, ibid.; Jerome Redding to Frank Wardlaw, 1929, Pioneers Bio.; Edison's testimony, 29 Nov. 1880, p. 7, *Nicholson v. Edison.*

2. See Docs. 85 and 86; "The Printing Telegraph," *Telegr.* 6 (1869–70): 269; Reid 1879, 621; and Edison's testimony, 29 Nov. 1880, pp. 7–8, *Nicholson v. Edison.* A photograph album in the West Orange, N.J., archives contains two photographs of a laboratory mock-up of this instrument (Cat. 551:12, 14, NjWOE).

3. See Doc. 12, n. 1.

4. See Doc. 54. This method was adopted in other machines as well. Reid 1879, 619.

5. The breakwheel was a standard piece of printing telegraph equipment. A wheel turned by a crank made intermittent electrical contact by one of several means. Edison, in an unpublished manuscript, left the only description we have of the varied breakwheel mechanisms em-

Hand-cranked breakwheels like this one were used on printing telegraph circuits. This example is labeled "0" to "20," with each number except "0" divided into "R" and "L."

ployed in the early 1870s. NS-74-002, Lab. (*TAEM* 7:240–82); Edison's testimony, 29 Nov. 1880, p. 8, *Nicholson v. Edison*.

6. Edison's testimony, 29 Nov. 1880, pp. 8–9, *Nicholson v. Edison*. This printing telegraph was sometimes used with more than one wire. Reid 1879, 621.

–76–

*And Franklin Pope
Patent Model: Printing
Telegraphy*[1]

[Jersey City, N.J.,[2] September 16, 1869?[3]]

M (10 cm × 4 cm × 6 cm), MiDbEI(H), Acc. 29.1980.1298. Frame of
the instrument slightly bent; one spring missing.

1. See headnote above.
2. See Doc. 85.
3. Edison executed the covering patent application on this date. Pat.
App. 102,320.

N York   Sept 17 1869

Friend Hanaford

I received your Letters, but owing to things being so mixed
by the Consolidation of S S Laws Gold reporting Telegh and
Callahans Stock reporting Telegh and the my[a] Consequent
dismissal has upset all my calculations.[1] The Printers are
being made abut were delayed, by the removal of Phelphs
Shop,[2] and a rush of work = I expect to receive pay soon[a] for
a job I am doing now and I will remit you sufficient money to
purchase Skillings Battery although, if you will read the Contract, you will see we did not Contract to furnish supplies, but
only to keep the Line in good working order. If I was you I
should inform Mr Skillings that his battery needed supplies,
and That you would purchase for him and hand in the bill for
same   My P Office address is ~~6138~~ 6010[a] Now—[3]

A barber told me yesterday that the roots of my hair were
all coming out grey, and that I would be gray in 10 months[4]
my hair is now all sprinkled with them by the trouble which
my Double Transmitter give me and Skillings & other things.
I shall forward Printers soon as done   Your

Edison

ALS, NjWOE, DF (*TAEM* 12:28). Letter begins on reverse of letterhead of the Gold and Stock Reporting Telegraph. [a]Interlined above.

1. Edward Calahan served as superintendent after the merger, so
Edison's services were no longer required. See Chapter 4 introduction.
2. George M. Phelps (1820–1888), inventor and manufacturer, su-
pervised Western Union's factory, which had recently been moved from
Brooklyn to larger quarters on Fifty–fifth St. in New York. He had earlier managed the machine shop of the American Telegraph Co. Phelps
invented in several areas but was best known for his work in printing
telegraphy. Reid 1879, 640–42; "Removal," *Telegr.* 5 (1868–69): 304;
"The Death of George M. Phelps, Sr.," *Elec. W.* 11 (1888): 268.
3. The number 6138 was the post office box of Laws's Gold and
Stock Reporting Telegraph Co.; 6010 was the box number of the *Telegrapher.* James Ashley, editor of the *Telegrapher,* soon became Edison's
business partner (see Doc. 78).
4. Edison's hair was not conspicuously gray until the 1890s.

*George Phelps, superinten-
dent of Western Union's
manufacturing shop in New
York City.*

New York, October 2, 1869.[1]

POPE, EDISON & CO.,[2]

ELECTRICAL ENGINEERS,

AND

GENERAL TELEGRAPHIC AGENCY,[3]

OFFICE:

EXCHANGE BUILDINGS,

Nos. 78 and 80 BROADWAY, Room 48.[4]

A necessity has long been felt, by Managers and Projectors of Telegraph Lines, Inventors of Telegraph Machinery and Appliances, etc., for the establishment of a Bureau of Electrical and Telegraphic Engineering in this city. It is to supply this necessity that we offer facilities to those desiring such information and service.

A LEADING FEATURE

will be the application of Electricity to the Arts and Sciences.

INSTRUMENTS

for Special Telegraphic Service will be designed, and their operation guaranteed.

CAREFUL AND RELIABLE TESTS

of Instruments, Wires, Cables, Batteries, Magnets, etc., will be made, and detailed written reports furnished thereon.

CONTRACTS

for the Construction, Re-construction and Maintenance of either Private or Commercial Telegraph Lines will be entered into upon just and reasonable terms.

VARIOUS APPLICATIONS OF ELECTRICITY.

Special attention will be paid to the application of Electricity and Magnetism for Fire-Alarms, Thermo-Alarms, Burglar-Alarms, etc., etc.

TELEGRAPHIC PATENTS.

We possess unequalled facilities for preparing Claims, Drawings, and specifications for Patents, and for obtaining prompt and favorable consideration of applications for Patents in the United States and Foreign Countries.

EXPERIMENTAL APPARATUS.

Attention will be paid to the construction of experimental apparatus, and experiments will be conducted with scientific accuracy. Parties at a distance, desiring Experimental Apparatus constructed, can forward a rough sketch thereof, and the same will be properly worked up.

DRAWINGS, WOOD ENGRAVINGS, CATALOGUES, ETC.

prepared in the best and most artistic manner.

PURCHASING AGENCY.

Telegraph Wire, Cables, Instruments, Insulators, Scientific

and Electrical, and Electro-Medical Apparatus, Telegraph Supplies of all descriptions, Telegraphic and Scientific Books, etc., will be purchased for parties favoring us with their orders, and forwarded by the most prompt and economical conveyance, and as cheaply as the same could be purchased by our customers personally. Our facilities for this business are unexcelled.

Letters and orders by mail should be addressed to Box 6010, P.O., NEW YORK.

PD, *Telegr.* 6 (1869–70): 45.

1. The advertisement appeared in the *Telegrapher* from this date until 23 April 1870.

2. Edison, Franklin Pope, and James Ashley formed the firm of Pope, Edison & Co. between the time of Pope's return from his Midwest trip (see Doc. 72) and the appearance of this announcement. James N. Ashley (n.d.) was at the time editor of the *Telegrapher*. He had begun his career at midcentury as an operator with the New York and Boston Telegraph Co. He helped introduce the House printing telegraph in the United States and Europe, managed several telegraph companies, and served as army correspondent for the *New York Herald* during the Civil War. In February 1868 he replaced Pope as editor of the *Telegrapher* and in 1877 became editor of Western Union's *Journal of the Telegraph* when that publication absorbed the *Telegrapher*. Reid 1879, 409–10.

3. The company leased private lines to businessmen and, acting as the Financial and Commercial Telegraph Co., used Edison's and Pope's first printer to provide merchants with quotations from the gold and commercial exchanges. See Docs. 76 and 86; and "The Gold and Stock Telegraph Company.—Its Inception, Development, Business, etc.," *Telegr.* 11 (1875): 127.

Many of the provisions of this announcement reflect Pope's expertise. His *Modern Practice of the Electric Telegraph* included long sections on testing wires, he was an experienced writer and designer, and he later became a patent attorney. Although Edison's biographer Matthew Josephson claimed that this advertisement "was the first announcement of a professional electrical engineering service in the United States," Chester, Partrick & Co. had been advertising themselves as "Telegraphic and Electrical Engineers" in the *Telegrapher* since 16 November 1867. Josephson 1959, 78.

4. This was the address of the *Telegrapher*.

–79–

To E. Baker Welch

New York Oct. 3, 1869.

E. B. Welch.

Please send me a copy of my assignment to you of double transmitter.[1] I need it to show to other parties so I can go ahead with it. Respy.

T. A. Edison.

TL (transcript), MWalFAR, *Welch v. Edison*. See Doc. 36 textnote.

1. Doc. 61.

Electrical and Telegraphic Engineering.

Anonymous Article in the Telegrapher[1]

THE necessity for a more general application of scientific knowledge and experience to the construction and operation of telegraph lines, and to all matters connected therewith, has become very apparent.[2] In Europe this necessity was earlier appreciated than in this country, and many able scientific, electrical and telegraphic engineers are there fully and profitably employed.

In this country, with the exception of Mr. MOSES G. FARMER, of Boston, whose great ability and attainments as a scientific electrician have placed him in the front rank of modern scientists, we know of no one who has heretofore devoted himself to this important work. As will be seen from the advertisement of POPE, EDISON & CO., in this paper, this firm offer their services for such practical and scientific services in this city. Mr. POPE's superior ability and acquirements as a practical telegrapher and electrician, are too well known to need any extended commendation from us. Although young in years he has, by careful study and experiment, fully qualified himself to give valuable assistance to such as may desire his services. As an operator of several years' experience he has acquired that knowledge of the *practical* working of the electric telegraph which will render his services of peculiar value in all matters pertaining to the construction and operation of telegraphic lines, the due adjustment and proportions of instruments, insulation, batteries, etc., which render telegraphs accurate and reliable.

Mr. EDISON is a young man of the highest order of mechanical talent, combined with good scientific electrical knowledge and experience. He has already invented and patented a number of valuable and useful inventions, among which may be mentioned the best instrument for double transmission yet brought out.

Their united genius and science cannot fail to render their services most valuable to all who may have occasion to employ them, and must ensure their constant and profitable occupation.

Connected with their other business they have established a purchasing bureau, which cannot but prove a great accommodation to persons at a distance, who may desire to purchase telegraph material, supplies, books, etc.

We commend this new firm to the favorable consideration of the telegraphic public, with entire confidence that they will afford to those who may desire their assistance complete satisfaction.

PD, *Telegr.* 6 (1869–70): 52.

1. Editor James Ashley probably wrote this article.

2. Ashley had previously argued the need for "scientific" expertise in telegraphy: "In no line of business can searching scientific investigation and study be so profitably applied as in the practical management and improvement of the telegraph. . . . Every telegraph company should employ a competent and scientific *Engineer,* to whom should be confided the entire care of, and responsibility for its wires and apparatus." "Errors in Telegraphic Management," *Telegr.* 5 (1868–69): 336.

–81–

To the Editor of the
Telegrapher

New York, October 16, 1869.
Queries.

TO THE EDITOR OF THE TELEGRAPHER.

DOES the continuous "jump spark" of an induction coil conduct low tension currents, and in what degree?

At what ratio does the effective power of the several layers of wire upon an electro-magnet decrease or increase as they recede from the core?

Who was the first discoverer of the insulating properties of paraffine?

Is the iron ray in the solar spectrum deflectable by a magnet?

Is there any known device for rendering a galvanometer perfectly astatic?[1]

Is the current of a voltaic battery intermittent? if not, the proof.

Was Voltaire the inventor of the Voltaic pile—*vide* Madison Buell,[2] in the *Journal of the Telegraph*[3]—or was it Volta?

E.[4]

PD, *Telegr.* 6 (1869–70): 58.

1. That is, insensitive to the Earth's magnetic field.

2. Madison Buell was chief operator and manager for Western Union in Buffalo. He frequently contributed to the technical press and was the inventor of a widely used switchboard. Taltavall 1893, 107–8.

3. "Scintillations from Scientific Authors," *J. Teleg.* 2 (1869): 197.

4. Authorship attributed to Edison. See Doc. 34, n. 10.

–82–

To Joel Hills and
William Plummer

New York, November 12 1869[a]

Gentlemen—

As the conditions upon which I conveyed to you my printing Telegraph instrument—namely the issue of the Patents and the instrument proving to be as represented[1]—have been fulfilled, the ~~conditions~~ consideration of two hundred and fifty dollars has become due   you will therefore please forward me a check for that amount on receipt of this Letter Respy Yours—

T. A. Edison

ALS, MiDbEI, EP&RI. Letterhead of Pope, Edison & Co. ᵃ"New York," and "18" preprinted.

1. See Doc. 51.

–83–

To Frank Hanaford

New York, Nov 30 1869ᵃ

Friend Hanaford =

I cannot do ~~a~~everything without money but to show you that I have not been Idle read this affidavit

    This is to certify that Mr Edison Is having one set of Private Line instruments constructed—[1]

Frank L Popeᵇ

These instruments will be done probably in 10 days—and I will ship then on to you with full ~~instruments~~ instructions how to put them up = They cost me all the money I have earned, ~~but~~ and I think you are a little hard on me   If I came to Boston I never could be able to earn enough to get them Respy

Edison

ALS, NjWOE, DF (*TAEM* 12:31). Letterhead of Pope, Edison & Co. ᵃ"New York," and "18" preprinted. ᵇSignature.

1. These instruments may be the "Printers" referred to in Docs. 73 and 77.

–84–

To E. Baker Welch

New York Nov. 30, 1869.

Mr. Welch:

I have not been able to see Laws yet, but hear he is in town to-day   will try and see him.[1]

Please send on a copy of my assignment to you for double transmitters. Respy.

Edison.

TL (transcript), MWalFAR, *Welch v. Edison*. See Doc. 36 textnote.

1. After his company merged with the Gold and Stock Telegraph Co., Samuel Laws began studying law. He received the LL.B. degree from Columbia College in New York in 1870. He was, however, still connected with telegraphy as one of the directors of Gold and Stock. *DAB*, s.v. "Laws, Samuel Spahr."

85

New York, December 1, 1869.

Dr. Bradley's Clock and Works.[a]

We had great pleasure a few days ago in visiting the shops of Dr. Bradley,[1] of Jersey City, and examining his electric clock, a sketch of which we have given on our fourth page. It is exceedingly simple and ingenious. The governor or pendulum is its chief feature. Acting as a part of the circuit by which motion is communicated, the centrifugal tendency is checked the instant it leaves its contact with the arm which propels it, and is, in like manner, sustained by its contact. The description is interesting and excellent.

Dr. Bradley has a large force manufacturing the new printing instruments of Pope & Edison,[2] which are remarkable for their ingenuity and simplicity. A single wire is used. In the circuit are two magnets, one more responsive than the other. The primary motion of a crank ratchets up the type wheel to its place by the more responsive magnetic arm. Arresting the motion of the crank allows the second magnet, which makes the stroke, to act. The return of the striking arm moves the paper forward. Thus three motions are made by a single wire and by the same current. The less responsive magnet is made so by lightness and quickness of contact in the primary motion. It acts when time is given by detention to connect. We are indebted to Mr. Ecklin, the intelligent foreman, for the inspection of this minute but effective instrument.

In an adjoining room we were allowed to see the Bradley process of filling the wire spools, which is executed by Miss Knight, an intelligent and skillful lady, to whom Dr. Bradley has assigned this duty.

PD, *J. Teleg.* 3 (1869–70): 5. [a]Followed by centered rule.

1. Leverett Bradley (1798–1875), educated as a physician, was a merchant, a civil engineer, and then a newspaper publisher before turning to electrical invention and manufacture in the late 1850s. He gained a reputation for his patented magnetic helices and electrical measuring instruments. "Obituary. Dr. Leverett Bradley," *Telegr.* 11 (1875): 222. See also App. 1.A30–31.

2. See Doc. 76.

86

New York, December 11, 1869.

The Financial and Commercial Telegraph.

THE first section of the above company's line, designed for furnishing reports of the prices of gold and exchange to merchants and others, has just been put in successful operation in this city. This enterprise, although in many respects similar to others now engaged in the same line of business, is intended to occupy a somewhat different field—that of furnishing quotations at low rates to mercantile and importing houses, who have only an indirect interest in the rate of gold, not sufficient to warrant them in employing the more expensive instruments in such general use among the brokers of Wall and Broad streets.

The line has been constructed, and the instruments furnished by POPE, EDISON & CO., contractors, who will also remain in charge of its working until fully completed. The line is constructed in the best manner, with No. 7 compound wire and BROOKS insulators,[2] and the single wire printing instruments of POPE & EDISON are employed. They are of elegant workmanship and appearance, and their simplicity and effectiveness will ensure them a large popularity.

The whole work has been done in the short space of seven weeks from the time the order was placed in the hands of the contractors, a fact which speaks well for the business enterprise of the new firm.

PD, *Telegr.* 6 (1869–70): 124.

1. Editor James Ashley probably wrote this article.

2. The Brooks insulator, patented by veteran telegrapher and inventor David Brooks, was considered a major advance in telegraph technology. Made with paraffin-saturated cement, it consisted of a hook cemented into an inverted glass bottle that was in turn cemented into a cast-iron shell. The materials in ordinary insulators allowed current leakage because they were hygroscopic, that is, their surfaces became moist in humid air. The cement in Brooks's insulator prevented the formation of a continuous water film, even in rain. Brooks' Patent Paraffine Insulator Works in Philadelphia manufactured the insulator. "David Brooks," *J. Frank. Inst.* 132 (1891): 75; Pope 1869, 61–62; "Brooks' Improved Insulators," *Telegr.* 5 (1868–69): 320; "Brooks' Improved Paraffine Insulator," ibid., 360; "Brooks' Patent Paraffine Insulator Works," ibid. 6 (1869–70): 127.

–87–   [1869][1]

Memorandum:	Diggers 8.	$10.80	Labor per Mile $	17.58
Telegraph Construction	Setters[2] 5.	6.75	Poles per Mile $	50.00
Estimate	Nailing Insulators. 1	1.35	Wire per Mile $	60.00
	Trimmer.[3] 1	1.35	Insulators per	
	Reel Men[4] 2.	2.70	Mile $	7.00
	Wire pullers 2.	2.70	Cost per Mile $	134.58

Climbers 2.	3.08	Labor to NY	$ 4,395.00
Reel Horse & Wagon	2.00	Cost to Build	
Labor per three Miles	$30.73	Lines to NY	$33,645.00[c]
Board per three Miles	$22.00		
" " "			

$17.000. for an additional wire and by use of double Transmitters on each 4 complete wires could be had to N York

HD, MiDbEI, EP&RI. [a]To this point, in an unknown hand; remainder in Edison's hand.

1. Docketed on back: "Tho's A. Edison's. Estimate Cost of Tel Line to N.Y. 1869."
2. Diggers and setters erected the poles.
3. The trimmer cut tree branches away from the line.
4. Reel men handled the reels of wire.

TELEGRAPH CIRCUITS   Doc. 88

Edison made the drawings of telegraph circuits shown in Doc. 88 in a pocket notebook sometime in 1869. Although the circuits depicted were for printing telegraphy (cf. Docs. 66, 93, and 109), they embody ideas that Edison also used or tried to use in multiple telegraphy. He introduced the drawings in an 1880 patent interference to support his claim that in 1869 he was working on circuits that in principle were suitable for diplex and quadruplex telegraphy. He testified that he recognized this possibility at the time but had first had the idea years earlier.[1]

Sketch E is the earliest document to display the general form of Edison's later successful diplex and quadruplex designs. The lower loop in the diagram represents a telegraph main line (Edison left out transmitting devices in all of these sketches). Edison described "a polarized relay and a neutral relay being included in the same circuit, one operated by a reversal of the current, and the other by a rise and fall of tension. Both instruments were provided with local circuits containing a battery and electromagnet."[2] The local circuits are shown in the upper part of the drawing. In the main circuit the polarized relay is shown on the left side, the neutral on the right. Because this arrangement allowed both the reversal of the current and the rise and fall of tension to be independent and simultaneous signals and because it did not require either the two local circuits or the two implied keys to be in the same location, this circuit could be used in multiple as well as printing telegraphy.

Sketch B depicts what Edison characterized as "a neutral relay and a polarized relay, both included in the same circuit." In this design the two signal types could be neither independent nor simultaneous. The lower loop in the diagram represents the main line of the telegraph system; the upper portion of the drawing shows a local circuit for the printing instrument. **P'** designates an electromagnet to move a printing device; immediately below that are the ratchet, the escapement lever, and two electromagnets of a typewheel circuit. Next below is a switch actuated by a neutral (standard) relay that is itself on the main line; the lowest device is a polarized relay, also on the main line, that switches between the local circuit branches. "The neutral relay served to open and close a local circuit and give rotation to a type wheel. The polarized relay served to open and close a local circuit and give motion to a magnet to effect the printing of a letter; a rise and fall of tension caused the neutral relay to move its armature back, and forward, opening and closing a local circuit at each reciprocation when the polarity of the current was in one direction, but when the polarity of the current was reversed, the tongue of the polarized relay closed a local circuit, and at the same time prevented the neutral relay from closing or opening the circuit operated by it by reason by opening the said circuit upon the polarized relay. . . ."[3]

Edison described sketch C as "nearly identical, except in the method of making the local connections."[4] The main-line wire is not shown. At center left is a paper tape that is pressed up against the round typewheel by the printing lever.

Edison did not discuss the other drawings. Sketch A is a printing telegraph circuit in which the polarized relay appears to switch the current between typewheel magnet **T** and printing lever magnet **P**, with the typewheel magnet partially cut in on the printing circuit. Sketch D is a combination neutral and polarized relay.

1. Testimony and Exhibits on Behalf of T. A. Edison, pp. 3, 22, 29, 147–54, *Nicholson v. Edison.* See also App. 1.A18, D42.

2. Testimony and Exhibits on Behalf of T. A. Edison, p. 29, *Nicholson v. Edison.*

3. Ibid., p. 28. Edison did not use this circuit in either of his successful 1869 printing telegraphs (Docs. 54 and 76).

4. Testimony and Exhibits on Behalf of T. A. Edison, pp. 28–29, *Nicholson v. Edison.*

Technical Drawings:
Telegraph Circuits[1]

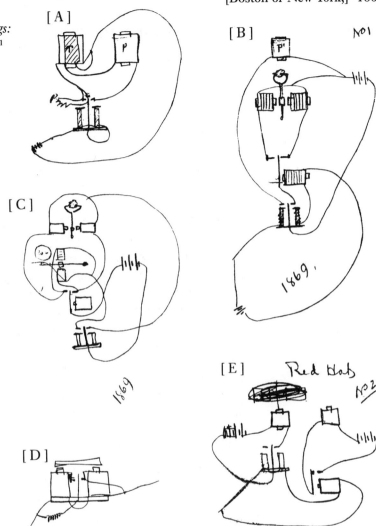

AX (photographic transcript), MdSuFR, RG-241, *Nicholson v. Edison*, Testimony and Exhibits on Behalf of T. A. Edison, pp. 197–98. The exhibit sketches appear to be photolithographic copies prepared for the printed record (Lemuel Serrell to TAE, 18 Dec. 1880, DF [*TAEM* 55:235]). During this testimony Edison inserted the labels "No 1" and "No 2" on the sketches. Apparently in the original document the two drawings labeled "1869" were on opposite sides of one piece of paper.

1. See headnote above.

2. Edison worked on many variations of such circuits in both Boston and New York in 1869.

New Alliances

January–June 1870

The first half of 1870 saw an important shift in Edison's business and personal relations as his work in printing telegraphy became closely tied to the Gold and Stock Telegraph Company and its officers. At the beginning of the year, he was working with Franklin Pope and James Ashley in connection with Pope, Edison & Company. Within four months he had moved his operations to Newark, New Jersey, and was developing facsimile and printing instruments for the Gold and Stock Telegraph Company and for himself. This activity demonstrated his unusual capabilities to the executives of that powerful company.

Edison's business relations with Pope and Ashley also involved the Financial and Commercial Telegraph Company. Established the preceding fall, it was doing well and attracted the attention of its powerful rival the Gold and Stock Telegraph Company. The patent covering Financial and Commercial's printer was not granted until April, but in the intervening months the inventors developed other devices—Edison a ratchet-and-pawl escapement mechanism for turning a typewheel,[1] Pope a combination Morse and polarized relay,[2] and the two together another printing instrument, the gold printer. Patent applications filed in April covered these inventions. Six days after submitting the application for the gold printer, the three partners sold to Gold and Stock partial rights to the four patent applications and to Edison's Boston instrument patent, which they had had to repurchase from Joel Hills and William Plummer. They also sold the other assets of the Financial and Commercial Company to Gold and Stock. Under the contract, Pope, Edison & Company retained all patent rights for use with private lines—lines that, for example, a business might use to communicate between an office and warehouse. Small printing telegraphs opened this new market area, which the three partners expected to cultivate.

Early in 1870 Edison's commercial horizons expanded as he began to work with Gold and Stock. Telegraph entrepreneur and Gold and Stock executive Marshall Lefferts took Edison under his wing, encouraging him to sign two contracts with Gold and Stock officers George Field and Elisha Andrews on 10 February. One contract called for the development of a facsimile telegraph to rival the Morse system, and the other specified a printer like the Calahan ticker. Lefferts, head of the Commercial News Department of Western Union, then succeeded Field as president of Gold and Stock in March. Gold and Stock, the sole transmitter of information from the New York Stock Exchange, commanded a market of great potential. In order to maintain its control, the company carefully hoarded printing telegraph patents, obtaining reissues when it needed to broaden or clarify claims. It owned the patents Edison had executed for Samuel Laws, and in the early months of 1870 the company's officers gained personal knowledge of Edison and his talent for inventing.

Edison completed his ticker for Gold and Stock in early April, and although it operated slowly, it satisfied the company.[3] Development of the facsimile telegraph took considerably longer, but the resources provided under the contract allowed Edison to rent a shop, furnish it, and hire a mechanic to assist with experiments. Taking advantage of this provision, Edison entered into a partnership with machinist William Unger, and in February they established the Newark Telegraph Works in a small shop across the street from Newark's principal railroad depot.[4] Soon Edison was boarding in the city as well.

Establishing himself in Newark was an important step for Edison because most of his needs could be met there. Newark was a generally prosperous city, its growth having accelerated during the Civil War. In practical terms, it was a major source of iron and brass castings, hand and machine tools, chemicals, and other supplies any manufacturer of telegraph and electrical devices would need.[5] It also boasted a concentration of precision metalworking industries, such as toolmaking and jewelry manufacture, which provided the type of skilled labor Edison needed.[6] As a place of residence it retained an almost

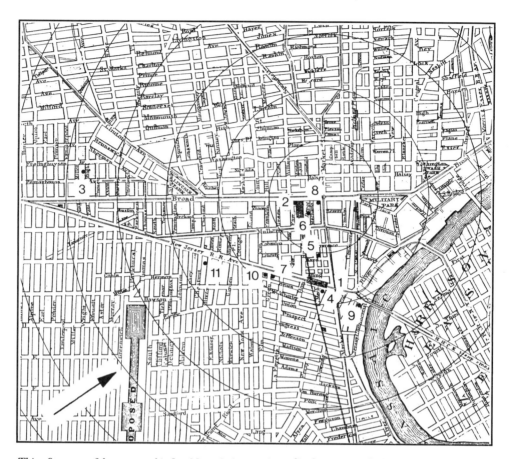

This 1875 map of downtown Newark, N.J., shows the locations of Edison's businesses and residences between 1870 and 1873. The concentric circles are spaced at quarter-mile intervals. (**1**) Edison boarded in this area of Market St. sometime in 1870–71. (**2**) Edison was boarding at 854 Broad St. before he married in December 1871. (**3**) Edison purchased a house at 53 Wright St. (later renumbered 97) in December 1871. (**4**) Edison set up the Newark Telegraph Works, his first Newark shop, with William Unger at 15 New Jersey Railroad Ave., in February 1870. (**5**) The Newark Telegraph Works was moved to 4–6 Ward St. (renumbered 10–12) in May 1871, and renamed Edison and Unger. (**6**) Edison used a small shop at 24 Mechanic St. as an annex to the Ward St. shop during the fall of 1871. (**7**) Edison set up the American Telegraph works with George Harrington at 103–109 New Jersey Railroad Avenue in October 1870. (**8**) Edison established a short-lived enterprise, the News Reporting Telegraph Co., in the Daily Advertiser Building at 788 Broad St. in the fall of 1871. (**9**) Edison had a small experimental shop in White's Building on the Morris Canal sometime in 1871–72. (**10**) Edison and Joseph Murray set up a small shop, Murray and Co., at 115 New Jersey Railroad Ave. in February 1872. (**11**) Murray and Co. moved to 39 Oliver St. in May 1872.

pastoral aura. Observers emphasized the broad streets, abundant greenery, and general beauty of the city's site along the clear Passaic River. Newark's flourishing manufacturing and transportation facilities, pleasant environment, comparatively low taxes and rents, and proximity to New York City made it a desirable residence and workplace for the ambitious Edison.[7]

In this desirable manufacturing environment, Edison began to expand his inventive horizons into the field of automatic telegraphy. Marshall Lefferts may have prompted Edison's first patent application in automatic telegraphy: an automatic telegraph transmitter. Lefferts had long-standing and deep interests in automatic telegraphy and is thought to have made Edison aware of problems that plagued the system of George Little, a system in which Lefferts was then interested. In automatic telegraphy an operator perforated a strip of paper with holes to represent the dots and dashes of Morse code and then fed the perforated strip into a transmitter, where electrical contact was made through the perforations, causing the circuit to close intermittently and thereby transmitting signals. A receiver at the other end of the line recorded the signals on paper. By using automatic machinery, such systems could send messages at higher speeds than hand-actuated instruments.[8]

Lefferts knew that practical limitations prevented automatic telegraphs from achieving high speed, however. One problem was that the inertia of the mechanism slowed the speed with which the signal could be transmitted. The most difficult problem was what Edison and his contemporaries called "tailing," a combination of electromagnetic induction and capacitance that lengthened each electrical signal in a message. Thus, dots often appeared as dashes, and dashes sometimes became long enough to cause the entire message to appear as a solid line. Edison designed his first automatic transmitter to overcome tailing and executed the patent for it in June 1870.[9] This was his initial step into an area that continued to occupy his inventive and entrepreneurial efforts for several years.

1. U.S. Pat. 103,035. See Doc. 97, n. 12.
2. U.S. Pat. 103,077. See Doc. 97, n. 11.
3. U.S. Patent Office 1872, 80; Henry van Hoevenbergh to Ralph Pope, 24 July 1908, Meadowcroft; G&S Minutes 1867–70, 126–30.
4. How Edison met Unger is unknown. See Doc. 92, n. 1.
5. Newark was noted for the wide range of its manufactures (Hirsch

1978, 16–21; Armstrong 1873; Atkinson 1878; Ford 1874; Van Arsdale & Co. [1873?]; and Popper 1951). Of the thirty suppliers of the Newark Telegraph Works in 1870 whose locations have been identified, twenty-one were in Newark and seven were in New York City. 70-006, DF (*TAEM* 12:155–201).

6. Foreign-born residents constituted about one third of Newark's population in 1870. Nearly 45 percent of these were from Germany, including the Unger family, and a large number were skilled workers with some capital (Popper 1951, 63–64, 131–33; Drummond 1979, 141–45; Urquhart 1913, 2:1059–60; U.S. Census 1964, roll 447). For a time Edison and Unger employed as a machinist one of William Unger's older brothers, George.

7. Atkinson 1878, 314; Drummond 1979, 21–30, 62–64, 128–34, 148, 241, 250–52; Ford 1874, 5; Lamb 1876; Popper 1951, 65–68; Sikes 1867a and 1867b.

8. See Doc. 33. Hand operators averaged 25–40 words per minute; ink recording automatic telegraphs in use in England averaged 60–120. Experimental tests of George Little's electrochemically recording automatic system in 1869 yielded speeds as high as 100–130 words per minute. Pope 1869, 118; Prescott 1877, 711; National Telegraph Co. 1869; "Little's Automatic Telegraph System," *Telegr.* 6 (1869–70): 150.

9. Each time the circuit was broken in the process of transmitting a dot or dash, the line discharged and induced a current. The induced current prolonged the signal, an elongation that was imperceptible at moderate speeds but noticeable at higher transmission speeds. The effect was compounded by capacitance, which increased with transmission distance. Telegraphers often referred to the cause of tailing as "electrostatic line induction." See Doc. 101, n. 3.

–89–

To Frank Hanaford

New York, [January?][1] 26 1870[a]

Frank

Did you receive my Letter enclosing money[2]   you said nothing about it in Letter = Am making Skillings instruments now— will forward with description how to put them up— They will go all right if the Line even lays on the ground = keep your Courage up and it will come out all right   Think I can get you a red hot Situation here where you Can make some stamps—[3] I may come on with the insts myself if I do will tell you about it

Answer if you got my Letter with money   My hair is damnd near white— Man told me yesterday I was a walking churchyard—[4] Your

Edison

ALS, NjWOE, DF (*TAEM* 12:42). Letterhead of the *Telegrapher.* [a]"New York," and "18" preprinted.

1. Conjecture is based on the content of the letter and its relationship to other documents.

2. Edison offered to send money to Hanaford at least twice, and later

did send money at least once. See Docs. 70 and 77; and Frank Hanaford to TAE, 12 July 1871, DF (*TAEM* 12:224).

3. "Stamps" meant money. Farmer and Henley 1970, s.v. "Stamps."

4. See Doc. 77.

–90–

From Dewitt Roberts

Ellenburgh N.Y. Feb 4 1870

Dear Sir—

I have just directed Mr Williams[1] to send the Voting Ins't[2] here by Express with facilities for working it—

I wish you would prepare the neccessary articles for the solution[3] and send it to me by the National Express Company as soon as possible. I would also like to know if you are willing I should proceed to put the Instrument before the public

Your early attention to this matter will greatly oblige   Truly Yours

D. C. Roberts

Am sorry the Boys had to give up as they did.[4] I would rather have given $50. myself than had them do that

ALS, NjWOE, DF (*TAEM* 12:220).

1. Charles Williams, Jr.

2. Doc. 43.

3. The chemical solution for soaking the recording paper; see Doc. 43 headnote.

4. Western Union operators engaged in an unsuccessful strike 3–18 January 1870. Ulriksson 1953, 23–28.

–91–

Agreement with George Field and Elisha Andrews

[New York?,] February 10, 1870[a]

This Article of agreement made and entered into this tenth (10th) day of February 1870 by and between Thomas A Edison of Elizabeth, New Jersey, party of the first part, and Geo B Field[1] of the city county and State of New York and Elisha W. Andrews[2] of the town of Englewood Bergen County New Jersey parties of the second part.[3] Witnesseth that the party of the first part agrees to invent and perfect a Printing Telegraph Instrument to be worked practically upon one wire, and to perform the same kind of work as the stock Instrument known as "Calahans ~~Invention~~ Patent" ~~As~~now used by the "Gold and Stock Telegraph Company" of New York City.[4]

Said instrument to be made, shall be constructed not to have more than thirty Six pieces and Six screws in its working parts, all parts to be made of iron and steel, the size to be smaller than the Calahan instrument and the speed to be as great if not greater.[5]

It shall be constructed to print two lines on the strip of paper, and all the parts to be clearly seen, and easily approached and the instrument to be easily taken out or put in an office, to have a superior inking apparatus, and the instrument to be worked with the same amount of battery as those used by the Gold & Stock Tel Co" if not less. All parts of the said instruments to be made interchangeable, and the said invention to be clearly patentable in its combinations, and upon the completion of two instruments fully demonstrating the perfection of the instruments for the purpose designed, to the satisfaction of the parties of the second part, the said part of the first part, agrees to prepare or have prepared an application for Letters Patent from the United states Government, and convey by an assignment at the Patent office of the United states to the parties of the second part— Nine tenths (⁹/₁₀) of the interest of the party of first part in said invention[6] upon the following conditions to wit:

The parties of the second part shall well and truly pay to the said party of the first part the sum of Seven (7000) thousand dollars in full payment for the said nine tenths (⁹/₁₀) when the Patent for the instrument above described shall have been allowed. by the Patent office.[7] And the said party of the first part ~~of the first part~~ further agrees that he will lease to the said parties of the second part from the date of said Patent, the remaining one tenth interest in said invention for Seventeen years from the date of the patent for a sum not exceeding one dollar per annum. And the parties of the second part shall by a special agreement agree not to re = issue said patent during its validity or Seventeen years with out the consent of the party of the first part, and shall pay the bills for the construction of the two instruments above referred to not to exceed the sum of two hundred & fifty dollars.

This Contract is assignable, and the agreements are obligatory upon the heirs or assigns of the parties here~~with~~unto.

In witness whereof the parties have hereunto set their hands & seals the day and year aforesaid

Thomas. A. Edison   Geo B. Field
                                                     Elisha W Andrews[b]
In presence of   A M Kidder[c]

In addition to the above agreement .I. agree to include my services as C Electrician for one year from date of the application for Letters patent for the above named instrument,[8] for the consideration named in the above writing. I also agree to assign the above named parties any future improvement in mechanical Printing Telegraph instruments which are applicable ~~to~~ directly to quoting the prices of Gold & Stocks ~~or any [scheme?] of telegraphing where a number of instruments are worked in connection with one another~~[d] in the United States which I may devise

Thomas A. Edison

DS (copy), NjWOE, LS (*TAEM* 28:931). In George Field's hand. [a]Date taken from text, form altered. [b]Representation of wax seal drawn next to each signature. [c]After this, text is in Edison's hand. [d]Canceled words interlined above.

1. George Baker Field (n.d.) succeeded Elisha Andrews (see n. 2) as president of the Gold and Stock Telegraph Co. on 13 July 1868. He became a director of the company on 7 September 1869 and was replaced as president by Marshall Lefferts in March 1870. Reid 1879, 607–8; Hotchkiss 1969, 437.

2. Elisha Whittelsey Andrews (n.d.), a New York stockbroker and original stockholder in Gold and Stock, served as president of Gold and Stock from 19 September 1867 to 13 July 1868, when he was elected to the company's board of directors. In 1870 he traveled to England, seeking to introduce Gold and Stock instruments there (see Chapter 7 introduction). Two years later he helped found the American District Telegraph Co., which provided alarm and messenger services (see Doc. 226, n. 2). Incorporation agreement of Gold and Stock Telegraph Co., exhibit in *G&S v. Pearce;* Reid 1879, 607, 635; Hotchkiss 1969, 434; Calahan 1901c.

3. Although Gold and Stock is not mentioned in this document, Field and Andrews were acting in the company's interest. They assigned this agreement to the company in late spring. G&S Minutes 1867–70, 126–30.

4. Edward Calahan's printing telegraph (U.S. Pat. 76,993), which was introduced in brokers' offices in New York at the end of 1867, was the first to have two independent typewheels—one for letters and one for numbers. It was the mainstay of the Gold and Stock inventory of printers until Edison's universal stock printer superseded it in the early 1870s (G&S Minutes 1870–79, 86; William Orton to S. G. Lynch, 22 Oct. 1872, LBO). Pope and Edison, like many other inventors, sought unsuccessfully to find anticipations of Calahan's design with two independent wheels so that they could circumvent his patent. All litigation concerning infringements of Calahan's design was decided in favor of Gold and Stock. The principal advantage of two typewheels was the ready legibility of the printed tape, which was considered "extremely important" by Gold and Stock and its customers. Because each wheel had a blank space (or a dot), there was no need for an extra mechanism to block one while printing with the other (as Edison later provided for his cotton instrument and universal stock printer). In addition the lower inertia and smaller circumference of the two independent typewheels allowed greater speed than either a single, large typewheel—which was Edison's solution in this case—or two coupled typewheels. See Doc. 116 for sketches of two other attempted solutions. Affidavit of Frank L. Pope, 5 Nov. 1883, *G&S v. Pearce;* George Scott's testimony, 12 Feb. 1884, ibid.; *G&S v. Wiley.*

The Edward Calahan stock printer used by the Gold and Stock Telegraph Co.: (left) *side view;* (right) *front view.*

The instrument Edison designed to satisfy his 10 February 1870 contract with George Field and Elisha Andrews.

5. The number of parts in Calahan's and Edison's printers is uncertain. The "iron and steel" qualification was probably economic. Brass had the advantages of being nonmagnetic and nonrusting but was at the time about six times as expensive as iron and 33 percent more expensive than steel. See invoices, 70-006, DF (*TAEM* 12:156–98).

6. The patent for the printer designed under this contract was executed on 24 May 1870 and issued as U.S. Pat. 128,608 (G&S Minutes 1867–70, 126–30). The entire right to the patent was assigned to Gold and Stock on 30 October 1872 (Digest Pat. E-2:209). This machine was

never used. Edison made an amendment to the patent application that placed the application in interference with another machine, and the application was not allowed for a year. After finally amending the application to remove the interference, Edison neglected it for another year before paying for its issuance. U.S. Patent Office 1872, 80–81; Pat. App. 128,608. See also Doc. 124, n. 1.

7. This payment was fulfilled in different terms under the agreement of 26 May 1871 (Doc. 164).

8. Edison filed the patent application (see n. 6) on 27 May 1870. The next year on May 26 he signed an agreement making him the company's "Consulting Electrician and Mechanician" for five years (Doc. 164).

[New York?,] February 10, 1870[a]

This article of agreement made and entered into this tenth (10)[b] day of February 1870 by and between Thomas A Edison of Elizabeth New Jersey, party of the first part, and George B Field, of the city county and State of New York and Elisha W Andrews of the town of Englewood Bergen County New Jersey[c] parties of this second part witnesses as follows, that the said party of the first part hereby agrees with the said parties of the second part, to invent and perfect an <u>Autographic or Fac Simile</u> Telegraph instrument which shall

1st	Equal the average speed of Morse
2nd	Be, simple, reliable, practical
3d	Transmit hieroglyphical characters, short hand, and messages in any written Language
4h	Transmit outline Photographs.
5h	Have a speed greater than the average Morse o~rn~ on[d] Press Telegrams.
6h	Have no elaborate preparation of the message to be sent.
7h	Drop copies at any point on circuits not excessively long.
8h	Be able to retransmit from chemical strip.
9h	Not conflict with any existing patent valid in the United States.
10h	Have a United States patent, clear, and po[s]itive.

And upon the completion of two instruments fully demonstrating the[e] practical working of said invention, to the satisfaction of the parties of the second part, the said party of the first part agrees to convey by an assignment to the parties of the second part <u>Two-Thirds</u> ⅔ds interest in his invention, and to have the same secured by letters patent of the United States. The parties of the second part agree to pay to the party of the first part, in consideration of said assignment the sum of Three Thousand $3000. dollars. And farther the parties of the second shall furnish a good comfortable room at a rent not exceeding ten dollars per month, and all the necessary tools, and machinery to make experiments for and <u>Construct</u> said apparatus, the tools not to exceed a cost of four hundred dollars $400., and to pay a first class mechanic to be employed by said party of the first part until said instruments shall be completed, not to exceed six months working time, and for the stock to be used in their construction and the incidental expenses pertaining thereto.—[1]

All bills to be presented to the party of the second part, by the party of the first part. And no bill to be contracted amounting to over the sum of Ten (10)[b] dollars without the consent of the parties of the second part. The parties of the second part shall pay all patent fees for securing Letters patent and upon fulfilling the conditions above recited, and by securing by special agreement to the party of the first part, one third of all profits oaccurruing from said invention, the parties of the second part, shall become owners of two thirds of said invention and the parties of the first the owner of one third ⅓ interest in said patent or patents.[2] It is hereby agreed that the obligations in this contract between the parties herinto are obligatory upon their heirs and assignees

In witness whereof the parties have hereinto set their hands & seals, the day and year aforesaid[f]

Thomas. A. Edison

Geo B Field

Elisha W. Andrews[g]

In Presence of   A. M. Kidder[3]

DS (copy), NjWOE, LS (*TAEM* 28:928). [a]Date taken from text, form altered. [b]Circled. [c]To this point, written in Edison's hand. [d]Interlined above in Edison's hand. [e]"of two . . . the" repeated upside down, overstruck, at bottom of page. [f]"hereby agreed . . . aforesaid" written in Edison's hand. [g]Representation of wax seal next to each signature.

1. Under this provision, Edison established the Newark Telegraph Works in conjunction with William Unger (1850–1878), a Newark machinist. Unger and Edison were partners in the Newark Telegraph Works until 3 July 1872 (see Doc. 264). Unger then moved to New York City and manufactured telegraph instruments, light machinery, and models at a shop over the New Haven Railroad Freight Depot on Franklin St. After a fire at this shop in February 1873, he went into partnership manufacturing electrical and telegraph instruments and other machinery with Hamilton Towle at 30 Cortland St. Unger later joined his brothers—Herman, George, Frederick, and Eugene—in the firm of Unger Brothers in Newark, manufacturing pocket knives and hardware specialties. In 1878 George, Frederick, and William Unger died. William Unger advertisement, *Telegr.* 8 (1871–72): 445; "Scorched But Not Destroyed," ibid. 9 (1873): 59; Rainwater 1975, 175–76.

On 15 February, Edison and Unger rented a room for $13.50 at 15 Railroad Ave. in Newark (70-008, DF [*TAEM* 12:206]), and by the end of April they had spent about $400 on machinery and tools (70-006, DF [*TAEM* 12:159–66]). The firm's accounts for experimental work on the facsimile telegraph began in May, after the shop had been equipped (70-005, DF [*TAEM* 12:144]). The establishment of the American Telegraph Works in October 1870 (see Doc. 109) provided Edison with a much larger and better-equipped machine shop, but he and Unger continued to operate the Newark Telegraph Works, where much experimental work was done.

2. Apparently Andrews, Field, and Marshall Lefferts (who acquired an interest in the autographic telegraph) continued to pay for these experiments for over a year, although Edison claimed that he bore the expenses after the six months stipulated here. As late as 1876 Edison still sought to collect money he claimed was due him. He continued to experiment periodically on facsimile telegraphs, finally patenting a system in 1881. Cat. 1183:2–13, Accts. (*TAEM* 20:861–66); Andrews to TAE, 20 June 1870, DF (*TAEM* 12:618); TAE to Andrews, 21 June 1876, DF (*TAEM* 13:1136).

3. A. M. Kidder, of the banking and brokerage firm of A. M. Kidder & Co., specialized in the trading of railroad stocks and bonds. He became a member of the newly consolidated New York Stock Exchange on 1 May 1869. *New York Times*, 27 Apr. 1903, 7; New York City 417:193, RGD; membership records, NNYSE.

–93–

Franklin Pope to Montgomery Livingston[1]

NY Mar 17 1870.

Dear Sir,

I have examined the patent of E. A. Calahan[2] referred to by the Examiner in our case,[3] and in reference to the same would remark as follows:

In working a printing telegraph by electricity, two distinct operations are involved.— 1. That of bringing the type wheel into a proper position, and 2. That of taking an impression from said wheel after having been thus moved.

In the House, Hughes, and a similar class of instruments, a single line wire is uemployed, in conjunction with local mechanical power applied separately to each recieving instrument, and this is an indispensable condition of their action. The movement of the type wheel is governed by a combination of mechanical and electrical power, while the impression is taken exclusively by mechanical power, or the reverse. In Calahan's patent Apr 18 1868[4] the type wheel is moved by one main circuit, and the impression taken by a second main circuit. He also describes arranging a number of of instruments in this way,— all the type wheel magnets being placed in one main circuit, and all the printing magnets placed in another main circuit. G. M. Phelps, (Pat June 22 1869) uses a main and local circuit,[5] the latter for giving the impression   Others use a single main circuit, combined with locals, or with mechanical power or both.

We therefore wish to restrict our claim[6] to the arrangement of a number of automatic printing instruments working simultaneously ion one wire, or in one circuit, when the power is derived exclusively from the main battery.[7]

We were the first to <u>practically</u> work instruments in this way by the main circuit on any number of wires, and believe ourselves to be the first and only inventors who have accomplished it upon <u>one</u> wire.

The value of this improvement over previous methods, is we think sufficiently obvious. Yours truly,

F. L. Pope for Pope Edison & Co.

ALS, Pat. App. 102,320. Letterhead of Frank Pope.

1. Montgomery M. Livingston (n.d.), a New York attorney specializing in patents, served as solicitor for the two Pope and Edison joint patents and for Edison's U.S. Patent 103,035.

2. See Doc. 91, n. 4.

3. The application executed on 16 September 1869 by Pope and Edison had been filed at the U.S. Patent Office on 27 October. The Patent Office failed to act on it, however, and on 3 February 1870 Livingston wrote to the commissioner of patents to inform him that Pope and Edison were "quite anxious concerning it, an early decision will therefore much oblige." In reply, on 17 February the examiner in charge of the case declared that Calahan's patent—owned by the Gold and Stock Telegraph Co.—and "the Dial Telegraph of Siemens & Halskie" anticipated the application. Consequently, several of Pope and Edison's claims were withdrawn, including all references to multiple-wire machines. The interference with Calahan resulted from the exaggerated breadth of Pope and Edison's original claims. Pat. App. 102,320; for the Siemens and Halske instrument, see Sabine 1872, 156–63.

4. Actually 21 April.

5. George Phelps's patent of 22 June 1869 (U.S. Pat. 99,662) was his fourth relating to the combination printer.

6. As finally amended, the application included several claims for new mechanical arrangements. Here Pope is amending the first claim of the original application and thereby removing the interference with Calahan.

7. This reliance on the main battery eliminated the need for local batteries and for contending with acid spills that could sometimes be damaging and dangerous. Because local batteries had already proved to be a nuisance to subscribers, Gold and Stock had replaced them with a single, large battery in a centrally located building. Hotchkiss 1969, 435.

–94–

To Joel Hills and William Plummer

NEW YORK, March 23rd 1870[a]

Gentlemen

This will introduce Mr. J. N. Ashley ~~who is~~ who proposes to visit Boston on business, and will call upon you.

I have authorized him to arrange with you in regard to the Two Hundred and Fifty Dollars which you were to pay me when the patent for the printer assigned to you was issued.[1] Any arrangement that he may agreed to and any release or other legal document that he may sign on my behalf will be recognised by me as good and valid as if signed or executed by myself.[2] Yours truly

Thos. A. Edison

LS, MiDbEI, EP&RI. Letterhead of the *Telegrapher.* [a]"NEW YORK," and "18" preprinted.

1. See Docs. 51, 53, and 82.
2. The Gold and Stock Telegraph Co. had been talking with Pope, Edison & Co. about patent infringement since February (see Doc. 93, n. 3). On 15 March, Gold and Stock's directors had requested newly elected president Marshall Lefferts "to take prompt action in relation to" a recent communication. On 12 April, Gold and Stock resolved to "purchase the lines, instruments and property of Pope, Edison & Co." Repurchase of this patent claim from Hills and Plummer was part of the settlement. See Docs. 95 and 97; and G&S Minutes 1867–70, 101–2, 115, 118.

–95–

James Ashley Draft Agreement with Joel Hills and William Plummer

[New York, March 1870?][1]

Agreement[a]

Whereas, Letters Patent of the United States for certain new and useful Improvements in Printing Telegraph Instruments, were, on the twenty second day of June, one thousand eight hundred and sixty nine, issued in due form of law to ~~Joseph~~ Joel[b] H. Hills and William E. Plummer as sole assignees of Thomas A Edison the inventor of said improvements; which said Letters Patent are numbered 91527.

And Whereas, the said Joseph H. Hills and William E. Plummer, by an instrument in writing under their hands and seals and duly executed and delivered, and bearing even date herewith did assign sell and set over unto James N. Ashley of the city of Brooklyn, in the state of New York, and unto his executors administrators and assigns the entire right title and interest in and to the said invention and the Letters Patent therefor, for and during the entire residue of the term for which the said Letters Patent are granted.

Now therefore this Agreement made and entered into by and between the said ~~Joseph~~oel[c] H. Hills and William E. Plummer parties of the first part, and the said James N. Ashley party of the second part

Witnesseth:— That for and in consideration of the above referred to assignment the said party of the second part covenants and agrees for himself and for his executors, administrators and assigns that he will promptly pay to them the said parties of the first part the sum of Twelve Hundred Dollars in quarterly installments in the following manner, to wit: The said party of the second part will pay unto the said parties of the first part the sum of Three Hundred Dollars on the First day of May one thousand eight hundred and seventy, and the sum of Three Hundred Dollars on the first day of August of the same year; and the sum of Three Hundred Dollars on the First day of November in the same year; and the sum of Three Hundred Dollars on the First day of February, one thousand eight hundred and seventy one; making a total of Twelve Hundred Dollars as aforesaid.

And the said party of the second part further covenants and agrees for himself and for ʰis executors, administrators and assigns that he will satisfy all just and legal claims which the said Thomas A Edison may have against them the parties of the first part for or on account of the purchase of the said invention by them the said parties of the first part from him the said Edison.

And the said party of the second part further agrees that in default of the payment of the said sum of Twelve Hundred Dollars the right and interest in and to the said invention and the Letters Patent therefor shall revert to them the said parties of the first part.

In Witness Whereof we the parties hereto hereunto set our hands and affix our seals this   day of   one thousand eight hundred and seventy.ᵈ

Df (copy), MiDbEI, EP&RI. ᵃ"Copy" written in left margin above. ᵇInterlined above in a different hand. ᶜ"oel" interlined above in a different hand. ᵈWax seals affixed for signatures.

1. This document may have been prepared for negotiations before Ashley's first visit to Boston (see Doc. 94).

GOLD PRINTER   Doc. 96

The gold printer,[1] so called because the Gold and Stock Telegraph Company used it to report gold sales, was the second joint invention of Edison and Franklin Pope. Although at the time of application they made no assignment of rights, within a week they sold partial rights to Gold and Stock, which employed it on gold-reporting circuits at least through the 1880s.[2]

The gold printer reflected Edison and Pope's experience with printing telegraphs. They modified elements from both

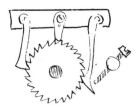

An 1872 notebook containing Edison's drawings of escapement mechanisms includes this design used in the Edison-Pope gold printer.

the Boston instrument and the Financial and Commercial instrument, and they also designed a unison stop to synchronize all the printers on a circuit. Edison had worked with a unison stop when he redesigned Samuel Laws's printer in 1869. That one, however, was mechanical, whereas the new unison was electrical. Like the Boston instrument, the gold printer required only one line and incorporated a polar relay as a switch. Unlike the Boston instrument (whose polar relay cut out either the typewheel or the printing electromagnet), the gold printer used its polar relay to cut out either the typewheel and printing mechanisms together or the unison stop. As with the Financial and Commercial instrument, the length of the transmitted signal determined whether the typewheel advanced or the instrument printed. This arrangement had led to problems because the available breakwheels could not be depended upon to generate impulses sufficiently short to avoid activating the printing mechanism.[3] Edison and Pope hoped to prevent that difficulty in the gold printer by shunting the current around the printing electromagnet during the first part of an impulse, thereby rendering the printing mechanism insensitive to all but deliberately lengthened signals.

The production instrument shown in Doc. 96 has been modified somewhat from the design specified in Edison and Pope's U.S. Patent 103,924. It has four binding posts instead of two, and there is no polar relay mounted on the instrument's base. These alterations suggest a variety of possible circuits. The polar relay could have been placed underneath the base,[4] designed as a separate piece of apparatus,[5] or simply eliminated because of its fallibility.[6] In the last case the machine would have required two circuits. Alternatively, although the original design did not call for a local battery, one may have been used with this particular machine. This artifact also exhibits two minor mechanical deviations from the patent specification: the retracting spring for the printing lever has been moved away from the lever's fulcrum, and there is a crudely cut slot in the base through which the paper strip for printing emerged.

1. An annotated duplicate of the album in which the photographs in Doc. 96 are found identifies this instrument as such. Cat. 66-42, Box 44, WU Coll.

2. Reid 1879, 614, 621.

3. See Doc. 76 headnote, n. 5, for a discussion of breakwheels.

4. Edison placed a polar relay under the base of the Chicago instrument, which he developed at the end of 1870.

5. Pope's U.S. Patent 103,077 combined a Morse and a polar relay and was used by Pope, Edison & Co. later in 1870 with the instrument of the American Printing Telegraph Co.

6. See Doc. 54 headnote. James Reid's 1879 description of the gold printer omits any mention of the unison and details a circuit similar to that of the Boston instrument. He notes that the polarized relay was eliminated by 1879, the gold printer then employing two wires. Reid 1879, 620–21.

–96–

*And Franklin Pope
Production Model:
Printing Telegraphy*[1]

[New York?,[2] April 12, 1870?[3]]

M (historic photograph) (est. 12 cm dia. × 18 cm), NjWOE, Cat. 551:16, 31. See Doc. 54 textnote. On design alterations in this instrument, see headnote preceding this document.

1. See headnote above.

2. No manufacturing information has been found.

3. Edison and Pope executed the covering patent application on this date.

-97-

And Franklin Pope and James Ashley Agreement with Gold and Stock Telegraph Co.

New York, April 30, 1870[a]

This Memorandum of an agreement made and entered into this 18th day of April 1870,[1] by and between "The Gold and Stock Telegraph Company of the City of New York" party hereto of the first part, and Frank L. Pope, James[b] N. Ashley, and Thomas A. Edison, parties hereto of the second part, jointly and severally witnesseth

That, for and in consideration of the sum of one dollar to them in hand paid by said party of the first part (the receipt whereof is hereby acknowledged) and of the covenants and agreements of the party of the first part hereinafter expressed, the said Francis L. Pope, James[c] N. Ashley, and Thomas A. Edison, parties hereto of the second part, jointly agree and each of them separately for himself agrees, as follows:

First: to sell, assign, transfer and convey (and they hereby do sell, assign, transfer and convey) to "The Gold and Stock Telegraph Company of the City of NewYork," each, every, and (except as is hereinafter excepted and reserved) all the right, title and interest, in, to, and under, each any and all of the inventions and improvements by said parties of the second part, either or any of them heretofore made or hereafter to be made, and of the patents therefor heretofore issued to and secured by, or hereafter to be issued, or re-issued to and secured by, said parties of the second part, either or any of them, for printing telegraph instruments, apparatus, and methods— and also each, every, and (with the same exceptions and reservations as are above referred to) all their right, title, and interest, and the right, title and interest of each and any of them, as the assignee or assignees of any patent, or license under any patent, to any other person or persons issued, or to be issued, or reissued, for any printing telegraph instrument apparatus or method.

Second: to sell, assign, and transfer, (and they hereby do sell assign and transfer) to said party of the first part, each, every, and all the right, title, and interest, in, and to, the lines of telegraph with the wires, poles, insulators, instruments, batteries and fixtures, and every thing to said lines of telegraph pertaining, now in the possession and under the control of said parties of the second part—either, or any of them (covenanting hereby that they, or either or any of them who may have the same in possession or under control, has, and have a good and absolute right and title to the same.) and to surrender and turn over (and they hereby do surrender and turn over) the same in working order, and free and clear of all incumbrances and liens whatsoever, to said party of the first part with all subscribers to quotations of gold, stocks, and commercial news by said lines of telegraph, and all subscriptions due, and to become due from said subscribers on or after the 18th day of April 1870.[2]

Third: To make, execute, and deliver to said party of the first part, their successors or assigns, on demand, any and all proper instruments in writing which may be required to effectuate or express the sales, assignments, and transfers, either or any of them, hereby made or agreed to be made.

Fourth: That they will not for a period of ten years from the date of execution of these presents, engage, either directly or indirectly, as individuals, or by association, in any telegraphing which is in effect in rivalry or opposition to the business of the party of the first part, and that they will not infringe any patent rights of party of the first part, but will by all means in their power, discourage others from such rivalry, opposition and infringements, and will assist the party of the first part in maintaining its rights and business against any and all persons who may attempt such rivalry, opposition or infringement.

Fifth: that said parties of the second part hereby covenant that they, or some or one of them, has, or have a good and absolute right and title in and to, each and all of the property, and rights of property, hereby assigned and transferred to the party of the first part;— and further, that they (and any particular one or two of them as the case may be, and as may be required) will make and complete any and every assignment and transfer hereby agreed to be hereafter made to the party of the first part, their successors or assigns, free and clear of any and every incumbrance, claim and lien whatsoever.

For and in consideration of said covenants, sales, assignments and transfers, and of the full and faithful performance of the agreements above expressed, said party of the first part agrees:

First: to pay said Francis L. Pope, J   N. Ashley, and Thomas A. Edison on order, the sum of fifteen thousand dollars, in payments as follows: to wit: Five thousand dollars at and upon the execution and delivery of these presents: Five thousand dollars as soon as said Pope and Edison shall have obtained and assigned to said party of the first part letters patent of the United States (four new-patents and one re-issue)[3] covering claims substantially as per memorandum this day furnished by said Pope to party of the first part (hereto annexed) and the further sum of Five thousand dollars in three equal payments to be made on the first days of September, October, and November 1870, respectively.

Second: to perform the agreements heretofore made by the parties of the second part with their present subscribers to furnish quotations of gold, for one year from the date of each of such agreements, at the rate of two dollars per week. It is however mutually understood and agreed that the parties of the second part have and reserve to themselves the right to manufacture and use, and to sell to others to be used, each, any, and all of the apparatus, instruments and methods for which patents are hereby assigned, or agreed to be assigned, for the specific purpose of transmitting over what are known as "private" lines of telegraph, messages other than commercial quotations, and quotations of the prices of gold and stocks[4]—but that said parties of the second part will not sell to be used, or allow to be used, any of[d] such instruments or apparatus, except upon receipt of an agreement in writing, to be made, executed and delivered, by each and every of the persons who may purchase the same for any such "private" line of telegraph, that they will not use or permit others to use the same for any other purpose than that for which said particular "private" line, for which the purchase is made, was originally intended. And that there may be no misapprehension as to what is herein meant by a "private" line of telegraph, it is mutually understood and agreed that the following is an example of one.

A.B. being either an individual, firm or corporation doing business, has a line of telegraph connecting AB's offices[d] shops, ~~offices~~, or factories, or any of them, where ABs proper or[e] usual business is carried on, such line is a "private" line, provided that the messages transmitted over it are for the use of, and used by A.B. only, and not by the public generally, or persons other than A.B., his or their agents and employees in said business.

It is further mutually agreed that should any of the instruments, apparatus or methods, for which patents or licenses have been obtained, or are to be obtained, and which are hereby assigned or agreed to be assigned to said party of the first part, prove available for use upon long lines of telegraph between distant cities, the same may be so used by either of the parties hereto, but only after thirty days notice in writing first to be given by the party so desiring to use the same, to the other party, accompanied in each particular case by a statement of the nature and extent of the proposed enterprise and of the estimated expense thereof, and also by an offer of [-]option to such other party to come in and contribute one half of the actual cost or expense of such enterprise as a joint undertaking, and that, in case such contribution is, within said[d] thirty days, made, or satisfactorily secured, the party so contributing shall be entitled to one half the profits of the enterprise—the object of the parties hereto being to make if possible, a joint enterprise of each and every undertaking attempted under the terms of this clause of their agreement, and to conduct the same in friendship and harmony as partners.

It is further mutually agreed that in case any disagreement or dispute shall arise between the parties hereto, or their assigns as to the true intent and meaning of this agreement, or of any clause or provision thereof, such disagreement or dispute shall be referred to the arbitration of three arbiters, of whom one shall be the president for the time being of the Chamber of Commerce of the City of New York City, and one shall be chosen by each of the parties hereto—and that the decision of such arbiters, or any two of them shall be conclusive.[5]

In witness whereof the parties hereto have sealed and delivered these presents this 30th day of April 1870.
Gold & Stock Telegraph Co by M. Lefferts[6] Presdt[f]
Frank L Pope[f]   James N. Ashley[f]
                                                         Thomas. A. Edison[f]

Witness   T F Goodrich[7g]

(1) F. L. Pope and T. A. Edison   Imp't in
Printing Telegraphs[8]
We do not claim distinctively the placing of an electro magnet for operating the type wheel of a printing instrument in the same circuit as the electro magnet that actuates the printing hammer, as this is not new, neither do we claim the use of a unison stop in general. We claim

Principle of Closed ckt instrument[h]

  1. In a printing telegraph instrument, the arrangment of

two electro-magnets in the same electrical circuits, ~~an~~one being employed to rotate the type wheel, and the other to actuate the printing mechanism, when the action of the latter is controlled by that of the former, by means of a branch circuit and mechanical shunt or cut off, or its equivalent constructed and operated substantially as described.

Unison Cut off[b]

2. The electrical unison cut off, whereby at a given point in the revolution of a rachet or type wheel a shunt or branch circuit may be brought into action and the electrical current diverted from the electro magnet operating the said rachet or type wheel so that the movement of the latter will be arrested at said given point, the same being constructed and operated substantially as specified.[i]

Polarized switch[b]

3. The Electro magnet R'.R',[9] soft iron bar T polarized steel bar N.S. and stops zz', combined arranged, and operated as set forth.

Paper feed[b]

4. The standard M. screws o.o', pawl arm N, Spurs q, and bed plate O, or their equivalents combined and arranged substantially as described, and for the purposes specified (paper feed)

Escapement[b]

5[d] The combination of the lever E, pawls F and F', stops J and K and toothed wheel G or their equivalents, arranged and operated substantially as specified.

Combination of polarized switch (3) with unison stop (2)[b]

6. The electro magnet RR', soft iron bar T. polarized steel bar N.S, or their equivalents, in combination with the spring X, insulating collar V and pin or stud w, or their equivalents, arranged and operated in the manner set forth, and for the purpose specified.

7 The three duplicate characters arranged upon the type wheel in the manner described and for the purpose specified.

(2) F. L. Pope and Thos. A. Edison   Improvement in printing Telegraphs[10]

Working instruments in main circuit on one wire.[b]

1. The combination of a number of automatic printing telegraph instruments, arranged in one main circuit, and operating simultaneously in unison when the electro motive power used in operating the same, is derived exclusively from one or more main batteries placed in such main circuits without the aid of secondary or local batteries or of mechanism, actuated by springs, or weights, or otherwise, sSubstantially in the manner and for the purpose set forth.

2. The combination of a polarized <u>magnet</u> with an electro magnet placed in the same electrical circuit and operated substantially as described and for the purpose set forth.

Escapement[b]

3. The combination of the rachet wheel I, bar F', pawls hh', stops i.i' and type wheel T, ~~bar~~ arranged and operating substantially as and for the purpose herein specified

4. The combination of an electro magnet with the rachet wheel, bar pawls, stops, and polarized magnet substantially as and for the purpose specified.

5. The arrangement of the permanent magnet N.S. polarized magnet E, electro magnet M, tongue C, arm D, bar E, pawls hh', stops ii' springs jj', rachet wheel H, type wheel T, and standards A.K. all constructed arranged and operated substantially as and for the purpose herein specified.

Paper feed[b]

6 The roller t   serrated wheel g, paul s, rachet wheel r, click x, and standard p, in combination with the polarized magnets E and the electro magnet M, and their appurtenances for the purposes set forth.

7. the screw stops dd' upon the standard A, in combination with the type wheel T substantially as herein specified

8 The arrangement of the tongue G with slot O, of the permanent magnet N.S, by means of a pivot a and screw o for the purpose specified.

3)T. A. Edison   Improvement in Telegraphic Apparatus[11]

1 the switch <u>n's</u> or its equivalent in combination with two electro magnets $\overline{E}$ and M constructed arranged and operated substantially as described and for the purpose specified.

2 The keys P and N, constructed as described, in combination with the switch N.S. or equivalent and the Electro-magnets[j] E and M for the purpose specified[k]

4)— T. A. Edison   Electro
motor escapement[12]

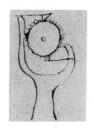

Edisons patent sold to Hall and   of Boston Mass, purchased by Pope & Co, is dated June 22/69, and is referred to
in Edisons patent for polarized switch No 968681—assigned
to SLaws—dated 9 Nov. 1869[13]

Also patent of [-]T. A. Edison June 22, 1869 and assigned
to [-]Joel E. Hills and William E. Plummer of Boston, and
purchased by Pope Edison & Co. to be reissued[l]

The above is a description of the patents referred to in the
above agreement.

Frank L. Pope.[m]

ADDENDA[n]

New York 7. May 1870[o]
Recd from the Gold & Stock Telegraph Co. Five thousand
dollars, being the first mentioned sum in the forgoing agreement[14]

$5000—   Pope Edison & Co.

New York 25. May 1870[o]
Received from the Gold & Stock Tel Co. Two thousand
dollars ~~be~~ on within agreement—

$2000.   Pope Edison & Co

New York 1t July 1870—[o]
Received from the Gold & Stock Tel Co.—Three thousand dollars—on within Contract—[15]

$3000.00   Pope Edison & Co.

[Washington, D.C.,] July 2 1870[p]
U.S. Patent Office   Recd for record July 2 1870 and recorded in Liber G 13 page 113 of Transfers of Patents. In
testimony whereof I have caused the Seal of the Patent Office
to be hereunto affixed. ⟨Exd. C. E. L.⟩[q]

Saml. J. Fisher.   Comr. of Patents

New York 8. November 1870.[o]
Received from the Gold & Stock Tel. Co.—Sixteen hundred & fifty dollars—being the last payments—(others not

endorsed) and in full for all claims and demand under this agreement—

Pope Edison & Co.

D (photographic transcript), MiDbEI, EP&RI. Each sheet of agreement has a canceled 5¢ Internal Revenue stamp affixed in the upper left corner. [a]Place taken from notarization; date taken from text, form altered. [b]Written in left margin. [c]Extra space on either side of name. [d]Interlined above. [e]"proper or" interlined above. [f]Seal affixed next to signature. [g]Followed by notarization of signatures of both parties. [h]Written in left margin; brace spans following paragraph. [i]Followed by "(over)" to indicate page turn. [j]"and the Electro-" interlined above. [k]Followed by line drawn across page. [l]Followed by centered horizontal line. [m]Signature; preceding sentence written in Pope's hand. [n]First three addenda have canceled 2¢ Internal Revenue stamps affixed in left margin. [o]This addendum is a D; apparently written by Lefferts, company name written by Pope. [p]This addendum is a DS; written in an unknown hand; date taken from text. [q]Marginalia written in an unknown hand.

1. Although Pope, Ashley, Edison, and Lefferts reached an oral agreement on 18 April, they did not sign this memorandum of agreement until 30 April. The parties had been negotiating since at least mid-March. G&S Minutes 1867–70, 101–2, 115, 118.

2. This paragraph refers to the property and business of the Financial and Commercial Telegraph Co.

3. See n. 13.

4. These reserved rights were the basis for the 1 July formation of the American Printing Telegraph Co. See Doc. 130.

5. It was common practice to refer commercial disputes to the New York Chamber of Commerce for arbitration. The Chamber had established arbitration procedures as early as 1768; from 1861 to 1873, by legislative order, decisions of the Chamber's Committee on Arbitration could be entered as judgments in courts of record. Bishop 1918, 120–27, 262.

6. Marshall Lefferts (1821–1876), who introduced the Bain automatic telegraph system to the United States in 1849, was a leading telegraph engineer and entrepreneur. From 1861 to 1866 he was general manager and engineer of the American Telegraph Co., on whose lines he tested John Humaston's automatic system for possible commercial use. Franklin Pope served as his assistant between 1862 and 1864. When this company merged with Western Union in July 1866, Lefferts assumed the posts of engineer and superintendent of Western Union's Commercial News Department. He became president of Gold and Stock in March 1870 but did not resign from his position as a Western Union engineer until the following January. Gold and Stock grew rapidly under Lefferts, and he remained president when Western Union took over the company in 1871. *DAB*, s.v. "Lefferts, Marshall"; *NCAB*, 10:243; "Obituary. Marshall Lefferts," *Telegr.* 12 (1876): 168; "Resignation of Gen. Lefferts," *J. Teleg.* 4 (1871): 47; Reid 1879, 421–22, 426, 566, 608, 653.

Lefferts was widely known and respected among inventors and investors interested in telegraphy. He became president of the American Printing Telegraph Co. (see Doc. 130), adopted Edison as a protégé,

Marshall Lefferts, president of the Gold and Stock Telegraph Co.

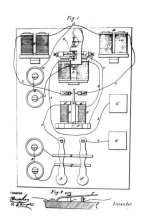

Patent drawing of Franklin Pope's switch for using local batteries with a polar relay (U.S. Pat. 103,077).

and backed him in several ventures over the next few years, including Edison's work on the electric pen. Henry van Hoevenbergh to Ralph Pope, 24 July 1908, GF; "Edison's Electric Pen" folder, Telegraphy Series, Lefferts.

7. Thomas Goodrich was a notary public and insurance agent with an office at 135 Broadway. Wilson 1870, 423.

8. This section of the Pope memorandum reproduces the claims in U.S. Patent 103,924, the design for the second Pope-Edison printer (Doc. 96). The letters identify elements in the patent drawings.

9. "RR′" in printed patent.

10. This section summarizes the claims in U.S. Patent 102,320, the design for the first Pope-Edison printer (Doc. 76). In claim 5, bar E should read F according to the patent specification; slot O in claim 8 should be slot b. Claim 8 is abbreviated here.

11. This is actually Pope's U.S. Patent 103,077. Pope's relay switch allowed the incorporation of a local battery into the Boston instrument circuitry, which made the operation of the individual printers more dependable by having the main circuit transmit only signals to trigger the mechanism and not the power to move it. Later in 1870 Pope and Edison used this design in the American Printing instrument. Pope executed the application on 5 April and filed it on 11 April, the same day on which Edison filed his escapement application (see n. 12).

12. Edison's U.S. Patent 103,035. The novelty of this escapement lay in the fact that the stops were an integral part of the lever arm, which obviated problems of adjustment that had plagued earlier devices. The patent was granted on 17 May 1870. Edison executed the application on 5 February, assigned it to Ashley as trustee on 15 March, filed the application with the Patent Office on 11 April, and registered the assignment on 24 June. Pat. App. 103,035; Digest Pat. E-2:143, 144.

(Left): *Patent drawing of Edison's electro-motor escapement used in printing telegraphs (U.S. Pat. 103,035).* (Right): *The model submitted with Edison's application for a patent.*

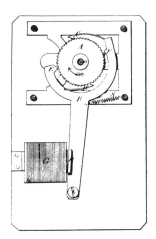

13. This paragraph and the next one refer to the Boston instrument (Doc. 54). Pope, Ashley, and Edison did not in fact repurchase the patent until 11 May (see Doc. 100, n. 1). This was the patent referred to in the payment schedule as the "one re-issue." Reissue 4,166 was executed on 6 September and granted on 25 October 1870.

14. Edison later claimed that Ashley had tried to keep all the money from him and that Lefferts had personally given him his share (App.

1.A31). The date of this payment coincides with the date of the first entry in the accounts for the Newark Telegraph Works. 70-005, DF (*TAEM* 12:144).

15. The agreement called for a second payment of $5,000 upon assignment of all patents. U.S. Patents 103,035, 103,077, and 91,527 were assigned to Gold and Stock on 1 July. Digest Pat. E-2:144.

–98–

Financial Statement from Mrs. Ebenezer Pope and Franklin Pope

[Elizabeth, N.J.,] May 7 1870.

T. A. Edison   To Mrs E L Pope[1]   D[ebto]r.

To. 32 weeks board, Sept 12 1869 to Apr 23 1870,
   at $5 per week. $160
Deduct $10 paid———   10
                                           $150.

Also to F. L. Pope for sundry small sums of money as follows:[2]

October	15	.50	December	4	5.00
"	23	2.00		10	.50
"	28	1.00		11	50
"	30	.50		15	1.00
November	5	1.00		20	1 00
"	9	.25		22	.50
"	12	2.00		28	2.00
"	15	50		30	.50
"	18	50			
"	20	25			$ 20.50
"	21	1.00			

                                                 $170.50

D, NjWOE, DF (*TAEM* 12:40). Apparently written by Franklin Pope.

1. Mrs. Ebenezer L. (Electra Wainwright) Pope was Franklin Pope's mother. Other members of the household in 1869–70 included Franklin's younger brothers, Henry and Ralph, both of whom were telegraph operators. Boyd and Boyd 1868, 207; *Elizabeth Directory* 1869, 164; ibid. 1870, 166–67.

2. These small loans from Pope to Edison began shortly after the establishment of their partnership in October 1869. Pope rendered this bill on the date of the first payment by the Gold and Stock Telegraph Co. under the agreement of 30 April (Doc. 97).

–99–

To Samuel and Nancy Edison

NEW YORK, May 9 1870[a]

Dear Father and Mother =
   I sent you another express package Saturday—enclosed you will find the receipts for same =

J C Edison[1] writes me that mother is'not very well and that you have to work very hard. I guess you had better take it easy after this— Dont do any hard work and get mother anything She desires = You can draw on me for money—write me and Say how much money you will need in June and I will Send the amount on the first of that month =[2] give Love to all the Folks—and write me the town news—What is Pitt[3] doing— Did you sell J. C. a Lot = is Truey[4] still with you & How is she— Your Affec Son

Thos A

ALS, MiDbEI, EP&RI. Letterhead of the *Telegrapher.* [a]"NEW YORK," and "18" preprinted.

1. Probably Jacob Edison, Thomas's cousin. Jacob Edison to TAE, 13 Apr. 1874, DF (*TAEM* 13:20).

2. Edison wrote this letter two days after the first payment to Pope, Edison & Co. under the agreement of 30 April (Doc. 97). The next payment from Gold and Stock was due at the end of May.

3. William Pitt Edison.

4. Truey was the Edisons' maidservant. Memorandum of Nellie Poyer, n.d., EBC.

–100–

Release from Agreement for Joel Hills and William Plummer

New York May 9th 1870

I hereby release Messrs. Joel H. Hills and Wm. E. Plummer from any claim that I may have upon them arising under my agreement with them for the assignment[a] ~~transfer~~ of the patent for a Printing Telegraph Instrument issued to them as my assignees.[1]

Thos. A. Edison

DS, MiDbEI. [a]Interlined above.

1. James Ashley may have taken this note to Boston to conclude negotiations with Hills and Plummer. Ashley came to terms with Hills (to whom Plummer had signed over his interest the week before) on 11 May. At the time this note was written, neither Edison nor Ashley appears to have been aware that Plummer was no longer involved. Libers Pat. E-2:141.

–101–

Patent Application: Automatic Telegraphy

[New York?,][1] June 22, 1870[a]

1[b] To all whom it may concern.

Be it known that I, Thomas A. Edison, of Newark in the County of Essex and State of New Jersey, have invented and made an Improved Telegraphic Transmitting Instrument, and the following is declared to be a correct description of the said Invention.

2   In telegraphing, a perforated strip of paper has been employed to make and break the electrical circuit in transmitting the message.[2]

3   In transmitting instruments adapted to said paper there is a small disk or wire brush that closes the metallic circuit through the perforations, and the circuit is broken by the paper when the unperforated portion intervenes between the roller or plate and the disk or wire brush.

4   The transmission of pulsations of electricity being very rapid in this system of telegraphing, there is a difficulty that sometimes arises from the wire not clearing itself, and the pulsations are attenuated and do not distinctly reach the distant station.[3]

5   My invention consists in arranging the connections and portions of the instrument in such a manner that a reverse current shall be thrown upon the wire of the circuit by a motion derived from the thickness of the paper when the same is drawn in between the plate or roller and the brush or disk.

6   In the drawing the device in question is represented by a side view.

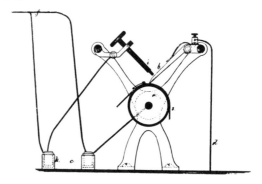

Thomas A. Edison[c]

Witnesses,   Chas. H. Smith[4]   Geo. D. Walker[5]

7   Let a. represent a plate, roller or metallic surface over which the strip of perforated paper s, is drawn, and b. represents a wire brush, stilus, or roller, these parts being of any known character for sending pulsations of electricity to a distant receiving instrument.

8   The battery is represented at c. and the ground wire at d. and the line wire at f. The current will therefore be sent

when the circuit is closed through the perforation of the paper, and when the unperforated portion of the paper is beneath the brush or stilus <u>b</u>. the end is lifted sufficiently to touch or nearly so the point <u>i</u>. that is adjustable and mounted in any convenient manner.

9   By the said movement the battery k is brought into action by closing the circuit between <u>i</u> and <u>b</u>. and a reverse current is thrown upon the telegraph line, thereby preventing the attenuation of the previous pulsation, clearing the wire and causing the mark at the receiving station to be clear and distinct.

I claim as my invention a circuit closer operated by the movement of the perforated paper in a telegraph transmitting instrument to throw a reverse circuit on the line, substantially as set forth   Signed by me this 22nd day of June AD 1870[6]

Thomas A. Edison

Witnesses   Chas H. Smith   Geo. T. Pinckney[7]

DS and PD, MDSUFR, RG-241, Pat. App. 114,656. Petition and oath omitted. [a]Date taken from text, form altered. [b]Section numbers written in left margin. [c]Drawing and accompanying signatures from printed patent.

1. This application was prepared at Lemuel Serrell's office, 119–21 Nassau St., New York; however, it may have been based on an original drafted by Edison in Newark.

2. On automatic telegraphy, see Doc. 33.

3. If the electrical potential of a telegraph line did not drop sufficiently between signals, the signals elongated and caused the recorded marks to run together. This effect was known as tailing. It was most noticeable either on long lines, which required more time to discharge, or when sending signals very rapidly. Because one of the principal advantages of automatic telegraphs was the greater speed of transmission, it was necessary to develop a method to discharge the line more rapidly (Prescott 1877, 689; Maver 1892, 288). Marshall Lefferts had encouraged the development of George Little's system of automatic telegraphy and may have mentioned to Edison the importance of solving the tailing problem. At about this time both Lefferts and Little applied for patents on devices to prevent tailing (U.S. Pats. 114,692 and 108,495). Their schemes also used reverse currents to neutralize the charge of the preceding signal.

4. Charles Smith witnessed and notarized many documents in Serrell's office.

5. George Walker witnessed many documents in Serrell's office.

6. The U.S. Patent Office approved Edison's application on 11 July 1870, but he did not pay the final fee within the prescribed six months and therefore had to reapply for the patent on 16 January 1871. He was issued U.S. Patent 114,656 on 9 May 1871. Pat. App. 114,656.

7. George Pinckney witnessed many documents in Serrell's office.

–6– Expansion and Diversification

July–December 1870

During the second half of 1870, Edison strengthened his relationship with Marshall Lefferts and the Gold and Stock Telegraph Company, an association that he maintained for some years, and intensified his efforts to design, develop, and manufacture instruments for automatic telegraphy. Having executed his first automatic telegraph patent in June, he soon contracted with the promoters of George Little's automatic telegraph to make improvements in Little's system. He also joined in partnership with one of these promoters to establish his second manufacturing company in Newark.

After their sale of partial rights to their printing telegraph patents to Gold and Stock in April, Edison, Pope, and Ashley continued their association solely on the basis of their reserved rights to engage in private-line telegraphy. With the capital gained from the sale, the partners fashioned a small printing machine and, with William Allen and Marshall Lefferts's son, established the American Printing Telegraph Company, "the first company for the erection of Private Telegraph Lines for individuals and business houses."[1] While a few large urban businesses had built their own private lines in the late 1860s, all early lines used either dial telegraphs (such as Edison's 1869 magnetograph) or Morse instruments—each requiring an attendant at the receiving end.[2] The success of private lines awaited a reliable printer—a low-maintenance and simple-to-operate machine that received and recorded messages unattended. The American Printing instrument (also known as the Pope and Edison printer) met that need and found a ready market. On 30 November, the American Printing Telegraph Company purchased the patent rights reserved by Edison, Pope, and Ashley on 30 April. The next day, having parceled out its assets, Pope, Edison & Company ceased to exist. Moreover, Edison no longer worked directly with the American Printing Telegraph Company after developing the instrument.

The nature and intensity of his work in printing telegraphy drew him closer to Gold and Stock and its individual officers. Between July and the end of the year, Edison filed sixteen caveats relating to telegraphy.[3] In at least one instance he agreed to assign to Gold and Stock officers any future patent derived from a specified caveat. He began developing his universal private-line printer, which looked sufficiently promising by mid-October to elicit an offer from Gold and Stock, and he improved his stock-ticker design enough to have Gold and Stock order 150 instruments in December.[4] He also executed five successful patent applications, one of which described a printing telegraph instrument.[5] Edison assigned two of the five applications in part to Marshall Lefferts and one in part to Lefferts, Elisha Andrews, and George Field.[6] Field and Andrews obtained two British patents in the same period: one identical to Edison's U.S. application of 24 May, which Edison held alone, and the other similar to an American patent Edison had assigned to Lefferts and himself.[7]

Induced by the entrepreneurial ambitions of Daniel Craig and George Harrington, Edison's inventive and commercial interests in automatic telegraphy burst into full bloom during the summer and fall of 1870. Daniel Craig, former head of the Associated Press and the leading American proponent of automatic telegraphy, had in 1869 obtained a financial interest in George Little's automatic telegraph system and had contracted with the National Telegraph Company to develop it for press transmission.[8] National's reluctance to fund technical development led Craig to approach Edison independently in August with a request to devise a faster, simpler perforator.[9] However, because the perforator was only one of several parts of Little's system that required improvement and because such technical efforts required additional financial resources, Craig sought new investors. He secured the support of George Harrington, former assistant secretary of the treasury. Harrington and some of his associates formed the Automatic Telegraph Company in late November for the purpose of developing automatic telegraphy and making it a commercial success. In return for a two-thirds interest in Edison's current and subsequent automatic telegraph inventions, Harrington independently began to provide substantial support for Edison to design his own automatic telegraph system as well as improve Little's. As an accompaniment to this effort, Harrington and Edison founded in October the American Telegraph Works. This enterprise primarily built machinery for the Little system, also made printing telegraphs for Gold and Stock, and undertook other outside jobs. Significantly, both the American Telegraph Works and the Edison and Unger shops provided Edison with experimental as well as manufacturing facilities.

1. Reid 1879, 621. Reid is incorrect in citing 1869 as the date of the company's formation.

2. "A Telegraph Line for Business Men," *J. Teleg.* 2 (1868–69): 154; "Private Telegraph," ibid., 194. A firm in Chicago also attempted to introduce dial telegraphs into private homes. "Too Much Electricity," ibid., 205.

3. Quad. 72.11, pp. 44–45 (*TAEM* 9:187).

4. Gold and Stock's order illustrates the need for caution when trying to reconstruct Edison's inventive activities from dates related to patenting. Edison did not execute a patent application covering the design—the cotton instrument—until 7 January 1871.

5. The Chicago instrument, U.S. Pat. 113,033.

6. Edison assigned U.S. Pats. 114,657 and 114,658 to himself and Lefferts; he assigned U.S. Pat. 111,112 to himself, Lefferts, Andrews, and Field.

7. Brit. Pat. 1,657 (1870) is identical to U.S. Pat. 128,608; Brit. Pat. 2,578 (1870) is identical to U.S. Pat. 114,657.

8. Under the terms of the agreement that Craig negotiated with National Telegraph, he became the company's general agent to secure stock subcriptions from editors and news agents; Lefferts became the company's superintendent, although he continued as chief engineer of Western Union; and Little became the company's electrician. The company's promoters had reincorporated it in 1866 in order to take advantage of a congressional act allowing state-chartered telegraph companies to construct lines along national post roads and across public lands. However, the company had failed to sell all its stock or erect any lines by 1869. Although adoption of the Little automatic telegraph system temporarily revitalized the company, the directors were unwilling to continue spending money to improve the system, and in 1871 they sold the rights to the Little patents. Agreement of 9 Sept. 1869, Respondents' Exhibit 4, 1:517–27, Box 735A, *Harrington v. A&P;* Lindley 1975, 47–59; New York City 371:782, RGD.

9. Aside from electrical and mechanical impediments to high transmission speeds, the ancillary activities of perforating tapes and translating messages from the receiving tapes were slow and labor-intensive. Because Craig advocated the use of children, particularly girls, as a means of reducing labor costs, it was necessary to devise simple and easily operated machinery. Craig apparently hired Edison because he was dissatisfied with the performance of the Little perforator in com-

parison with the Wheatstone perforator used by the British Post Office. He probably learned of Edison's inventive talents from Marshall Lefferts. "The 'Little' Automatic Telegraph System," *Telegr.* 7 (1870–71): 28; "National Telegraph Company," ibid. 3 (1869–70): 60–62; Craig to Lefferts, 1 Mar. 1871, Lefferts; Doc. 146.

–102–

Technical Drawing:
Dial Telegraphy

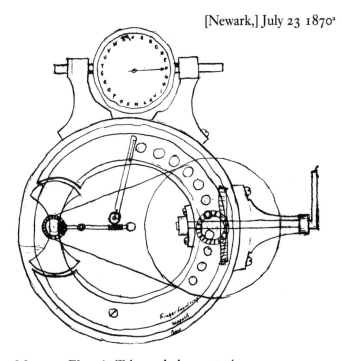

Magneto Electric Telegraph Apparatus[1]
Invented[2] July 23 1870
Witnesses   William Unger[3] Martin Heger[4] Andrew Hyde[5]

AX, NjWOE, Lab., Cat. 298:70 (*TAEM* 5:131). [a]Date taken from text.

1. Edison's sketch shows a transmitter for either a dial telegraph or a small printing telegraph. According to the letters in the accompanying diagram, rotating the geared crank **A** turned wheel assembly **B**. This spun the armature **C** between the poles of magnet **D**, which both generated pulses of electricity and mechanically turned shaft **E**. At the same time, the worm gear on shaft **E** turned arm **F**. The keys **G**, **G'**, **G''**, etc., when depressed, served to block the motion of arm **F**. When arm **F** was stopped by a key, armature

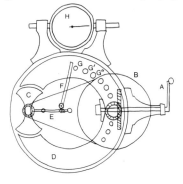

C was likewise stopped after a particular number of pulses had been sent over the line (not shown). The local indicator dial H could have worked either electromechanically, like the receiving devices, or through a direct mechanical connection. On dial telegraphs see Prescott 1877, 562–602.

The words on Edison's sketch are, top to bottom: "Fingerboard ring", "magnets", and "Base". The key at which arm F is stopped is labeled F in faint pencil; correspondingly, the hand on dial H points to F.

2. No caveat or patent application has been found for this device, nor did Edison receive a patent for it.

3. See Doc. 92, n. 1.

4. The accounts of Newark Telegraph Works list "Hager" and "Haeger," 16–26 May and 1–4 June 1870, 70-005, DF (*TAEM* 12:144, 145, 149).

5. Andrew Hyde worked at the Newark Telegraph Works from 18 June to 3 September 1870. From 26 September 1870 until 25 February 1871 he was superintendent of the American Telegraph Works.

–103–

*Agreement with
Daniel Craig*

[New York?,] August 3, 1870[a]

Memorandum of an Agreement between T. A. Edison, of Newark, andNew Jersey, and D. H. Craig,[1] Agent,[b] of Peek-skill, New York, entered into on the 3d of August, 1870.

Whereas Mr. Edison has invented or is about to invent and construct a Machine for the purpose of Perforating Paper, in connection with Automatic or "Fast" Telegraphy;

And whereas Mr. Craig, for himself and as agent for other parties,[2] is desirous of purchasing the control of and an interest in said Perforating machinery, and in the Patent or Patents which may be obtained for the same, together with all improvements upon said machinery:—

Now, therefore, in consideration of the sum of one dollar, the receipt of which the said T. [-]A. Edison hereby acknowledges, and other considerations to be named, he agrees to and with the said D H Craig, Agent, as follows, to wit:—

To devise and perfect a Perforating Apparatus, to be constructed of Steel and Iron, not to occupy a space [---][b] exceeding twelve inches square, to have an extremely small number of pieces, to bring its reliability to the highest point possible with mechanism, to be capable of a speed of twenty-five words or more, per minute, according to the expertness of the writer, to space, write & punch mathematically accurate, to weigh not more than twenty pounds, to require no extraneous power, motor or force, but to work by the mere act of working. The cost, when manufactured in numbers, not to exceed fifty dollars: The machine to be perfectly adapted for the work to be performed, and to leave nothing further to be desired in tha direction.[3]

Mr. Edison also agrees to procure a good substantial United States Patent, to cover every patentable feature of the said machine or invention, and will assign the same to the said D H Craig, Agent[b] together with all improvements which the said Edison may make in machinery for Perforating paper for Automatic or Fast Telegraphy, upon the folowing terms and conditions, to wit:

The said Craig to pay to said Edison within thirty days after the completion of a perfect machine, as above, and application for a Patent for the same, the sum of thirteen hundred dollars in cash, and thirty seven hundred dollars of the paid-up, or non-assessable stock of the National Telegraph Company, together with one-half of whatever sum or advantage which said Craig may receive from the National Telegraph Co. or from any other party or parties, over and above the sums above named—the said Craig hereby promising and agreeing to make the most advantageous arrangements possible for the sale of the invention and patents, free of all charges for commissions or services—and hereby promises & agrees to pay over to said Edison, forthwith, one half of whatever sum or sums may be received over and above the aSums herein before[b] provided to be paid to Mr. Edison—and said Craig also promises and agrees to make no sale or bargain for sale, of the said machine or patents without the full knowledge of said Edison—and said Craig also agrees that he will not make any sale of the machine or patents to any party without the consent of said Edison, except upon the distinct agreement or understanding that said Edison is in all cases to have the preference in all contracts or agreeements for the manufacture of the said machines.

It is mutually understood and agreed between the said Edison and the said Craig that for all acceptable improvements which the said Edison may devise for facilitating the Perforating of Paper, after the completion of the before mentioned machine, he is to be paid by said Craig a fair and reasonable sum for all the time, labor or expense which said Edison may expend in perfecting such improvements. Provided however, that in fixing the measure of such payments by said Craig, due regard is to be had to the fact that said ~~Craig~~Edison is to ~~be~~participate in the advantages of the improvements to the same extent as said Craig.

It is also mutually agreed between said Edison and said

Craig, that in the event of any differences in construing their obligations to each other under this agreement, the matter or matters in dispute shall be referred to the President of the Board of Trade of New York City and his decision shall be conclusive and be binding upon the said Edison and the said Craig.

T. A. Edison[c]   D H Craig[d]

DS, NjWOE, DF (*TAEM* 12:107). In Craig's hand. [a]Date taken from text, form altered. [b]Interlined above. [c]Seal accompanies each signature. [d]"stamps" written in left margin; canceled 5¢ Internal Revenue stamp affixed; scrap of paper affixed with circuit diagrams on one side, "T. A. Edison 15 Railroad Avenue Newark N.J." in Edison's hand on the other.

1. Daniel H. Craig (1814?–1895) began his career as a journalist in New York, Baltimore, Boston, and Halifax. Before the laying of the first successful transatlantic telegraph cable in 1866, Craig gained notoriety by sending news-laden carrier pigeons ashore from incoming European ships. He worked for various press telegraph services, serving as president of the Associated Press from 1861 to 1866. Craig claimed that Western Union forced him from his position with the Associated Press. Subsequently, he sought to wrest control of press telegraphy from that company through pursuit of automatic telegraphy and related telegraph enterprises. Reid 1879, 362–70; *DAB*, s.v. "Craig, Daniel H."

2. These other parties were either the National Telegraph Co. or George Harrington and his associates.

3. Meeting these specifications required major improvements in perforator design. The fastest perforation speed that had been achieved by 1870 was 10 words per minute, and all perforators patented in the United States up to this time (except Leverett Bradley's U.S. Patent 48,479) required an extraneous power source. Prescott 1870, 292.

Daniel Craig, the principal proponent of automatic telegraphy in the United States.

–104–

From Daniel Craig

New York.[a] Aug. 12/70

Confidential. Dear Mr. Edison:

As I have not heard to the contrary, I suppose your financial wants have been met. At the end of a few days I hope to be in funds beyond my immediate necessities, & if you should then be in want, I will most cheerfully aid you. It evinces a mean disposition and great want of common sense, for any one who has control of your very superior talents, to keep you in trouble about a few paltry dollars.[1] When you & I get into close business relations I will assure you of better treatment.

In all of my calculations & estimates, I have assumed that we must employ a copyist to write out our messages, at the rate of one copyist to every 500 words per hour. In talking upon this subject with Prof. Farmer, yesterday, he told me it was perfectly practicable to have a little writing (or printing) machine by which at least 3,500 words per hour could be written in plain Roman characters—that he had worked the thing out years ago, but had never had time to finish the machine, and had no time now, but would give me all his ideas & help any competent machinist to perfect the thing, and then I could give him one of the machines for his own use, and anything more that I could afford.[2]

Now, you can see that such a machine could be made of great value to any company or any person, having much copying to do, and I am sure I could find a market for 10,000 of them in a very short time.

Can I not prevail upon you to take hold of this idea of Farmer's & work it up into a practical machine.—You to have the exclusive manufacture of the machines and you Farmer & myself sharing equally in all expenses & all profits. I really wish you could oblige me in this matter. I have been a sufferer in this direction already, to the extent of some $5000 six years ago—but my inventor was'nt blessed with Farmer's or your head. You can work it out, I am sure.

Farmer is here, to test our Line,[3] and I suppose he will stay till Saturday. I wish you could come in early Saturday morning and meet Farmer—say at 8 a.m. at the Astor House, and have an hour's talk with him, and then see me before you go home. If you cannot come so early as 8 or 9, then fix the hour at 1 o'clock at 66 Broadway.[4] Yours faithfully

D. H. Craig

☞ If the note to Farmer is satisfactory, say so at the bottom, before you hand it to him.[5] C[b]

ENCLOSURE[c]

New York Aug. 12/70

Prof. Farmer   My dear Sir:

I have arranged with the bearer, my friend, J. A. Edison, a genius only second to yourself, who will give us the benefit of his brains and machine shop, to work up your ideas for the writing machine, as you have proposed to me. I propose to pay all necessary expenses in the development of the ideas—to debit the invention accordingly, and when necessary advances shall have been re-embursed, then the profits to be divided equally between you, Mr. Edison and myself. I to manage the whole business to the best of my ability without any charge whatever for commissions.

Mr. Edison assents to this programme, & will unite with us

in the ~~signature~~ execution of a suitable memorandum covering these points.[6] Yours truly

D. H. Craig

ALS, NjWOE, DF (*TAEM* 12:112). [a]"P.O. Box, 3237" written above. [b]This postscript written in left margin of first page. [c]Enclosure is an ALS.

1. Craig is probably referring to James Ashley. See Doc. 97, n. 14.

2. This machine was probably a form of typewriter. Craig proposed that such a machine be used for copying messages received by the automatic system. He later negotiated with Christopher Sholes for the rights to Sholes's typewriter.

3. This line, established by the National Telegraph Co. between Washington, D.C., and New York to test the Little automatic system, employed Farmer's compound wire. Farmer 1870, 138; "The Little Automatic Telegraph System," *Telegr.* 6 (1869–70): 192.

4. The offices of the National Telegraph Co. were located at 64–66 Broadway. A back room on the fourth floor of 66 Broadway was being used as a "testing office" to experiment with the Little system. Later these offices became the headquarters of the Automatic Telegraph Co. Craig 1870.

5. Edison did not write anything on the enclosed note to Farmer.

6. No arrangement seems to have been made.

–105–

Draft Agreement with Daniel Craig[1]

New York, August 17, 1870[a]

Memorandum of an agreement between J. A. Edison, of Newark, N.J. and D H Craig, of Peekskill, N.Y., entered into at the city of New York, August 17, 1870.

Whereas Mr. Edison has invented a new and valuable Repeater,[2] for Automatic or Fast Telegraphy—

And whereas Mr. Craig desires to become interested in the ownership of said invention and in any patent or patents which may be granted to the said Edison in connection with the same, or any improvements upon the same.—

Now, therefore, in consideration of one dollar paid to said Edison by said Craig, the receipt of which is hereby acknowledged, the said Edison and the said Craig promise and agree as follows, to wit:

Mr. Edison will, if possible, improve the said Repeater and will endeavor to secure all the valuable points of the same by Letters Patent from the United States Government.

Mr. Edison will recognise Mr. Craig's interest in the invention and in any patent which may be received for the same, to the extent of one-half of the whole, and will, if desired by said Craig, execute all necessary or proper papers connectied with the title or ownership

Mr. Craig will furnish to Mr. Edison all necessary money to pay the expense of a working model of said Repeater and also all patent fees and expenses.

Mr. Craig is to have the management of all business connected with the sale and use of said Repeater and Patents, but is in no event to make a sale of the same without the full knowledge of Mr. Edison, and is in no event to sell the invention except with a proviso that Mr. Edison is to have the exclusive manufacture of the machines, provided he so desires, and also provided he will make them well and as cheap as they can be made by other first class machinists

Mr. Craig is to charge no commissions for his services as manager of the said business, & hereby promises and agrees to hand over to said Edison one-half of whatever sum or sums he may receive for the said invention or patents less only actual and necessary disbursements connected with the Patents or business.

Df, NjWOE, DF (*TAEM* 12:118). In Craig's hand. [a]Place and date taken from text.

1. Two copies of this document were enclosed with Doc. 106.

Patent drawing of Edison's repeater adopted for automatic telegraphy (U.S. Pat. 114,657).

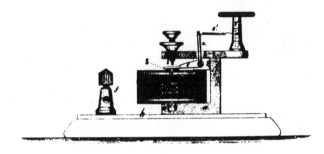

2. Edison used a modified polarized relay for his automatic repeater. He applied for the patent on 6 September 1870 and assigned it to Marshall Lefferts that same day. He amended the application in November, and the patent issued as U.S. Patent 114,657. TAE to Commissioner of Patents, 14 Nov. 1870, Pat. App. 114,657.

–106–

From Daniel Craig

New York,[a] Aug. 19/70

Dear Mr. Edison:

Inclosed, please find the balance due to you on the model Relay[1]—$6.—

Mr. Little[2] says he does not want half the evident speed of your Relay,[3] and hopes you will give all possible care to make the works substantial, & so that the machine can be <u>relied upon</u> in Repeating offices. He takes to it kindly and decidedly, and I can rely upon him for the strongest testimonials, if needed.

If the inclosed memorandum is satisfactory, please sign & return one to me, through the P.O. If not satisfactory, change it to suit yourself.[4]

I keep all my arrangements with you <u>private</u>—strictly so— & yet, I suppose it will be impossible to prevent a few people around us from suspecting that we are <u>putting our heads together</u>.

I have not been able to get off to Washington yet, but hope to do so today, and on Monday, if you will come in, I think, you will see 400 words per minute from Wash'n.[5] Yours faithfully

D. H. Craig

George Little, a British telegraph inventor who introduced an automatic telegraph system into the United States in 1868–69, which Edison later improved.

ALS, NjWOE, DF (*TAEM* 12:116). a"P.O. Box, 3237" written above.

1. This probably refers to a patent model. Edison executed his patent application for a repeater (i.e., relay) on 6 September 1870 (U.S. Pat. 114,657).

2. George Little (b. 1821) was best known for developing automatic telegraphy between 1868 and 1875. In 1849, while employed by the Electric Telegraph Co. of England, he invented the inverted cup or "umbrella" telegraph line insulator, which was widely adopted in Europe and America. He also invented batteries, chronographs, duplex indicators, lightning arresters, secret telegraphing systems, and visual code dials. In 1868 he demonstrated his chemical recording pen to Marshall Lefferts and others in the offices of Western Union. The applicability of this pen to automatic telegraphy may have led Lefferts to finance Little's work in exchange for a half interest in his patents. Little served as company electrician to the National Telegraph Co. from 1869 to 1870. During the years he was working on automatic telegraphy, Little resided in Bergen County, N.J. Little 1868; Pope 1868b; "George Little's Inventions," *J. Teleg.* 2 (1869): 150.

3. That is, the repeater could work much faster than the rest of the system.

4. Two copies of Doc. 105 were enclosed with this letter. Neither copy was signed.

5. A test of automatic telegraphy conducted between New York and Washington in late 1870 achieved a speed of only 250 words per minute. Farmer 1870.

AMERICAN PRINTING TELEGRAPH   Doc. 107

This printing telegraph, known as the American printing telegraph or the Pope and Edison printer, was the third and last collaborative effort of Edison and Franklin Pope. They developed this instrument for private lines and established the

American Printing Telegraph Company in the summer of 1870 specifically to exploit it.[1] The instrument "was ingenious and exceedingly simple," although "its action was comparatively slow."[2] It became quite successful, and Gold and Stock purchased the American Printing Telegraph Company in May 1871.[3] The polarized relay continued to be the hallmark of Edison's work in printing telegraphy. The American printing telegraph used the polarized relay to switch the current to either the printing or the typewheel magnet, in the fashion of Edison's first patented printing telegraph,[4] and thus avoided the subtleties of operation associated with having the typewheel and printing electromagnets in series.[5] The polarity of a signal, rather than its length, determined whether the typewheel advanced or the instrument printed.

In the American printing telegraph, Edison and Pope separated the polarized relay from the printer itself. In May 1870, Pope had received U.S. Patent 103,077, "Improved Electro-Magnetic Movement," which combined a polarized relay and a conventional relay to operate a receiving instrument on a local battery (see photograph below). This reduced problems caused by current variability on the main circuit. By the spring of 1871 Edison had replaced Pope's combination relay with the style of polarized relay used in the Chicago instrument.[6]

In addition to Pope's relay patent, this instrument incorporated part or all of the patents covering the Boston instrument and the gold printer, as well as U.S. Patent 103,035, "Improvement in Electro-Motor Escapements," an Edison design for an improved typewheel-ratchet mechanism. An article in the *Telegrapher* assured interested parties that "suitable arrangements are provided for bringing the typewheels of the two instruments together in case they should be accidentally thrown out of correspondence."[7]

1. See Doc. 130.
2. Reid 1879, 621, 623.
3. See Doc. 130, n. 2.
4. Doc. 51. In the Boston instrument, the polar relay short-circuited one or the other magnet, rather than positively switching the current, but the effect was the same.
5. See Docs. 76 and 96.
6. The Chicago instrument embodied U.S. Patent 113,033. See Doc. 164, p. 288, "No. 4."
7. "Pope & Edison's Type Printing Telegraph, for Private Lines," *Telegr.* 7 (1870–71): 65. This long, descriptive article about the American printing telegraph—probably written by James Ashley or Franklin Pope—appeared on 22 October 1870. The same account appeared sub-

stantially unchanged in subsequent editions of Pope's *Modern Practice of the Electric Telegraph.*

Opposite a blank page in a pocket notebook from October 1870, Edison wrote: "The unison that Ehlerich put on the first Model of Pope & Edison," but no sketch exists (PN-70-10-03, Lab. [*TAEM* 6:808]). A later unison prompted an infringement claim by George Phelps. William Orton to Marshall Lefferts, 18 Mar. 1871, LBO.

–107–

And Franklin Pope Production Model: Printing Telegraphy[1]

[Newark or New York, Summer 1870]

M (historic photograph) (est. 17 cm dia. × 17 cm), NjWOE, Cat. 551:6, 18, 39. See Doc. 54 textnote. The machine frame is stamped "G. C.

WESSMAN & SON NEW YORK 91". Gustave and Charles Wessman were brass dealers located on Centre St. in New York City.

1. See headnote above.

–108–

To George Harrington[1]

[Newark, September 1870[2]]

Dear Sir

We need "lease" papers immeddiately, as the parties who own itBuilding[a] seem to desire it before we put our shafting up,[3] which approaches completion.

Please embody following in Lease   That:—

We shall have the right to run nights—

Shall have the right to use the power nights by paying for its cost.

We shall Have[a] any amount of power we desire above the four horse power which is included in the rental, by pay one hundred and fifty dollars per horse power per year. The total amount of power shall not exceed 10 horse =

That he will heat the shop by steam

That he shall belt on our main shaft at his own expense

That he shall keep the belt and pulleys that run the main shaft in order—

That we shall not be limited as to the amount of machinery used in the room[b]

The rent is One Thousand dollars per year for the room, which includes four full horse power payable monthly in advance = Commencing Oct 1st 1870—

I Think it advisable to lease one year with the privelege of 2 or 3 as you think best = IPlease have Lease made out and sent to

T A Edison   Care Ritchie and Boyden[4]   No 15 Railroad Ave   Newark N Jersey

Please write and Let me know what bills you have received.[5] One machine is in the freight house now = The reason that I write instead of coming to see you is that I am working a night gang onf men on the perforator which I hope to have done by tuesday evening or wednesday 10 oclock = ItI cannot have all the keys in but will have as many as possible = I am compelled to stay here and watch men and give instructions— Mr Hyde[6] is attending to the Belts pulleys, shafts and Carpenter work all right— I may possible come in to see you tomorrow at 11.30. = Yours Respy

T A Edison

ALS, NjWOE, DF (*TAEM* 12:47). ᵃInterlined above. ᵇFollowed by a line drawn across the page.

1. George Harrington (1815?–1892) had been a clerk in the U.S. Treasury Department during the Polk administration (1845–1849). He became chief clerk under Secretary of the Treasury Salmon Chase and was appointed assistant secretary of the treasury in 1861. From 1865 to 1869 he served as U.S. minister to Switzerland. While in Europe, he observed government-controlled telegraphy, and upon his return home he became interested in the commercial possibilities of automatic telegraphy. *NCAB*, 12:337; Quad. 72.10, pp. 1–5 (*TAEM* 9:149–51).

2. This date is indicated by Edison's noting that the rent is to be "payable monthly in advance = Commencing Oct 1st 1870."

3. Edison and Harrington leased space from Ezra and Roscoe Gould in their factory building at the corner of Green St. and Railroad Ave. in Newark. One of Newark's leading tool and machinery manufacturers, E. & R. J. Gould produced machine tools, woodworking machinery, steam engines, and fire-fighting apparatus. "Shafting" refers to the system of overhead shafts, pulleys, and belts used to transmit power to the machines on the floor. Holbrook 1870, 889; idem 1871, 300, 891; Ford 1874, 73–74.

4. E. A. Ritchie and G. Boyden manufactured padlocks at 15 Railroad Ave., which was also the location of the Newark Telegraph Works.

5. There are bills and correspondence regarding machinery for the shop addressed to Harrington and dated 27 September 1870. DF (*TAEM* 12:45–46).

6. Andrew Hyde.

–109–

*Agreement with
George Harrington*

New York, October 1, 1870ᵃ

This Indenture, made this first day of October, one thousand eight hundred and seventy, by and between Thomas A. Edison, of Newark, in the State of New Jersey, of the first part, and George Harrington, of the City of Washington, District of Columbia, of the second part,

Witnesseth, That for and in consideration of one dollar, paid in hand, one to the other, the receipt of which is hereby acknowledged, and of the mutual trust and confidence which said parties have in each other, do each covenant and agree with the other as follows:

First—That the said parties as above named will be partners as inventors and as manufacturers of all kinds of machinery, instruments, tools, battery materials, and all and whatsoever may be required by the various systems of telegraphy, and of all such other machinery, instruments, tools or articles or things, the manufacture of which may be offered to or obtained and accepted by them, the said parties to be interested as owners in all original inventions and improvements invented, purchased or obtained by them, or either of them, and in all the interests and profits arising therefrom, and in the profits and losses arising from the business of manufacturing, in the proportions as hereinafter set forth.

Second.—That the business of said firm shall be known and conducted under the name and style of The American Telegraph Works.

Third.—The place of manufacture shall be in the City of Newark, State of New Jersey, until such time as it may be mutually agreed to select some other locality.

Fourth.—The capital of the firm shall be nine thousand ($9,000) dollars, of which the party of the first part shall furnish the sum of three thousand dollars, in the manner hereinafter set forth, and the party of the second part shall furnish the sum of six thousand dollars in cash.[1]

The capital to be furnished by the party of the first part shall consist of the stock, machinery, tools and inventions owned wholly or in part by him, of which an inventory shall be made, without reservation; but so much of the stock, machinery, tools and fixtures partly owned by said party of the first part, and in part owned by one William Unger,[2] as are now located and in use at the former place of business, at No. 15 Railroad Avenue, Newark, New Jersey, shall be allowed to remain there for use by the parties hereto and the said William Unger, under the unexpired partnership as existing at this date between Edison, party of the first part, and the said William Unger; but said shop, machinery, tools and fixtures, known as No. 15 Railroad Avenue, shall not be used as a place of general manufacture upon orders to the detriment of the interests of the manufactory to be established and known as the "American Telegraph Works," under the auspices of and to be owned by the parties to this indenture, it being understood and stipulated that the general manufacture, as heretofore carried on, is to[b] transferred to the American Telegraph Works, to be established under this agreement; and the transfer of the title to the stock, machinery, tools, fixtures and inventions owned wholly or in part by the party of the first part to the parties of the first and second parts jointly, to be held by them in the proportions respectively, according to the amount of capital furnished, as herein stipulated, shall be taken and received as full payment of the proportion of capital to be supplied by the party of the first part.

Fifth.—The party of the first part shall give his whole time and attention, talents and inventive powers, to the business and interests of the firm, and shall admit no other parties to any direct or indirect interest in and to any inventions or improvements made or to be made by him, except as hereinafter set forth; but all such shall enure and belong to the parties of the first and second parts as above set forth, in the proportion as set forth in section sixth of this indenture. Provided, however, that the inventions made exclusively for the Gold and Stock Company, which, under a contract between said party of the first part and Mr. Marshall Lefferts are to be the sole property of the Gold and Stock Company, and^c not to be included in this agreement.[3] But the said Edison, or party of the first part, binds himself not to invent under said contract any machinery that will militate against automatic telegraphy, nor to sell, transfer or convey to any parties whatever, without the consent of the party of the second part hereto, any invention or improvement that may be useful or desired in automatic telegraphy. And provided further, that for any original inventions or improvements that the party of the first part may make, other than such as may be suggested or arise from the current work in the manufactory, there shall be allowed and paid by the firm to the party of the first part a reasonable and proper compensation therefor, according to its practical value, all things considered. Such payment to be in addition to and irrespective of the proportionate part of the profits of the business of the firm to which the party of the first part would be otherwise entitled.

And it is further agreed, that if any disagreement shall arise as to the sum which may be claimed as "reasonable and proper" to be paid for such original inventions, the question shall be referred to an arbitrator, or, if preferred by either of the parties, to three disinterested parties, one to be chosen by each, and a third by the two thus chosen, and whose decision shall be final and binding on both.

Sixth.—That all profits arising from the business of the firm, and from all inventions and improvements, and from the manufactory, shall be divided between the parties as follows: One third therof to the party of the first part, and two thirds to the party of the second part; and all taxes, rents, insurance and other expenses, and all losses or damages, if any such shall occur, shall be paid from the general receipts of the firm arising from its business; if there shall be insufficient receipts, the deficiency shall be supplied by the parties hereto in the ratio of one third and two thirds, or shall be taken from the capital of the company.

Seventh.—The partners shall be allowed and paid from the gross revenues arising from the business a sum equal to fifteen per cent. upon the capital per annum, to be divided into monthly payments, and a like per centum on moneys advanced by either party over and above their proportionate parts of the capital as above set forth, and all excess of profits shall remain in the treasury of the firm, to be appropriated to the enlargement of the works and manufactory, and extension of the business, as may from time to time be agreed upon. Otherwise than as set forth in this section, there shall be no moneys or property belonging to the firm withdrawn, taken or used by either partner except upon the written consent of both partners.

Eighth.—The party of the first part shall have the control and direction of the manufactory, and shall employ and dismiss all workmen, as he shall deem best for the interests of the firm; shall purchase at lowest cash prices without commission, the machinery, tools, stock and other necessaries required in the manufactory, and generally shall be responsible for the careful preservation of the machinery and property of the company, and the economical conduct of the manufacturing part of the business. But the manner of keeping the accounts and books of the firm and manufactory, and the employment of persons required in keeping such accounts and books, and all that relates to the financial affairs of the firm and business, and the disposition of the proceeds[d] of the manufactory, shall be performed, or approved, controlled and directed, at his option, by the party of the second part.

Ninth.—There shall be no notes given nor any liabilities created by any member of the firm without the previous assent of both partners.

Before contracts shall be entered into for the manufacture of any given number of articles, it shall be the duty of the party of the first part carefully to estimate the whole amount of moneys that will be required to fulfil such contract if made, and the length of time that will be required to produce the articles wanted; and such estimate shall be submitted to the party of the second part in order to ascertain if the financial condition of the firm is such as to justify the outlay, and whether, when making the contract, it should not be provided in such contract for advances to be made by the parties for whom the work is to be done, in proportion as the work progresses, and before completion.

Tenth.—Full accounts shall be kept of all business done by the firm, and all transactions of the purchase, manufacture, sales, receipts and payments shall be clearly and fully recorded, together with a detailed account of all expenses of whatever character incurred, and the books and accounts shall at all times be open to the inspection of either partner.[4]

Eleventh.—Each partner shall give a true account of all moneys, property, matter and things that may come into his hands, or to his knowledge belonging to or concerning, or in any wise affecting said partnership on said business.

Twelfth.—It is further stipulated, agreed and understood, that the manufacture of all machinery, instruments, tools and other articles, other than so much as may be necessary to develope inventions and improvements, and make experiments, arising out of or from any inventions and improvements heretofore made, or that may hereafter be made by the party of the first part, or orders for machinery and instruments, or any part thereof, that may be obtained by either of the parties hereto, shall be manufactured, made and filled at and from the manufactory to be set up, created or established under this copartnership, and at no other place, shop or manufactory, without the consent of all the parties to this indenture.

Thirteenth.—It is further stipulated and agreed, that the party of the second part may, at his own option, admit a third party into the firm upon terms of equality with him and with the party of the first part, that is to say: To an equal third part or interest in all the inventions, stock, machinery, tools and all other property of the firm and in the business, with one third share of the profits and losses arising therefrom, and one third benefit, and an assumption of one third of all the liabilities of the firm. Provided, that by the admission of such third party the interest of the said party of the first part in the property and business of the firm shall not be lessened thereby, nor the stipulations and agreements and provisions of this indenture changed or modified, except in so far as must necessarily follow the admission of a third partner upon an equal footing in interest, and in all other respects, with all the rights and privileges, and subject to all the restrictions to be enjoyed or as imposed upon the parties to this indenture.[5]

Fourteenth.—This partnership shall continue for a period or term of five years from the first day of October, eighteen hundred and seventy, unless sooner dissolved by mutual consent of all the parties.

Fifteenth.—At the expiration of the partnership, or on its final dissolution, the property and assets, after paying all liabilities of the firm legitimately created in the course of the business, shall be divided among the respective partners, according to their respective interests; and in case any one of the partners shall die before the expiration of the partnership, the surviving partner or partners, if there shall be more than one, shall account for and pay over to the executors, administrators or other legal representatives of such deceased partner, his proportion of the moneys and of the proceeds of all property and assets owned by said partnership or firm.

Sixteenth.—The provisions of this indenture may be altered or modified from time to time, upon the agreement and written consent of all parties.[6]

In witness whereof, the said Thomas A. Edison, and the said George Harrington, have hereunto set their hands and affixed their seals, in the City of New York, on the day and date first above written.

GEORGE HARRINGTON,[e]   THOMAS A. EDISON,[e]

In presence of   J. W. TREADWELL, CHAS. S. HIGGINSON.[f]

PD (transcript), NjWOE, Quad. TLC.2, p. 83 (*TAEM* 10:861). Ten different versions of this agreement exist, including eight printed transcriptions (seven in Quad. and one in *Harrington v. A&P*) and two handwritten versions (one in LS and one in Libers Pat.). The original apparently had 5¢ Internal Revenue stamps on each sheet. All the printed copies were recorded or entered in legal records as accurate copies, yet they disagree in many minor points, none of which were noted in court proceedings. This text was selected because it has clear punctuation and only a few obvious errors. [a]Place and date taken from text; form of date altered. [b]In most copies, "to be". [c]In most copies, "are". [d]In some copies, "products". [e]Followed by indication of a seal. [f]Some copies transcribe the subsequent notarization, dated 31 December 1870.

1. See Doc. 159.
2. See Doc. 94, n. 1.
3. See Doc. 91.
4. Detailed accounting records were kept by the firm, including an extensive set of labor and cost accounts. American Telegraph Works Accounts, Accts. (*TAEM* 20:126–603).
5. On 10 May 1871 Harrington sold half of his interest to a group of investors. Daniel Craig later claimed that he was to have been the third party admitted to the firm under this clause. See Doc. 159; and Quad. 70.8, pp. 53–55 (*TAEM* 9:790–91).
6. See Doc. 155.

POCKET NOTEBOOK[1]   Docs. 110–123, 125, and 128

Edison made most of these notes and drawings at the office of patent attorney Lemuel Serrell[2] on 3 October 1870. He had constructed many of the devices described one to four months previously and probably was making these notes to establish priority for patent purposes.[3] The partnership agreement made with George Harrington two days before (Doc. 109) may be related. Under its terms Edison and Harrington became partners in inventing as well as manufacturing, and this notebook establishes that Edison had developed certain inventions prior to the agreement.

Most of the entries made by Edison in this notebook concern printing telegraphs. Other descriptions include a perforator invented two months before (see Doc. 103); a battery; a double-tongued polarized relay; improved electromagnets; a machine to punch out the figures in a check; a device to adjust the reaction time of an electromagnet's armature; and a complex mechanical linkage. Edison used the notebook for another drawing on 10 October (Doc. 125) and an agreement at the end of the month (Doc. 128).

Edison also used this notebook in October 1870 to record lists of machinery, tools, office furniture and supplies, and personal and business accounts. One of the accounts contains entries also found in the accounts of the American Telegraph Works.[4]

Lemuel Serrell, Edison's principal patent attorney from 1870 to 1880.

1. The notebook is disbound and the pages have been trimmed and laminated.

2. Lemuel Wright Serrell (1829–1899) became Edison's patent attorney in May 1870 and continued to act in that capacity until the early 1880s. He began his career working as a clerk in the office of his father, William Serrell, who was a civil and mechanical engineer and who also acted as a patent solicitor. In 1848 William Serrell's office handled the automatic telegraph patent applications of Alexander Bain, and this experience initiated Lemuel's specialization in telegraph patents. Among his clients were many of the leading telegraph inventors in the New York area, including Marshall Lefferts and George Little. Lefferts may have suggested that Edison switch from patent solicitor Montgomery Livingston to Serrell. Quad. 71.1, p. 95 (*TAEM* 10:52); Ricord 1897, 448–49; "Lemuel Wright Serrell," *New York Times*, 2 Aug. 1899, 7.

3. On each 3 October entry (except Doc. 112), Edison added as a postscript the place and date of composition; and he usually noted the date on which he constructed the pictured device.

4. Cat. 30,109, Accts. (*TAEM* 20:127–28).

110

Notebook Entry:
Automatic Telegraphy[1]

[New York,] Oct 3 1870
Drawn at L Serrells office   Constructed 2 months before[a]

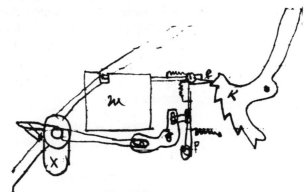

Machine for perforating paper for Telegraphic purposes[2]
M is punch which is sheared upon its end so that it will shear
the paper like shears instead of pressing out   X is paper driving apparatus which derives its motion by the forward & backward motion of the upright P   e is a knife edge fixed in such
a manner that it will move down but not upwards   K is a Key
provided with Cam Teeth of any number or required shaftpe
when the Key K is depressed it gives a number of back &
forward motion to the punch & paper driving apparatus which
produces a letter

AX, NjWOE, Lab., PN-70-10-03 (*TAEM* 6:809). On numbered leaf
18. [a]See headnote above, n. 3.

  1. See headnote above.
  2. This machine was constructed about the time Edison contracted
with Daniel Craig to develop an improved perforator for automatic telegraphy (Doc. 103).

111

Notebook Entry:
Printing Telegraphy[1]

[New York,] Oct 3 1870
Drawn at L Serrells Office   Hyde[2] & myself 2 months [be-
fore?][a]

Plan for making type wheels for Printing Telegraph & other purpose[3]   g is the wheel Divided and rotated at a great velocity by steam   X is a hardened steel die, M′ & M are geared wheel[s] which mash into each other and keep the speed of g & X equal   now by feeding slowly the die wheel against the wheel g when in motion the die wheel containing sunk letters impresses or shapes the Letters upon the edge of the teeth in wheel g etc

AX, NjWOE, Lab., PN-70-10-03 (*TAEM* 6:810). On back of numbered leaf 18. [a]Last word cut off by page trimming. See headnote, p. 196, n. 3.

1. See headnote, p. 196.
2. Andrew Hyde.
3. Apparently this idea failed. Between 1 October and 30 December 1871, Edison and Unger bought about 400 pairs of typewheels for printing telegraphs. 71-014, DF (*TAEM* 12:546–76 passim).

–112–

Notebook Entry:
Battery[1]

[New York,] Oct 3 1870 =
Grind carbon or coke and Black Oxide manganese to a flour and make into sticks for a Leclanche Manganese Battery—[2]   Written at L Serrells Office[a]

AX, NjWOE, Lab., PN-70-10-03 (*TAEM* 6:810). On numbered leaf 19. [a]Followed by crossed-out drawing of toothed wheel.

1. See headnote, p. 196.
2. The Leclanché battery used carbon for the cathode and manganese peroxide for an electrolyte. The carbon could be derived from coke.

–113–

Notebook Entry:
Battery[1]

[New York,] Oct 3 1870
Drawn at L Serrells office   Described to Carter[2] 2 weeks before[a]

Rotating Battery to prevent polarization of the electrodes = The Two elements are fastened to a shaft [---]their perpheries running in a trough containing the exciting fluids which can if necessary be seperated aby a porous plate   The Shaft can be of great length & contain any desired number of elements and be rotated by any power constantly or at intervals[3]

AX, NjWOE, Lab., PN-70-10-03 (*TAEM* 6:810). On numbered leaf 19. [a]See headnote, p. 196, n. 3.

1. See headnote, p. 196.
2. Possibly Robert Carter, an employee of the Gold and Stock Telegraph Co. Reid 1879, 626; Wilson 1871, 193.
3. Edison's "Rotating Battery" was a variation on the so-called trough battery, a wooden box holding sulfuric acid or some other electrolytic solution into which two gangs of electrodes were immersed. When the standard trough battery was not in use, the electrodes were lifted out of the liquid to preserve them from contamination by electrolytic action. By using only a part of each large circular plate at any moment, Edison's battery would retard corrosion of the surface and require a less frequent replacement of the electrodes.

–114–

Notebook Entry:
Regulating Device[1]

[New York,] Oct 3 1870

Drawn at Serrells office   Made 1 month before[a]

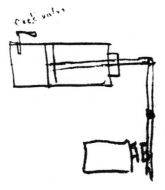

To produce a adjustable slow motion by means of a piston worked by a magnet or other power acting in a cylinder or air chamber and a cock valve   The cock being regulated and the power applied to the piston   it goes forward as fast as the air can pass through the cock which can being so regulated that an infinitesmal amount passes only or a large amount   which takes time & gives a slow motion to the piston[2]

AX, NjWOE, Lab., PN-70-10-03 (*TAEM* 6:811). On numbered leaf 20. [a]See headnote, p. 196, n. 3.

1. See headnote, p. 196.
2. Edison later used a similar device to slow the printing lever on a printing telegraph (U.S. Pat. 128,604).

Notebook Entry:
Printing Telegraphy[1]

[New York,] Oct 3 1870
Drawn at L[a] Serrells office[b] Commenced building Month before[c]

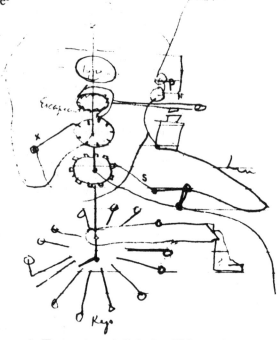

automatic Transmitter & Printing Telegraph[2] combined[d]

the shaft K has 4 wheels upon it   the first is a type wheel the second the escapement ~~the~~Wheel   [t]he[c] third a brass wheel will[3] thin insulated places [u]pon[c] its periphiery upon which the spring X Rubs   [T]his[c] wheel is so connected with the printing [m]agnet[c] P that when the shaft has been stopped [b]y[c] a continuous closure of the current the spring X is upon one of the insulated indents which allows the current to enter the Printing magnet to print   but when the wheel is rotated quite rapidly the c[ontact] spring X being in conta[ct][f] with the brass portion of the wheel 6 to [1?]o[c] times as much as the insulated parts   the current is of too short a duration to effect the printing magnet   The 4th wheel is a break wheel having a spring [--] rubbin[g][f] upon the teeth upon its periphiery this wheel & spring is connected directly with the line and by the action of the escapement it is made to make & break [t]he[c] current of the magnet which gives it motion   Consequently it automatically [i]ntermitts[c] upon the same principle as the pointer Dial Telegraph of Krammer illustrated & described by Sabine & Dub—[4] upon the [e]nd[c] of the shaft is a detent which is rigid & revolves with the shaft around [t]he[c] path of a number of key— which when depressed stop the shaft its its wheel a given positions etc

AX, NjWOE, Lab., PN-70-10-03 (*TAEM* 6:812). Text on numbered leaf 21 and drawing on back of numbered leaf 20. [a]Interlined above. [b]Interlined below. [c]See headnote p. 196, n. 3. [d]Followed by "over" to indicate page turn. [e]Cut off by page trimming. [f]Word runs off the edge of page.

1. See headnote, p. 196.
2. This is an early stage of Edison's universal private-line printer. See Doc. 126.
3. Probably means "with."
4. Sabine 1867, 70–72; Dub 1863, 329–39.

Notebook Entry:
Printing Telegraphy[1]

[New York,] Oct 3 1870

Drawn at L Serrells o[ffice][a]   Page [3][b]

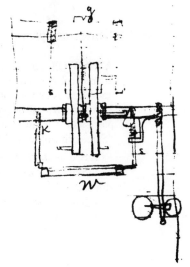

Fig 1[b]

Arrangement for rotating two type wheels upon one shaft by one escapement & one Magnet[2] So that one type wheel can be rotated when the other is Locked & vice vers[a][c]   This is done by a keyed moveable lock on the shaft of the type wheel which is moved back by being cammed by the cammed detent g and cammed forward by the upward movement of the paper Lever m [and] cam fork s when the shaft is in a certain position. One of the type wheels will be always carried arround except when the operator brings the wheel into a given position   then by an upward movement of the printing lever the type wheel hise desires is locked to the shaft & carried arround but will be cammed back by g unless he raises the printing Lever again etc   K is the unison apparatus for bring the shaft at every revolution at [--] a given point & releasing by the Printing Lever etc

Fig 2[3] Page 4[b]

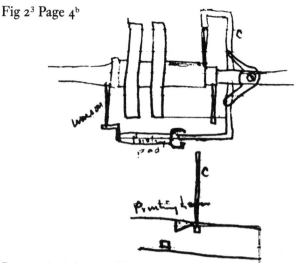

Same principle as in Fig 1 Pag 3[b]

AX, NjWOE, Lab., PN-70-10-03 (*TAEM* 6:813). Fig. 1 and some text are on back of numbered leaf 21; remaining text is on numbered leaf 22; and Fig. 2 is on back of numbered leaf 22. [a]See headnote, p. 196, n. 3. Last word cut off by page trimming. [b]Probably added at same time as place and date information: "Page [3]"; "Fig 1"; "Fig 2 Page 4"; and "Same principle as in Fig 1 Pag 3". [c]Letter cut off by page trimming.

1. See headnote, p. 196.

2. See Doc. 124. Neither of the designs sketched here was patented, although Edison used a similar idea in U.S. Patent 126,534.

3. The labels in the drawing are "unison", "Printing pad", and "Printing lever".

–117–

Notebook Entry: Relay[1]

[New York,] Oct 3 1870

Double Tongued[a] Polarized Relay[b]
Drawn at L Serrells office   Constructed 3 months before[c]

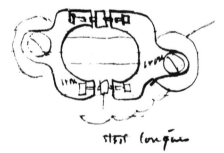

Polarized Magnet with two tongues[2] but with but one pair of magnets   These tongues are both pivoted in the back of the magnet and by giving them [-]or putting them closer to one side of the other one tongue will work on a positive cur-

rent the other remaining still[d] every time that Current is intermitted like a common Relay and vise versa when a negative
[current] is intermitted with the other tongue or they can be
so arranged between the forks of the magnets that they will
only change when the current is changed but wont act when it
is intermitted—[3]

AX, NjWOE, Lab., PN-70-10-03 (*TAEM* 6:814). On numbered leaf
23. [a]Interlined above. [b]This line was apparently added at same time as
reference to Serrell's office. [c]See headnote, p. 196, n. 3. [d]"the other remaining still" interlined above.

    1. See headnote, p. 196.
    2. Polarized relay armatures. The labels in the drawing are "iron",
"iron", and "steel tongues".
    3. Edison sought a single device that could act as both an ordinary
(neutral) and polarized relay.

–118–

Notebook Entry:
Printing Telegraphy[1]

[New York,] Oct 3 1870
Escapement for Printing Teleghs & other apparatus[2]
Constructed Sept 15 1870   Drawn at L Serrells office[a]

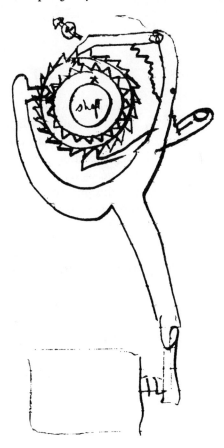

K a stop   X wheel V shaped 3teeth [&?][b] wedge acting [o]n[b] Ratchet & pawl acting[c]

AX, NjWOE, Lab., PN-70-10-03 (*TAEM* 6:815). On numbered leaf 24. [a]See headnote, p. 196, n. 3. [b]Cut off by page trimming. [c]Numbers "225", "550", and "515" scattered in a rough column at right.

1. See headnote, p. 196.

2. Edison used this ratchet-and-pawl design in U.S. Patent 113,033 and a similar one in U.S. Patent 113,034.

–119–

Notebook Entry:
Switch[1]

[New York,] Oct 3 1870[a]
Constructed Sept 145 71870[b]   Drawn Oct 3 1870 at L Serrells ofs[c]

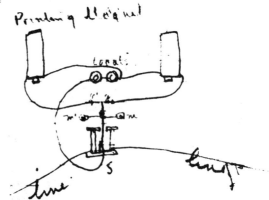

Means of working two magnets upon one wire[2]

S is a polarized Magnet   it tongue is centered equidistant between ~~neither~~ the points g g′ by the springs m m′— When say 6 pulsation of negative Electricity is sent through line the tongue of relay S is attracted to point g′ & back by spring m   this put local battery into type wheel mag[3] 6 times & vise versa when positive

AX, NjWOE, Lab., PN-70-10-03 (*TAEM* 6:816). On back of numbered leaf 24. [a]Date taken from text. [b]Numbers "5 1870" are in much darker pencil than preceding or following words. [c]See headnote, p. 196, n. 3.

1. See headnote, p. 196.

2. The labels at the top of the drawing are "Printing Magnet" and "Type wheel Mag". Edison used this switch with the American printing telegraph and incorporated it into the Chicago instrument (U.S. Pat. 113,033). See Doc. 164, p. 288, "No. 4."

3. If the tongue moved to g′, it would charge the printing magnet, not the typewheel magnet.

Notebook Entry:
Electromagnet[1]

Electro Magnet

Drawn at L Serrells office   Constructed 4 or 5 months before[a]

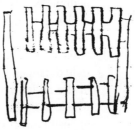

Constructed with washers in this form or turned out of solid stuff & planed in like a key seat. the wire by means of these slats in the Cores can be wound from one end of the

Entire Magnet to other getting all or nearly so wire in contact with the iron[2]

AX, NjWOE, Lab., PN-70-10-03 (*TAEM* 6:816). On numbered leaf 25. [a]See headnote, p. 196, n. 3.

1. See headnote, p. 196.

2. Edison used magnets with slotted cores two years later in U.S. Patent 138,869, in which he claimed they released the armature faster than did solid-core magnets. The patented magnet cores had only one longitudinal groove, however.

121

Notebook Entry:
Mechanical Movement[1]

Drawn on Oct 3d 1870 at ML Serrells office   Constructed 5 or 6 weeks before[b]

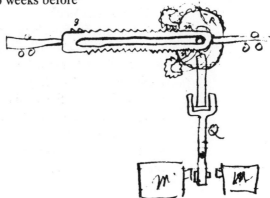

Mechanical Movement. to To transmitt a continuous rotary Motion into a back and forward motion quite long which shall stop itself at each pof the point between which it plays   This is done by having teeth at g & none at x & teeth at n & none at flange which has a milled periphery   by the side of these wheels is a arm [-]which are rigid upon the Shaft   upon the ends of these arms are [-] clicks moveable which rub along the milled flanges   now if the rack goes forward the click engages in the milled teeth and the gear wheel is locked to the shaft which is the[n] rotatined but the other wheel [that] goes in opposite direction is not locked   now when a back motion of the rack occurs othe other wheel locks & carries shaft forward in same direction = [2]

AX, NjWOE, Lab., PN-70-10-03 (*TAEM* 6:817). Starts on back of numbered leaf 25 and concludes on numbered leaf 26. [a]Date taken from text. [b]See headnote, p. 196, n. 3.

1. See headnote, p. 196.
2. Extended reciprocating motion, the goal of this design, was a feature of facsimile telegraphs, where typically the transmitting stylus moved back and forth over the message while the receiving stylus moved synchronously over the paper on which the message was recorded. See Docs. 46 and 92.

–122–

Notebook Entry:
Embosser[1]

[New York,] Oct 3 1870[a]

Done Oct 3 1870 at Serrells office[b]

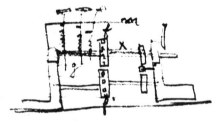

Gear teeth Deep
Apparatus for Punching out the figures for the amount of a check etc   x is shaft with ½ to ¾ Diameter having 10 to 20 Ribs cut in it Lengthwise   These ribs in the form of gear teeth   m is a plate & g is a plate   These plates guide keys with Racks upon them   the teeth of the Racks engaging in going down & back with the ribs on x =   f & f′ are two wheels one containing letters in relief the other sunk in   they act like a punch & die   paper goes between them to be punched the shaft x has a sliding bearing

3 ~~to~~Any number of keys can be used   on the key after the rack is a pin to press the wheel f down to wheel f′ & punch.[2]

AX, NjWOE, Lab., PN-70-10-03 (*TAEM* 6:818). Begins on back of numbered leaf 26 and concludes on numbered leaf 27, the bottom half of which was torn off and is missing. [a]Date taken from text. [b]See headnote, p. 196, n. 3.

1. See headnote, p. 196.
2. The pin is on the key shaft immediately *above* the rack.

–123–

Memorandum[1]

[New York?, October 3, 1870?][a]

all new inventions I will here after keep a full record[b]

AX, NjWOE, Lab., PN-70-10-03 (*TAEM* 6:819). This, the inside back cover of the notebook, is numbered leaf 28. [a]Place and date taken from Docs. 110–22. [b]Numbers "3,051 50" are written above and to the left.

1. See headnote, p. 196.

–124–

From Marshall Lefferts

New York, Oct 7 1870[a]

D Sir

Please go ahead and finish as soon as possible the arrangement of two wheels for your Printer = and I propose as follows. If you by means of a new transmitter or in any other manner succeed in making your Instrument comply with your Contract as to speed[1] we will pay you for the two wheel improvement[2]—such an amount as you may fairly claim under all the circumstances, as right for us to pay. if we cannot agree no harm can come to you, for if you cannot complete your contract without the device of two wheels then ~~we are to have it~~ you are bound to give it to us[b] without charge— It is also understood that if I wish it: we can go ahead & use the two wheel improvement without waiting for your further experiment with [-] transmitters—[3] Yours

M. Lefferts, Presdt.

ALS, NjWOE, DF (*TAEM* 12:137). Letterhead of the Gold and Stock Telegraph Co. [a]"New York," and "1870" preprinted. [b]"you are . . . us" interlined above.

1. Edison had fulfilled this contract (Doc. 91) to the satisfaction of Gold and Stock in May, "so far as the finishing of the instrument" was concerned. G&S Minutes 1867–70, 126–30.
2. Doc. 91, n. 4. See Doc. 116 for the two-typewheel designs Edison was working on at this time.
3. Probably the universal private-line printer. See Docs. 115 and 126.

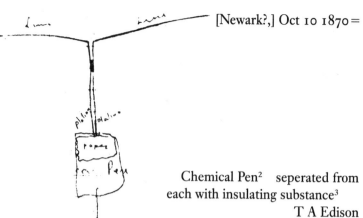

–125–

Notebook Entry:
Automatic Telegraphy[1]

[Newark?,] Oct 10 1870 =

Chemical Pen[2]   seperated from
each with insulating substance[3]
T A Edison

AXS, NjWOE, Lab., PN-70-10-03 (*TAEM* 6:808). On numbered leaf
17.

1. See headnote, p. 196.
2. The labels in the drawing are "Line", "Line", "platina", "platina",
and "paper".
3. In electrochemical telegraph instruments, current usually passed
from a single pen or stylus, through chemically treated paper, and thence
to a metal roller over which the paper passed. Here, Edison used two
platinum ("platina") pens electrically insulated from each other. Current
passed from one pen across the surface of the chemically treated paper
and back through the second pen, thus avoiding the varying resistance
of unevenly saturated paper. The use of platinum lessened the corrosion
of pen points by the chemical in the paper. Cf. Docs. 46, 180, and 290.

–126–

Draft for Marshall
Lefferts[1]

New York, October 19, 1870
DSir
 I have no intention of doing ~~anything~~ the Gold & S Te Co
any injury if I can help it— they will however recognize my
right as also my necessities to invent new[a] and improve existing Telegraph Machinery, but in doing so I am willing they
should be benefited by such inventions as I may devise   And
as regards the small Printer now in course of construction,
which I call the "Universal Printer"[b] I will finish it complete,
and give the Gold & Stock Te Co. ~~Three~~ Two[a] months in
which to experiment, ~~for the~~ and[a] test ~~of its~~ it[a] and I state it as
my belief that it will prove the best and most acceptable[c] private line ~~In~~ Printing Inst., yet devised and introduced.—[2] It is
also specially intended by me for working a number of the[a]
Inst in one circuit, (not as stock quoting Inst. although it can
be so made to work), but for general information—such for
instance as your projected "Lawyer lines"[3]— after this
examination if they desire to purchase it, I will dispose of

the Patent Improvements[a] to them—for a Maximum sum of ~~thirty~~ ~~forty~~ $30,000[d] thousand dollars—payable in Stock of your Company—supposing the Capital not to be over $1,250,000[4]—but ~~as the capt~~ if[a] it can be shewn that the instruments is not what I claim for it—or that it is not patentable ~~in all its parts~~—I shall be willing to discuss the ~~price~~ Maximum price named, and will do what is fair in its proper adjustment—giving the Compy the right to accept or reject—as they please[e] It seems to me that this is doing what is right—

~~Should they purchase this Instrument, I will enter into an agreement binding myself~~. The above sale to be subject however, to a reserved exclusive[a] right for myself and associates—for the ~~use~~ Manufacture and use of the instruments and its[a] improvements ~~and~~ for all Municipal and governmental[f] purposes, ~~that is.~~ including[a] Fire Alarm—Police—&c—[5]

Should the Company purchase the Instrument above refered to, I ~~am~~ will enter into an agreement, binding myself, to offer to the Gold & Stock Te Co. all ~~of my~~ future Printing Telegraph Inst. or improvements ~~in~~ of those already in use which I may devise and which are not already owned or controlled by the Company,[g] and[h] that I will complete all such inventions or improvements, ~~by~~in a good workable instruments which shall be placed in ~~their~~ the Companys[i] hands for experiments for ~~3~~2 months—and if they then desire to purchase the invention ~~and~~ or improvements, I will sell and assign the same, for a fair and reasonable—Compensation, taking into consideration what the Company, have already purchased and our mutually friendly relations, which I w~~a~~ish to preserve—

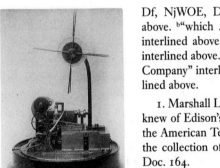

George Phelps's private-line printer known as the "financial instrument."

Df, NjWOE, DF (*TAEM* 12:138). Written by Lefferts. [a]Interlined above. [b]"which . . . Printer" interlined above. [c]"and most acceptable" interlined above. [d]"~~forty~~ $30,000" interlined above. [e]"as they please" interlined above. [f]"and governmental" interlined above. [g]"and which . . . Company" interlined above. [h]Written in margin. [i]"the Companys" interlined above.

1. Marshall Lefferts emended this document as he wrote it. He likely knew of Edison's contractual involvement with George Harrington and the American Telegraph Co. No signed copy of this document exists in the collection of Gold and Stock papers at Columbia University. Cf. Doc. 164.

2. The universal private-line printer (Docs. 115 and 165)—still unfinished—was signed over to Gold and Stock as part of Edison's 26 May 1871 contract (Doc. 164). In 1870, Western Union's acquisition of George Phelps's new printing telegraph (U.S. Pat. 110,675) threatened Gold and Stock's business (see Doc. 128, n. 3). The universal private-line printer was Edison's response to Phelps's printer. After Western Union and Gold and Stock merged in May 1871, the Phelps printer

Edward Calahan's printing telegraph known as the "lawyers' instrument."

became their "financial instrument." The universal, in its several forms, enjoyed moderate success on private lines. As late as 1877, 150 remained in service. Reid 1879, 613–14, 624.

3. Although an 1872 photograph album of Gold and Stock telegraph instruments contains a picture labeled "Lawyer's Instrument" showing a modified form of Edward Calahan's bank printer (U.S. Pat. 99,401), telegraph lines were not run from New York City courts to law offices until 1874. Cat. 66-42, Box 44, WU Coll.; "Telegraphic Communication Between the Courts and Lawyers' Offices," *Telegr.* 10 (1874): 179.

4. See Doc. 164.

5. This reservation is not included in Edison's May 1871 contract with Gold and Stock but is similar to that made by Pope, Edison & Co. in its agreement of 30 April 1870 with Gold and Stock (Doc. 97), which led to the establishment of the American Printing Telegraph Co. Edison's "associates" were probably the active members of the American Printing Telegraph Co.—Pope, Ashley, and Lefferts himself.

–127–

Agreement with Elisha Andrews and Marshall Lefferts

New York Oct 24. 1870

It is mutually agreed between the undersigned E. W. Andrews, M. Lefferts[1] and T. A. Edison, that the said Edison, having invented or discovered a new form of Electro-Magnet, and for which he has lodged a Caveat and proposes to apply for a Patent for the same.[2] Will assign and doth hereby assign said patent and all improvements thereof upon the same principle[a] to the said Andrews, Lefferts and Edison, one third to each party, but to be enjoyed so far as profits are concerned as in common.

The said Edison further agrees that he will give his best endeavors, to improve discover and invent Electro-Magnets for the purpose of use and application for motive power and that all such future ~~inventions~~ improvements upon this magnet[b] shall belong to the parties above mentioned, in the proportion

The expenses of making any instrument for testing any invention which the said Edison may make, shall be borne equally by the (3) three parties hereto—but such expenditure shall only be incurred upon the consent ~~to~~ of two parties to this agreement

Patents if taken out in Europe shall be subject to the same conditions as herein agreed to for the United States.

The said Andrews, and Lefferts agree that they will use thier best exertions to introduce and make profitable such inventions as may be made by the said Edison under this agreement.

T. A. Edison.

E. W Andrews
M. Lefferts.

DS, NjWOE, LS (*TAEM* 28:947). Probably written by Andrews. [a]"upon the same principle" interlined above. [b]"improvements upon this magnet" interlined above.

1. The officers of the Gold and Stock Telegraph Co. commonly contracted agreements as proxies for the company. See, for example, Doc. 91, n. 3.

2. Edison's caveat, titled "Electromagnets," was filed 22 October 1870; no patent resulted. Quad. 72.11, p. 44 (*TAEM* 9:187).

–128–

Agreement with Samuel Ropes, Jr.[1]

Newark Oct 29 1870

In consideration of the sum of one dollar the receipt of which is hereby acknowledged I promise to pay to T A Edison his heirs or legal Representatives one twelfth $\frac{1}{12}$ undivided ~~interest~~ of all: profits arising directly or indirectly from ~~his~~My[2] Connection and relations and introduction of instruments etc with the Gold & Stock Telegraph Co. of N York[3] for a period of Ten 10 years from date after ~~paying~~ receiving from them full payment of outlay by me on my own printer[4] which is $3,556—

S. W. Ropes Jr.

DS, NjWOE, Lab., PN-70-10-03 (*TAEM* 6:804). Written by Edison. On numbered leaf 10.

1. See headnote, p. 196.

2. This was probably a slip of the pen resulting from Edison's writing about himself in the third person.

3. In the fall of 1870, the Gold and Stock Telegraph Co. extended its lines to cities outside New York at the same time that Western Union planned to introduce George Phelps's printing telegraphs and to expand the operation of its own Commercial News Department lines in Chicago. Marshall Lefferts, then president of Gold and Stock, sent Samuel Ropes surreptitiously to Chicago in early or mid-October with Edison's Chicago instrument (U.S. Pat. 113,033) to test the Chicago market for stock and grain quotations. Ropes acquired fifty subscribers within a month. After telling Western Union president William Orton of this new threat to both companies, Lefferts purchased the lines, instruments, and subscribers from Ropes. Ropes became the Chicago agent for Gold and Stock, remaining there until his death in April 1871. William Orton to Anson Stager, 12 Nov. 1870, 24 Feb. 1871, 22 Mar. 1871, LBO; G&S Minutes 1870–79, 27–28.

4. Edison failed to correct this lapse into the first person. The phrase refers to Edison's instrument; Ropes was not an inventor.

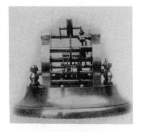

Edison's printing telegraph known as the "Chicago instrument."

–129–

To Samuel and Nancy
Edison

Dear Father and Mother

Why dont you write to me and tell me the news. You spoke in you last Letter that you had a good chance to buy a good peice of property very cheap   If you have your eye on it still write me, describing it, and why you think it valuable. I can send you the money for it. How is mother getting along you wrote the last time she was getting along nicely = I am in a position now to Let you have some Cash, so you can write and say how much = I may be home some time this winter = Can't say when exactly for I have a Large amount of business to attend to. I have one shop which Employs 18 men,[1] and am Fitting up another which will Employ over 150 men =[2] I am now—what "you" Democrats call a "Bloated Eastern Manufacturer"   Is the Buchanans still Live in Port Huron, and is Carrie[3] Married yet = Give my Love to all   Your Son

Thomas A

ALS, MiDbEI, EP&RI.

1. Newark Telegraph Works.
2. American Telegraph Works.
3. Carrie Buchanan, whom Edison later called "my old flame when a boy." See Chapter 1 introduction, n. 12.

–130–

American Printing
Telegraph Co.
Agreement with Pope,
Edison & Co.

New York, November 30, 1870[a]

This Memorandum of an Agreement made and entered into this —first—[b] day of July[b] One thousand eight hundred and seventy[1] By and Between "The American Printing Telegraph Company,"[2] a Corporation incorporated and organized under General laws of the State of New York party hereto of the first part, and Frank L. Pope, James N. Ashley and Thomas A. Edison heretofore carrying on business as copartners under the firm name of "Pope, Edison & Co" parties hereto jointly and severally of the second part Witnesseth

That for and in consideration of the sum of One Dollar to each of them in hand paid by said party of the first part (the receipt whereof is hereby acknowledged by each of them) and of the covenants and agreements of the party of the first part hereinafter expressed, the said Francis L. Pope, James N. Ashley and Thomas A. Edison, parties hereto of the second part, jointly under their said firm name of "Pope, Edison, & Co" and each of them separately and severally for himself, in and under his own name, agree and agrees as follows.[c]

First. to sell, assign, transfer, set over and convey (and they do, and each of them does, hereby sell, assign transfer, set over and convey) to "The American Printing Telegraph Company" each, every and all the right, title and interest of said parties of the second part, and of each of them, in, to and under the business heretofore carried on by them under said firm name of "Pope, Edison & Co," and in and to the rights, interests and credits of said firm, the property, rights of property and claims therof, and in and to the good will of said firm in said business, for and as of the date of the first day of July one thousand eight hundred and seventy, to the same force and effect as if this assignment transfer and conveyance had been made and executed on said day.

Second. That Whereas there was made and entered into on the eighteenth day of April one thousand eight hundred and seventy an agreement by and between "The Gold and Stock Telegraph Company of the City of New York" party thereto of the first part and said Frank L. Pope, James N. Ashley and Thomas A. Edison parties thereto jointly and severally of the second part, of which Agreement a Memorandum was executed and delivered in duplicate by the said parties thereto on the thirtieth day of April one thousand eight hundred and seventy,[3] and of which memorandum a copy is hereto annexed[d] marked "A" and made a part of this Agreement: and Whereas by and under said agreement of April eighteenth one thousand eight hundred and seventy said parties of the second part sold, assigned, transferred and conveyed to said "The Gold and Stock Telegraph Company of the City of New York" among other things certain rights, titles, and interest in, to and under certain inventions and improvements theretofore made, and in, to and under the patents therefor theretofore issued and secured, and agreed to sell, assign, transfer and convey to said "The Gold and Stock Telegraph Company of the City of New York" certain, rights, titles and interests in, to and under certain inventions and improvements thereafter by said parties of the second part or some or one of them to be made, and which said inventions and improvements have heretofore been made and in, to and under patents therefor thereafter to be issued or reissued and which have heretofore been issued or reissued for printing telegraph instruments, apparatus and methods:[4] and Whereas by and under said agreement of April eighteenth, one thousand eight hundred and seventy said parties of the second part expressly reserved to themselves certain rights therein mentioned and referred to in and about said inventions improvements and patents and printing telegraph instruments, apparatus and methods.) said parties of the second part have, and each of them has, sold assigned, transferred and conveyed, and hereby do, and each of them does, sell, assign, transfer and convey to said "The American Printing Telegraph Company" each, every and all the rights, title and interest which are in and by said agreement of April eighteenth one thousand eight hundred and seventy reserved to said parties of the second part, in, to, under and about said inventions and improvements heretofore made, and in, to and under said patents heretofore issued or reissued to the same extent and effect as the same are therein reserved to said parties of the second part, and to be exercised, used and enjoyed by said "The American Printing Telegraph Company" upon and subject to the same conditions as are therein imposed upon the exercise use and enjoyment thereof by the parties of the second part—and particularly the right to manufacture and use, and to sell or lease to others to be used, each, any and all of the apparatus, instruments and methods for which patents are thereby assigned or agreed to be assigned, for the specific purpose of transmitting over what are therein designated as "private" lines of telegraph messages other than commercial quotations and quotations of the prices of gold and stocks—and also particularly the right which the parties of the second part have under said agreement (and subject to the same conditions of exercise, use and enjoyment) to use the instruments, apparatus and methods for which patents or licenses are thereby assigned or agreed to be assigned, upon long lines of telegraph.

Third: That said parties of the second part hereby covenant that they have not, and that no one or two of them has, sold assigned transferred or conveyed or by any instrument or agreement purported or agreed to sell, assign transfer or convey to any person or persons other than "The American Printing Telegraph Co" either or any of, the rights, titles, interests or claims herein and hereby sold, assigned, transferred and conveyed or purported or agreed to be sold, assigned, transferred or conveyed to "The American Printing Telegraph Co"—and to that covenant the said parties of the second part hereby bind themselves, their heirs executors and administrators, and each of the parties of the second part binds himself, his heirs, executors, and administrators.[c]

Fourth: To make, execute, and deliver to said party of the first part, their successors or assigns, on demand, any and all proper instruments in writing which may be required to effectuate or express the sales, assignments, and transfers, either, or any of them, hereby made, or agreed to be made. And that: For and in consideration of said covenants, sales, assignments conveyances and transfers and of the full and faithful performance of the agreements above expressed said party of the first part agrees.[c]

First: To issue to said parties of the second part certificates of full paid shares of the Capital Stock of "The American Printing Telegraph Company" party hereto of the first part as follows i.e certificates for 510[e] full paid shares[f] of said Capital Stock to said Frank L. Pope and certificates for 510[e] full paid shares of said Capital Stock to said James N. Ashley, and certificates for 180[e] full paid shares of said Stock to Thomas A. Edison said certificates to be transferrable upon the books of the Company and said shares to be free and clear of and from any liability to calls or assessments by "The American Printing Telegraph Company." [5c]

Second: That said "The American Printing Telegraph Company" will not sell or lease to be used, or allow to be used by any other person or persons any of such instruments, apparatus or methods except upon receipt of an agreement in writing to be made executed and delivered to party hereto of the first part by each and every of the firms or persons who may purchase or lease the same from said party of the first part for any such "private" line of telegraph, that such purchaser or lessee will not use or permit others to use the same for any other purpose than that for which said particular "private" line for which the purchase or lease is made was originally intended[c]

It is further mutually agreed that in case any disagreement or dispute shall arise between the parties hereto or their assigns as to the true intent and meaning of this agreement, or of any clause or provision thereof, such disagreement or dispute shall be referred to the arbitration of three arbitrators, of whom, one shall be the President for the time being of the Chamber of Commerce of New York City, and one shall be chosen by each of the parties hereto, and that the decision of such arbitrators or any two of them shall be conclusive[g]

In Witness Whereof the parties hereto have sealed and delivered these presents this thirtieth—[b] day of November—[b] in the year of our Lord one thousand eight hundred and seventy.[c]
The American Printing Tel. Co. by M. Lefferts President
James N. Ashley[h]   Frank L. Pope[h]
                                                          Thomas. A. Edison.[hi]

Witnesses[j]

DS, NNC, Edison Coll. Each sheet of document has 5¢ Internal Revenue stamp affixed to upper left corner, front. ªPlace and date taken from text; form of date altered. ᵇWritten by Lefferts. ᶜFollowed by line drawn to right margin. ᵈAnnexation missing. ᵉIn an unknown hand, in a large space. ᶠ"510 – 180 = 330" (written in vertical column) and "Cliff 180 Allen 180" written in left margin in same hand as numerals. ᵍFollowed by double horizontal rule. ʰWax seal affixed next to signature. ⁱFollowed by notarization dated 1 December 1870. ʲWritten in left margin; corporate seal embossed below.

1. Negotiation of the terms of this agreement began at the first directors' meeting of the American Printing Telegraph Co. on 18 October (see n. 2). APT.

2. The American Printing Telegraph Co. was incorporated in the summer of 1870 to exploit the Pope and Edison printing telegraph patent rights reserved for the development of private lines (see Doc. 97). Edison, Pope, Ashley, Marshall Clifford Lefferts (son of Marshall Lefferts), and William H. Allen (of New Haven, Conn.; otherwise unknown) signed the incorporation papers on 1 July and filed them on 3 September. However, by mid-October control of the company had effectively passed to the Gold and Stock Telegraph Co. At the first directors' meeting (18 October) Marshall Lefferts was elected president and Tracy Edson and Joseph Cook (both Gold and Stock directors) were present and involved. The next month Gold and Stock acquired a controlling interest in the company stock. Neither Edison, Pope, nor Allen ever participated at any company meetings. American Printing Telegraph Co. Certificate of Organization, NNYCo; APT; G&S Minutes 1870–79, 3–14.

This illustration in the Telegrapher *promoted the American Printing Telegraph Co.'s Pope-Edison printer.*

The American Printing Telegraph Co. was aggressive and successful. They began advertising in the 5 November 1870 *Telegrapher,* and Ashley used the paper to tout the enterprise whenever possible. At least one set of printers had been leased by late summer, and the company was setting poles and laying cables for several weeks before signing this agreement. Receipt of first prize for the "Best Electric Printing Telegraph Instrument" at the American Institute fair in New York provided an additional opportunity to promote the business. During December the company's business expanded dramatically in the New York region, and the firm introduced instruments in textile mills, steamship company offices, and other businesses on the East Coast and in San Francisco. "Close of the American Institute Fair," *Telegr.* 7 (1870–71): 89; "The New Printing Telegraph Instrument," ibid., 100; "New Telegraph Cables across the

Edison's ticket to the American Institute industrial exhibition, where the American Printing Telegraph Co. printer was shown.

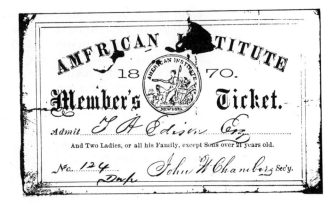

North and East Rivers," ibid., 108; "The New Printing Telegraph Instruments," ibid., 123; "The New Enterprise," ibid., 194; John 1870.

3. Doc. 97.

4. The patents covered under the 30 April agreement with Gold and Stock had all been issued by the end of October. U.S. Patent 102,320 was granted on 26 April; the polarized relay and escapement patents (U.S. Pats. 103,035 and 103,077) were both issued on 17 May; the second Pope-Edison printer patent (U.S. Pat. 103,924) was issued on 7 June; and the reissue of the Boston instrument patent (4,166) was granted on 25 October.

5. At the time of incorporation, American Printing issued stock worth $125,000 as 2,500 shares of $50 each. No further stock was issued. The controlling interest, bought by Gold and Stock, included 180 shares each from Edison, Pope, and Ashley. When Gold and Stock purchased the remaining company stock in May 1871, Pope and Ashley each held 330 shares. The certificate of incorporation proposed a distribution of 716 shares each to Pope and Ashley, 384 shares each to Lefferts and Allen, and 300 shares to Edison, but the actual distribution was as listed in this agreement, with 180 shares each to Lefferts and Allen (see textnote f). APT; G&S Minutes 1870–79, 5, 13–14, 78.

–131–

From Marshall Lefferts

New York Dec [c. 3,] 1870[a]

D Sir

You will please make and deliver to this Company, one hundred & fifty (150) Instruments like what we call the Cotton Instrument—[1] We will pay you at the rate of ~~Seven~~ Fifty eight dollars each—[2] Yours

M Lefferts Prest

ALS, NjWOE, DF (*TAEM* 12:141). Letterhead of the Gold and Stock Telegraph Co. [a]"New York" and "187" preprinted.

1. Doc. 136. Under Lefferts's direction, Gold and Stock was expanding its services to include a wide range of commercial information. When the New York Cotton Exchange opened in 1870, Gold and Stock

provided instruments to report on cotton prices, futures, production, and shipments for traders in the new exchange. Indeed, commodity exchanges in general were a growing market for telegraphy in 1870. Reid 1879, 613–14; "Changes in the Cotton Trade," *New York Evening Post*, 27 Oct. 1871; DuBoff 1983, 259–61.

2. The accounts of the American Telegraph Works show that by the end of May, Edison had delivered at least 131 cotton instruments. The first shipment of forty "Printing Instruments with Unisons attached at $58. Each" left the shop on 1 April 1871. Five other shipments followed (on 5, 10, 21, and 25 April and 30 May), costing Gold and Stock a total of $7,598. 71-007, DF (*TAEM* 12:415–17).

–132–

To Daniel Craig

Newark Dec 7, 1870

Friend Craig

Yours just received = It appears that Mr Harrington came over here and engaged in conversation with Mr Hyde and the draughtsman regarding the perforator and by his fidiggity manner got them so excited that they didn't know which end they stood on, as neither one of them ever saw the perforator until the day you was here, when it was brought from the other shop, and neither of them have worked or attemped to work it or even understand it = I had taken one of the parts out when experimenting with it & did not replace it and when Mr Harrington came over the draughtsman attempted to work it and couldn't, and from this Mr Harrington conceived that it was a failure = It was nothing but an experiment to determine if the principle and details were correct. they are correct and you will have your perforator in good time and beyond your most[a] sanguine expectations. I have eight men working on a pair night and day and it is an[a] utter impossibility to hurry it more than I am. I and the draughtsman and men were up nearly all last night = Galieo discovered the principle of acurate Hologlogy in the swinging Lamp of Pisa— It wouldnt be a very sage remark to say—why damn it that lamp aint a clock = You must remember that I am making your copying printer[1] which will be very rapid also your Transmitting and receiving motors,[2] which includes nearly all your machinery for the present you must know that it takes considerable thinking and brains to carry on all these machines at once = Besides a thousand small details in the other room in the manufacture of 150 printing Machines =[3] I have also in process of manufacture 12 Universal Printing Machines,[4] a Regulating Temperature Machine[5] Two other perforating Machines[6]— a New screw slotting machine, A wire straightning Machine =[7] Polishing

Machine, and other things which also takes thinking= Regarding an excess of expenditure I got his permission Every time that I made an overpurchase and gave satisfactory reasons at the time=   I cannot take Ten thousand dollar contracts on Three thousand or four thousand worth of Machinery— The Manufacture of Mechanism is a Slow operation being legitimate, and all Legitimate Businesses they are slow but sure, and the slower the more sure. If Mr. H. feels dissatisfied, with the Expenditures (every cent which is represented in good sound solid and acurate machinery) I will sell some of my Stock coming to me from the American Printing Co[8] and pay him for what he has spent or the excess over what he expected to pay= Mr H says that some of our experiments were useless= But after he has had more experience in this business, he will find that No experiments are useless. I had thought of coming over to see you today but cannot find time—

Of course I expect you (having had experience) understand this. My men here say that Mr H understood them the exact opposite of what they meant= Yours Truly

Edison

ALS, NjWOE, DF (*TAEM* 12:122). ªInterlined above.

1. Probably a typewriter.

2. These motors were probably based on those described in George Little's U.S. Patent 96,332. Little used them to synchronize the transmitters and receivers of his automatic telegraph.

3. Cotton instruments. See Doc. 131.

4. These were experimental models of Edison's universal private line printer.

5. Dr. G. M. Sternberg's electromagnetic regulator for furnace dampers and valves, which the American Telegraph Works manufactured. Bill to G. M. Sternberg, Mar.? 1871, 71-007, DF (*TAEM* 12:413); 71-008, DF (*TAEM* 12:423).

6. The accounts of Newark Telegraph Works for 7–14 January 1871 show work on one small and two large perforators. 70-005, DF (*TAEM* 12:152).

7. On 28 September 1870, Edison had purchased two screw-making machines from Pratt Whitney and Co., at least one of which used heavy-gauge wire as feed stock. The wire, generally wound on reels, required straightening before being fed into the screw-making machine. The screw-slotting machine cut slots in the screwheads. Bill from Pratt Whitney and Co., 28 Sept. 1870, 70-002, DF (*TAEM* 12:51); Knight 1876–77, s.vv. "Screw-cutting Machine," "Wire-straightening."

8. See Doc. 130.

Advertisement showing Dr. Sternberg's electromagnetic regulator, which was constructed by Edison and Unger.

To George Harrington

Mr Harrington

Mr Gould would like some Cash on account = I think $200. would satisfy him. I shall be ready by Saturday I think to Show you The Copying Printer[1] if everything goes smoothly with it = I shall also be able to show you the Transmitting and Receiving Motor or Apparatus by Next Tuesday I Think = [2] The Two Punching Machines[3] will probably be done inside of Two weeks from today =

You quite misunderstood my men here; The Machine that they attempted to show you work had been taken apart by me in Experimenting, and put together again but with not all the parts in. That is the reason it did not work = Mr Craig saw it operate— But it is of no consequence whether it worked or not It was an experiment as I told you once before, not made to show but to Satisfy me that I was all right. The Machines that I am making now will be made well & Complete, and if they dont perforate more than 80 words a minute[4] then there will be a funeral over here pretty quick = I shall need considerable money saturday to pay hands off as I am running about 45 hands day and night <u>Sixteen</u>[b] of them are on the Perforators =

I insured Today for $10,000 on Machinery and $3000 on Stock in N York Cos at 1½ per cent = Is that right = insurance commences tomorrow 7 oclock

<u>You must remember that the money you are now paying is not for Machinery But for work on the Automatic System and on the Gold and Stock Contract of $7,500</u> = Yours very Respy

T. A. Edison

P.S. I am attending to the shop from seven7 in the morning till one & two oclock next morning— so you cannot blame me for lLoossing time   E

ALS, NjWOE, DF (*TAEM* 12:122). Letterhead of American Telegraph Works. [a]"Newark, N.J.," and "187" preprinted. [b]Underlined three times.

1. Probably a typewriter.
2. See Doc. 132, n. 2.
3. Perforators.
4. Edison's perforators achieved average speeds of 30–35 words per minute. A skilled operator using the Siemens perforator could perforate approximately 40 words per minute. "Edison's System of Fast Telegraphy," *Scribner's Magazine* 19 (1879): 842; "Automatic Telegraphy—The Wheatstone and Siemens System," *Telegr.* 7 (1870–71): 399.

New York,[1] December 17, 1870[a]

The object of this invention is to produce two distinct movements at a distant station ~~upon~~[b] by[c] one wire and is especially applicable to that class of Printing Telegraphs which work upon one wire

Description[d]

A polarized switch operated by ~~th~~ an electro magnet in the main line becomes a shunt, the electro magnet a resistance and the diverted portion of the current is sent through one of two electro magnets and thence to the main line, the polarity of the current sent causes the switch to direct the current through one magnet or the other.[e]

A′ is the Printing lever magnet and A is the type wheel magnet. S. and S are the Main line wires and ground circuit[f] within which is the Polarized Magnet X X[2]

O and O′ are the keys arranged with the ~~main~~[g] battery C in the usual manner for reversing the Current upon the Main Line—one current throwing the Magnetized bar B ~~forward~~[b] to the point p and the ~~other Current throw~~[g] reverse Current throwing it to the point p′ These points are both Connected with wires to the Printing & Type wheel Magnets thence join together and are Connected to the Main Line at D. The Polarized Magnet has a very high resistance Compared to that of the ~~o~~Magnets A′ and A being wound with very fine wire. The bar B is also Connected with the Main line by the wire g—It will be seen that when the bar or tongue B is equidistant from the points p′ and p and touching neither, the Printing Magnet A′ and Type Wheel Magnet A is not within the Main circuit ~~being~~.[g]

Operation

If by depressing the key O a positive current be sent over the Line the bar B will be deflected to the point p, and the Current will then split at c The greater portion passing through the wire g bar B point p through the Magnet A Thence to D & back to the transmitting Station. ~~But~~[b] Only[h] a small portion of the Current will pass through the Polarized Magnet X X to D owing to the resistence it encounters through the fine wire but sufficient to actuate Bar, B, ~~But~~[b] and[h] not enough to prevent the Magnet A from working Now if the key O′ be depressed a Current of opposite polarity will be transmitted over the wire and the Bar B will be attracted to the point p′ Leaving the magnet A Entirely out of the Circuit and allowing the Current to pass through the Printing Magnet A′ thence to D in the same manner as before

My claims will probably be The Two Magnets A & A′ placed within a^b secondary^j circuits^j ~~independent of the Main Curcuit Current~~^b and operated by a Shunt Current from The Main Curcuit by means of a polarized ~~Magnets~~. ~~w~~Constructed as described & for the purpose set forth^k

Signed by me this 17th day of December AD 1870[3]

Thomas A. Edison^l

Witnesses   Geo D Walker Geo T. Pinckney

AD (copy), NjWOE, DF (*TAEM* 12:214). Letterhead of the American Telegraph Works. ^aPlace and date taken from petition; form of date altered. ^bOverstruck by same hand as in textnote c. ^cInterlined above in unknown hand. ^dUnderlined twice, surrounded by the inserted text referred to in textnote e, and overstruck in same hand. ^e"A polarized . . . other" in second unknown hand. ^f"and ground circuit" interlined above in first unknown hand. ^gOverstruck by Edison. ^hInterlined above in first unknown hand and traced over by Edison. ⁱWritten in the margin in first unknown hand. ^j"s" written in first unknown hand. ^kRemainder of document written in second unknown hand. ^lName preceded by symbol for "signed" and followed by petition and notarization in second unknown hand.

1. Edison's preliminary text was probably written in Newark.

2. The original drawing is missing; the following diagram is an editorial reconstruction.

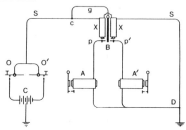

3. No patent was issued. According to the docket, Edison filed this caveat on 22 December 1870 and renewed it 6 January 1872. A 13 January 1873 notation reads, "Model will be furnished."

–135–

From Marshall Lefferts

[New York?, 1870?[1]]

of that particular case, was simply, that I had to agree to give Mr. Craig ⅓ of the interest.[2] After this you went on experimenting with "Perforators" and as you now inform me, you have several and one or two first rate,[3] but they belong to M Craig & Harrington, with an interest in the patents for yourself—[4] This you have also informed me, will all be made right, but I see no evidence of it—but I do see most clearly, that I shall <u>through you</u> be a very heavy looser— My outlay in this branch of Telegraphy, <u>in Cash</u>, has been over $25,000. Twenty five thousand dollars—of which amount between $6 & 7000, was expended last year—[5] You have said to me, on several occasions, that no man in this country, had worked as hard and as long to perfect this system as myself, and that I should make a fortune by it— All I ~~want~~ ask[a] is, that you do what is right. You should recognize our early understanding. A mans word should be as binding as his bond— Your try

M. Lefferts.

ALS (fragment), NjWOE, DF (*TAEM* 12:142). Only the last page of this letter has been found. [a]Interlined above.

1. Lefferts's large outlay of money in automatic telegraphy suggests that this letter dates from 1870 (see n. 5). Edison made no successful working perforators until 1871, but he did devise a perforator that Harrington saw in December 1870, and he was experimenting on both small and large perforators by the end of the year.

2. Craig's one-third interest may have been in George Little's patents in automatic telegraphy. In 1869 Lefferts obtained a half interest in these patents in exchange for supporting Little's inventive work. Later that year, rights to the Little patents were transferred to Craig for purposes of negotiating an agreement whereby the National Telegraph Co. would acquire the patent rights to the system and build a line to test it. Quad. 70.8, pp. 35–36 (*TAEM* 9:781–82); Agreement between Little, Craig, and the National Telegraph Co., 9 Sept. 1869, Exhibit 4, Defendants' Exhibits, Box 17B, 2:67–78, *Harrington v. A&P.*

3. See Doc. 132.

4. See Doc. 109, n. 5.

5. Lefferts had invested in both the Bain and the Humaston systems of automatic telegraphy, and in 1869 he provided substantial sums for Little's work. He paid for patents, apparatus, and experiments, and he may have provided as much as $200 per month in salary to Little. Craig to Lefferts, 20 Sept. 1869, Lefferts; Quad. 70.8, pp. 35–40 (*TAEM* 9:781–84).

–7– New Resources

January–June 1871

Edison's work in printing telegraphy was crowned with success during the first half of 1871. He was manufacturing cotton instruments at the American Telegraph Works for the Gold and Stock Telegraph Company, and at the end of May he signed a lucrative five-year contract with Gold and Stock that gave him a substantial lump-sum payment and an annual salary for his inventive work. During this period Edison also turned increasingly from improving George Little's automatic telegraph system to designing his own, and his backer in automatic telegraphy, George Harrington, enlisted several associates in purchasing a one-third interest in the American Telegraph Works. In May the Newark Telegraph Works moved from New Jersey Railroad Avenue to Ward Street in Newark and changed its name to Edison and Unger. Both shops were kept busy with Edison's inventive work—including a foray into sewing-machine design—and with Gold and Stock orders for printing telegraphs. However, in the midst of this activity came sad news; on 9 April, after a long illness, Edison's mother died.

In printing telegraphy Edison's inventive activity continued unabated in 1871. He continually modified the new ticker—his cotton instrument—while he labored on his universal private-line printer. He and Marshall Lefferts had discussed the value of the private-line printer the previous fall; in May, Edison completed a patent model, and Gold and Stock bought patent rights to the machine for $15,000 worth of stock. In the first six months of 1871, he filed ten caveats, at least five of which described improvements in printing telegraphs.[1]

Meanwhile, Edison's relations with the men who ran Gold and Stock and Western Union grew closer. In January, Tracy Edson, a major stockholder and director of Gold and Stock, called Edison to New York and gave him a letter of introduction to William Orton, president of Western Union. Gold and Stock directors George Field and Elisha Andrews applied in May for a British patent (which they later received) covering what would eventually be three of Edison's U.S. patents, including that for the cotton instrument.[2] In early June the men set out to introduce Edison's ticker on the London stock exchange.[3] Lefferts also applied for a British patent describing Edison inventions, but he never completed the process, and the devices he specified were covered in a British patent Field and Andrews filed the next year.[4]

Three related events occurred as May ended, the cumulative effect of which was to draw Edison closer to Western Union. First, on 25 May, Western Union assumed control of Gold and Stock.[5] As Gold and Stock had become increasingly successful, Western Union had moved to enter the New York market for private telegraph lines and financial information by contracting with George Phelps to design a printer that was competitive with Calahan's.[6] Tracy Edson had reacted to the threat, negotiating a deal that gave Western Union control of Gold and Stock. The agreement left intact Gold and Stock's corporate structure and ceded Western Union's increasingly important Commercial News Department to Gold and Stock. For Edison, the new arrangement meant closer contact with the powerful men in the Western Union organization and a wider sphere of operations for his machines.

The second event was the signing of a contract by Edison with Gold and Stock on the day after the Western Union–Gold and Stock agreement. Edison became a "Contract Electrician and Mechanician" for a period of five years, agreeing to assign to the company all of his patents in printing telegraphy. In return, he received company stock valued at $35,000 and an annual salary of $2,000, plus $1,000 each year Gold and Stock used any of his new inventions.

The last of the three events was Gold and Stock's purchase of the American Printing Telegraph Company. In addition to the Commercial News Department, Gold and Stock had taken over Western Union's private lines, and buying the American Printing Telegraph Company assured it complete control of all important patents covering private lines.[7] On 27 May, the *Telegrapher* ran a paragraph in which Frank Pope and

James Ashley announced that they had "disposed of their interest in the business of the [American Printing Telegraph] company, resigned their official connection with, and retired from its service."[8] This marked the termination of their business association with Edison.[9] Their personal relations with him seem to have been deteriorating for some time as well. Although the breach was not publicly mentioned at the time, many in the telegraph community knew of it. Edison later claimed Ashley tried to prevent him from getting his fair share of the 1870 Financial and Commercial Telegraph Company settlement,[10] and—as is clear from subsequent invective—Ashley's opinion of Edison's talent was considerably lower than Edison's own. Edison's name virtually disappeared from the pages of Ashley's *Telegrapher* for three years, and when it reappeared (in connection with the labyrinthine patent struggle over the quadruplex telegraph) it was as the "professor of duplicity and quadruplicity."[11] Pope, who had been Edison's friend, became a willing supporter of claimants against Edison.[12] One man later recalled, "Frank [Pope] took up Mr Edison when he (Edison) was poor and even ragged, and . . . Edison treated Frank with the grossest ingratitude."[13]

Originally hired by Daniel Craig and George Harrington to help solve problems with the Little system of automatic telegraphy, Edison increasingly turned his attention to designing his own system during the first months of 1871. Although the Automatic Telegraph Company acquired the rights to the Little system in January, company president Harrington saw great advantage in providing Edison with the necessary resources to design his own. Harrington's October 1870 agreement with Edison made Harrington a partner in Edison's inventions; between October 1870 and May 1871, Harrington advanced Edison $11,000 for expenses.[14]

As 1871 began, Edison focused his attention in automatic telegraphy on the problem of perforator design, building prototype perforators throughout the winter and spring. Edison and his backers considered large perforators preferable because their typewriter-like keys required no special knowledge on the part of the operator. However, large perforators were complex, slow, and liable to get out of order, so Edison also designed small perforators, which had fewer keys but which required familiarity with the Morse code in order to punch combinations of dots, dashes, and spaces. He explored the possibility of electrically powered perforators and made numerous experiments to determine the best arrangement of perforations to differentiate dashes and dots. By early June, the Automatic Telegraph Company had begun to use his perforators, but their operation proved troublesome and Edison kept working to improve them.

Unfortunately, much less is known about the other automatic telegraph components Edison devised early in 1871. Account records indicate that he constructed prototypes of a number of devices used in automatic telegraph systems, including various forms of transmitters, an ink recorder, and a "chemical paper machine."[15] Edison's backers encouraged his work on a copying printer—a typewriter. They believed that an operator could transcribe a message from telegraphic code to roman letters much more rapidly with such a printer than by hand. Craig, who saw the printer as a key component in a successful automatic system, negotiated with inventor Christopher Sholes for the use of what became the first commercially successful typewriter and, at the same time, encouraged Edison to develop his own. By the end of March, Edison's various devices were far enough along that he and Harrington laid plans for a demonstration system. Delays in devising a working printer, however, caused them to postpone the demonstration until early June, when they decided to proceed without the printer.

The expenses of supporting Edison's inventive activities apparently led Harrington in May to invite five others to join him in an agreement that gave them half of his two-thirds interest in the American Telegraph Works and in Edison's automatic inventions in exchange for $19,500 cash and the agreement to provide 50 percent of any new funds required. The support of these new backers—particularly Josiah Reiff—enabled Edison to continue his work. Reiff, secretary of the Automatic Telegraph Company, later provided Edison with an annual salary of $2,000 to work on automatic telegraphy.

1. Quad. 72.11, pp. 44–45 (*TAEM* 9:187).

2. Brit. Pat. 1,400 (1871). Under British patent law, the importer of an invention could be granted a patent regardless of whether the importer was the inventor. Davenport 1979, 26.

3. "Agreement by and between George B. Field, Elisha W. Andrews, William A. Camp, and Edward A. Calahan," 1 June 1871, Lefferts. The Exchange Telegraph Co. began operation in London in June 1872.

4. Brit. Pats. 1,444 (1871) and 828 (1872).

5. Reid 1879, 611–13; Marshall Lefferts to stockholders, 27 May 1871, DF (*TAEM* 12:614); G&S Minutes 1870–79, 27–77 passim.

6. U.S. Pat. 110,675.

7. Reid 1879, 621–22.

8. "The American Printing Telegraph Co.," *Telegr.* 7 (1870–71): 316.

9. The only exception seems to be some entries in a Murray and Co. daybook on 25 May, 7 June, and 11 July 1872 indicating orders placed by Pope and Ashley for the manufacture of some of Pope's Nonpareil telegraph apparatus. At that time Edison had not formally joined with Murray as a business associate. The entries are peculiar in that they are the only ones in the book without associated dollar amounts. Cat. 1214:18–19, 22, Accts. (*TAEM* 21:573, 575).

10. See Doc. 97, n. 14.

11. The vituperation began on 18 July 1874 ("'The Dutch have Taken Holland!'" and "'More Startling Inventions for Rapid Telegraphy.,'" *Telegr.* 10 [1874]: 169, 172) and continued for years. Ashley first called Edison the "professor" on 16 February 1875 ("The Telegraphic Situation," ibid. 11 [1875]: 34).

12. For example, Pope 1889.

13. Henry van Hoevenberg to Ralph Pope, 24 July 1908, Meadowcroft.

14. By 10 May 1871, Harrington had advanced $16,000 for tools, machinery, and property used in establishing the American Telegraph Works and another $11,000 for experiments. Doc. 159.

15. The "chemical paper machine" may have been a device for saturating paper with chemicals for use in Little's automatic receiver. The surviving records for the first half of 1871 show that Edison worked on ink recorders rather than chemical receivers, although he did begin working on chemical receivers later in the year. See Doc. 153, n. 6; 70-015, DF (*TAEM* 12:577); and Cat. 30,108, Accts. (*TAEM* 20:188).

POCKET NOTEBOOK   Docs. 135A–D, 138A, and 139A–B[1]

Edison began making entries in this pocket-sized notebook about 1 January 1871 and continued to use it for some months.[2] One entry considers the possibility of constructing a flying machine (Doc. 135B); several others concern the design of an electric sewing machine (Docs. 135D, 138A, and 139A–B). Also included are some of Edison's earliest technical drawings and notes regarding perforator designs for automatic telegraphy (Doc. 135C). Many of the miscellaneous sketches scattered throughout the book show components for the electric sewing machine or for perforators. Other designs include a paper feed for a printing telegraph (Doc. 135A), an ink recorder that was probably intended for the automatic telegraph system, a circuit diagram for testing magnets, a punch for cutting letters on printing-telegraph typewheels, and two sketches of what may be hand-cranked transmitters for automatic telegraphy.

1. The notebook (PN-70-12-25, Lab. [*TAEM* Supp. III]) came to light as this volume was in press; hence the intercalary numeration of the documents. It will be filmed at the end of Part III of *TAEM* as a supplement.

2. The notebook is crudely constructed of folded sheets of blank white paper stitched together and may have been homemade. The final dated entry is from 15 January 1871, but one of the last entries—a list of the time required to construct Edison's new worm unison for printing telegraphs—was probably made in late winter or early spring. There is also a drawing that Edison appears to have retrospectively dated 25 December 1870 (he first wrote "1871" and then overwrote the second "1" with a "0").

[Newark,] Jany 1st 71

Printing Telegraph Paper feed[2]

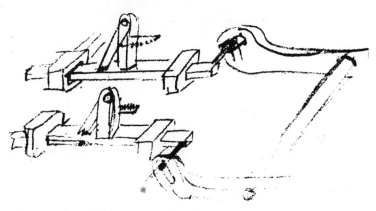

AX, NjWOE, Lab., PN-70-12-25 (*TAEM* Supp. III).

1. See headnote above.

2. The arms at right are extensions of the printing lever. Their up-and-down motion will make the paper feed mechanism reciprocate.

[Newark, c. January 1, 1871]

A Paines engine[2] can be so constructed of steel & with hollow magnets so as to obtain the resquisite strength and still be of extreme lightness—and combined with suitable air propelling apparatus wings parchoutte etc. so[-] as to produce a flying machine of extreme lightness & tremenduous power—

AX, NjWOE, Lab., PN-70-12-25 (*TAEM* Supp. III).

1. See headnote, p. 228. No other contemporary references to this idea have been found.

2. Henry Paine, a Newark inventor, announced in early 1871 a new type of electric motor that aroused great interest. He was soon exposed as a fraud. The controversy can be followed in the 1871 *Scientific American* (11 [n.s.]: 167, 370–71, 375, 404; 12 [n.s.]: 21, 84, 101, 180).

–135C–

Notebook Entry:
Automatic Telegraphy[1]

[Newark, c. January 1, 1871]

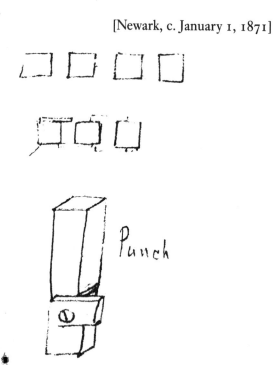

Perforating Apparatus made with ~~peculiar~~ ordinary punch & dies and knifes upon ~~these~~ punches extending over the intermediate space between each punch hole so that if six punch & dies a suitably combined with proper mechanism six punch holes can be made thus

but if it is desired to perforate a dash ~~th~~and 4 dots thus:

it can be done by depressing the first two keys ~~st~~further than is necessary to perforate the dots but sufficient to bring the knife edges down to the little slot ~~beat~~ the ends of the blank space between the dots & cut out this space leaving a dash—[a]

A perforator can be made by embossing paper by suitable mechanism & with Morse dots & dashes and after they are raised a revolving knife may be made to pass over the surface of the paper & cut the raised portion off, le~~ve~~aving a holes

~~eithe~~ [th]rough[b] the paper   this paper will perform the same work as on perforated—[2]

AX, NjWOE, Lab., PN-70-12-25 (*TAEM* Supp. III). [a]Followed by centered horizontal line. [b]Paper damaged.[*]

1. See headnote, p. 228.
2. Cf. Doc. 151A.

–135D–

Notebook Entry:
Sewing Machine[1]

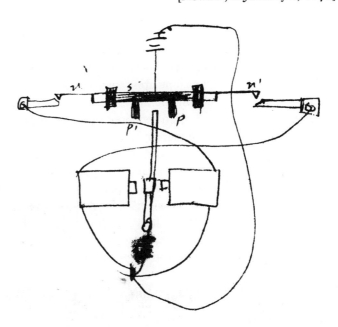

Positive Self Make and Break   s is a square slide which runs in bearings   n n′ are springs   p p′ are pins upon the slide which the armatur of the magnet hits & changes alternately as will be seen throwing the current through one magnet & then through the other = This be~~reeak~~ it will be seen prevents the armature from making strokes of unequal length the parts being so arranged that the armature must complete its full incursion before a change of current takes place etc etc—[2]

AX, NjWOE, Lab., PN-70-12-25 (*TAEM* Supp. III).

1. See headnote, p. 228.
2. Edison used this mechanism to generate a reciprocating motion for his sewing machine. See Doc. 138A; and "Sewing Machine Break," Cat. 298:43 (*TAEM* 5:83).

The cotton instrument was the telegraph printer employed by the Gold and Stock Telegraph Company to report prices from the New York Cotton Exchange and later from the New York Produce Exchange.[1] Although Edison did not begin manufacturing the instruments until the spring of 1871, the design was sufficiently complete by December 1870 that the company ordered 150 of them.[2] Edison executed a patent application that covered the new features of the device in January 1871 and patented several modifications during the next twelve months. By the end of the summer he had abandoned the cotton instrument in favor of a new basic design, which he incorporated into his universal stock printer.[3]

With the cotton instrument Edison solved the problem of manipulating two typewheels—one containing letters, the other numbers—without infringing Edward Calahan's patent.[4] Edison mounted both wheels on the same shaft and fitted a sliding cover over the paper strip that moved through the machine and a sliding platen under it. Activating the printing lever at one of two points in the typewheels' revolution shifted the cover and platen to one side or the other, allowing either the numbers or the letters to print alone.[5] Edison devised a unison stop, but this was soon superseded because it required an operator at the receiving end.[6] Incorporating the polarized relay and other circuitry of the Boston instrument, the cotton

Detail of Edison's cotton instrument showing the shifting mechanism for the paper cover and platen that allowed only one typewheel to print at a time.

instrument originally worked on one wire. In practice, Edison abandoned the single wire for the two-wire circuitry of the universal stock printer.[7] The general form of the cotton instrument followed that of the American printing telegraph and the Chicago instrument, both of which resembled the Calahan printer.

1. Healy 1905, 101.

2. U.S. Pat. 113,034. The invoices for instruments made in April and May of 1871 by the American Telegraph Works for Gold and Stock reflect the fulfillment of this order. To judge by existing manufacturing records, these 150 were the only ones made. See Doc. 131, n. 2.

3. The photographs reproduced in Doc. 136 are the only extant record of a production model of a cotton instrument. The Edison Institute (Dearborn, Mich.) holds the patent models for each of the modifications patented (U.S. Pats. 123,006, 126,531, 126,534, and 126,535). The models are unpainted instruments altered to illustrate the patent claims. The universal stock printer incorporated improvements that made the cotton instrument obsolete. When Edison began manufacturing universal printers in the summer of 1871, he numbered them sequentially with the cotton instruments of the December Gold and Stock order (PN-71-09-06, Accts. [*TAEM* 20:974–83]). Edison made explicit the evolution of the instrument in the specifications for U.S. Patents 123,006 and 126,532.

4. U.S. Pat. 76,993, Reissue 3,810. See Doc. 91, n. 4.

5. Pierre Dujardin, a French physician and inventor, had in 1868 designed a shifting platen mechanism for a printing telegraph, but Edison's version was sufficiently different to be patentable. It is likely that Edison had seen Dujardin's patent. The first claim in Edison's patent application actually described Dujardin's device (covered by U.S. Patent 82,502), and Edison deleted it from his application on 15 February 1871. The application was examined the next day and allowed. James Reid denied Edison credit for this design and for the screw-thread unison (Doc. 158). In neither case, however, did the original design to which Reid referred bear much resemblance to Edison's. Gold and Stock purchased the rights to Dujardin's patent on 28 March 1876, rewrote the specifications, and obtained Reissue 7,627 on 24 April 1877. Pat. App. 113,034; Reid 1879, 617–18; Dujardin patent assignment to Gold and Stock, Edison Coll.

6. By May 1871 Edison had developed the screw-thread unison (Doc. 158).

7. See Doc. 136. On the shortcomings of the polarized relay, see Doc. 51 headnote.

Production Model:
*Printing Telegraphy*¹

M (historic photograph) (21 cm dia. × 13 cm), NjWOE, Cat. 551:7, 17, 22 (see Doc. 54 textnote). This artifact displays two important changes from the original cotton instrument design (U.S. Pat. 113,034). First, the polarized relay switch has been by-passed, converting the machine from a one-wire circuit to a two-wire circuit (note the four binding posts)—that is, one wire for advancing the typewheel and one for actuating the printer lever. Second, a screw-thread unison has replaced the original, manually operated unison.

1. See headnote above.
2. Edison executed the covering patent on this date. Pat. App. 113,034.

New York Jan 12/71

Dear Mr. Edison

Hurrah for us!

Hurrah for we!

Your notes, like your confident face, always inspire us with new vim[1]

I intend to see you Friday or Saturday.

I have recorded your perforations & you will see that your arrangement of dots, dashes and spaces areis not quite right, but of course you can easily, I hope, fix that.[2] Yours, all the time

D H Craig

ALS, NjWOE, DF (*TAEM* 12:440).

1. The Edison notes that inspired this confidence are not extant. However, Craig may have been referring to Edison's claim that he had reduced to paper his plans for an automatic printer. See Doc. 140.

2. The particular arrangement of holes used by Edison in these tests of automatic telegraph apparatus is unknown.

New York Jany. 12th 1871[a]

My Dear Sir,

I would like to see you the first time you come to New York, in regard to a matter that may be of interest to you   Yours Very Truly

Tracy R. Edson[1]

ALS, NjWOE, DF (*TAEM* 12:613). Letterhead of the Gold and Stock Telegraph Co. [a]"New York" and "187" preprinted.

1. Tracy Robinson Edson (1809–1881) began his career as an engraver of bank notes and eventually initiated the consolidation of major American engraving companies in order to maintain standards against counterfeiters. From 1858 to 1875 he served as president of the American Bank Note Co. He made a large part of his personal fortune from royalties on patent rights he purchased for a permanent green ink used to print U.S. currency.

Shortly after the formation of the Gold and Stock Telegraph Co. in 1867, Edson became a major shareholder. In 1869 he was elected a director and served as a member of the executive committee thereafter. He played a significant role in obtaining for Gold and Stock the printing telegraph patent rights held by Samuel Laws in August 1869 and in transacting the sale of Gold and Stock to Western Union in May 1871. *NCAB* 19:394; Reid 1879, 607–8, 612–13, 626, 632, 799, 807.

Break for Sewing Machine Electric[2]

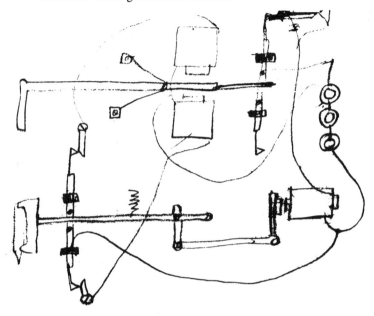

Break for Electric Sewing Machine   Jany 12 = 71

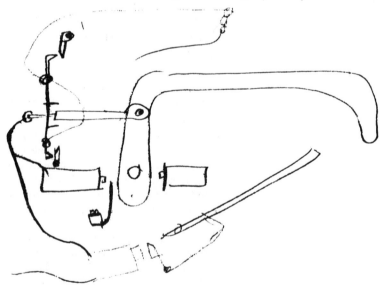

AX, NjWOE, Lab., PN-70-12-25 (*TAEM* Supp. III).

1. See headnote, p. 228.

2. Both designs in this entry use the mechanism described in Doc.
135D. Edison is not known to have built any models of motor-driven

sewing machines; however, the following sketch appears later in the pocket notebook that contains this entry (PN-70-12-25 [*TAEM* Supp. III]).

Edison's rough sketch of a sewing machine and motor.

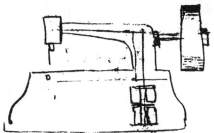

–139–

Tracy Edson to William Orton[1]

New-York, Jany. 13th, 1871[a]

Dear sir,

I have the pleasure (with your permission) of introducing my ingenious friend Mr. Thos. A. Edison, the bearer of this, who is worthy of the kind consideration which I am sure you will accord to him.

Mr. Edison is prepared to confer with you on the subject to which I alluded on Thursday last.[2] Very Respectfully sir, Your Obt. servant

Tracy R. Edson

ALS, NjWOE, DF (*TAEM* 12:613). Letterhead of the Gold and Stock Telegraph Co. [a]"New-York," and "187" preprinted.

1. William Orton (1826–1878) graduated from Albany Normal School in 1846 and taught for several years. He was successively a partner in a publishing firm, a member of the New York City Common Council (the city's governing body), an Internal Revenue collector, and commissioner of Internal Revenue under President Andrew Johnson in 1865. That same year he left government service to become president of the United States Telegraph Co. When Western Union bought that company in 1866, Orton became Western Union's vice-president. In July 1867 he succeeded to the presidency, a position he held until his death. He established the *Journal of the Telegraph* in 1871 to disseminate information throughout the Western Union system. One of Orton's most notable accomplishments as Western Union's president was his long, successful fight against a government takeover of American telegraph companies. *DAB*, s.v. "Orton, William"; Reid 1879, 520–38, 781–812.

2. Orton, Edson, and Marshall Lefferts (but not Edison) met several times during the following week to discuss the disposition of the local market reporting telegraph business. These negotiations led to the merger of Western Union and Gold and Stock. It is not known whether Edison subsequently met with Orton. Orton to Anson Stager, 24 Jan. 1871, LBO.

Notebook Entry:
Sewing Machine[1]

alternating apparatus[2]   Jany 13—71—   E[a]
Have a difrential-gear etc on Sewing Machine for alternating Battery in cellar = or use a Daniells[3]   big crocks—to get
small Resistence & put a positive break on engine & neutralize
the perm mag by winding last layer seperate & throwing a
quick current in

E[a]

AXS, NjWOE, Lab., PN-70-12-25 (*TAEM* Supp. III). [a]Edison initialed
each of the two pages of this entry.

1. See headnote, p. 228.
2. This mechanism uses a modification of the reciprocating break described in Doc. 135D. The short, wide electromagnet to the right of the
escapement lever might have as its core the "perm mag" mentioned below. The circle at upper right is a breakwheel for transmitting intermittent electrical impulses.
3. See Doc. 336, n. 3.

Notebook Entry:
Sewing Machine[1]

Magnet with armature attached at its upper end to a projection of the attracting core of The Magnet =

This gives greater strength for the Reason that the armature is magnetized by actual contact.

Edison

AXS, NjWOE, Lab., PN-70-12-25 (*TAEM* Supp. III).

1. See headnote, p. 228.

-140-

From Daniel Craig

New York,[a] Jany 18/71

Dear Mr. Edison

Yours of yesterday rec'd.[1]

Two weeks ago you told me you had, in a rough way, reduced to paper, your ideas for an Automatic Printer that would transmit two or three hundred words per minute, and you promised to have the man make full drawings for your caveat or patent.[2]

I then told you that a friend had caveated an invention for the same idea, and that as he explained it to me, it seemed to me very likely to work, and that I had thought it best to arrange to control the invention—but that I did not want to have it developed until you had thought over the subject and completed your own drawings and that then I should want you to sit down with my friend[3] and examine your plans with him and after making up your mind as to the best plan to build the machine go ahead & develop the same, and try, at least, to get ahead of Farmer and beat him, too

There, I reckon, I have complied with your request and given you my ideas in a very "explicit" way

I agree with you, fully as to "Dry Goods clerks [-]or Digger Indians," but you captivate my whole heart when you speak of making machines which will require "No Intelligence." That's the thing for Telegraphers.

I should say that, if possible, we ought to start with a Perforator that will punch any letter when the key is properly struck, and I would say that such a machine, that could be driven up to 25 words per minute, and costing $50, would be better for us than one which should require thought in the making of dots & dashes, & working even up to 40 words and costing $25[4]

How soon can you give us, (to enable girls to begin to practice) two perfect Perforators, and what rate of speed may we hope for from experts?

When, <u>in your best judgment</u>, can we have all the necessary machinery for, say 4 offices (N.Y., Phila. Balto. & Wash'n) When shall I go to Newark again? Yours truly

D H Craig

ALS, NjWOE, DF (*TAEM* 12:441). ª"P.O. Box 3237" written above.

1. Not found.

2. The "Automatic Printer" mentioned here is not the typewriter, but rather a printing receiver for the automatic telegraph system (see Doc. 143).

3. This may be Charles Jones. See Doc. 163.

4. See Chapter 7 introduction for a discussion of perforator designs.

–141–

National Telegraph Co. Agreement with Automatic Telegraph Co.

[New York,] January 18, 1871ª

MEMORANDUM OF AN AGREEMENT made this eighteenth day of January, A.D. 1871, between the National Telegraph Company, a corporation created under the laws of the State of New York and duly organized pursuant thereto, and The Automatic Telegraph Company,[1] also a corporation created under the laws of the State of New York and duly organized pursuant thereto;

WITNESSETH: that whereas the said National Telegraph Company has this day sold, transferred, assigned and set over unto the said Automatic Telegraph Company, its successors and asssigns, an Agreement between the said National Telegraph Company and Daniel H. Craig and George Little, dated September 9, 1869, and all extensions and modifications thereof, together with all the property rights and interests of the said National Telegraph Company under the said contract or to arise therefrom subject to the performance by the said Automatic Telegraph Company of all the conditions, stipulations and undertakings in the said Contract contained which are to be hereafter performed by the said National Telegraph Company and liable to all damages and forfeitures for non-performance thereof, and placed the same in the hands of Erastus Corning[2] in escrow—

AND WHEREAS the said transfer and assignment has been made in pursuance of an agreement between said Companies, that the Automatic Telegraph Company shall within two months from the 20th of December, 1870, unless prevented by circumstances beyond the control of said Automatic Telegraph Company, complete the line with at least two more wires between New York and Washington and prepare it with offices, instruments and all things necessary to open it to the public for business; that it shall when so prepared be opened for business by the said Automatic Telegraph Company and the system covered by said Contract so assigned and transferred shall be thoroughly tested in practical telegraph work between New York and Washington and the intermediate principal cities, for such period of time not exceeding four months from and after the completion and opening for business as aforesaid as shall be sufficient to satisfy the said Automatic Telegraph Company as to the value of the said system; that if such test shall satisfy the said last named Company, that it will be justified in going on with the construction of lines connecting the principal cities, and applying thereto the said system, then the said Company shall, within six months from the expiration of the period of four months during which the system is to be tested as aforesaid, from among the first monies received from the further sale of its bonds and stock, pay to the said National Telegraph Company the sum of One Hundred and Five Thousand Dollars in cash, and the further sum of Four Hundred and Seventeen Thousand Dollars in the Capital Stock of the said Automatic Telegraph Company—

And upon the payment to Hon. Erastus Corning of Albany with whom said Instrument or Agreement has been deposited as a Trustee, of the sum hereinbefore mentioned and the delivery to him of the Stock, as aforesaid, he is hereby authorized and directed to deliver said Contract and the transfer and assignment thereof to the said Automatic Telegraph Company and deliver the money and Stock to the National Telegraph Company.

But it is expressly understood and agreed that in case the results of the said tests shall be unsatisfactory to the said Automatic Telegraph Company then it shall be under no obligation to receive the transfer and assignment of the said Contract nor to pay anything on account thereof, nor assume any of the liabilities or undertakings in the said Contract contained, but the said Contract shall remain the property of the National Telegraph Company.

In Witness Whereof the said National Telegraph Company has caused its corporate seal to be affixed hereto and these presents to be executed in triplicate by its President and Secretary on this eighteenth day of January, A.D. 1871.

ROBERT SQUIRES,[3] Vice Prst. National Tel. Co.[b]
Signed, Sealed & delivered in the presence of
GEO B WALTER[4]   Secy Nat Tel Co

WHEREAS, by the terms of the foregoing agreement,[5] the Automatic Telegraph Company stipulated to complete the line between New York and Washington with two more wires within two months from the 20th of December, 1870, and Whereas, an unexpected delay has occurred and prevented the execution of the transfer papers at the date contemplated when the Agreement was made, and, therefore, the fulfillment of the said stipulation has become impossible within the period named, and, whereas, the snow, ice, and frost of the inclement season now upon us will prevent the resetting of the poles, and the making of other repairs necessary to enable the poles to sustain the strain of additional wires until the rigors of winter are passed.

NOW THEREFORE, in order to remove any further causes of delay in making the necessary tests of the system, the foregoing Agreement is hereby modified so as to permit the Automatic Telegraph Company to acquire other lines or wires on and over which to make the test of the system; and the stringing of additional wires upon the poles of the present line is left to await the results of such tests over other lines or wires and to the discretion of the Automatic Telegraph Company.[6]

IN WITNESS WHEREOF, the National Telegraph Company, has caused its corporate seal to be affixed hereto, and these presents to be signed by its President and Secretary on this eighteenth day of January, A.D. 1871.
ROBERT SQUIRES, Vice Prst. National Tel. Co.[c]

GEO. B. WALTER, Secy. Nat. Tel. Co.

The Automatic Telegraph Company hereby accents the within contract.

GEO HARRINGTON,[d] President Aut. Tel. Co.
Signed sealed and delivered in presence of
JOSIAH C. REIFF,[7] Secretary.[e]

PD (transcript), NjBaFAR, *Harrington v. A&P*, Box 735A, Respondents' Exhibit 2, 1:499–503. Internal Revenue stamps affixed to pages of the original. [a]Date taken from text, form altered. [b]"SQUIRES," followed by representation of seal; "Co." followed by representation of corporate seal. [c]Followed by representation of corporate seal. [d]Followed by representation of seal. [e]Followed by representation of seal and by notarization dated 7 February 1871.

1. The Automatic Telegraph Co. was incorporated on 28 November 1870 by George Harrington, Daniel Craig, Josiah Reiff, William Mellen, and John Elliott. Harrington became president, and Reiff secretary, of the new company, which had a capital stock of $13 million divided into 130,000 shares valued at $100 each. Harrington held 129,960

shares, and each of the others held 10 shares. By this agreement Automatic Telegraph obtained the rights to George Little's system from National Telegraph. Certificates of incorporation of the Automatic Telegraph Co., 28 Nov. and 2 Dec. 1870, NNYCo.

2. Erastus Corning was a banker, iron manufacturer, and railroad president and promoter. A prominent figure in Albany politics, Corning served several terms as mayor and congressman. His son, Erastus Corning, Jr., also was involved in Automatic Telegraph. *DAB*, s.v. "Corning, Erastus"; Quad. 71.8, pp. 45, 49 (*TAEM* 9:786, 788).

3. Robert Squires was president of the Third Avenue Railroad in New York as well as vice-president of National Telegraph. N.Y. 371:782, RGD.

4. George Walter, secretary and one of the principal promoters of National Telegraph, was later involved with Marshall Lefferts in promoting Edison's electric pen in England. Electric Pen folder, Lefferts; N.Y. 350:1135, RGD.

5. That is, the agreement between the companies mentioned in the third paragraph.

6. See Doc. 261, n. 1.

7. Josiah C. Reiff (1838–1911), a railroad financier, became associated with William Palmer in the development of the Kansas Pacific Railroad in the late 1860s and then joined Palmer in supporting Automatic Telegraph. Reiff later claimed that he provided the principal financial support for the development of the automatic system (roughly $175,000). After Jay Gould and the Atlantic and Pacific Telegraph Co. acquired Automatic Telegraph in 1875, Reiff initiated a lawsuit, seeking reimbursement for the automatic telegraph interests of Edison, George Harrington, and himself. Reiff also held extensive interests in copper mining and was a partner in the banking firm of Charles Woerishoffer. *New York Times*, 2 Mar. 1911, 9; "Reiff Family History," BC; Reiff's testimony, 1:52, Box 17A, *Harrington v. A&P*.

–142–

From Daniel Craig

Peekskill,[1] Jan. 31/71[a]

Dear Mr. Edison

I had your perforations (a & b)[2] telegraphed today, and send you the result. At low speed (say, about 300 words per minute) the writing came pretty well, but even then, the dashes were too short for the dots. When run up to 6 or 7 hundred words per minute, many of the dashes were <u>nearly</u> in two parts, and the shortness of the dashes, as compared with the dots was even more apparent than when the speed was less.

Of course, you will know how best to remedy this trouble, but my way would be to increase the size of the <u>dash</u> holes about ⅜ths and so arranging as to have the dash <u>actually</u> <u>lengthened</u>, say ⅛″ or ¼ and throwing the other fraction in to <u>assure continuity</u> thus ▫   You now have it thus ▫

and the result, in rapid recording is you get it thus[3] ⬥⬥⬥ which, I am sure, you will say is not exactly according to Gunter[4] or Edison.

I hope that not a single hour is to be lost in giving us a complete Perforator for practice, and all possible speed should be applied to the completion of enough for the Washington line. I presume Mr. Harrington would agree with me that 15 Perforators would do nicely, and 10 might answer, & should, if the added 5 would prolong the completion of the 10 to any extent.

Please explain to Mr. Harrington that you can make a note form[5] printer at short notice, and that (working ~~print~~ a type wheel it must print faster & much better than can possibly be done by the levers or arms plan of Sholes.[6] My idea is that it is a decided object for us to have the control of the Sholes machine, provided it costs little or nothing—but if not, not— and of course, relying upon your inventive head to work us through when and as may be necessary.

I hope you will make sure that our friend Unger don't take the least amount of push off of the Puncher. That fully & successfully completed is a thousand fold more important, pecuniarily and in every other way, than all the Gold machines ever made or dreamed of.[7]

The next big thing is the little automatic "Recorder"[8] (I think that will do, "for high"[9] till you find a better name) and pray don't fit[b] out the opposition so nicely that we cannot do better. I rely upon you upon this point, and if you do not let us slip up, you will assure to yourself a big pile of profits every year—and a good deal more than the Callahan hounds[10] would pay you for your brains through your natural life

☞ I have not yet had an opportunity to talk to Mr. Harrington, but I shall not forget nor neglect the point, as my heart and judgment is in the arrangement I suggested. It shall be   Truly &c

D. H. Craig

ALS, NjWOE, DF (*TAEM* 12:444). ᵃ"P.O. Box 3237 = New York." written above. ᵇInterlined above.

1. Peekskill, N.Y., is located on the Hudson River 39 miles north of New York City.

2. The "perforations (a & b)" were probably two separate trial runs of Edison's perforated tape.

3. Edison later adopted a form of overlapping perforations similar to that suggested here by Craig.

4. Edmund Gunter was an English mathematician and inventor of

several mathematical instruments. The phrase "according to Gunter" meant that something was correctly done according to rule. *OED*, s.v. "Gunter"; Farmer and Henley 1970, s.v. "Cocker."

5. "Note form" means a sheet of paper as opposed to a paper tape.

6. Christopher Sholes, Wisconsin printer and journalist, invented a typewriter in 1868. The Remington Arms Co. acquired the patent rights in 1873 and, after making improvements, introduced it as the Remington typewriter. In 1871 Craig proposed using Sholes typewriters with the automatic telegraph system and had a number of them made for that purpose. Doc. 186; *DAB*, s.v. "Sholes, Christopher"; Current 1954.

This illustration of Christopher Sholes's typewriter appeared in the 10 August 1872 issue of Scientific American.

7. Craig here urges Edison to make one of his manufacturing partners, William Unger, spend more time working on the automatic system's perforator ("Puncher") and less time on the apparatus Edison and Unger were making for the Gold and Stock Co. ("the gold machines").

8. See Doc. 143, n. 2.

9. Abbreviated form of "How's that for high?"—a common slang expression meaning "What do you think of it?" Farmer and Henley 1970, s.v. "High."

10. That is, Gold and Stock and its superintendent, Edward Calahan.

–143–

From Daniel Craig

Peekskill,[a] Feb. 13/71.

Dear Mr. Edison—

I gave, a few days ago, a letter of introduction to you, and when my big-bellied (but none too big for his <u>head</u>) friend calls upon you, I shall feel personally obliged if you can devote a few moments to him.[1] He is a great admirer of good machines and <u>good work</u>, and I believe he has very few equals in the iron foundry business. He carries in to that business a good deal more Brains than would be necessary to run the White House in first class style and effect. If you should see fit to give him your casting business, I am sure he will do it to your satisfaction, so far as you may give him your wishes.

I do not want you to work, or think of work, over 10 hours out of the 24, but I do want to have my heart put at rest in the matter of the "Automatic Writer"[2] (How'll that do for a name?) desperately, and if you should ever feel as though you wanted to be prayed for, you can secure a first class journeyman if you will tell me you have done it— I fully believe you will do it, if you live long enough,—indeed, if you should tell me you could make babies by machinery, I should'nt doubt it.

As the machine was invented and patented (at least, in part, I presume) years before you had your first dealings with L. or the G. & S. Co.,[3] and as we all want to be at peace with L., why would it not be legitimate for you to ante-date a bill of sale to some suitable person & then let him convey to Harrington for the Am. Tel. Works Co. Of course, you are to be paid specially for the invention.

If we could commence on the Domestic business in Competion with Ashley, very soon, I would like it very much.[4]

Will you write me as soon as you have anything to say, and on no account, I beg of you, delay the machines for the full test of Automatic Telegraphy. As we discard Sholes,[5] we shall, of course feel a deep interest in your proposed Printer, and I beg you will consider whether it is possible to use the upper & lower case letters of about the size of Pica type, or a size larger. I would say that for general purposes a machine that would print thus, nicely, 35 words per minute would be more valuable than one printing all caps that worked up to 50 words per minute. Of course, caps alone will answer our immediate purposes nicely. Yours truly

D H Craig

ALS, NjWOE, DF (*TAEM* 12:447). a"P.O. Box 3237 = New York." written above.

1. Unidentified.

2. The following paragraph suggests that Edison had proposed using some printing telegraph as a receiver in the automatic system. See Doc. 151A, n. 4.; Doc. 194; and Cat. 1182:17–21, Accts. (*TAEM* 3:58–60).

3. Marshall Lefferts and the Gold and Stock Telegraph Co.

4. James Ashley was the agent for a burglar alarm company. American Burglar Alarm Telegraph Co. advertisement, *Telegr.* 7 (1870–71): 118.

5. The Automatic Telegraph Co. was still using Sholes's printer at the end of July. See Doc. 186.

–144–

Daniel Craig to Marshall Lefferts

Peekskill,[a] Feb. 13/71.

Dear General—

We spiked Bryan. Baldwin is played out—the dirty thief.[1]

A nice backer, you, truly, if[b] you told Harrington, as he understood you, that we could not compete successfully with the W.U. Co. with the Little Perforator, which now averages over 600 words per hour, reliably. 25 Perforators, at 600, give 150,000 words per day.

25	Girl Perforators, at $400 pr. anm.		$10,000
2	Telegraphers	400 "	800
5	Copyists (printers)	400 "	2,000
	Total cost of Labor, yearly,		$12,800

Now, for Morse—I give the Liars 800 ws. per hour, though they don't average 600.[c]

~~1938~~	Morse 1st class ops $1,200	$~~22800~~45,600
19	2d class[2] to attend Repeaters, in every circuit over 300 m.	15,200
		$60,800

More than 4 to 1, in Labor, against the Morse people. How's that for high?

Harrington decided, against my judgment, that it was best to deal with the Corning party & avoid all differences.[3] I am not sorry on one account, as it will render less doubtful my ability to raise money on our interests as soon as we can get a full test with regular offices at New York, Phila Balto & Washington— and when that happy time will come, now depends upon Edison, and he says you keep him stirred up all the time with a sharp stick—just as though your trumpery gold reporting was of any importance along side of our operations. I respectfully protest against your going over the North River,[4] or sending a Letter in that direction for the next four weeks. I never have any trouble with Edison except when you stick in your oar—keep it out.

There will, I am satisfied, be no trouble in consolidating all outside lines upon a fair basis, within 60 days after we can make a regular business test between Wash'n & New York, & intermediate offices—and then, if not before, you must come over to us, boots & breeches.

The Automatic Telh. Co. provide in their articles of Association for a News Department, & have elected me to manage it for five years. My idea is that the discredit of News-peddlers & News speculators, should not attach to the Automatic Telh. Co. directly or indirectly. What would you think of the idea of a News Co. who should arrange, in the way of a Lease with the Automatic Co. for the control of one and ultimately of two wires, to be used exclusively for News business—the Automatic Co. fixing its rates for this class of business so^d as to give the News Co. special advantages—and if this scheme should be thought wise, what Lease money should be paid to the Automatic Co., for one good compound (No 5) wire, & Brooks insulators—wire costing at factory about $80 per mile, & insulators about 35¢ each? including right to use automatic machines & system.

I should propose to let Editors in to the News Co. if they wished, as stockholders, and should expect to send a General News report as now, to the whole press, & then let the first class journals get over as many specials as they pleased, at a very low tariff—their reporters perforating & copying their own reports.

As I know we can send at least 30,000 words per hour, in any circuit, by the aid of the Edison Repeater, I think all the Press business can be well done over one wire, and the Commercial News Reporting can also be well done with the occasional use of a second wire between a few chief points.

Have you any objection to tell me, confidentially or otherwise, what arrangement we could probably make for the use of your Universal Printer (I believe that is what you call it) for the distribution of Commercial & General news in cities? I don't mean the exclusive use, as I don't care how many competitors I have of the Simonton[5] and other jumping jack order.— Little says he has new ideas, and that he can beat Edison & the whole world—but you and I will only take that covered all over with salt.

Harrington having advanced over $20,000 to help along the main thing, at Newark, I am ashamed to ask him to now let Little go in, & if, as I suppose, you have personal rights in the Universal Printer, I should rather see you getting extra profits than Little or other parties.

Will you do me the favor to explain to Mr. Hoadley[6] how impossible it is for Mrs. Craig to do more than pay the interest on what is due to him, till we get the Automatic system at work, and tell him how immensely advantageous has been the delay in getting the improved machinery—about a month more will put us all right. <u>I am sick</u>   Truly &c

D. H. Craig

ALS, NNHi, Lefferts. ᵃ"P.O. Box 3237" written above. ᵇUnderlined twice. ᶜA line separates "600." from the following column of numbers. ᵈInterlined above.

1. Bryan not identified. "Baldwin" may refer to A. J. Baldwin, a "well-known backer of opposition lines" competing with Western Union. He was an incorporator and president of the Pacific and Atlantic Telegraph Co. in 1866 and an incorporator of the Southern and Atlantic Telegraph Co. in 1869. Reid 1879, 451.

2. Telegraph companies had no clear criteria for separating first-class and second-class operators; however, first-class operators generally could send or receive about 40 words per minute without mistakes or breaks and could transcribe messages in neat, clear handwriting. Second-class operators possessed inferior skills and therefore often serviced way wires that ran through small towns and had only limited traffic. Gabler 1986, 69–71; Abernathy 1887, 38–41.

3. This may be a reference to Erastus Corning, Jr., and the National Telegraph Co. See Doc. 141, n. 2.

4. The Hudson River was sometimes referred to as the North River.

5. James Simonton replaced Daniel Craig as general agent of the Associated Press in 1867 and held that post for fourteen years. *DAB*, s.v. "Simonton, James William."

6. David Hoadley, president of the Panama Railroad, had financial dealings with Craig and Lefferts outside their interests in telegraphy. Material on these financial relationships can be found in the Lefferts Papers. Sterling 1922, 70–71.

–145–

From Daniel Craig

Peekskill, Feb. 18/71

Dear Mr. Edison:

As you have not given me your promised note I suppose you have not <u>yet</u> got out of the woods on the <u>Automatic Writer</u>— and the thought makes me feel very sad.

I have been quite sick with inflammation of the Lungs, and at one time thought I should have a very <u>serious</u> time—but I am pretty well, today, & hope to see you soon.

The Turkish Consul has talked to somebody, who thought it of sufficient interest to have it repeated to me, in a way to lead to the conclusion that some of the Gold and Stock people were using you to set the Gold, Exchange and Stock business in operation in Constantinople, London, Paris, &c &c[1]

Of course, it is none of my business what you do for such a venture, but I would like to be kept posted. <u>We of the Automatic Telegraph people</u>, shall, as soon as we get fairly afloat here, go vigorously at work to utilize all we may have on the other side, and I only hope that as Harrington fully agrees with me that you are to be <u>with</u> us and <u>of</u> us, in all Telegraph engineering, you will keep us as carefully in your mind as we shall certainly keep you.

I think we shall want to patent your <u>Repeater</u> and your Perforator and Printer (note paper), in Europe

If you are not fully posted as to what is going on by the Turkish Consul, Laws,[2] &c I wish you would get booked up, & let me have the story. The <u>Turk</u> says the machine will record Chinese, Japan or any other <u>Language</u>. I take it there is no other machine that will do that, but yours.[3] Yours truly

D H. Craig

ALS, NjWOE, DF (*TAEM* 12:450).

1. Elisha Andrews, George Field, Edward Calahan, and Marshall Lefferts were arranging to introduce Edison's cotton instrument, when finished, in a stock-reporting system for the London Exchange in 1871. See Chapter 7 introduction.

2. Probably Samuel Laws.

3. The machine Craig describes would most likely be a facsimile telegraph, such as the one Edison was devising for Field and Andrews under the terms of their 10 February 1870 agreement (Doc. 92).

–146–

Daniel Craig to Frank Scudamore[1]

New York,[a] Feb 20, 1871

Dr. Sir:

I have just read Mr. Hubbard's[2] Communication to the Speaker of the House of Representatives and your full & interesting Letter to the former, in relation to Telegraph matters.

I learn from Mr. Hubbard's statements, at p.p. 8, 9, that you are able to perforate your messages at the rate of about thirty words per minute—1.800 per hour—and that you transmit, by the Wheatstone Automatic System Press report, dropping at 5 stations, at the rate of about 2.000 words per hour, and that for 5 days' general business, from London, in circuits of about 200 miles, the Wheatstone machines averaged about 3.600 words per hour.

It seems to me that we could do a good deal better than this for you, if you felt inclined to give our new system a practical test.

Since you wrote me that you could Perforate 40 words per minute, and were kind enough to send me samples of your beautiful work, I have been quite disgusted with our slow process by which we could accomplish only 10 or 12 words. Immediately on receipt of your Letter, I set a very ingenious man to work,[3] and since last July,[4] he has invented & made the models of fourteen different Perforators, and out of the whole, we have selected two that seem to me to be perfect, in their way. One, very simple, reliable & inexpensive, perforates about 30 to 35 words per minute, and can be manufactured for about $20. The other is equally simple & reliable, fitted with piano-shaped keys, and may be worked up to 50 or 60 words per minute, and can be manufactured for about $30.[5]

A girl of ordinary intelligence and a few hours or days of practice, works either machine with perfect accuracy.

We found that where a single wire was tumbling messages into the office at the rate of sixty thousand[b] words per hour, that we must have something more rapid than the hand-pen to copy the messages—and as necessity is the mother of invention, we set to work to get up a hand key Printing machine,[6] which is worked by a girl at the rate of about 3.000 words per hour, and prints as neatly as can be done with a regular press in a printing office. The messages are printed on note paper (that is, not on a paper ribbon) and thus one girl copies, in a very elegant manner, (and in manifold with, 10 copies, if you wish, for the Press) as many words in an hour as four or five girls could copy with a pen.

Now, then, I am satisfied, and now we will go on with our new system, which I laid on the shelf six months ago, with a feeling (for which I have, in the main, to thank you) that we had not done all that was possible to do. Meantime, our friend Orton,[7] and hosts of other deeply interested parties,—not excepting Hubbard—have had months of the keenest delight in circulating reports that we had failed to develop anything that could compete with the Morse system.

The facts are as follows:

With 10 girls & 10 inexpensive and durable machines, we can perforate 30.000 words in one hour, and 300.000 words in 10 hours.

With 1 girl to transmit & 1 girl to receive, these 300.000 words can be transmitted, certainly 500, and we fully believe 2000 miles, over one wire, and still not have the wire fully employed, by from 25 to 50 per cent.

At the receiving end of the line, 10 more girls, with 10 inexpensive & durable printers, can copy, in plain Roman print, the 30.000 words per hour and the 300.000 words per day of 10 hours.

Now, this I know[b] we can do with our new machinery, which we are driving ahead as fast as possible, & shall have enough completed to stock the Wash'n Balto, Phila & New York offices within the ensuing month. Meantime, I like to have my opponents and enemies here believe that I have abandoned all hope of destroying the Western Union Co.

Orton and Morse have testified that 600 words per hour, for 10 hours per day, is the full average speed of the Morse system, but, of course, there are a few operators who can average 1000 words per hour—10.000 per day of 10 hours. <u>Thirty</u> of such operators (who, here, would command $1.200 or $1.500 per year) with 30 good wires, in a circuit of 200 or 250 miles, would transmit 300.000 words in 10 hours, and <u>Thirty</u> more, equally good, would record the same at the other end.

60 Morse operators, at $1.200 per an.	$72.000
22 Automatic girls, at $400 " "	8.800
Difference against Morse	$63.200

Now, if the circuit is 600 miles, Morse-people must use <u>Repeaters</u> half way & thus require 30 more 2d rate operators, at $800 per an. $24.000

Difference against Morse $87.200

Here, you see, we beat Orton in the matter of <u>Labor</u> alone, at the rate of about 10 to 1, and of course, in the matter of Capital, expense of Lines, battery, depreciation, &c we shall beat our opponents nearly 20 to 1, and yet they console themselves that we are of no account & their <u>worthless</u> stock has been run up from 29 to 47.

When we get fairly to work, I should be pleased if you would send some friend, in whose judgment you can rely, to me, & I will take pleasure in exhibiting our system to him, & I have no doubt his report will be even more favorable then this Letter. <u>We can send 30.000 words per hour with absolute correctness, (by the aid of our Automatic Repeater)ᶜ to every city in the union, direct & at one writing—dropping copy at every place, with one wire.ᵈ</u> Truly &c

D. H. Craig.

ALS, UKLPO, ATF, item 21. ᵃ"<u>P.O. Box 3237</u>" written at top. ᵇUnderlined twice. ᶜ"Repeater" underlined twice. ᵈ"One wire" written in bottom right margin.

1. Frank Ives Scudamore (1823–1884) was at this time second secretary in the British Post Office. He was instrumental in the 1869 nationalization of the telegraph system and its subsequent management. *DNB*, s.v. "Scudamore, Frank Ives"; Kieve 1973, 128ff.

2. Gardiner Hubbard, a former lawyer from Cambridge, Mass., was a central figure in efforts to nationalize American telegraphy. His daughter later married Alexander Graham Bell. *DAB*, s.v. "Hubbard, Gardiner"; Lindley 1975, esp. chaps. 3–5.

3. Edison.

4. See Doc. 103.

5. These two perforators cannot be identified with certainty. The former probably resembles Doc. 255; the latter may be similar to an extant three-key perforator (Cat. 1523, NjWOE).

6. Typewriter.

7. William Orton.

–147–

From Daniel Craig

Peekskill, Feb. 25, 1871

Dear Mr. Edison:

Mr. Grace[1] wrote me that Mr. Harrington was agoing to Washington yesterday, so I reckon you are having no <u>bothering</u>, & I should'nt wonder if you wished I might be sick a month longer and Harrington be put into the Washington Penitentiary for at least that length of time.

On Tuesday next, weather permitting, I propose to bring a townsman, Mr. Anderson,[2] to you for you to examine what seems to <u>me</u> (very poor authority) a good beginning for a first rate Perforator, for Merchants' use. Harrington has been negotiating with Anderson for several weeks or months, & has got to a point where, if you think the machine of real merit, or that it could be made of any use to a party who should undertake to run <u>opposition</u> to us, we are to put it into your hands to develop, for our benefit & under our[a] control. We shall try to get out in the 11 o'clock train, and will go direct to your shop, and if you think best to let Unger see the machine (a full working model, made very rough) let him call there on his way to dinner. Anderson is an artist (painter) but belongs to a family of very ingenious people—his father a machinist—and as he had never seen or even talked with any person knowing anything about Perforators, I think you will agree with me that he has produced a <u>very creditable machine</u>. No doubt you can greatly improve it, and as we shall get the benefit of whatever merits it possesses, I hope you will be able to find time to give it a careful examination, and tell Anderson kindly & <u>moderately</u> what you think of his invention. I hope you are getting along nicely. Truly &c

D H Craig.

ALS, NjWOE, DF (*TAEM* 12:452). [a]"under our" interlined above.

1. Probably George Grace, superintendent of the Automatic Telegraph Co., who was assisting Craig in testing the Little system. Grace began his career as a telegraph messenger in 1851 and by 1860 worked as chief operator in the Springfield office of the American Telegraph Co. under his brother Frederick Grace. He then served as superintendent for the Franklin, the Bankers' and Brokers', and the National telegraph companies, before becoming superintendent of Automatic Telegraph. In 1874 Grace became general superintendent of the Southern and Atlantic Telegraph Co. Reid 1879, 454; "Automatic Telegraphy," *Telegr.* 6 (1869–70): 408.

2. Frank Anderson was an artist who, like Craig, was from Peekskill, N.Y. Operation of Anderson's perforator (U.S. Pat. 122,098) required neither knowledge of the telegraph code nor an external power source. By 1888 Anderson had developed a system of automatic telegraphy capable of clearly transmitting 3,000 words per minute. The perforator of this system had the capacity to punch 60 words per minute. Craig established the Machine Telegraph Co. to exploit the Anderson automatic system. *Elec. W.* 17 (1891): 183; Maver 1892, 291–94; Craig 1888.

–148–

*Daniel Craig to
Marshall Lefferts*

Peekskill, March 1/71

Dear General—

You are indebted to my inflamed Lungs and a fearful Cold, for your undisturbed serenity during the past three or four weeks.

I write you now to say that since Mr. Harrington got all the papers, executed, from that dirty Scallowag, Walter,[1] all has been going along as nicely as anything ever goes without my push; and Harrington seems to expect to be in full operation this month, but, of course, he cannot be until Edison says the word. Nothing could look better than the future does, and if nothing breaks, we (you and I) will soon be in clover—that is, we will be able to pay our debts.

I suspect you have not said a word to Mr. Hoadley for if you had, and had told him, as you might, that I was agoing to bring Automatic Telegraphy out and make it the top of the heap, very soon, he would not have written me as he did, the other day, a very sharp dun for his money.

Will you not go and tell Mr. Hoadley the situation of things, and tell him that it is utterly impossible for him to get his money until we get Automatic Telegraphy born. Mrs. Craig has paid & will pay the interest, and with that, on her side, you must aid me in keeping that old Christian quiet, and the sooner you get to work on this line the better it will be for your digestion.

I was told by Edison that Orton had seduced that drunken loafer—[-]Edison's late foreman[2]—who bulled[3] your 150 machines, and it is surmised that Orton, Prescott[4] and Phelps are agoing to "see," now, through a very dark lantern, all there is in your machines and in ours. Orton must be getting desperate.

In a private talk with Bennett[5] the other day, he intimated to me that he had a good understanding with Vanderbilt,[6] in Telegraph matters, and he said, with considerable emphasis—"all your old enemies will be out of the W.U. Co. within six months, and then if you and Lefferts have really got anything of value in your new system, I don't see why you should object to let the control fall into Vanderbilt's & his friends hands any more now than two years ago.

To my mind, this signifies that Bennett and the Vanderbilt people are one in Telegraph aspirations, and that there is to be a clean sweep, from Orton down, and that you[a] are the coming man. Now, tell me, on the square, is not this conjecture substantially correct? You know I have always had a sort of a hanker to see you Boss at 145,[7] and have said to you that then an arrangement satisfactory to that Co. and to all on our side, would be easy. Of late, I have felt as though that old hope was about played out, and I have been at work creating machinery which would be even now, hard to manage in that direction, and there should be no time lost in throwing an anchor to windward, if there really is anything at work looking to your becoming the manager of the W.U. Co.

What would be the harm of your asking Clark[8] for a private talk, and then tell him what you know of Automatic Telegraphy, and the utterly unreliable character of the disparaging statements put forth by Orton, Prescott and the other fools of the W.U. Co.

Then, for any good reason that may occur to you, arrange for me to see Clark, & let me tell him what we are prepared to prove over any good Line, and let him understand that you are the only person qualified to introduce Automatic Telegraphy upon the lines of the W.U. Co. and also that you are the only man who could exert sufficient influence to arrest Automatic Telegraphy as an opposition and hostile element and bring it in to harmony with the interests of the W.U. instead of the opposition Companies' interests.

The result of this little management, which can be confined to you & myself, would be to assure to you the position you ought to hold, as manager of the W.U. Co. and then it would be an easy step for you and me to make all parties see their interests in harmonizing with our views.

What I want, & all I care for, is the control of a News Co.

with one or two Automatic wires leading to all news-centres, and I will guarantee to bring every newspaper in the country into harness, and drive every Telegraph news shyster out of the business in less than 12 months.

With our new machinery, we shall certainly Perforate, with 1 girl, at three or four hundred dollars per year, 2000 words per hour—Edison says 2500 to 3000.

One of our copying (printing) machines has been worked by one who is over 50 years old & who has not had equal to 60 days practice upon the machine, 57 words per minute, for messages of over 500 words, and any girl or Boy can certainly average 2000 words per hour.

We have telegraphed, with accuracy, over 282 miles of wire, one thousand words per minute, and our Repeater works perfectly above 600 words per minute, and when we worked, for several days, from Washington to Albany (over a fearfully poor A. & P. wire nearly half the way) we really could discover no difference in the writing at Albany and at New York, at 600 words per minute, working direct, & dropping at New York.

Now then, here (on next page) is Automatic Telegraphy as Morse people have got to encounter it:—

Automatic = 600 miles Line.

5 girls, perforate 30.000 ws. pr. h @ $400.	$6.000
5 girls copy (plain Roman print) 400	6.000
2 girls to tend Motors.	800
Total, for operators, 1 year	$12.800[b]

Morse = 600 Miles Line.

43 wires, averaging 700 words per hour, will require 86 first class operators, at $1.200 per year.	$51.600
43 Repeater operators, @ $800	14.400
Total, for operators, 1 year,	$66.000

More than 5 to 1 against Morse, on Labor, and more than 20 to 1 on Capital & incidental Line expenses.[b]

We shall not require Repeater instruments on 1000 mile circuits, & probably not on 2000 miles. Our careful tests go to prove this, as they clearly do to telegraph at almost any speed where there is a regular flow of current, much or little.

Medill,[9] you know, said we could beat Morse out of sight on great lines or routes, but not on small lines or routes. He was green or fool enough to be taken-in by the Morse blowers.

Take a Line 100 miles long, & 10 offices, each averaging

100 words per hour: A girl will Perforate 100 words in 3 minutes & copy the same in 3 more, and will transmit the same in 1—Total 7 minutes. 3 minutes to Perforate & 1 to transmit, each office on the Line could be called every 40 minutes.

No Morse way-line with ten small stations, could call through & pass 1000 words over the wire, in less than 2 hours—so Morse offices would be served five times in 10 hours, and Automatic offices would be served fifteen times in 10 hours.

Morse Line would send 5000 words in 10 hours—Automatic Line would send 10,000 words or more, if offered, within the 10 hours.

The greater the business, the greater the Comparative money advantages—but the Automatic system, on small way-lines will show more than 50 per cent of business advantages—for, all other things being equal, telegraph business will go to the Line having the best facilities. We shall have a vastly more correct system than Morse can ever be even with first class operators, which, of course, are never found on way-lines.

I suppose, of course, that you know all these things much better than I do—but then, you have been running along in the Morse rut so long, that it will not do you any harm to get an eye-opener, occasionally.

Somebody told me that the G. & S. Co. was agoing in to the reporting of general news in all the great cities. Is this true? I hope not. I aggitated this feature of the news business, and started it as far as the Ass'd Press would permit me, 14 years ago. It is only a News organization like the Associated Press that can do that business legitimately, and whoever gets in to that business will have to bury me or I shall them. I should be very sorry to bury you or even your old shoes—so I hope your Co. will not force you in to a position where they cannot sustain you. Yours, all the time,

D. H. Craig

ALS, NNHi, Lefferts. ᵃUnderlined twice. ᵇFollowed by two centered horizontal lines.

1. This is a reference to the agreement between the National Telegraph Co. (of which George Walter was secretary) and the Automatic Telegraph Co. concerning the use of Little's automatic telegraph (Doc. 141). Harrington to Walter, 2 Feb. 1871, and Robert Squires to Harrington, 3 Feb. 1871, Respondents' Exhibits 2, 1:454, Box 735A, *Harrington v. A&P*.

2. Probably Andrew Hyde, Edison's foreman at the American Telegraph Works. The last date on which Hyde's name appeared on the

company payroll was 25 February. Cat. 30,108, Accts. (*TAEM* 20:226).

3. To "bull" meant to make a mistake or a blunder. Farmer and Henley 1970, s.v. "Bull."

4. George Bartlett Prescott (1830–1894), chief electrician of Western Union, authored a number of important works on electricity, including *History, Theory, and Practice of the Electric Telegraph* (1860) and *Electricity and the Electric Telegraph* (1877). He began his career in 1847 as an operator with the New York and Boston Magnetic Telegraph and soon became assistant manager of the company's Boston office. He later served as chief operator, manager, and superintendent for a number of telegraph companies, including the American Telegraph Co., which merged with Western Union in 1866. In 1870 William Orton appointed him chief electrician of Western Union and placed those electricians formerly serving as assistants to the company's general superintendents under Prescott's supervision. Prescott also assumed responsibility for all improvements to the company's lines and for evaluating the importance of new inventions. He continued to participate in the telegraph industry until 1882, when he retired and devoted his time to writing. *DAB*, s.v. "Prescott, George B."; Orton 1870; "A New Order of Things," *Telegr.* 7 (1870–71): 92.

5. Probably James Gordon Bennett, publisher of the *New York Herald*. *DAB*, s.v. "Bennett, James Gordon."

6. Cornelius Vanderbilt, railroad president and financier, was also a major shareholder and a director of Western Union. *DAB*, s.v. "Vanderbilt, Cornelius Henry."

7. Western Union's headquarters were located at 145 Broadway.

8. Horace Clark, railroad financier and chairman of the Western Union executive committee, was William Vanderbilt's brother-in-law. *DAB*, s.v. "Clark, Horace Francis."

9. Joseph Medill, publisher of the *Chicago Tribune*, was mayor of Chicago in 1871. Two years earlier he had witnessed the test of Little's automatic telegraph made by George Hicks, agent of the Western Associated Press. *DAB*, s.v. "Medill, Joseph"; Little 1874b.

–149–

From George Harrington

[New York?,] 4 Mch [1871]

My dear Edison

Hurrah! Now push to immediate completion one if not more of the hand transmitter[1] & one at least of the 3 keyd perforators—and that large perforator—the two former at once.[2]

I want to confer with you about switches[3] and the size of our tables & what kind will be best. at the proper time I will fix up one room for the tests in a manner suitable for gentlemen & in a manner not to disgrace our pretty machinery.[4]

Harrington

ALS, NjWOE, DF (*TAEM* 12:244).

1. On 4 March, Edison finished constructing a hand transmitter,

which probably operated by turning a hand crank. Just such a device is illustrated in the accompanying undated drawing labeled "Hand Transmitter Motor". Cat. 298:106, Lab. (*TAEM* 5:198–200).

Edison's sketch of an automatic telegraph transmitter operated by a hand crank.

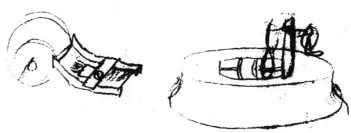

2. Account records show that Edison had worked on two large perforators as early as January 1871 and a three-key perforator in early February. On 17 February he had filed a caveat (not found) for a perforator; on 6 March he executed another, for a large perforator. Bills to Automatic Telegraph Works, Jan. and Feb. 1871, 71-015, DF (*TAEM* 12:578–82); Cat. 30,108, Accts. (*TAEM* 20:205–20); Quad. 72.11, pp. 44–45 (*TAEM* 9:187–88); caveat of 6 Mar. 1871, DF (*TAEM* 12:643 and *TAEM* Supp. III).

3. Switches diverted current from one circuit to another, thereby saving the time otherwise used in removing instruments from one circuit and rearranging them in another. Because a variety of switches existed, Edison and Harrington had to evaluate which kind would best serve their needs. Pope 1869, 37–42.

4. On 5 May, Edison had "four men working day & part night—on fitting up" a sewing machine table with instruments and batteries for the purpose of exhibiting the automatic telegraph. Bill to Harrington, 5 May 1871, DF (*TAEM* 12:586).

–150–

From Daniel Craig

Peekskill, Mar. 16/71

Dear Mr. Edison:

By doubling the Perforated paper you will see that ordinary pens or rollers <u>would not</u>, by considerable, make <u>dashes</u>.[1] What I send is your <u>M</u> on "Mississippi."[2] To make a dash, the upper hole would have to be at least ⅓d larger. To make sure, it should <u>overlap</u> the two lower holes as much as the thickness of three thicknesses of thick writing paper.

Would it not be possible to have the 3 keys so arranged that you could depress the whole <u>three</u> at once?[3] I think you struck <u>two</u> at one time & <u>one</u> the next, to make the dash. I did not see you make the long dash at all, but I suppose it is <u>thar</u>.[4]

If you don't let me see your Perforator and Printer <u>working</u>, soon, I think I shall hang myself—for things are getting too <u>hot</u>. I am too much a subject for <u>commiseration</u> on the part of my friends.

I shall try & see you on Saturday   Truly &c

D. H. Craig

ALS, NjWOE, DF (*TAEM* 12:454).

1. By doubling the thickness of the paper, Edison hoped to prevent it from tearing.

2. Transmission of the letter "M" consisted of two consecutive dashes.

3. In his first perforator patent (U.S. Pat. 121,601), Edison described only overlapping holes of equal size. He developed an arrangement of punched holes to represent a dash similar to that recommended here by Craig, but he did not claim it until his second perforator patent (U.S. Pat. 132,456). See Doc. 255.

4. A long and a short dash, in addition to the dot, were elements of the telegraph code used in nineteenth-century America. For example, a single short dash meant the letter "T," and a sing' long dash represented the letter "L." Pope 1869, 99–101.

–151–

From George Harrington

New York, 18 Mch 1871[a]

Sir;

I accept your proposition to make fifty of your perforators with the various improvements referred to,[1] for one hundred dollars each and hereby authorize and request you to manufacture them forthwith

Geo Harrington   Presdt Aut Co

ALS, NjWOE, DF (*TAEM* 12:245). Letterhead of the Automatic Telegraph Co. [a]"New York," and "187" preprinted.

1. Edison continued to experiment on designs for perforators, and on 11 May he filed two covering caveats. One included modifications of his large perforator; nothing is known of the other. Account records show that by the end of April he completed one large perforator and one three-key perforator. Quad. 72.11, pp. 44–45 (*TAEM* 9:187–88); caveat of 11 May 1871, PS (*TAEM* Supp. III); Cat. 30,108, Accts. (*TAEM* 20:236–37).

–151A–

Technical Note: Automatic Telegraphy[1]

1[3]

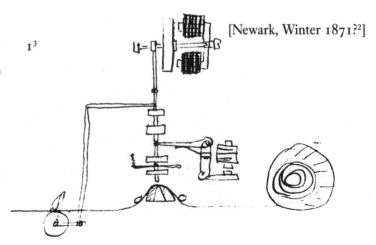

[Newark, Winter 1871?[2]]

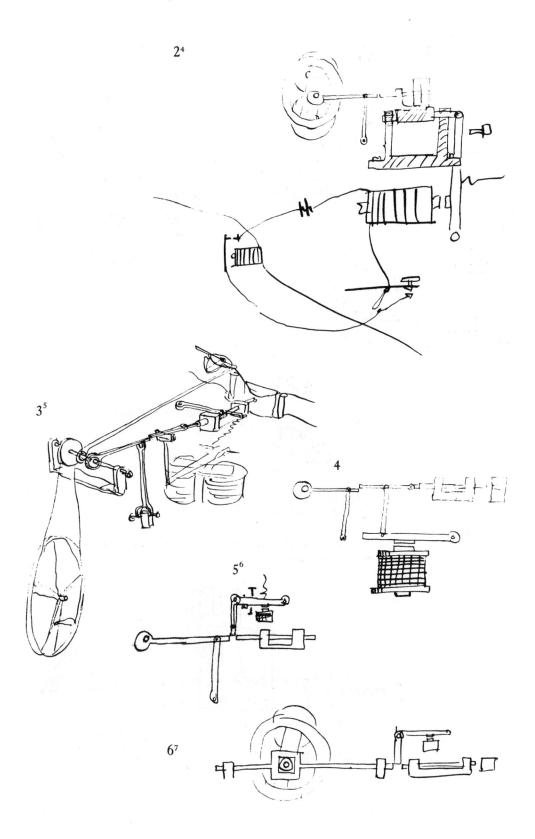

2⁴

3⁵

4

5⁶

6⁷

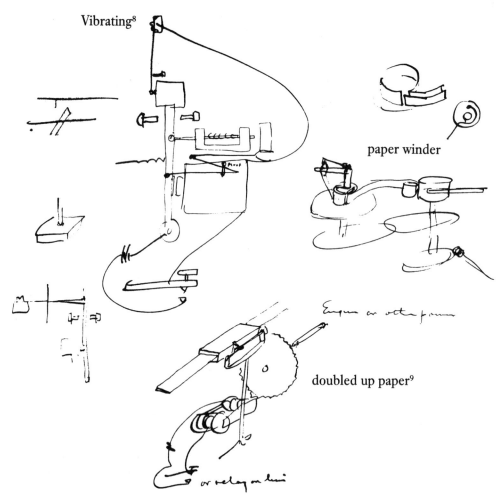

Vibrating[8]

paper winder

doubled up paper[9]

emboss paper deeply & cut off top of embossed dot or dash
by a revolving knife = [10]

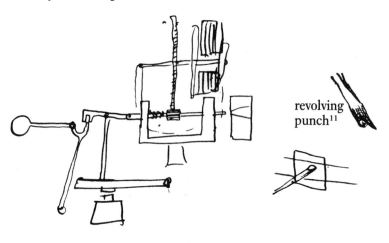

revolving
punch[11]

magnet brings revolving punch forward and it acts as a shear revolving with great reapidity. use continuous feed =

moveable die = = wheel & ratchet for paper feed = clock-work for greciiprocating = self[a] vibrating weighted lever = cutting wheel = revolving cutting punch = = reciprocating transmitter = for sending reverse currents bring second punch in only when other punch not wkg =

mention that in local perferater magnet may be dispensed with & key & hand used =

AX, NjWOE, NS-Undated-005, Lab. (*TAEM* Supp. III). [a]Interlined above.

1. This material, which appears to be preliminary notes for a caveat, came to light as this volume was in press; hence the intercalary numeration. It will be filmed at the end of Part III of *TAEM* as a supplement.

2. Edison might have made these notes as early as January. On 26 July 1871 he executed a caveat describing an electric perforator, but it was unlike these designs. See Doc. 289 headnote.

3. In each of the six numbered sketches, Edison is trying to harness a constantly reciprocating rod to punch holes in the paper tape, either by moving a punch in line with the rod (3 and 4), by closing a gap between the rod and the punch (2), or by introducing a tongue of metal into that gap (1, 5, and 6). In each case the sketch shows only one punch; a perforator would have several. Depressing a key would activate the appropriate punches for that letter or dot/dash signal. Several of the punchers drawn have electric motors like the one Edison was using in his universal private-line printer.

In this design the motor (at top) drives an eccentric (as shown in elevation in sketch 2). At lower left, driven by the eccentric, is the "wheel ratchet for paper feed" mentioned at the end of the document. When the electromagnet at right center is charged, levers connected to its armature drive a tongue between the reciprocating rod and the punch lever. The object at lower right is the roll of paper.

4. This design incorporates the "moveable die" mentioned near the end of the document. When the electromagnet at center is charged and attracts its armature, the die (top center) slides to the left and the reciprocating rod (driven by the electric engine at top left) hits the punch, perforating the paper. The line wire (with electromagnet) slanting from lower right to middle left indicates that Edison intended this to be used remotely (perhaps as the "local perforator" mentioned later). Cf. Doc. 194.

5. Here the operator supplies power with the hand wheel at lower left. The long rod attached to the rotating shaft at left center is the paper feed; the other rod supplies power for the punch. Each is on an eccen-

tric. When the electromagnet at lower center is charged, the bar hinged to the punch (center) descends and is struck by the reciprocating rod. Sketch 4 shows the basic punch mechanism from the side.

6. In this variation, charging the electromagnet pulls down the tongue (drawn vertically at center) between the reciprocating rod and the punch.

7. This design is essentially the same as sketch 5; here the power supply is an electric motor.

8. This is the "self Vibrating weighted lever" mentioned later in the document. Closing the key at bottom charges the electromagnet at right center. The armature lever is weighted at the top and has the punch attached (extending to the right just above the electromagnet). As the lever moves toward the magnet, the circuit breaks (top); as the lever is pulled from the magnet, contact is made and the magnet charges again. The oscillation will continue until the key is released.

9. Here the paper tape is folded and cuts are made in the fold. When the paper is opened, the perforations will appear in the middle. The labels on the figure are "or relay on line", which refers to replacing the key, and "Engine or other power".

10. Cf. Doc. 135C.

11. The motor at top drives the punch shaft at center through a gear wheel. When the magnet at left center is charged, a lever engages the reciprocating piece at left, pushing the punch through the paper. At right is a detail of the punch.

–152–

From George Harrington

New York, 22 Mch 1871[a]

My dear Edison

I telegraphed to day to know when you will be over.

I want your "sounder" to try.[1] We can have the circuit to Charleston for the purpose. If successful 'twill be a card for you. Also I want a roll or two of your chemical paper for per-manent recording which shall have a full trial. I will give you the rolls.[2]

I hope to see the sounder on Monday. very truly

Harrington

I think we can get the Charleston circuit at between 12 & 2 o clock.[3]

ALS, NjWOE, DF (*TAEM* 12:246). Letterhead of the Automatic Telegraph Co. [a]"New York," and "187" preprinted.

1. The "sounder" may be Edison's relay (U.S. Pat. 114,657), as those two terms were sometimes used interchangeably. "Western Union and Automatic Telegraphy—Mr. Little in Reply to Mr. Orton," *Telegr.* 10 (1874): 259; Pope 1869, 32.

2. Depending upon the chemicals employed and the metal constituting the stylus, the marks recorded in automatic telegraphy were either transitory or permanent. A platinum pen and a solution of potassium iodide (KI), for example, left transitory marks called "fugitive printing."

An iron pen used with paper soaked in an aqueous solution of ammonium nitrate (NH_4NO_3) and potassium ferrocyanide ($K_4Fe(CN)_6$) gave a permanent record. Sabine 1872, 202, 212.

3. This is probably a reference to the line of the Southern and Atlantic Telegraph Co. that ran between Washington and Charleston, S.C. Between 12:00 P.M. and 2:00 A.M., telegraph traffic was light, allowing Harrington to test Edison's relay.

-153-

List of Machinery

[Newark,] March 28—71

Machinery ordered by Mr Harrington[1]

4[a]	Large perforators[2]
34[a]	small hand perforators[3]
I	Perforator with compressed air—hand—[4]
3	Pocket Correctors[5]
I[b]	Paper wetter[6]
I[a]	Power Transmitter[7]
3[a]	Repeaters[8]
3[c]	Ink Receivers—Electric[9]
I	" " Weight[d]
2[b]	Hand Transmitters straight[10]
2[e]	" " Aperture Pen devices
3[a]	weight hand Transmitters, with Automatic thrown in[11]
3	Magnetic[12] " " " "
3	Polarized Relays
3	Mechanical Printers[13]
	Assortment of Pens.

AD, NjWOE, DF (*TAEM* 12:247). [a]Entry preceded by "x" and followed by Harrington's "—". [b]Entry preceded by "x" and followed by Harrington's "d[e]l[ivere]d—". [c]"x" precedes entry. [d]Followed by Harrington's "—". [e]"x" precedes and follows entry.

1. As seen in the illustration on p. 266, Edison also wrote a slightly different, undated version of this list, which he entitled "List of Machinery necessary to exhibit a perfect 'Fast system' [i.e., automatic telegraphy] and all its ramifications." 71-003, DF (*TAEM* 12:256).

2. According to American Telegraph Works accounts, employees worked on a large perforator between February and April 1871. Foundry work on ten other large perforators was completed by 18 May, and several perforators were constructed during the summer. Cat. 30,108, Accts. (*TAEM* 20:188); Doc. 162; agreements between Frederick Spannenberg, William Parkinson, and the American Telegraph Works, 17 and 26 July 1871, DF (*TAEM* 12:301-7).

3. The small hand perforators were most likely the manually operated, three-key machines for which Edison and Unger billed the American Telegraph Works on 3 April 1871 (71-015, DF [*TAEM* 12:583]). Employees of the American Telegraph Works also worked on a three-

Edison's list of the machinery that was needed to operate an automatic telegraph system.

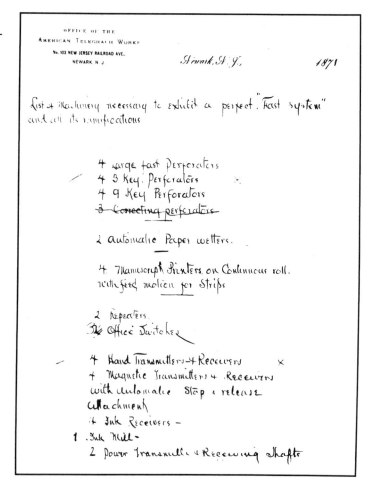

key perforator between February and April 1871. Cat. 30,108, Accts. (*TAEM* 20:188).

4. This device may have been powered by a compressed-air cylinder similar to that described in Doc. 196. "Hand" may refer to the manner in which the device was operated, perhaps like the perforator of Richard Culley and Charles Wheatstone, the valves of which were worked by finger keys. "Wheatstone's Automatic Telegraph," *Telegr.* 4 (1868–69): 273.

5. A pocket corrector could fix faulty perforations such as those caused by operator error, the punched-out portion clinging to the paper tape, or the punch cutting out only a partial opening. Ibid.

6. The paper-wetter was probably the "chemical paper machine" listed in the payroll accounts of the American Telegraph Works for 18 and 25 March 1871. It may have saturated the chemical paper for the Little automatic receiver. Cat. 30,108, Accts. (*TAEM* 20:188).

7. The accompanying drawing shows a device with a number of transmitters operated by a belt-driven pulley. Cat. 298:21, Lab. (*TAEM* 5:38).

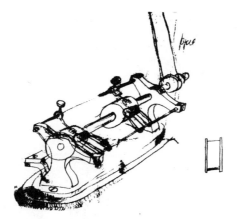

8. Edison and Unger charged the American Telegraph Works $50.00 for each of three repeaters on 5 May 1871. These were probably Edison's patented repeater (U.S. Pat. 114,657). 71-015, DF (*TAEM* 12:586).

9. Edison did not execute a patent for an ink recorder until 12 August 1871 (U.S. Pat. 124,800). Edison and Unger billed the American Telegraph Works on 6 April 1871 for experimental work on a model of an ink recorder, on 24 April 1871 for polishing an ink recorder, and on 2 May 1871 for a Patent Office model of an ink recorder. As seen below,

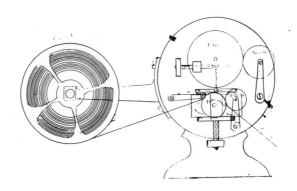

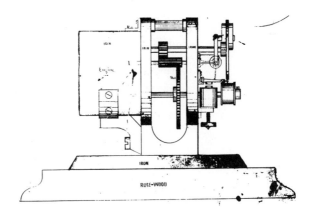

two undated notebook entries show a slightly different form of ink recorder from that found in Edison's patent. Bills to American Telegraph Works, 71-015, DF (*TAEM* 12:583–85); Cat. 298:9, 64, Lab. (*TAEM* 5:19, 123). See also the drawing in PN-70-12-25, Lab. (*TAEM* Supp. III).

This drawing from Edison's patent for an automatic telegraph ink recorder differs from his notebook drawings (U.S. Pat. 124,800).

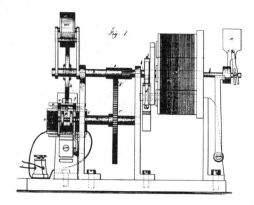

10. The hand transmitter (see Doc. 149, n. 1) was operated by a hand crank. Cat. 298:106, Lab. (*TAEM* 5:199).

11. The accompanying drawing of a weight transmitter does not show the drive mechanism, but it was probably powered by a descending weight. The "automatic" may have been an automatic paper-feeding device. Cat. 30,094, Lab. (*TAEM* 5:350).

Edison's drawing of an automatic transmitter operated by a weight.

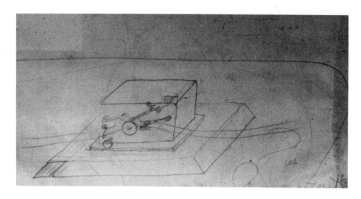

12. Account records show that Edison and Unger charged the American Telegraph Works for work on a magnetic transmitter on 24 April 1871. An electromagnetic (hence "magnetic") motor probably drove the apparatus. 71-015, DF (*TAEM* 12:584).

13. Typewriters.

–154–

From Daniel Craig

Peekskill, N.Y. March 29/71

Dear Mr. Edison:

Hall[1] made for us, in March, 1868[9], a working model of the Little Motor and Puncher, and Tablet.[2]

In Jan. 1870, after Little had considerably changed all the

mechanism,[3] Hall & two other men, went to work and made Ten of each in 13 weeks, and Hall himself did not work on the machines over ½ of the time.

I asked your foremen today, how long it would probably take to make 4 more Perforators, supposing the present one was found to be all right, and he almost took my breath away by saying, "about three months!"[4]

Surely, there must be some grievous mistake about this.

He said working drawings could be made from the machine, in a week or less, so that we could take the machine away for practice—and I hope you can confirm this.

Are you absolutely sure, from your own personal observation, that all is being done that can be, to expedite the completion of our machines? The 3 weeks of January have run in to three months, and as I know we are all suffering terribly by the delay, I feel as though I must speak to you about it—but, of course, only to be sure you are doing all you can to help us out of hot water. Truly &c

D H Craig

ALS, NjWOE, DF (*TAEM* 12:456).

1. Probably Thomas Hall, Boston electrical manufacturer.

2. Regarding George Little's motor, see Doc. 132, n. 2. His puncher (U.S. Pat. 96,331) was a multikeyed perforator that used electromagnets to punch the holes and advance the paper. Little's tablet (U.S. Pat. 96,330) used essentially the same perforating mechanism, but the operator indicated the letter to be punched by sliding a metal stylus the length of a groove corresponding to that letter. A unique arrangement of contacts at the bottom of each groove differentiated the letters.

3. It is not known what alterations Little made.

4. This is probably a reference to the four large perforators ordered by Harrington on 28 March. See Doc. 153, n. 2.

–155–

Agreement with George Harrington[1]

Newark, April 4, 1871[a]

Whereas I, Thomas A. Edison, of the City of Newark, State of New Jersey, for certain valid and valuable considerations, to me in hand paid, and in further consideration of certain covenants and stipulations to be fulfilled by George Harrington, of Washington, District of Columbia, did stipulate and agree to invent and construct for the said Harrington full and complete sets of instruments and machinery that should successfully and economically develop into practical use the Little or other system of automatic or fast system of telegraphy, and subsequently to improve and perfect such instruments and machinery by adding thereto from time to time such further inventions as experience should demand, and my ability as an inventor and electrician might suggest and permit; and furthermore, to prepare, or cause to be prepared, the necessary descriptive[b] papers, the models[c] and drawings requisite and necessary[d] to obtain patents for all such inventions and improvements, to[e] be the joint property of the said Harrington and myself, and the patents to be issued to the said Harrington and myself in the proportionate interest of two thirds to said Harrington and one third to myself, the whole to be under the sole control of said Harrington, to be disposed of by him for our mutual benefit in the proportions hereinbefore recited, in such manner and to such extent as he, the said Harrington, should deem advisable, with power to sell, transfer and convey the whole or any part of the rights and titles in and to any or all of said inventions and improvements, as also of the patent or other rights arising therefrom; and the said Harrington having faithfully fulfilled all of the covenants and stipulations entered into[f] by him: Now, therefore, be it known, that in consideration thereof, and of the sum of one dollar to me in hand paid, I, Thomas A. Edison, of the City of Newark, State of New Jersey, do[f] by these presents hereby assign, set over and convey to him, the said Harrington, two thirds in interest of all my said inventions, including therein all my inventions of mechanical or copying printers,[2] and of all the patents for all such inventions and printers, whether already issued, applied for, or to be hereafter applied for, and of all and whatsover of my inventions and improvements made or to be made, and of all the patents that may be issued therefor, that are or may be applicable to automatic telegraphy or[3] mechanical printers.

And whereas, I am desirous of obtaining the coöperation and assistance of the said Harrington[g] in disposing of my said one third interest, as before recited, and for the purpose of united and harmonious action in negotiating for its use or its sale and transfer by or to others in conjunction with his own, and in such free and unrestricted manner as will tend to success, and for the sum of one dollar to me in hand paid, the receipt whereof is hereby acknowledged: Now, therefore, be it known, that I, Thomas A. Edison, of the City of Newark, State of New Jersey, have constituted and appointed, and by these presents do constitute and appoint George Harrington, of the City of Washington, District of Columbia, my true, lawful and only attorney, irrevocable,[4] with power to substitute for me and in my name, and in such manner as he may think best, to sell, transfer and convey all my rights, titles and interest in and to any and all of my said inventions, and the improvements thereto, whether made or to be made, and to sell, transfer and convey all of my rights, by patent or otherwise, arising therefrom, already made and obtained, and all such as may hereafter be made or obtained, and to execute in full any or all necessary papers and documents requisite for the transfer of title, and to invest in other parties full and legal ownership therein; hereby divesting myself of and investing him, the said Harrington, with all the powers necessary in the premises fully and completely to carry out the purposes and intentions herein set forth, hereby fully confirming all that my said attorney may or shall do in the premises as fully as if done by me in person; and request the Commissioner of Patents to recognize him as such attorney.

In witness whereof, I have hereunto set my hand and affixed my seal, in the City of Newark, this fourth day of April, eighteen hundred and seventy-one.[5]

T. A. EDISON,[h]

In presence of   A. D. COBURN, A. B. CANDEE.[6i]

PD (transcript), NjWOE, Quad. TLC.2, p. 89 (*TAEM* 10:864). Ten versions of this document are in the Quadruplex Case, two more are in *Harrington v. A&P*, and another is in Libers Pat. U-13:412. Among the versions there are dozens of minor textual differences, most of which were not noted in the court proceedings. However, a few discrepancies were observed and one became an issue (see n. 3). This copy was selected because it best represents all the legal copies, its obvious errors are few, and it is clearly punctuated. [a]Place and date taken from text; form of date altered. [b]"description" in some copies. [c]"model" in some copies. [d]"and necessary" missing in some copies. [e]Preceded by "the said inventions and improvements" in some copies. [f]Not present in some copies. [g]Preceded by "George" in some copies. [h]Representation of seal follows. [i]Followed by representation of 50¢ Internal Revenue stamp; other copies indicate a different placement and another stamp, both canceled by Edison.

1. This document amplifies the agreement of 1 October 1870 (Doc. 109) by explicitly assigning to Harrington partial rights to certain patents and granting to him Edison's power of attorney in regard to those patents. Quad. 70.1, pp. 1–2 (*TAEM* 9:293–94).

2. Typewriters.

3. In the Quadruplex Case, Western Union claimed that this word had been forged on the original; it does not appear on some copies. For evidence and some argument regarding this and other disputed textual variations, see Quad. 71.1, pp. 1–24 (*TAEM* 10:5–17).

4. The duration of this arrangement was governed by the five-year term of the agreement reached on 1 October 1870 (Doc. 109).

5. This copy has an addendum (dated 6 May 1871) indicating that it had been copied into the Libers of Patent Assignments (U-13:412).

6. Unidentified.

–156–

To George Harrington[1]

[Newark?, April 1871?][2]

Had Printer[3] operating, but broke it. am making some alterations, may be done tomorrow

Perforator getting on finely, will probably have it within the time promised Mr Craig.

Tools all done for Gold & Stock Contract[4]—and we are pushing right ahead on it =

In the Other Shop[5] I am having made the "Little Distributer"[6]   Expenses may be a Little heavy for the present but the results will be tremenduous!

Edison

ALS, NjWOE, DF (*TAEM* 12:255).

1. The discussion of the printer, perforator, and "the other shop" indicates that Edison wrote this to Harrington.

2. Edison was working on both a perforator and a typewriter (printer) in the spring of 1871, and in April he began production of cotton instruments for Gold and Stock at the American Telegraph Works. The universal stock printer, manufactured in the fall, was produced at the Newark Telegraph Works.

Advertising card for the American Telegraph Works, established by Edison and George Harrington.

AMERICAN

TELEGRAPH   WORKS,

No. 103 New Jersey Railroad Ave.,

NEWARK, N. J.

Manufacturers of Telegraph Machines, Small and Accurate
MECHANISM.

T. A. EDISON, Agent.

3. Typewriter.

4. The contract was for 150 cotton instruments; see Docs. 131 and 136.

5. The Newark Telegraph Works.

6. Unidentified; possibly some component of George Little's automatic system.

*And William Unger
Lease from William
Kirk*

LEASE.[b]

This is to Certify, That I have this First[c] day of May 1871 LET and RENTED unto[d] Wm Unger & Thos A Edison the third and fourth stories of my new factory corner of Ward street & Pear Alley in the city of Newark N.J.,[1] with the privilege of four horse steam power   The power in the third story not to be charged for until applied with Appurtenances, and the sole and uninterrupted use and occupation thereof for the term of one year to commence the first d[c] day of May 1871 at the yearly rent of Fifteen hundred[c] Dollars payable[c] monthly in advance.

WITNESS,[b] Wm. S. Vliet[2]   Wm H Kirk[3]

ADS, NjWOE, DF (*TAEM* 12:480). Written on a standard form with "LANDLORD'S AGREEMENT" preprinted at left. [a]Place and date taken from text; form of date altered. [b]Preprinted. [c]Followed by dash to fill in extra space. [d]To this point, "This . . . That", "have this", "day of", "18", and "LET . . . unto" preprinted. [e]"with . . . of", "to commence the", "day of", "at . . . of", and "Dollars payable" preprinted.

1. The legal address was changed to 10–12 Ward Street under a new numbering system adopted by the Newark City Council about 1869, but the old address, 4–6 Ward Street, continued to be used for some time.

2. William Vliet was a Newark carpenter and builder. *Holbrook's Newark City Directory* 1871, 632.

3. William Kirk was a Newark builder and landowner. Ibid., 403; Hopkins 1873, Fourth Ward map.

SCREW-THREAD UNISON   Doc. 158

Printing telegraph systems required that the sending and receiving instruments be precisely synchronized. A problem at the sending end, on the line, or in the printer could make the printer "throw out"—that is, fall behind the transmitter by one or more letters. A unison device whereby the transmitting operator could reset all the printers on a circuit to the same character was therefore much desired. Sometime in the spring of 1871 Edison devised a unison stop mechanism that was one of his two most significant inventions in printing telegraphy (the other was the typewheel-shifting mechanism invented for the cotton instrument). By mid-May, Edison had sent a draft of the U.S. patent specifications for the unison stop to his patent attorney, Lemuel Serrell. George Field and Elisha Andrews also included the unison in a British patent application they communicated to their London solicitor, William Lake, in early May.[1]

The fundamental feature of Edison's unison was a small peg seated in a screw thread on the typewheel shaft (see Doc. 158, photograph B). As the typewheel rotated, the thread moved the peg to one side. Every time the printing-lever magnet was charged, the peg lifted and was drawn back to its starting point by a spring. If the printing magnet was not charged, however, the peg continued to move until the arm holding it hit a "stop" attached to the typewheel, which stopped the rotation of the shaft. By sending enough impulses to rotate a typewheel the requisite number of turns, a transmitting operator could bring all the instruments on a circuit to unison. The operator then charged the printing magnet, which freed all the typewheels at the unison point. The operators on reporting circuits were supposed to run the machines to unison every five minutes or so, but in practice they were often too busy. For private lines, Edison recommended resetting the printers before each communication.[2]

These undated sketches appear to show Edison's redesigned screw-thread unison: the mounting for the release arm on the printing lever (top); the thread on the shaft (center); and the top of the release arm (bottom).

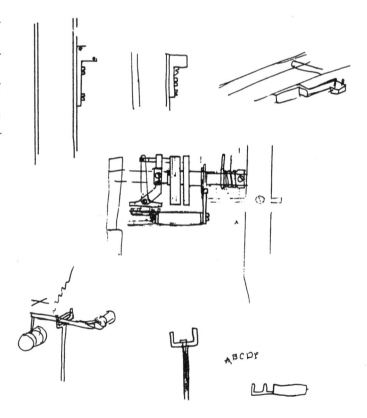

In its earliest form, the screw-thread unison was actuated by the printing magnet but was mechanically independent of the printing lever.[3] Edison soon modified it, however, by attaching the release mechanism to the printing lever, and the arm holding the peg to the ink-roller shaft.[4] Subsequently he attached the arm itself to the shaft of the printing lever.[5]

1. James Reid called the screw-thread unison "the invention of . . . Theodore M. Foote . . . as constructed by Edison," but the only similarity between the Edison and Foote unisons was that both were actuated by the printing magnet (Reid 1879, 618). Foote's unison was specified in U.S. Patent 105,060.

A 13 February 1871 Newark Telegraph Works invoice to the American Telegraph Works included a pattern for a "unison post" for the Gold and Stock Telegraph Co. (71-015, DF [*TAEM* 12:580]). Edison's agreement with Gold and Stock in May (Doc. 164) included the screw-thread unison under "List of patents applications for which are pending and in progress at the solicitors," even though the unison application was not executed until 13 November (U.S. Pat. 126,535). Field and Andrews's application, filed in England on 24 May, became British Patent 1,400 (1871). An undated notebook entry lists the time needed to make and install one screw-thread unison. PN-70-12-25, Lab. (*TAEM* Supp. III).

2. NS-Undated-005, Lab. (*TAEM* 8:258); Healy, 1905, 121; Gold and Stock 1872, 5–6; idem n.d., 14.

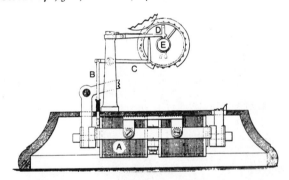

3. According to the diagram shown above, when printing magnet **A** charged, it pulled down rod **B**, which pivoted arm **C** and lifted tooth **D** out of the screw thread on typewheel shaft **E**. Photograph B of Doc. 158 shows the screw thread and tooth; rear-view C shows the lifting mechanism.

4. U.S. Pat. 126,532.

5. George Phelps later adopted the screw-thread unison on his stock ticker, which was widely used toward the end of the century. In Phelps's version, a wire was wrapped in an open spiral around the typewheel shaft, and a flat bar rested on edge in the groove thus created. The New York Stock Exchange Archives houses a representative machine.

[A]

Patent Model: Printing Telegraphy[1]

[Newark,[2] c. May 1, 1871[3]]

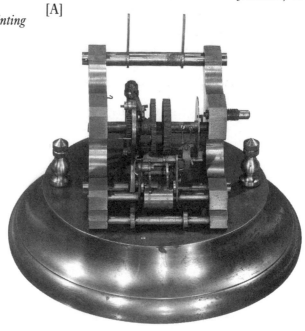

[B]

[C]

M (21 cm dia. × 13 cm), MiDbEI, Acc. 29.1980.1313. The ink roller and the spring between the frame and unison arm are missing.

1. See headnote above.

2. Edison modified a cotton instrument (serial no. 29) to create this patent model. Although the instrument is stamped "P. Kenny 35 Ann St. N.Y.," the design and modifications were probably made in Newark.

3. See headnote above, n. 1.

*George Harrington
Agreement with
William Palmer,
Samuel Parsons,
John McManus,
Josiah Reiff, and
William Mellen*

[New York?,] May 10, 1871[a]

Memorandum of an Agreement made the tenth day of May, A.D. 1871, between George Harrington, party of the first part, and William J. Palmer,[1] Samuel B. Parsons,[2] John McManus,[3] Josiah C. Reiff and William P. Mellen,[4] parties of the second part. Witnesseth:

That whereas the said Harrington entered into a contract with Thomas A. Edison, dated October 1st, 1870, a copy of which is hereto annexed,[5] on account of which he has paid out the sum of six thousand dollars, and advanced the further sum of ten thousand dollars for machinery, tools, fixtures and other property, situated at corner Green street and Railroad avenue, in the City of Newark, an abstract of the inventory of which is hereto attached; and also the sum of eleven thousand dollars, for work, materials, etc., in inventing, making and perfecting machines connected with the development of automatic telegraphy, and a machine for mechanical printing, some of which have been perfected and some of which are but partially completed:

And whereas said Harrington has also received a power of attorney, from said Edison, dated April 4th, 1871, by and in virtue of which he has acquired certain rights in connection with said contract and with inventions made and to be made by said Edison, and services to be performed by him, a copy of which is also hereto annexed.

And whereas still further money is required for the purpose of purchasing more machinery, etc., and for continuing to invent, make and perfect machines as aforesaid, and for doing all things contemplated by the contract with said Edison:

And whereas the said parties of the second part desire to become interested with said Harrington in the said contract and in the prosecution of the pursuits and business aforesaid and said Harrington desires that they should become so interested:

Now, therefore, it is mutually agreed by all the parties hereto as follows:

Said Harrington agrees to sell, transfer, assign, and set over unto the said parties of the second part, one, full, equal, undivided half part and interest[6] in the said contract with said Edison, and in all property, rights, and interests of every name and nature which have been acquired under, or have grown out of the said contract or otherwise in connection therewith, or which may result therefrom, and the said parties of the second part agree to purchase and receive the same, upon the terms and conditions following, viz.:

Said parties of the second part agree to reimburse and refund to said Harrington the sum of twelve thousand dollars, that being the amount which he has paid as aforesaid over and above the sum of fifteen thousand dollars in connection with the contract, pursuits and business aforesaid and also the sum of seven thousand five hundred dollars, in consideration of the greater risk of loss incurred by him in originally making the said contract with said Edison, and in advancing and paying the money under it as aforesaid.[7]

And the said parties mutually agree that after the date hereof all further moneys which both parties shall consider necessary for use in prosecuting the said pursuits and business shall from time to time be paid and furnished one-half by said Harrington and one-half by said parties of the second part.

And it is further mutually agreed that if either or any of the said parties shall desire to limit the whole amount invested and to be invested under this agreement to thirty thousand dollars he or they may do so, and nothing herein contained shall be construed as obligating him to pay any more than his pro rata share thereof, and further, that in case said Harrington or said parties of the second part shall wish to terminate the said business after the thirty thousand dollars shall be invested, and before undertaking to prosecute it upon a larger scale, it may be done upon written notice to that effect, and thereupon the property, rights, interests and effects aforesaid shall be sold and the net proceeds therof shall be divided pro rata, or either party desiring to continue may purchase the same by refunding to each of the other parties the amount paid in by him. And it is further stipulated that no debts shall be contracted in carrying on the said business or in connection therewith.

In testimony whereof, we have signed this agreement the day and year first above written.

GEO. HARRINGTON,
WM. J. PALMER, by Wm. P. Mellen,

SAM. B. PARSONS,
JOSIAH C. REIFF,
WM. P. MELLEN.

ADDENDUM[b]

Newark[c] [10 May 1871][8]

Abstract of Inventory of property, fixtures, etc., at Newark:[9]
1061 Files, assorted sizes, and of all kinds.
44 Vises of different kinds and sizes.

826 Lbs. Cast iron anvils and 2 larger anvils.

19 Gongs of various kinds.

63 Hammers of various kinds and sizes.

10 Hand punches.

72 Chisels.

9 Oil cases, various kinds and sizes.

112 Turning and planing tools.

4 Engravers.

46 Lathe tool rests and fixtures.

426 Drills of various sizes.

19 Set brass vise clamps.

48 Figs[10] of different kinds.

9 Gauges.

50 Reamers.

668 pieces and articles, comprising tape measure, clamps, brushes, emery wheels, funnels, wrenches, oil stones, nippers, shears, mandril, planes, drill chucks, saws, bitts, hand-turning tools, etc., comprising all the tools necessary for a well furnished machine shop.

Machinery. Consisting of shafting, pulleys, 18 engine lathes, punch machine, milling machine, belting of various sizes, etc.

Office Furniture. Comprising tables, chairs, carpet, matting, and generally all necessary office furniture.

Fixtures. Comprising gas pipe of different dimensions, gas burners, shelving, rubber hose, window shades, work bench, ash and pine plank, etc.

Stock. Consisting of iron wire, sheet iron, nails, cast iron, cast steel, screws, bases, printer No. 1, copper wire, etc.

A detailed inventory of all the articles in this abstract is entered in the Property Book.

PD (transcript), NjBaFAR, *Harrington v. A&P*, Box 735A, Defendants' Exhibits, 2:1434–38, 1444–47. [a]Date taken from text, form altered. [b]Addendum is a PD (transcript). [c]Place taken from text.

1. William Jackson Palmer (1836–1909) was a railroad executive and financier. Prior to the Civil War, he served as private secretary to J. Edgar Thompson, president of the Pennsylvania Railroad. During the war he organized the 15th Pennsylvania Cavalry, eventually earning the rank of brigadier general. In 1894 he received the congressional Medal of Honor for his service during the war. After the war Palmer was treasurer of the Eastern Division of the Union Pacific Railroad, which became the Kansas Pacific Railroad before merging with Union Pacific. In 1870 he founded and headed the Denver & Rio Grande Railroad. He was also involved in Mexican railroads, including the Mexican National Railway, of which he was president. Palmer helped found the city of Colorado

Springs, Colo. His wife was the daughter of William Mellen (see n. 4 below). *DAB*, s.v. "Palmer, William Jackson"; *New York Times*, 14 Mar. 1909, 1.

2. Samuel Brown Parsons (1819?–1906), a nurseryman from Flushing, N.Y., introduced the seedless orange into the United States in 1859. *New York Times*, 5 Jan. 1906, 11; Parsons's testimony, 1:156, Box 17A, *Harrington v. A&P.*

3. John McManus (n.d.) was a partner in the Reading, Pa., firm of Seyfot, McManus & Co., proprietors of the Reading Iron Works, Scott Foundry, Steam Forge & Sheet Mill. McManus to Thomas Eckert, 19 Apr. 1875, Complainants' Exhibits, 1:386, Box 17B, *Harrington v. A&P.*

4. William P. Mellen (n.d.), of Flushing, N.Y., was one of the original incorporators of the Automatic Telegraph Co. He served as an assistant solicitor to Secretary of the Treasury Salmon Chase. "Gen. William J. Palmer," *New York Times*, 14 Mar. 1909, 11.

5. See Doc. 109.

6. According to Daniel Craig, this interest had originally been reserved for him. Quad. 70.8, pp. 53–55 (*TAEM* 9:790–91).

7. This contract was endorsed as follows:

Endorsed—Contracts—Geo. Harrington with Wm. J. Palmer and others—dated May 10th, 1871—On account of this contract the following sums have been duly paid which are hereby acknowledged: Sam'l B. Parsons, one-fifth in full, $4,500; Wm. J. Palmer, one-tenth in full, $2,350; J. C. Reiff, $1,000; all of which was expended in factory—Geo Harrington, April 17, 1875—Also received from J. J. Marsh, $1,500; Corning, $1,000; Morten, $1,000; Gouraud, $200—Geo. Harrington.

Marsh is unidentified. "Corning" probably refers to Erastus Corning, Jr. Alex Morten was connected with Edison's work on the electric pen and domestic telegraphs in the mid-1870s. George Gouraud (1842–1912) was associated with Josiah Reiff and William Palmer in several business and real estate ventures before becoming involved in the Automatic Telegraph Co. in 1873. Gouraud achieved the rank of colonel during the Civil War and received a congressional medal for his service. Following the war, he worked for a time at the Treasury Department, where he became associated with George Harrington. When Edison went to England in 1873 to introduce his automatic telegraph, Gouraud was there as an agent for the Pullman Palace Car Co. and acted on behalf of Automatic Telegraph. He remained associated with Edison's interests in England for many years. George Gouraud folder, BC.

8. A copy of the complete inventory, dated 10 May 1871, can be found at the Edison National Historic Site. 71-008, DF (*TAEM* 12:420–31).

9. For a description of tools and machinery, see Doc. 280.

10. Probably a misprint for "jigs."

Peekskill, May 11/71.

Dear Mr. Edison:

I have your note.

I expected & desired you to be interested in the process for Insulating Wires, if there was anything good about it, and I feel so now, and you can judge better than I can as to what answer I ought to make to you about giving up the wire covering machine.[1] If I was'nt at the moment so devlishly <u>hard up</u>, for money to keep half a dozen valuable things moving, I should not say a word but pay all expenses of machine & everything else, and push right on with the Insulation of wires.

I was sorry to have more than a few dollars expended to give me enough wire for a simple test, but as you acted as you thought for the best, I have not a word to say, except that I will, before a great while, pay all expenses provided you think our real interests will be promoted by having the use of the machine—but if you think that is not material, then, for your convenience as well as for mine, I would like you to get your pay as you suggest.

I am very busy in other matters just now. I would be delighted to know <u>from you</u>, just how automatic machines now stand. Truly

D H Craig

ALS, NjWOE, DF (*TAEM* 12:456).

1. An Edison and Unger account book entry for 6 May 1871 shows over one hundred fifty dollars' worth of work done for Craig on a wire-covering machine. Cat. 1183:3, Accts. (*TAEM* 20:861).

Peekskill, May 14/71

Dear Mr. Edison:

Our little philosopher, the <u>Great Little</u>, says our Paraffine insulation will not answer because it in a peculiar degree, operates to <u>retard the current</u>![1] As he also said your three-keyed perforator was a very complicated & expensive affair, and that your big Perforator was merely a poor imitation of Wheatstone's, his insulating story has'nt made a very deep indentation into my heart—still I would like you to tell me if there is a bit of truth in what [he?][a] says.

Do not fan[cy I am?][a] losing any interest in your operations because I do not bother you personally. I have been very busy about other matters, and am getting ready for active business as soon as you can give us the needed machines. I am not quite sure but that I shall have to take a hand in the News business even before we get the new telegraph system to work, as I am hard pressed by some of my friends.

I shall try to run out [to see?][a] you on Tuesday [ ][a] [y?][a]ou can tell me [ ],[a] if you have any, which [ ][a] [not?].[a] Truly &c

D H Craig

ALS, NjWOE, DF (*TAEM* 12:458). [a]One corner of the document is missing.

1. In 1887, the British mathematical physicist Oliver Heaviside discovered that minimum distortion of the transmitted message occurred when the product of resistance and capacitance of a circuit was equal to the product of leakage and self-induction. Insulating the wire increased the capacitance while decreasing the leakage, thus distorting the signal and causing it to elongate. George Little thought that a retardation of the current caused this elongation of the message. Brittain 1970, 39.

–162–

To George Harrington

Newark, N.J., May 18 1871[a]

Mr Harrington

Please give Murray[1] $265 and $70. The former for a Milling Machine[2] and the Latter for the Last of the Type wheels =[3]

Town[4] has taken stock of the whole Establishment, Has the Experimental Perforator Done,[5] and has the patterns in the foundry for the 10 perforators Which he is going right ahead to Build = The Mechancial Printer[6] Is nearly finished   Will be ready Next Saturday for visitors   We have had it together and it works Very Easy and Nicely = It is a good Looking peice of Mechanism

Saturday Night will settle the Success of all the Automatic Machinery = Hoping you aAre all right again   I remain Yours Automatically

Edison[b]

ALS, NjWOE, DF (*TAEM* 12:248). Letterhead of the Newark Telegraph Works. [a]"Newark, N.J.," and "187" preprinted. [b]At the bottom of the page Harrington wrote, "Received three hundred & thirty five dollars New York May 19 1871", and Joseph Murray signed it.

1. Joseph T. Murray (1834–1907) was a machinist who worked for Edison at the American Telegraph Works and assisted him with experiments. In 1872 he and Edison became manufacturing partners. After the dissolution of their partnership in 1875, Murray continued to operate his own electrical manufacturing firm. *New York Times*, 28 Jan. 1907, 7; *NCAB* (rev. ed.) 9:543.

2. Milling machines were used in machine shops to shape metal. See Doc. 280, n. 6.

3. The typewheels were for printing telegraphs being manufactured for the Gold and Stock Telegraph Co.

4. N. W. Towne was superintendent of the American Telegraph Works.

5. This was probably one of the perforator designs Edison covered in his 11 May caveats. See Doc. 151, n. 1.

6. Typewriter.

Albany May 25, 1871

To[a] T. A. Edison

At Mr Craigs request[1] I have sent you a ~~w~~Printing machine upon which I was working some years ago & nearly completed. I was obliged to take the train before I could have it weighed & charges fixed upon it from Buffalo, or I should have paid them in advance. I hope to be at your place in a few days & we will talk the thing over[2]   Yours &c

C. S. Jones[3]

ALS, NjWOE, DF (*TAEM* 12:458). Message form of the Western Union Telegraph Co. [a]Preprinted.

1. See Doc. 140.

2. Nothing further is known of Jones's machine or his relations with Edison.

3. Charles Jones, a telegraph operator, was later superintendent of telegraphs for the Illinois Central Railroad. Reid 1879, 322.

[New York,] May 26, 1871[a]

Agreement, made this Twenty sixth 26th[b] day of May 1871 by and between Thomas A. Edison of Newark, State of New Jersey, Inventor and Telegraph Instrument maker, of the first part, and The Gold & Stock Telegraph Company, a corporation incorporated under the laws of the State of New York, of the second part.[1]

Whereas the party of the first part is the inventor of various inventions and improvements in Electro-Magnetic Printing Telegraph instruments and appliances, applicable to the business of private lines and the simultaneous distribution of market reports and other information to a number of persons by means of instruments operated in the same circuit, some of which have been already patented, as to others of which applications for patents are now pending before the United States Commissioner of Patents, and as to others, of which caveats have been filed, by the said party of the first part, and all of which inventions and improvements, are intended for the party of the second part, a portion thereof being in actual use by the party of the second part, and the remainder not yet fully completed, but in process of development, and

Whereas, the amount of compensation to be made to the party of the first part therefor, and for the other inventions specified in the Schedule hereto annexed, has been left open for adjustment up to the present time, and a general settlement has now been agreed on between the said parties touching the said inventions, and improvements, and all future modifications and improvements thereof, and the patents to be issued therefor, including all reissues of said patents upon the basis hereinafter set forth.

Now the said parties have, and do in consideration of the premises, and each, in further consideration of the covenants hereinafter contained to be kept and performed by the other, covenant, and agree, to, and with each other, as follows, that is to say:

First— The party of the first part, agrees to sell, convey and assign, and hereby does bargain, sell, assign and transfer, unto the party of the second part, the full and exclusive right to all the inventions and improvements made by him, in Electro-Magnetic Printing Telegraph instruments, apparatus and appliances, for which patents have been already obtained, or as to which applications for patents are now pending, or embraced in caveats filed by the party of the first part, the same being more particularly described in Schedule "A" hereto annexed and the other inventions in said Schedule mentioned with the exception only of those embraced in the caveats in said schedule therein described as "not relating to printing telegraphs" (the intention being to embrace in this agreement all such inventions and improvements of the party of the first part, in Electro-Magnetic Printing Telegraph instruments, apparatus or appliances, whether specified in said Schedule or not) and all letters patent, granted, or to be granted, by the United States, for any and all said inventions and improvements, and all reissues, of said letters patent, granted, or to be granted, and all the right title and interest, which the party of the first part, has, or may, or can, or shall have, now or at any time hereafter, to, or in the said inventions, and improvements, and each and every of them, and all future improvements, and modifications thereof, and each and every of them, to have and to hold, to the party of the second part, their successors and assigns, to their sole use, benefit and behoof forever.

Second— The party of the first part further agrees, to go on and complete, and perfect, the said several inventions and improvements or so many of them as the party of the second part shall upon further examination and consideration in writing direct, with all reasonable dispatch, for the benefit of the party of the second part, and to prosecute with diligence, applications for patents for such inventions and improvements, as the same, shall respectively be matured and perfected; such patents to be issued to the party of the second part as assignee, and to make and deliver to the party of the second part free of all charge, complete working machines, embodying the said several inventions and improvements, one for each. As to the invention known as the "Universal Printer" it is expressly understood, and agreed, that the party of the first part, is to complete and deliver to the party of the second part, two complete and perfect working instruments, including the apparatus for the transmission and reception of messages, and also a modification of said instrument, without the transmitting apparatus, capable of being worked with a number of other instruments of the same character in the same circuit, from a common transmitter, so as to record the messages so transmitted.[2]

Third— The party of the first part further covenants and agrees, to and with the party of the second part, that he will on demand, from time to time, and as often as he may be required so to do, by the party of the second part, execute and deliver free of charge, to the said party of the second part, all such further conveyances, assignments, and instruments, as the party of the second part, their successors and assigns, or their counsel may reasonably devise, advise, or require, for the purpose of vesting in the party of the second part, the full and exclusive right to each and every of the inventions, and improvements, herein, or in the said annexed schedule, mentioned or intended to be comprised, and to each and every future modification or improvement therein, and in any Electro-Magnetic Printing Telegraph instrument, apparatus, or appliance whatever, applicable to the business of the party of the second part, which the party of the first part shall invent, design, or acquire, during the period of five years, succeeding the date of this agreement, and each and every patent for any such invention or improvement obtained or to be obtained during such period of five years succeeding the date of this agreement, and each and every reissue of each and every patent already issued, or which may hereafter be obtained therefor, and that he will during such period, give the said party of the second part, the benefit of his best skill and ability in the perfecting and improving of such inventions and improvements, and will to the best of his ability, co-operate with and assist the party of the second part, in obtaining such patents and in protecting the same, and the exclusive right of the party of the second part thereto.[3]

Fourth— The party of the first part further agrees, in consideration of the annual salary, hereinafter agreed to be paid, to enter the service of the party of the second part in the capacity of consulting Electrician, and Mechanician, for the period of five years from the date of this agreement, and during such period, to give the said party of the second part the benefit of his best skill and ability, in that capacity, in all matters relating to their business, and in all things to strive to promote the success of the said party of the second part.

Fifth— The party of the second part, in consideration of the premises, agrees to pay to the said party of the first part, for said several inventions and improvements, patents, and services, as follows, that is to say:

1st— For the whole of the said invention known as the "Universal Printer," complete and entire, and the exclusive right to the same, and the several modifications and improvements thereof, made, and to be made, at any time hereafter, and the patents to be obtained therefor, including any and all reissues of said patents, the sum of fifteen thousand dollars, in shares of the capital stock, of the party of the second part, at par.[4]

2nd   For all the other inventions, and improvements mentioned in this agreement, including as well those already made, as those to be hereafter made, and all patents issued, and to be issued therefor, and all reissues of such patents, and the good will of the party of the party of the first part, the further sum of twenty thousand dollars, in shares of the capital stock of the said party of the second part at par.

3rd   For the services of the said party of the first part as consulting Electrician and Mechanician, during the period of five years from the date hereof, a salary of two thousand dollars per annum, payable semi-annually in cash. And the party of the second part further agrees, that in each of said years, in which the party of the first part, shall invent and perfect, a new, useful, and valuable improvement in Printing Telegraph instruments, which shall be adopted and used by the party of the second part, the party of the second part will pay to the party of the first part, a further or additional compensation for that year, of one thousand dollars cash. It is understood, that the party of the second part, is also to pay the usual patent office fee and expenses, upon applications for patents under this contract.

In Witness whereof, the party of the first part has hereunto set his hand and seal, and the party of the second part has caused this contract to be signed by its President and Secretary, and its corporate seal to be hereto affixed, the day and year first above written.

Thomas A. Edison.[c]   The Gold & Stock Telegraph Co

H L Hotchkiss[5] Secretary[d]   by M. Lefferts   President

Witness[e]   A F Roberts[6]

Schedule A referred to in the preceding agreement.[f]

List of Patents issued and in the hands of the Gold & Stock Telegraph Company.[7]

Patent[g] No. 113033. Printing Telegraph, = called the "Chicago instrument"

Patent[g] No. 113034. Printing Telegraph = shifting Printing pad and devices.

Patent[g] No. 114658; Printing Telegraph magnets for working a Printing Telegraph instrument on one wire.[f]

List of Patents applications for which are pending and in progress at the solicitors.

Printing Telegraph No. 1. This is an instrument which works upon one wire and with one magnet the two distinct operations of Printing and Rotating The Type wheel being performed by actuating a polarized clutch and devices, by a reversal of the current. ⟨Model in Serrells   Patent No 126530⟩

Printing Telegraph No. 2. This invention consists of having two Type wheels worked independent of each other by two separate escapement levers both levers being operated by one magnet—either type wheel being grabbed at the will of the operator by means of a shuttle which is thrown in on either, by the printing current when both Type wheels are in certain positions. ⟨Patent applied for   Patent No 126534⟩

Printing Telegraph No. 3. Shifting Type wheel and its devices. ⟨Patent issued not in safe Jany 1872   No 123006⟩

Printing Telegraph No. 4. Universal Printing Telegraph with

its various devices, same as exhibited to the Company. It is doubtful if everything can be covered in one patent. ⟨applied for   Granted No 123005 131340 140488 140487 134866⟩

Printing Telegraph No. 5. This invention consists of working two type wheels independent of each other producing the same result as No. 2 but with entire different means, No. 2 being mechanical and No. 5 electrical. ⟨Applied for 131339⟩

Printing Telegraph No. 6. Shifting shield, Unison, and other devices. ⟨Applied for   126535⟩

List of Caveats in the Patent office.

No. 1. Not relating to Printing Telegraphs.

No. 2. Not relating to Printing Telegraphs.

No. 3. Printing Telegraph. A lengthy caveat describing several Printers, with unisons, paper drivers, escapements, Printing devices and modes of working on one wire. The claims are: ⟨128,608⟩

> 1st.   To an electro-magnet combined with a bar that is moved according to the polarity of the current, and cuts of [f] the electrical current, or admits it to its own magnet, substantially as described.

> 2nd.   To the escapement wheel with teeth combined with double acting pawls, and stops, arranged and acting as specified.

> 3rd.   To the polarity armature, either inert or performing the duty of a^h motor according to the direction of the current.

> 4th.   To the paper clamps constructed as specified.

> 5th   To the unison lever and stops, worked by the magnet that is the motor for the printing mechanism.

> 6th   To the yielding frictional unison for rendering the type wheel operative and inoperative according to the strength of the current.

> 7th   The expansion of a spring by the heat of an electrical current through a reduced wire as a means of disconnecting the unison mechanism.

No. 4. Relays for Printing Telegraphs, for working any Printing Telegraph upon one wire   Lately adopted by Amer Printing Tel. Co. ⟨Included in Chicago Inst ([Refused?])^i⟩

No. 5. Not relating to Printing Telegraphs.

No.^{g} 6. Universal Printer = relating thereto. ⟨See no ~~113~~ 123006⟩

~~No. 7. Not relating to Printing Telegraphs.~~

No. 8. Printing Telegraph = Shifting pad devices and Mechanical equivalents. ⟨Patent issued   No 113034⟩

No. 9. Electrical Rheotome, and apparatus for switching a main Trunk Line on to any number of lesser lines at a distance. Applicable to a system of Private Lines, as one main line may be made to do the work of twenty five firms and yet be independent of each other. ⟨No 131,334⟩

No. 10. Printing Telegraph magnets. A device for rendering the magnets of a Printing Telegraph more sensitive to rapid pulsations. The claims are   The helices arranged with the armatures operated substantially as described. ⟨found in 126532   ~~Sel~~ Sloting cores &c⟩

No. 11. Printing Telegraph. One wire, Switch working magnet, Local, Suitable for Private Lines. The claim is: The combination of polarized bars, with the type wheel and Printing magnet and local battery arranged and operated substantially as set forth. ⟨Abandoned⟩

No. 12. Printing Telegraph. One wire, Polarized magnet, Local, suitable for Private Lines.ʲ The claim is: The combination of two polarized magnets, adjusted as described in a main circuit and a Printing magnet and type wheel magnet in a local circuit, connected and operated substantially as set forth. ⟨~~Abandon~~ No 140489⟩

No. 13. Printing Telegraph. one wire, Mechanical Lock, Reversal, suitable for stock quoting   My claim is: The polarized bar and electro-magnet, combined with two other electro-magnets and connected substantially as specified so that the polarity bar will lock out of action, either of the armatures of the electro-magnets, according to the polarity of the current substantially as specified. ⟨Patent issued   No 114658⟩

No. 14. Printing Telegraph. One wire, Shunting magnet, Reversal, suitable for stock quoting   claims in caveat are: The two magnets A and A′ placed within a secondary circuit, and operated by a shunt current, from the main circuit, by means of the magnet constructed as described.

No. 15. Printing Telegraph. One wire, "Rheostat" total cessation of the current to Print, Relays, Local, suitable for Private Lines. The claims are,

> 1st. To a local electrical circuit at a distant station, open or closed by an electro-magnet, and pulsations from the sending station when combined with a connection in the Local circuit, that is self closing when the main line is deprived of an electrical current so as to pro-

duce an operation, that is distinct from that performed by the electrical pulsation.

    2nd. I claim increasing & decreasing a current over a constant current upon one wire, to effect the rotation of the Type wheel and the total interruption of the constant current to effect the impression of a letter, or vice versa, or to produce separate operations.

    3rd. A battery reversing device, or its equiv lent. ⟨~~Cannot be found~~ ~~Supposed to be issued September 1874~~ Covered practically by 139128⟩

No. 16. Printing Telegraph. One wire, Air Relay, Suitable for stock quoting. The claims are   The use of an air chamber, and two or more plungers or their equivalents, combined with two or more electro-magnets for the purpose set forth. ⟨Not taken out   experiments not satisfactory⟩

No. 17. Printing Telegraph. One wire, Double helice, Local, Suitable for Private Lines.[j] The claims are   The two compound electro magnets to perform different operations and connected to a local circuit and to a main line circuit, so that a positive current acts in one magnet to increase the attraction of the armature, and in the other the currents neutralize each other, and allow the armature to be moved by a spring, and when the negative pulsation is sent the operations are reversed as specified. ⟨No 128,605⟩

No. 18. Printing Telegraph. One wire, Residual Relay, Local, Suitable for Private Lines. claims are

    1st. The combination of the magnet G or its equivalent, operated as described with a Type wheel and Printing Lever magnet for the purpose set forth.

    2nd. The relay G and Local battery B, and its connections combined with a magnet whose lever is rendered in operative on account of residual magnetism during rapid pulsations, and a Type wheel & Printing lever magnet for the purpose set forth. ⟨Not taken out—Experiments not Satisfactory⟩

No. 19. Printing Telegraph. One wire, Air Retardation, suitable for stock quoting. claims are   The air cylinders used in connection with a Printing Telegraph to render one or the other magnet[h] inoperative, the same being controlled by the polarity of the current. ⟨No 128604⟩

No. ~~20~~. Not relating to Printing Telegraphs.

No. ~~21~~.   d[itt]o   do   do

No. ~~22~~.   do   do   do

No. 23. Unisons, for Printing Telegraphs. The claims are

1st. The retarding spring D or equivalent ~~magnet~~ for the purpose set forth.

2nd. The arm D. operated by an electro-magnet for the purpose set forth.

3rd. The combination of the arm D or its equiv lent with an armature on the Printing or type wheel magnet. ⟨The instrument which contains these devices was given to serrell but nothing has ever been heard of it, ~~nor can I~~⟩[k]

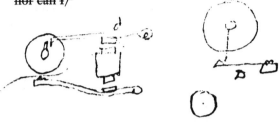

No. ~~24~~. Was an error and has been withdrawn.

No. ~~25~~. Not relating to Printing Telegraphs.

No. 26. Universal caveats. Long description of various devices in the universal   12 claims. ⟨No ~~126005~~ 123005⟩

Thomas A. Edison.

DS, NNC, Edison Coll. Canceled 5¢ Internal Revenue stamp affixed in left margin of each page. [a]Date taken from text, form altered. [b]"Twenty sixth 26th" written in Edison's hand. [c]Followed by wax seal. [d]Company seal embossed at left. [e]Written by Lefferts. [f]Followed by centered horizontal line. [g]Preceded by check mark in margin. [h]Interlined above. [i]Faded. [j]"Private Lines" underlined in an unknown hand. [k]Sketch accompanies marginalia.

1. This contract represents the culmination of the arrangement proposed by Marshall Lefferts in October 1870 (Doc. 126). Edison and Gold and Stock had begun formal negotiations in mid-April and settled the terms of this agreement on 9 May (G&S Minutes 1870–79, 44, 49–51). The Gold and Stock committee responsible for the contract considered it "a very advantageous one for the Company, receiving as it does the Co-operation and good will of Mr Edison for the future [and] acquiring the 'Universal Instrument' for the Company for . . . all purposes at one half the price originally expected" (ibid., 51). At the time of this agreement, Gold and Stock was merging with Western Union. See Chapter 7 introduction.

2. Because Edison's universal private-line printer was intended for point-to-point communication between unskilled users, it combined a thirty-character keyboard and transmitter with the printing mechanism for receiving. Gold and Stock also wanted to be able to use the printing mechanism independently as a ticker.

3. Edison assigned a total of forty-five patents to Gold and Stock, thirty-six after this agreement. They represent all of his work in printing

telegraphy. Documents recording the assignments and other material concerning Gold and Stock are at Columbia University, New York City (Edison Coll.).

4. Gold and Stock was not a publicly traded issue. There is no record of Edison's disposition of the stock described here and in the following paragraph.

5. Horace Leslie Hotchkiss (1842–1929) was a New York stockbroker and financier. He became secretary and treasurer of Gold and Stock in 1868. Reid 1879, 607; *DAB*, s.v. "Hotchkiss, Horace Leslie."

6. Unidentified.

7. At the time of application, Edison assigned U.S. Patent 113,033 to himself, U.S. Patent 113,034 to Gold and Stock, and U.S. Patent 114,658 jointly to Lefferts and himself. He then assigned U.S. Patents 113,033 and 114,658 to Gold and Stock on 5 June 1871. Digest Pat. E-2:169.

UNIVERSAL PRIVATE-LINE PRINTER, 1871
Doc. 165

The universal private-line printer, a combination telegraph transmitter-receiver, was Edison's most sophisticated device to date and represented many months of work by Edison and his co-workers.[1] The Gold and Stock Telegraph Company's purchase of rights to it formed the basis of Edison's May 1871 contract with that company. The universal printer was meant for the same private-business market as the American Printing Telegraph Company instrument and George Phelps's 1870 printer (U.S. Pat. 110,675).[2] Like those machines, it used a single typewheel with only letters.

The universal's primary innovations lay in its transmitter. Edison devised a continuously operating "pulsator" (breakwheel) powered by a small electric motor of his own design.[3] The rapid intermittent signal from the pulsator drove the typewheels of the sending and receiving instruments through ratchet-and-pawl escapements much like those on Edison's earlier printers. By depressing a transmitting key, the operator blocked the rotation of the typewheel on the sending instrument and at the same time halted the pulsator, thereby stopping the rotation of the typewheel on the receiving instrument. As in Pope and Edison's gold printer, the printing magnet was in series with the typewheel magnet and was too sluggish to respond to the rapid impulses that moved the typewheel. When the typewheel stopped, the printing magnet had time to reach full strength and then actuated the printing lever.

The universal private-line printer was not a finished machine when Edison patented it. For example, it had no unison stop.[4] Also, the speed of the motor-driven pulsator varied with the strength of the battery and was a source of trouble until Edison devised a governor to make it constant and reliable.[5] Still, Edison never managed to make this design practical. His notebooks show that he continued to work on it until March or April 1872, when he abandoned its printing mechanism and combined a modified form of its transmitter with his universal stock printer to create a successful private-line printer.[6]

1. The earliest extant mention of the universal printer is early October 1870, when William Unger and an unidentified draftsman received payment for their work on the universal. During the last week of that year, Charles Batchelor and James Eagan—two mainstays of Edison's crew—spent 61½ and 38 hours, respectively, working on it. PN-70-10-03, Lab. (*TAEM* 6:803, 812); Cat. 30,108, Accts. (*TAEM* 20:202).

2. Edison later testified that Gold and Stock might have used this model of the universal private-line printer in the summer of 1871. Testimony of Edison, p. 16, *Edison v. Lane.*

3. One of the earliest commercial applications of electric motors was for powering small printing telegraphs. Although several inventors—including Charles Page and William Davenport in the United States, Robert Davidson in Great Britain, Moritz Jacobi in Russia, and Gustave Froment in France—had used electric motors to drive industrial machinery, railroads, and boats as early as the 1840s, their creations were impractical because electric power produced by batteries was far too expensive to be used in large-scale work. George Phelps's large printing telegraph (U.S. Pat. 26,003) had been designed for any available power, including electric motors, but in practice was powered by human "grinders" (see Doc. 54 headnote, n. 4). In 1868 Phelps devised a smaller printer (U.S. Pat. 89,887) incorporating an electric motor much like the one Edison used in this printer. Western Union had the Phelps printer in service by the end of 1870, and Edison was no doubt familiar with the design. Although Edison's motor appears to be original, he did not attempt to patent it. Post, 1976, chaps. 4–5; Martin and Wetzler 1887, chaps. 1–2; William Orton to Anson Stager, 8 Oct. and 12 Nov. 1870, and 11 and 24 Feb. 1871, LBO.

4. Edison later adapted his screw-thread unison to this machine. See Doc. 243.

5. See Docs. 184 and 208.

6. Doc. 262. Edison received several patents on the parts and modifications of this model (U.S. Pats. 131,340, 131,343, 134,866, and 140,487).

Patent Model: Printing Telegraphy[1]

M (28 cm dia. × 15 cm), DSI-NMAH(E), Cat. 252,615.

1. See headnote above.

2. Although the patent application was not completed until July, Edison's patent attorney indicated that this patent model was finished by May. Lemuel Serrell to Commissioner of Patents, 14 Aug. 1871, Pat. App. 123,005.

–166–

From George Harrington

New York, June 4, 1871

My dear Edison

I can get no strength and am ordered to the Country for ten days, & I go immediately.

I have concluded that your suggestion that we make <u>one test</u> without waiting for the printer,[1] is the best & if you will have the Ink recorders, weight transmitters and stop or automatic attachment[2] completed we will make the test with Littles and the 6 Key'd perforators.[3] Towne says he will have one of the large perforators done "next week." I am sick of "next week," but on what is he expending—4 or 500 $ per week. What are his 20 odd men doing, as there are no orders nor likely to be unless I give them, would it not be well <u>to Stop all work</u> except upon the perforator. When one is completed if ever, as a model, I can understand that more men will be required to build a number. But there is no working model yet & until there is if I understand it, it is of no use to make jigs or special tools to build a [l]arge number.

All may be judicious but I cannot see what 20 men are doing. We shall see what progress has been made after 10 days hence   truly

Geo Harrington

ALS, NjWOE, DF (*TAEM* 12:250).

1. Although Edison originally claimed that his typewriter would be ready by 20 May, experiments continued after this note from Harrington. Bills to the American Telegraph Works, 10 June–23 Aug. 1871, 71-015, DF (*TAEM* 12:589–94).

2. For discussion of these instruments, see Doc. 153, nn. 9, 11.

3. Edison completed work on four 6-key perforators about 17 May. Bill to the American Telegraph Works, 17 May 1871, 71-015, DF (*TAEM* 12:587). See also Doc. 255.

–167–

From Daniel Craig

Peekskill June 8/71.

Dear Mr. Edison:

Being pretty busy and short-handed here, & nothing to do in the city, I don't go there much now-a-days, and did not see your small perforators till today, at 66.[1] Mr. Darrow tried to work them, but it was impossible, and he could not perforate two words consecutively to save his soul. I think, if those machines are to be at the office, you should go in & teach Darrow & Grace[2] how to use them, as they now reflect utter disgrace on all concerned. Even when, by possibility, Darrow could perforate 4 or 5 letters consecutively, the spacing was fearfully bad, as you will see by what I send you. I do not know what the word is, but I recollect Darrow said the two first letters were S A, and the last, M. If you cannot spend time to shew Darrow & Grace how to work the machines, do, pray, have them taken away, for they seem to justify Little's foolish spite when he says Harrington and Edison is simply fooling away time and money to no good purpose.

Why, the deuse don't you let me know how you are getting along. When are you agoing to have something to <u>show</u> in the way of the new Perforator & Printer.

I hear the W.U. Co. have swallowed the G.&S. Co. I hope you have sold your stock.

Harrington, I hear, has gone to Saratoga, to be back early next week.

Don't be rough on your poor friends, & let me hear from you upon all points, or, what I would like better, come here Saturday, and be <u>happy</u>, for one day of your life by breathing fresh, cool air, filled with the perfume of a thousand delightful odors. We are in the midst of Haying. Write me if you will come by the 4 p.m. train, Saturday & return when you like next week.

Capital place to Loaf! Truly yours,

D. H. Craig

ALS, NjWOE, DF (*TAEM* 12:460).

1. The main office of the Automatic Telegraph Co. was 66 Broadway, New York City.

2. W. E. Darrow was a shophand at the American Telegraph Works. George Grace was superintendent of the Automatic Telegraph Co.

–168–

Agreement with Marshall Lefferts and Elisha Andrews

New York[a] 8h June 1871

Copy[b]

It is hereby agreed and understood that, Geo B. Field, is after this date interested in the Telegraph Instrument, Known as the "Universal Printer"—to the full extent of one fourth (¼) of the English patents.[1]

Marshall Lefferts   (Signed)   E W. Andrews
                                  "   T. A. Edison

The original given to M Field. Private Line Printer

D (copy), NNHi, Lefferts. Letterhead of the Gold and Stock Telegraph Co.; written by Lefferts, including names. [a]"New York" preprinted. [b]Underlined twice; written above dateline.

1. One week before this agreement, Field, Andrews, Lefferts, and two other men had initiated a stock-reporting system on the London Stock Exchange (see Chapter 7 introduction). Edison executed the American patent application covering the universal private-line printer (U.S. Pat. 123,005) on 26 July 1871. William Lake, a patent solicitor in London, filed the provisional specification for the British patent to which this agreement refers on 19 March 1872 (Brit. Pat. 829 [1872]) as "a communication from George Baker Field and Elisha Whittelsey Andrews" and filed the complete specification on 18 September. The agreement between Edison, Lefferts, and Andrews to which this is apparently an amendment has not been found.

–169–

Stockton Griffin Memorandum[1]

[New York,] Saturday June 10 1871

8:35[a] AM at Test room[2]   Took The Inst that Prince[3] just fixed to Kent & Cos[4]   placed it in the Circuit   had considerable trouble with paper hangers[5] on account of one being longer than the other—imperfection in casting, fixed them OK— Inst worked all right— ~~Globe~~ Shade[6] wanted for this Inst—

9:40ª AM   Went to test room got a shade & at

10:15ª AM   at the request of Scott[7] Started over Cotton[8] wire   visited Kents first and fixed the shade— not being acquainted with the route of the Cotton wire it took me an hour & a half to find all of Them— Every Inst working satisfactory except Dunnells[9] which wanted Ink—which I put on & it went all right. Johnson & H[10] Complain that reports of transactions in NY are behind 15 to 30 Minutes & while I was standing by The Inst it commenced recording the 10½ AM Transactions which were 30 minutes late, it being just 11 AM.[11]

11:35ª at Test room   sent a msg to Ludwig[12] notifying him of above Complaint   Then went to see Prince & hurry him up on the 2d machine for Prod[13] wire

12:30ª Complaint from Fatman & Co[14]   went there found spring of printing lever weak and Inst needed oiling   fixed it so that it went allright

1 PM Dinner

1:30ª Prince still working at Inst— Everything going smoothly—

2:30ª got the Inst from Prince & placed it in C[ir]k[ui]t— works first rate   required no adjstmt   Battery made up fresh this AM—

3 PM   Visited Kent & Co— Inst had Thrown out[15] once— caused by Transmitting op[erato]r—Fatmans Inst going OK—

Remained at Test room till 34:30ª when all hands left— Callahan[16] said just before leaving would want me in the Long room[17] Monday & several days after to learn taking reports— said Scott would attend to Cotton & Prod wire meantime   So ends The first week

G

ADS, NjWOE, DF (*TAEM* 12:615). On message form of the Western Union Telegraph Co. ªColon supplied.

1. Stockton L. Griffin (n.d.), a telegraph operator, worked with Edison in Cincinnati in 1867. He later became Edison's secretary at Menlo Park. At this time he worked for Gold and Stock. "Personal," *Telegr.* 4 (1867–68): 164; Jehl 1937–41, 1:27, 35.

2. Gold and Stock stored spare instruments and parts, made repairs, and tested instruments in its central office. Gold and Stock 1873, 7–8.

3. Unidentified.

4. Albert E. Kent & Co., merchants, 76 Broadway. Wilson 1871, 643.

5. Paper hangers were the arms extending upward to hold the reel of paper.

6. The domed glass shade kept dust off the instrument.

7. In May of 1872 George Scott would succeed Edward Calahan as

general superintendent of Gold and Stock. "The Gold and Stock Telegraph Company," *Telegr.* 8 (1871–72): 307.

8. The circuit providing quotations from the New York Cotton Exchange.

9. Unidentified.

10. Johnson Hirsch & Co., merchants, 135 Pearl St.; or H. A. Johnson & Co., brokers, 66 Broadway. Wilson 1871, 610.

11. It was not unusual for the stock market ticker to be fifteen to thirty minutes behind the market. Healy 1905, 101.

12. Edwin Ludwig was manager of the Commercial News Department. Reid 1879, 610.

13. "Prod[uce] wire" refers to the circuit that provided quotations from the New York Produce Exchange.

14. Lewis & Solomon Fatman & Co., merchants, 70 Broadway. Wilson 1871, 372.

15. "Throwing out" meant failing to register a transmitted impulse, in which case the machine would print one letter behind in the alphabet. The instruction booklet for the stock ticker (Gold and Stock 1873) lists nine possible causes of throwing out.

16. Edward Calahan.

17. The main trading room at the New York Stock Exchange, from which quotations were transmitted.

–170–

From Lemuel Serrell

New York, June 10th 1871[a]

Dear Sir

I would like you to call at my office on Monday to look over some papers in connection with your telegraph cases, if you can make it convenient to do so, as I expect to go to Washington on Tuesday.[1] Yours Truly

Lemuel W. Serrell per H. Serrell.[2]

L, NjWOE, DF (*TAEM* 12:626). Letterhead of Lemuel Serrell. [a]"New York," and "187" preprinted.

1. When Edison had a device to patent, he drew up a description and sent it with a model to Serrell. Someone at Serrell's office wrote the patent specification and drafted the figures for the application. The cases Serrell refers to here are unknown. The U.S. Patent Office was located in Washington, D.C.

2. Harold Serrell was Lemuel's son and bookkeeper. Ricord 1897, 449; *Plainfield Merchant's City Directory* 1875, 76.

–171–

From Stockton Griffin

[New York?,] Dated, June 12 1871

To[a] Ed

How do you like the idea of my being in the Long Room[1] this week. Calahan wants to get a man to send reports but I suppose he wont offer an op[erato]r a living salary so he is going to have me learn so in case of necessity I can act there

Scott[2] runs his books in the office & runs your Insts and you know how they will be run   Is it a conspiracy of C's to fix things so you will not have a fair show[3]

Griff

ALS, NjWOE, DF (*TAEM* 12:617). Message form of the Western Union Telegraph Co. [a]"Dated," "187", and "To" preprinted.

1. See Doc. 169, n. 17.
2. George Scott.
3. Edward Calahan's stock ticker was still the principal instrument of Gold and Stock.

–172–

From Elisha Andrews

June 20, 1871

Dr Sir

Please do no[a] more work on the 5 Printers you are finishing until farther orders.[1] It may be some weeks before they are wanted, & you are so hurried on other work that you let this rest— Yours Truly

E. W. Andrews

How is Autographic?—[2] Let me Know as soon as you can get it to work.

ALS, NjWOE, DF (*TAEM* 12:618). [a]Interlined above.

1. It is not clear to what printers Andrews is referring.
2. The "autographic" is the facsimile telegraph Edison was developing for Andrews and George Field under the 10 February 1870 contract. See Doc. 92.

–173–

From George Scott

N.Y.[a] June 20/71

Dear Sir

Genl. Lefferts wished me to write you, & ask where are the Cotton inst (with new switches)[1] you promised to send along with the produce insts—[2] He wants them badly & hopes you will send them at once   Yours Truly

Geo. B. Scott

ALS, NjWOE, DF (*TAEM* 12:618). [a]"Gold & Stock Tel Co" written on date and place line.

1. Edison made two alterations to the cotton instruments used by Gold and Stock. He fitted them with his screw-thread unison and eliminated the polarized switch under the base (see Doc. 136). It is uncertain to which of these changes Scott is referring here.

2. Since Gold and Stock adopted Edison's one-wire ticker on both its cotton- and produce-reporting lines, the distinction between produce and cotton instruments could mean that the instruments of the two circuits had distinguishing typewheel characters. However, it is also possible that it referred to as trivial a difference as decoration.

-174-

From James Hammett[1]

Boston, June 26th 1871[a]

Dear Sir

I have received nothing from you yet. You asked me to state the amount due me on your note. I did so and have also written you a second letter in regard to it. in yours of the 12th you say you will send checks.[2]

On receipt of this you will immediately send the funds or write and give me some explanation of the matter   I am tired of writing about it and if you can not attend to it I propose to see if it can be collected.[3] Yours Truly

J Hammett

ALS, NjWOE, DF (*TAEM* 12:224). Letterhead of James Hammett. [a]"Boston," and "18" preprinted.

1. Hammett's letterhead indicates that he was a banker and broker.

2. No other letters to or from Hammett have been found.

3. An account book entry shows that Edison owed Hammett $100 in January 1869. The same entry indicates that he paid this sum, plus $25 interest, to Hammett on 27 April 1872. Cat. 1185, Accts. (*TAEM* 22:549, 563).

Innumerable Machines
in the Mind

July–December 1871

During the second half of 1871, increased stability and security marked Edison's life. He met and married Mary Stilwell of Newark; his association with Gold and Stock grew closer; and Edison and Unger became prime contractors for that company. New-found financial security allowed Edison to buy a house for his bride and to assist his father and his brother William Pitt with their enterprises. For a time he had five shops in Newark operating at once.[1] Although Edison's association with the American Telegraph Works became less intimate, the shop began manufacturing his large perforator for use on the Automatic Telegraph Company lines that were then being prepared for operation.[2] Edison's procedure for recording his designs for printing and automatic telegraphy became more systematic when in late July he started maintaining a series of laboratory notebooks.[3]

In printing telegraphy he had abandoned his cotton instrument in favor of the universal stock printer by early summer. He delivered two prototypes to the Gold and Stock Telegraph Company in June and then spent two months improving the machine. Manufactured by Edison and Unger, the first instruments were delivered in early September. By the end of the year Gold and Stock had bought 600, installing them on lines in New York and other cities.[4]

Edison continued his attempts to make the universal private-line printer commercially viable. Precision was the biggest stumbling block, both in arresting typewheels synchronously on transmitting and receiving instruments and in positioning them accurately. Edison also modified the printing lever in an attempt to increase the speed with which it recoiled after having made an impression. Toward the end of the year he designed a governor for the motor, a significant step in the machine's evolution. He also conceived how to make the printer's transmitting mechanism an independent instrument.

Sometime in the fall Edison became involved in starting the News Reporting Telegraph Company, a Newark enterprise that placed printing telegraphs in subscribers' homes and offices in order to provide "news of the world—financial, commercial, domestic, and foreign . . . hours before such news is published in the papers."[5] The business seems to have quickly failed, but through it Edison met the woman he would soon marry, Mary Stilwell.[6]

Automatic telegraphy demanded increasing attention from the young inventor. By the middle of July, the American Telegraph Works had begun production of his large perforator, while he, Joseph Murray, and William Unger prepared the New York office of the Automatic Telegraph Company for operation. Beginning in September, the American Telegraph Works began shipping perforators and other equipment to Automatic Telegraph's Washington office and the following month to its Philadelphia office.[7] Edison designed many of these instruments, his most important contribution being the large perforator. This instrument had individual keys for each letter, allowing operators with no knowledge of Morse code to work it. Among the equipment sent was a copying printer, an early form of typewriter. It is unclear, however, whether the printers used by Automatic Telegraph in 1871 were those designed by Edison or were Christopher Sholes's typewriter.[8] Edison most likely designed the ink recorder that was installed in the Washington office.[9] While no clear statement exists to explain the use of an ink recorder in place of the faster, chemical recorder, Daniel Craig did indicate in a July letter that the Automatic Telegraph Company planned to use a newly devised automatic "dot and dash printer" (i.e., ink recorder) on short lines.[10]

Although a considerable amount of equipment had been shipped to its various offices by the end of 1871, Automatic Telegraph did not actually begin transmitting public messages until December 1872. This suggests that critical technical problems remained unsolved. The automatic system installed on the company's New York to Washington circuit was still principally that of George Little; yet his chemical receiver, the key to high-speed automatic telegraphy, still was not in use,

*The Edison and Unger shop
on Ward St.*

and only one of his transmitters had been shipped. An examination of Little's many patent applications in 1871 indicates some of the technical difficulties that were plaguing the system.

Since the work of Alexander Bain, a continuing problem with automatic transmitters had been the adherence of fibers from the perforated paper to the transmitter's stylus, which prevented the stylus from making contact with the metal cylinder on which the paper rode. In April 1871, Little executed a patent on a technique to obviate this difficulty by replacing the wire brush in his earlier transmitters with a roller. By the end of the year, he had also designed an improved paper feed to lift the paper from the roller.[11] A second and more serious problem occurred in the recording of messages. Induced currents in the line elongated the chemically recorded marks. This problem, called "tailing," continued to plague the Little system. Little developed several techniques to clear the line of induced currents by using improved electromagnets and shunt circuits. Use of the automatic telegraph was also severely restricted by the system's inability to drop messages at intermediate stations and to record messages on two machines at the same station.[12]

Edison had his own ideas for improving automatic telegraphy. He recorded some of them in his notebooks and communicated some of them to Craig. Craig, however, wrote Edison that he considered Little's transmitting and receiving instruments to be "perfect" for their needs, although future improvements might be necessary. Not wanting Little to feel that Edison was attempting to supplant him as the inventor of the system, Craig suggested that Edison's perforator and copying printer would bring him sufficient glory.[13]

1. App. 1.A34–35, D155. The five "shops" were the Edison and Unger shop on Ward St.; a supplementary shop at 24 Mechanic St.; the American Telegraph Works on Railroad Ave.; an office at 788 Broad St. for the News Reporting Telegraph Co.; and, probably, a small experimental shop in White's Building on the Morris Canal, which Edison maintained sometime during the period 1870–1872. See gas receipts, 71-014, DF (*TAEM* 12:493); and Doc. 205. For the experimental shop, see Murray's testimony, Quad. 71.1, pp. 34–35 (*TAEM* 10:22); App. 1.A34; *Holbrook's Newark City Directory* 1871, 187, 580; and Hopkins 1873, Fifth Ward map.

2. Edison claimed that he walked out of the American Telegraph Works in a dispute with Harrington. Employment records show that most of the employees from that shop then began working instead at Edison and Unger in October 1871. However, because the American Telegraph Works was the principal supplier of apparatus for the Automatic Telegraph Co., Edison does appear to have continued some form of association with the shop. Cat. 30,108 and Cat. 30,107, Accts. (*TAEM* 20:304, 307; and 21:4–9). For Edison's claim see his testimony and that of Joseph Murray in Quad. 70.7, pp. 273, 277; and 71.1, p. 36 (*TAEM* 9:503, 505; and 10:23).

3. See Docs. 179, 181, 183, and 187.

4. Cat. 1183:10, 15, and PN-71-09-06, Accts. (*TAEM* 20:865, 867, 974–83).

5. Doc. 205.

6. Docs. 205 and 218.

7. Copy of an Affadavit by Edison, Murray, and Unger, 10 Feb. 1871, DF (*TAEM* 5:996); and agreements between Frederick Spannenberg, William Parkinson, and American Telegraph Works, 17 and 26 July 1871, DF (*TAEM* 12:301–7). By the end of the year there were eleven perforators, one transmitter, and two copying printers in the office in New York; six perforators, one transmitter, one printer, and one ink recorder in the Washington office; and six perforators, and one printer in the Philadelphia office. Each of the offices also received sundry other items such as paper reels and switches. PN-71-09-05, Accts. (*TAEM* 20:171–74).

8. In August, Craig indicated that the Automatic Telegraph Co. was using Sholes's typewriter as its copying printer but that Edison claimed he could improve on its performance. Since the printers shipped to the

company's offices were manufactured at the American Telegraph Works rather than at Sholes's shop in Milwaukee, they were probably Edison's. Craig to Lefferts, 8 Aug. 1871, Lefferts; Cat. 30,108, Accts. (*TAEM* 20:256–327 passim).

9. Although George Little also executed a patent for improved ink-recorder pens on 29 July (U.S. Pat. 120,289), the instrument made at the American Telegraph Works and shipped to the Washington office was probably Edison's (U.S. Pat. 124,800). Bill of Edison and Unger to American Telegraph Works, 19 June 1871, 71-015, DF (*TAEM* 12:590).

10. Craig to Frank Ives Scudamore, 20 Feb. 1871, ATF, UKLPO.

11. See Little's U.S. Patents 115,968, 120,288, 120,291, and 123,491.

12. See Little's U.S. Patents 122,266, 122,474, 123,490, 123,711, and 130,813.

13. Doc. 203.

–175–

Memorandum: Port Huron and Gratiot Street Railway Co.

[Newark?, c. July 1, 1871]

Recpts

June	1st	1871—	37.90
"	2	"	34.00
"	3	"	26.60
"	4	"	55 45
	5	"	28 90
	6	"	———
	7	"	39.15
	8	"	42.45
	9	"	33.00
	10	"	35 55
	11	"	50.25
	12	"	40.00
	13	"	42 00
	14	"	57 75
	15	"	34.80
	16	"	37.85
	17	"	34 45
	18	"	———
	19	"	41.70[a]

Cost of running road, ~~rent~~, of Buildings, Baggage Wagon, Freight Wagons—etc   Tkt ofs = $20. per day =

Conductors	Salary—		12.00
Drivers	"	Each	9 00
Mens	"		~~7~~8 00
Clerk			12 00
Supt			18 00

Owned by[b]

John[c] Miller[1]	70	shares
Wm[d] P Edison[2]	30	"
Wm Wastell[3]	70	"
Mrs.[d] M Cooper[4]	66	"
Wm Stewart[5]	35	"
Wm[d] Hartsuff[6]	30	
	301.	shares

Stock issued at 80. par 100. =
Company own—

2	Freight Wagons		
1	Baggage Wagons		
7[e]	Horses & Furnishings		
	New Stable just put up	costing	$1200.[f]
	V Property on which stables are	"	927.
	Car House = Costing	"	500.
2	Cars.		2670.
½	mile Spare iron		
1	Snow Plow		
1	Freight Car		
1	Large Sleigh for winter		330.00
1	Offices in Port Huron. Ticket office etc		
	Costing =		$450.

AD, NjWOE, DF (*TAEM* 12:238). Letterhead of the Port Huron and Gratiot Street Railway Co. [a]"17 days/671.80/40 say" and "~~17 days/644/36 pr dy~~" written below, probably in Edison's hand; "Average about $40 pr day, and this agrees with yearly return for this year 1871—If the running expenses—do not exceed $20. pr day—profit about 100 pr cent out of which must come equipment and dividends" written in left margin. [b]"100.", "115 00", and "65 00" added later in column at right. [c]Canceled check mark precedes this name. [d]Check mark precedes this name. [e]Ink badly smeared. [f]Figures in this column may be in another hand.

1. John Miller was the company treasurer and a banker in Port Huron. Jenks n.d., 4, 7; *History of St. Clair County* 1883, 585–86.

2. Pitt Edison was the company's superintendent.

3. William Wastell, a Port Huron druggist, was one of the original organizers of the line and served as the company's secretary. Jenks n.d., 2, 4, 7; *History of St. Clair County* 1883, 600.

4. These shares were actually bought by Gage M. Cooper, manager of the shops of the Grand Trunk Railway at Fort Gratiot (Jenks n.d., 2, 4, 7; *History of St. Clair County* 1883, 565–66). An account-book entry for 28 November 1871 shows Edison purchasing Cooper's 66 shares for $3,500. An entry for 1 January 1872 shows $3,100 still owed on the stock purchase (Cat. 1213:3, Accts. [*TAEM* 20:6]; see Doc. 266, n. 3).

5. William Stewart was a major hardware dealer in Port Huron. *History of St. Clair County* 1883, 596–97.

6. William Hartsuff was the postmaster of Port Huron and part of the family that had forced Samuel and Nancy Edison out of their original house in Port Huron. Ibid., 573–74; Stamps and Wright 1982, 10–14.

–176–

From Pitt Edison

Port Huron, Mich. July 12 1871[a]

T A E

I send you to day a Port Huron paper showing you the Amenmant to the St R R Charter which gives the Co the Exclusive right over the Park which is a big thing as the City will probeley [-]never open a public [-]St through it[1]   I am building the Car House and I am talking about building many other buildings and laying out a large amt of mony in building new road   also I tell them that it is nesesury to Extend our road to connect with the Port Hurn & Lak Michgan Depot[2] and thay all say thay cannot Stand it   it will cost to much   thay want more Dividends and not so much Rail Road   I sent you by mail[b] the maps on the 6th of this month and suppose you have got them[3]   if not let me know and I will send you more by Express   if you can get this stock taken I think the sooner the better as these parties are about as sick as they can be made[4]   I had some talk with Miller and I think par will fetch him[5]   the recets on the 4th of July was $169.40   as soon as the light House Reservation is sold[6] the Dutch agoing to buy and open a big Lager beer garden which will help the road considerable[c]

tell Charly[7] to be careful agoing rond the Engine or the machinery and not be careless   I hope he is not to much trouble to you   should he be let me know and we will have him come home   tell him that I say[d] to be a good Boy and mind his uncle   we are all well and send love to you & Charly

W. P. Edison

PS   tell Charly that Nellic[8] and Trofton[9] left last night on the Montgomery[10] for Chicago   WPE

ALS, NjWOE, DF (*TAEM* 12:236). Letterhead of the Port Huron and Gratiot Street Railway Co. [a]"Port Huron, Mich." and "18" preprinted. [b]"by mail" interlined above. [c]Followed by "over" to indicate page turn. [d]"that I say" interlined above.

1. When the land of the Fort Gratiot Military Reservation was divided into lots and sold, beginning in 1870, a large section was reserved for Pine Grove Park. Jenks n.d., 6; Bancroft 1888, 253–54.

2. The Port Huron and Lake Michigan Railroad was opened in 1871 as far as Flint, Mich., and was later extended to Chicago. The depot was

located on the river on the south side of Port Huron, near the center of town.

3. No maps have been found.

4. Probably a reference to the stock that Edison purchased on 28 November 1871. See Doc. 175, n. 4.

5. See Doc. 175, n. 1.

6. The sale was not authorized until 1873. Bancroft 1888, 254.

7. Pitt's son, Charles P. Edison (1860–1879).

8. Nellie Edison Poyer (1858–1947) was Pitt's daughter.

9. Unidentified.

10. Probably a steamboat.

–177–

To George Harrington

[Newark, July 22, 1871[1]]

Mr Harrington.[a]

Payroll this week is	577.86.
Unger[b] wants on acct	250.00
	827.86[c]

If That old Gentleman[2] is going to take Charge of The Books it is about time. That department is in a very slack Condition =

I cannot stand this worrying much longer about Bills—Pay roll etc = it engrosses my attention to the detriment of our machinery   You Cannot expect a man to invent & work night and day, and then be worried to a point of exasperation about how to obtain money to pay bills— If I keep on in this way 6 months Longer I shall be completely broken down in health & mind and the 30,000

AL (fragment), NjWOE, DF (*TAEM* 12:251). [a]"Credit Aut Tel Co" written by Harrington at top of document. [b]"828 − 577.86 = 250.14" written in column in left margin. [c]"828 paid 22 July" written below by Harrington.

1. On this date the payroll of the American Telegraph Works was $577.86 (later changed to $580.86). Cat. 30,108, Accts. (*TAEM* 20:255).

2. Unidentified.

–178–

To Samuel Edison

Newark N.J. July 24 [1871][1]

Dear Father

Rene[2] wrote me a Letter today about going into the Liquor Store, and wants me to put in from $500. to 1,000. and wants 200 or 300 cash down—

He says he will be satisfied with half the net profits =

He also says he will take the Liquor Business from the other store and put it in the new—

I send you \$300.[3] today for the purpose of going into this business. But you must be very careful what agreements you make   Do not agree under any circumstances to attend the store yourself= you want all ythe time you have to run around the Country and build houses, speculate etc

Have him advance his share of the Money and get every thing in <u>Black</u> and <u>White</u> which is the only possible Manner of doing business=

After investigation you do not think best to invest the money sent into the business—then invest in Land & houses to the best advantage

T. A. E

Enclosed is Renes Letter=[4]

ALS, MiDbEI, EP&RI.

1. Handwriting style and content, particularly regarding Edison's own financial condition, suggest that this letter dates from 1871.

2. Probably Rene Edison, a first cousin once removed (Samuel's grandnephew). "Rene" may be a familiar name for P. M. Edison, who opened a new store in Port Huron in 1871. Obituary of Anna Edison, *Detroit News*, 21 Nov. 1910, in Edison, Anna E., EBC; *History of St. Clair County* 1883, 630.

3. There is no account entry showing that Edison gave Samuel \$300 on 24 July 1871, although that amount is listed for 10 June (the placing of the entry suggests Edison may have meant July) and 15 August 1871. There is also a 4 July 1871 entry for \$400. Cat. 1213:3, Accts. (*TAEM* 20:6).

4. Not found.

–179–

Notebook Introduction: Automatic Telegraphy[1]

Newark N.J. July 28, 1871.[a]

Record of Ideas as they occur day by day applicable and for the Dot and Dash system of fast Telegraphy[2] invented for Geo Harrington and D H Craig= All of which are for them with the exception of one undivided one fourth interest in all patents caveats or ideas to myself applicable to that system and the proceeds from that one fourth interest either directly or indirectly=[3]

I do not record anbut fewe previously carried out inventions—but commence at this date to record all my ideas made previously to the opening of this book which I have not put in practical operation at the time of writing, and if possible all future ideas relating to this system while it shall be my interest to do so=

T A. Edison

AXS, NjWOE, Lab., Cat. 1182:1 (*TAEM* 3:50). ᵃ"Nothing on Tel" written at top by William Carman, a bookkeeper at Edison's later Menlo Park laboratory.

1. This entry appears on the first page of the notebook; Doc. 180 begins on the second page. This notebook also contains Docs. 190–91, 227, 231, 234–36, and 256–58.

2. Automatic telegraphy was commonly referred to as "fast telegraphy."

3. It is not clear why Edison refers here to either a one-fourth interest for himself or a continuing interest on Craig's behalf when the 1 October 1870 agreement between Edison and Harrington (Doc. 109) gave Edison rights to a one-third, not one-fourth, interest in his inventions, and a subsequent agreement, dated 4 April 1871 (Doc. 155), again specified one-third. On Craig's interest, see Doc. 109, n. 5; and Doc. 261, n. 2.

–180–

Notebook Entry:
Automatic Telegraphy

[Newark,] July 28, 1871.

Fig 1ᵃ

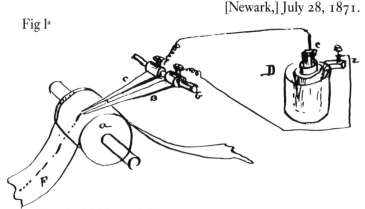

Chemical Telegraph Pen.

In this case I do not pass the current through the paper by using the paper carrying drum a as a conductor in the usual way, but pass the current from one pen to the other (pens—C.B) across the paper[1]   The pens C B are insulated from each other on the shaft G and one connected with Carbon and the other with Zince,[2] so if the paper a[3] be chemically prepared and the drum a rotated a The chemicals with which the paper is saturated will be decomposed as long as the current passes from C to B through the paper, but if the current is interrupted the mark will cease. The object of arranging the pen In this manner is to make the paper resistance variable, so that when the paper is too wet or the Line is short and there is too much battery on, the pens can be seperated which increases the resistance to the passage of the current, and a lighter mark is made which is too heavy or blurry when the paper is wet or the current too strong. Another advantage which this pen has is that, the distance which the current has to travel through the paper can be reduced to the ⅟₅₀₀ of an inch or even less, consequently reducing the resistance greatly and allowing a number of these receiving apparatus to be worked in one circuit without the electrical contrivances which has been found necessary on account of the tremendous resistance which the current encounteres in passing through the paper in some instances. paper properly wet has a resistance of 200 miles of 13 ohms eachs to a current passing through it. But by the use of this pen the pens A B C may be adjusted from 3 to 4 times as close as the thickness of the paper thereby reducing the resistance ~~from~~ to 50 tor 75 miles[4]

Continued—

Fig 2ᵃ

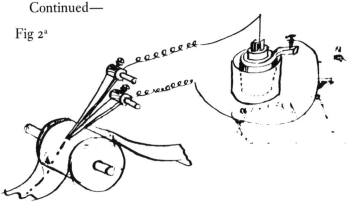

The pens may be arranged in this manner, Fig 2ᵇ one before the other, and a very close adjustment obtained—or what might be still better arrange them like that in fig 3. The pen C playing in between two prongs on the pen B the object of which is to get more surface on the non-recording pole or pen so as to reduce resistence.

Fig 3ᵃ

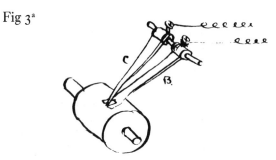

I have shown these pens with flat ends, but they may be provided with little rollers or wheels.

In sending positive and negative currents, for a chemical Telegraph The following idea has struck me =

Fig 4

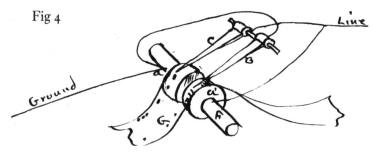

a and a′ are paper carrying rollers insulated from each other on the driving shaft F on these two rollers are two pens C.B also insulated from each other. it will be noticed by referring to the drawing that the main line is connected to the pen and the opposite roller and the ground wire to the other pen and the opposite roller = Now when paper perforated in two lines and sent through this following apparatus Fig 5, a positive Current will pass over the line when one set of holes passes under the pen and pass through the pen C through the paper thence through the roller to the ground decomposing the chemical solution with which the paper is saturated and leaving mark corresponding to the holes in the perforated paper at the other end of the Line. now the same current will pass through the roller a thence through the pen B and to the ground but in this case the current passes through the paper in an opposite direction. consequently the marks if any will be underneath the paper and not be noticed while it will leave no mark on top—if now a current of opposite polarity is sent over the wire the reverse action takes place and the pen B records, so that a dot on one line will represent a dot and a dot on the other line will represent a dash. Fig 5 show the apparatus for sending the double rowed perforated paper = [5]

Fig 5

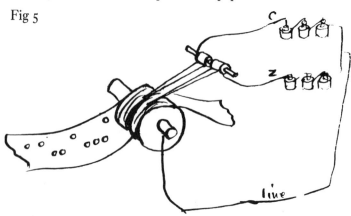

The upper row of holes throws a z̶iCopper battery on the Line and the Lower holes throw a Zinc battery on the Line

Syphon, chemical pen.—or device for recording chemically upon "dry paper"[6]

Fig 6

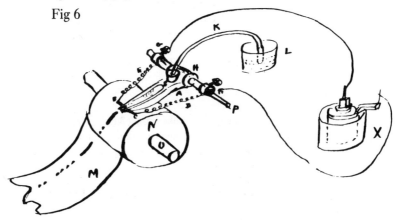

A is a hard rubber ~~or any~~ pen constructed after the manner of a drawing pen. it is secured to the shaft p and moves freely on it, and may be held closely on the paper by a spring. B and C are small platina points passing through each prong of the rubber pen of its extreme end and between which points the solution passes on to the paper   E and D are wires connecting these points to the binding screws F and G and thence to the battery X   S is a small funnel into which a syphon K leading from the resevoir L passes and keeps the pens supplied with the Chemical solution as fast as it is used   Now when the drum N is revolved carrying the recording paper M, The pen gives out a continuous line of the chemical Solution which being colorless leaves no mark and is quickly absorbed by the paper. But if the Current is closed for an instant all the solution passing the platina points at that moment will be decompossed and passing on to the paper will leave a mark. The syphon might be dispensed with and The solution supplied to the pen by dropping it drop by drop (according to the speed of the drum) into the funnel .S. One fact I will mention here is that only that portion of the Solution which is directly in contact with the platina points will be decomposed—

All written July 28 1871

Witness   Jos. T. Murray[c]   T A Edison

AXS, NjWOE, Lab, Cat. 1182:2 (*TAEM* 3:51). [a]Rectangle drawn around figure label. [b]"Fig 2" interlined above. [c]Signature.

1. The usual practice in chemical telegraphy was to pass the line current to ground through a metallic pen, chemically impregnated paper, and a metallic drum (here Edison represents the line circuit with a single battery wired directly to the pens). Edison's arrangement presented several potential advantages, as stated here.

2. Battery electrodes were commonly made of carbon and zinc, which here constitute the positive and negative poles, respectively.

3. Should be F.

4. See Doc. 34, n. 2.

5. On 10 October 1871, Edison executed a caveat incorporating several arrangements of this type (PS [*TAEM* Supp. III]). See also the second drawing of Doc. 194.

6. The drying of the chemically prepared paper used in automatic telegraphy was a serious problem and one that Edison had been working on since at least March. The "syphon" receiver had the advantage that chemical paper did not need to be prepared in advance because the machine deposited the chemical solution while in use. In March 1873 Edison executed a patent application for a "syphon" receiver operating on the same principle as outlined in this notebook entry (Doc. 290). Bill to American Telegraph Works, 16 Mar. 1871, 71-015, DF (*TAEM* 12:581); Cat. 30,108, Accts. (*TAEM* 20:236).

–181–

Notebook Introduction: Printing Telegraphy[1]

Newark N J July 28, 1871

Record of ideas and inventions relating to Printing Telegraphs Which work with a Type wheel, all of which I am under Contract with the Gold & Stock Telegraph Co of NY. to give them subject to the Conditions of the Contract within the next five years from the date of Contract Reserving for myself any ideas contained in this book which I do not see fit to give said G & Stock Telegraph = [2]

This will be a daily record, containing ideas previously formed some of which have been tried some that have been sketched and described, and some that have ~~b~~never been sketched tried or described

AX, NjWOE, Lab., Cat. 1174:1 (*TAEM* 3:7).

1. This entry appears on the first page of the notebook; Doc. 182 begins on the second page. This notebook also contains Docs. 193, 207–10, 212–13, 216–17, 219–20, 222–25, 230, 232–33, 238–39, 243–48, 250–51, and 253.

2. Edison assigned all of his subsequent inventions in printing telegraphy to Gold and Stock. See Doc. 164.

–182–

Notebook Entry: Printing Telegraphy

Fig 1ª

[Newark,] July 28 1871

Double link motion for a shifting Type wheel, description unnecessary

Fig 2

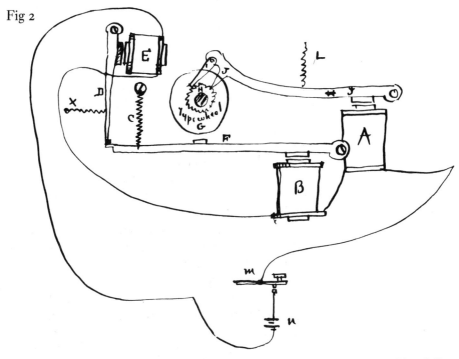

Printing Telegraph—to work upon one wire = The difference between this and most of the Printing Machines hitherto invented is the printing of the Letter by the spring instead of the magnet as it is usually done, and preventing the printing Lever from responding to the pulsations which rotate the type wheel, by causing a lever of a seperate magnet to lock the printing lever when the pulsations are of any degree of rapidity.[1] by referring to Fig 2 the principle upon which this printer work will easily be seen and will need but a few words of explanation =

IHaving found in the Course of my numerous experiments that when a magnets armature is adjusted so close to the face of the core that it nearly touches   that a current passed through it intermitted slowly it will stick or hug the core notwithstanding a considerable tension on a reacting spring. I have combined one of these magnets, adjusted in this manner with a printing lever arranged to print by a spring. E is the magnet which I will call the "Residual Relay"   D is the armature lever   X the reacting or adjusting spring   B is the printing Magnet   A The Type wheel rotating magnet. all these magnets are in one circuit in which is also the key m or any intermitting Current device & the battery n. J is the Escapement Lever   H the ratchet   G the type wheel.

Now when the Current is sent through ~~th~~All the magnets The Magnet B pulls the printing lever F away from the Type wheel. The Magnet E pulls the lever D towards its core and the end of the lever locks the printing lever and the Magnet A rotate~~d~~s the type wheel. if now the key is opened quickly & closed quickly the magnet A keeps rotating the Type wheel, but the lever of the Magnet E does not respond to these intermissions. consequently the printing lever remains locked   if now we wish to print we open the key for a moment Cut off all the circuit from the Line, the Magnet E has time to discharge, the Spring ~~D~~X pulls the arm D back releases the printing lever, and the Spring C pulls it up and effects the printing. if the Current is now closed   the lock action again takes place, and the Current can again be intermitted, the type wheel rotated, without effecting the Printing lever. In practice I think it will be necessary to have a seperate clutch, click or lever to lock the ~~type~~ Printing Magnet, and knocking it out and in by the lever .D. which will have dead play enough to acquire a Momentum, and Knock the clutch out. Thus

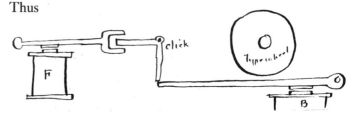

All foregoing written & drawn July 28 1871
Witness   Jos T Murray[b]   T A. Edison

AXS, NjWOE, Lab., Cat. 1174:2 (*TAEM* 3:8). [a]~~"Gold & Stock~~ Stock Quotation Printer" written by William Carman above drawing. [b]Signature.

1. There is another (probably earlier) version of this idea in a laboratory scrapbook. Cat. 297:7, Lab. (*TAEM* 5:60–61).

−183−

Notebook Introduction:
Automatic Telegraphy[1]

= Newark N.J., July 29, 1871

Automatic Printing. Telegraph.
Invented by and for myself exclusively.
Daily record of ideas as they occur applicable to this system, and a record of ideas already conceived, and account of experiments =

Dot and Dash and Automatic Printing Translating System, Invented for myself exclusively, and not for any small brained capitalist.[2]

AX, NjWOE, Lab., Cat. 1172:1 (*TAEM* 3:78).

1. This entry appears on the first page of the notebook; Doc. 184 appears on the second page. This notebook also contains Docs. 186, 194, 196, 198, 237, and 241–42.

2. Perhaps Edison felt Craig and Harrington (the "small brained" capitalists) would have no legal right to the system described in this notebook because it records the message in roman characters rather than dots and dashes. However, Edison later contracted with Harrington and Josiah Reiff to develop a roman-character system of automatic telegraphy (see Doc. 295).

–184–

Notebook Entry:
Automatic Telegraphy[1]

[Newark,] July 29–30,[a] [1871]

The general plan of this system is this =

I propose to perforate or punch in by in paper in strips Roman Characters, by dies and punches, a punch and die of each Letter of the Alphabet, and made on a the[b] stencil principle = The Machine which I am about to describe will be provided with 30 keys the depression of any one cutting out a Letter, after the manner of a stencil, and at the same time feed the paper ahead the distance of one Letter = ready for the next = After a strip of paper containing a message has been punched in this manner it is taken to a sending apparatus for transmission over the Line = This apparatus will be as follows = The Machine will have a paper carrying drum revolved by a weight or by magnetism, over the drum and revolving at right angles with it is another drum having a springs a number of springs in the same circle raidiating off of the drum and all Connected to the Main Line   When the Machine is in motion both drum revolve   the paper carrying drum very slowly and the drum carrying the springs very fast = The drum upon which the paper is carried being connected to the battery   every time a spring passes over ta portion of the Letter is closes the circuit through the aperatures of[c] the Letter   by the time one spring has passed over another Comes on in a different portion of the Letter as the paper drum feeds the paper ahead a given distance   Now another Machine precicely similiar to this receives the message at the other end of the Line, the perforated paper being replaced with Chemically prepared paper   It will be seen that if the two machines run in unison that an exact facsimile of the stencilized letters of the perforated slip will be record chemically at the other end of the Line. I propose to Constructio two perfect Machines as near alike as possible, and provide each Machine with a very fine electrical governor, and a Very convenient, deelicate, and quick adjustment for varying the speed, by the use of this contrivance and the attention of an expert to adjust instantly any differences. I hope to obtain 300 words legiably recorded every minute[2]

The great advantage of this system of any other is that is but one machine which requires intelligence of any amount and that is the adjusting of the receiving instrument =

Fig 1^d

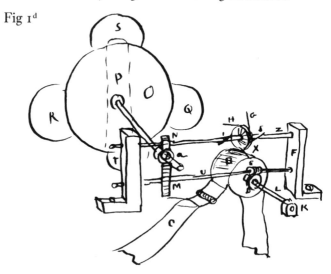

I will describe roughly the idea of my receiving and sending apparatus. B is the paper drum on the shaft L secured to the bearing K   D is a gear wheel   E a worm on the shaft U   M is a gear wheel driven by the worm a of the engine   O is the fly wheel of the engine   Q R S T the magnets   N is a gear wheel driven by the worm a   z the shaft carrying the wheel X with its sending or recording contact springs G H I J = C is the paper. When the Engine starts the Mechanism it all^c the parts revolve   the shaft Z and springs quite rapidly and the paper carrying drum slowly. While say the spring G is passing from one end of the drum to the other and recording a portion of the Letter, the paper carrying drum carries the paper ahead the $\frac{1}{64}$ of an inch so that when the spring H comes on the paper it records another portion of the Letter the $\frac{1}{64}$ of an inch behind that of G   And the Wheel X being provided with a sufficient number of springs to record a letter at every revolution—

It may be found in practice that moving the paper feeding

drum while the pens G H etc are crossing the paper may interfere with the record   in that case the wheel D may be made and rotated step by step and adjusted so that just at the moment one pen leaves the paper, it will feed the drum ahead the ¼₄ of an inch before the other pen comes on the paper. The device is illustrated in fig. 2.

Fig 2  Fig 3

F ͨ is the shaft carrying the pin wheel D provided with pins E E   a is a cam on the drum B which is on the shaft C   when the shaft C is rotated the cam a in passing between the pins E E cams the wheel D forward one pin   Fig 3 shows the cam a just in the act of passing between one of the pins and caming the pin wheel ahead

The rotary motion of the pens might be dispensed with and ~~The~~ a back and forward motion made by a cam wheel on the driving shaft   in this case but one pen would be necessary. This would give a motion to the pen much evener than a rotary, for when rotated the spring bears harder on the paper when in the middle than when it comes in contact. Fig 4 will give an idea of the devices by which I propose to accomplish the object =

Fig 4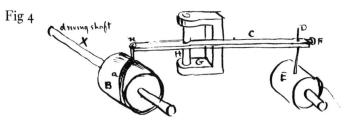

B is a drum with a grove planed or cut in it at an angle into which a pin n plays connected to the writing arm .C. and when the driving shaft X is rotated the pin following the Course of the slot gives a forward and backward motion to the p̶shaft or rather arm C which of course impells the writing pen D. back and forth on The paper & drum E = G is the support for the shaft H of the arm C =

The paper should be held down firmly upon the drum by a Shield   thus:—

Fig 5

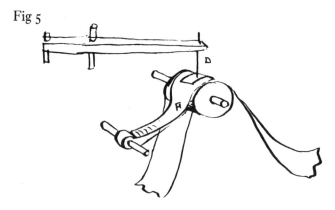

F is the shield; made preferably of an insulating substance.

The pen D may be provided with a small roller at its end for sending which I think will be found preferable to a flat pen, and it may be equally as good for receiving.[3]

I propose to try several forms of Governors   fig 6 will give one

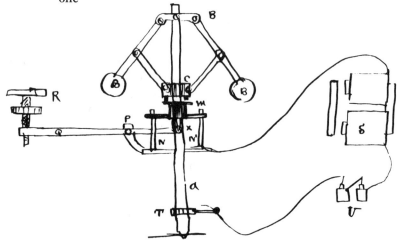

A is the shaft carrying the Governor. BBB is the governor[e]   C a collar which raises up on the shaft A as the speed increase and the balls spread out. M is another collars through which the shaft A[b] passes but does not touch,[e] ~~on~~the top part of this collar is platinized, and the bottom of the Collar C is provided with several platina points which rub on the platinazed surface of the Collar M[e]   N N' are rigid rods upon which the Collar M slides up and down,[e] Q is the adjusting rod for varying[c] the speed, pivoted at p, and adjusted up & down by the screw R[e]   This rod is secured to the Collar ~~by~~ at X. S are the Engine magnets   V is the battery   T is a collar upon which a spring rubs and is used as a means of connecting the battery with the shaft and its devices.[e] It ~~will~~ is obvious that if the collar is adjusted to a certain position and the battery power is increased the speed of the Engine will be increased also, and the balls will spread out by reason of the increased speed, ~~at~~and lift the Collar C and its platina points off from the platinized surface on M and break the Circuit, until the speed diminishes sufficient to allows The balls to collapse when the circuit will be again closed and so on the two collars will vibrate closely together alternately opening and closing the circuit and keeping the speed at the same rate In practice I apprehend that the governor shown[b] will not regulate the speed with sufficient acuracy[e] if this is the case, a Leverage can be abtained, and an extremely fine adjustment obtained by the following means

Fig 7.

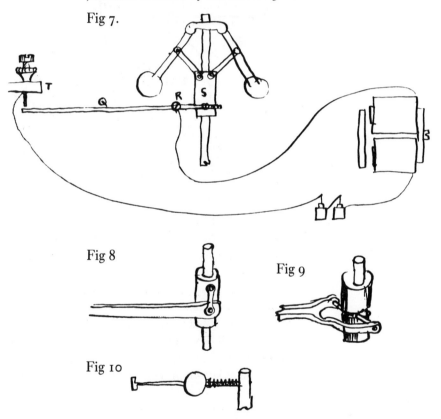

Fig 8

Fig 9

Fig 10

In fig 7 The electrical break as the speed increases is made by a lever Q secured to the Collar S by the devices shown in Figs 8 & 9 or in any usual manner[e] This Lever is pivoted at R, and the circuit is broken at .T. By using this Lever, and placing the fulcrum R in close proximity tho the Collar S greater leverage is obtained and the least rise of the governor

balls quickly shows itself in magnified proportions at the point
.T. Fig 10 shows a mode of arranging the governor balls. This
governor device may be worked on a different principle, instead of breaking the Current of the engine when it goes too
fast and thereby slowing it. A magnet is arranged with a break
upon the fly wheel and the governor made to Break and make
the circuit of this magnet and thereby put on and take off the
break from the fly wheel. Fig 11 will show this arrangement

Fig 11

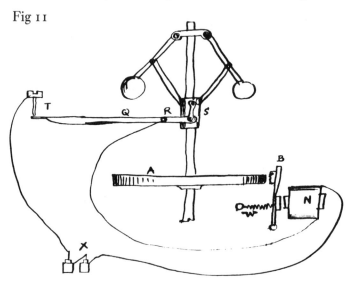

A. is the fly wheel secured to the carrying shaft, B is a
armature provided with a Leather pad which when there is no
circuit in the magnet N is held on the periphierry of the wheel
A by the spiral spring W, thus slowing it up, and the balls
dropping closes the circuit at^c T through the magnet N, and
pulls the break off from the^c fly wheel A, and doing this as
often as the speed increase beyond the point set at .T.

Instead of a worm and gear to drive the paper carrying
wheel ahead The idea has occurred to me that a cam motion
the same as shown in figure 4. could be used thus   fig 12

Fig 12

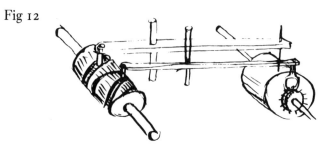

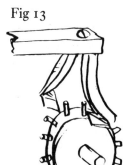

Fig 13

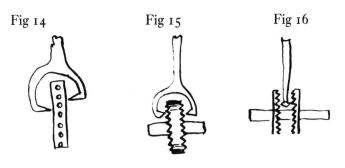

Fig 14   Fig 15   Fig 16

Fig 13 shows the manner of feeding the pin wheel ahead step by step—

Fig 15 shows another devicce for feeding the paper carrying drum ahead step by step twice for͏ᶠ at every revolution of the main driving shaft

Fig 14 shows another view of fig 13

Fig 16 shows another device = Fig 17 shows another device for getting motion from the main shaft and also for driving the paper drum

Fig 17

Fig 18 Is another device for feeding paper drum

Fig 18

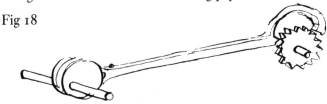

Fig 19 The same

Fig 19

Fig. 20 Another device for procuring a double motion for each revolution of the driving shaft for feeding the paper drum =

Fig 20.

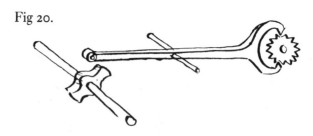

Fig 21 Represents a case for keeping the chemically prepared paper in to prevent evaporation, and to keep it wet and
ready for use the moment the machine commences operation.[c]
heretofore in Chemical Telegraphs considerable trouble has
been experienced in keeping the paper sufficiently damp to
give a clear record, and I find that no provisions for extermi-
nating this difficulty have been yet made to my knowlege. In
my machine more than any other it is absolutely necessary
That the paper should be in the machine and always ready for
use, and owing to the slowness of the paper in passing through
the machine it would quickly dry in coming from a roll exposed to the air.

Fig 21

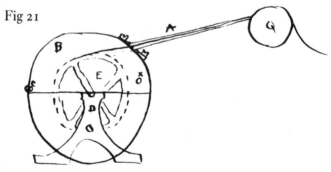

B is a metal case hinged at C provided with a handle X for
opening   This case is secured to the standard .D. A is a long
spout secured to .B. and leading directly to the paper carrying
& writing drum in which the paper passes   E is the paper reel
running within the Metal case upon a[b] tShaft the Centers
of which run in the standard D. The reel of paper which has
previously been prepared is set within the Case B and the end
led through the spout A, and then closed, and being almost
or airtight the paper will Keep moist and ready for use for
weeks.[c] it will be noticed that no part of the paper is exposed
to the air until it reaches the paper Carrying drum ready to be
used =

Fig 22

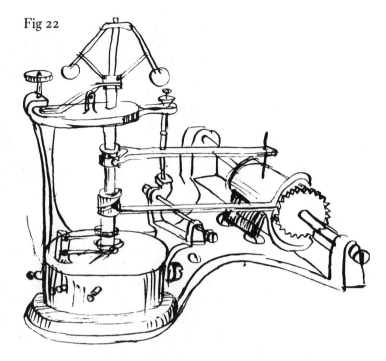

Proposed form for Constructing the Machine = Fig 22

In practice it may be found necessary to feed the paper drum by a seperate magnet, which will take an Intermittent work off from the Engine and perhaps add to the acuracy of the Unison of the two machine   It may be done in This manner   Fig 23

Fig 23

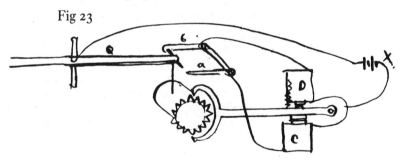

a and b are contact points, each connected with the magnets C.D, the connections arranged in the usual manner— When the writing pens comes over to one side it makes contact with one spring closes one magnet which feeds the paper ahead one tooth and going backs throws the current from the battery X through the other magnet and ~~closes th~~and feeds the paper ahead another tooth =

One magnet could be dispensed with and a spring used in its place

It also may be found necessary to make the writing pen travel straight across the paper and not perform an arc as I have heretofore shown   Thus   Fig 24

Fig 24

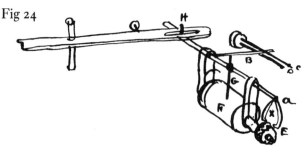

Q is the lever   a a sliding bar carrying the pen G and prevented from tipping by the wire B and guides C.D. It is connected to Q by the pin H and slot   E is the paper feeding ratchet = Q and a might be connected by a link which would reduce the friction. The ratchet wheel E might be a double Crown wheel with V shaped teeth or a pin wheel and ~~an~~ a fork X made to feed the paper by the back and forward movement of the writing shaft a.

From Page 3 to Page 14 Written, and illustrated on the 29 and 30th of July

Witness   Jos. T Murray[g]   T A. Edison

AXS, NjWOE, Lab., Cat. 1172:2 (*TAEM* 3:79). [a]Date taken from text, form altered. [b]Interlined above. [c]Immediately preceded by "x" in an unknown hand. [d]"x" written in an unknown hand next to drawing. [e]"(x)" interlined above in an unknown hand. [f]"twice ~~for~~" interlined above. [g]Signature.

1. This entry is continued in Doc. 186.
2. A roller was less likely to tear the perforations in the paper tape.
3. This is the first evidence of Edison's use of a flyball governor for regulating the speed of telegraph instruments. Although not used in his subsequent automatic telegraph instruments, this governor became an important element in Edison's universal private-line printer. See Doc. 208, n. 1.

–185–

Daniel Craig to Marshall Lefferts

Peekskill, July 31/71.

Dear General:

I have read your doleful note, and also that of our most excellent friend, who asks you if there is really any better ground for hope, now, as to our new system, than there was a year ago?

Whilst I do not admit that we had not advantages, a year ago, over the Morse system, ~~of~~ fully equal to 50 per cent. I do not hesitate to say, what I am sure, of your own knowledge, you can confirm, that within the past year, we have succeeded in making improvements, which, in the very material business of Perforating and Writing out (Copying) of messages, we are now at least three to five hundred per cent better off than we were 12 months ago.

I say, farther, and as the result of my closest observation and best judgment, that with out present <u>complete</u> machinery, we can and shall, very soon, prove our ability to beat the Morse system, in promptness and accuracy of transmission, <u>very decidedly</u>, and we shall prove our ability to beat all other systems, in case of labor, at least 4 or 5 to 1, & in cost of Lines, Batteries, &c more than 10 to 1, taking the whole country through.

There is another consoling thing, <u>to me</u>, in this matter, and that is that our inventors all say they have exhausted their wits, on our machinery, and they see neither the <u>necessity</u> nor the possibility of further improvement.

Mr. Harrington and the Automatic Co. are acting on this assumption and there will now be only so much delay as may be necessary to enable the operators to acquire a moderate degree of expertness in the use of the new machinery.[1]

Mr. Hoadley, and you and I are very near the end of our troubles. Yours truly

D. H. Craig

<u>Private</u>   Dr General: Of course you know it would be[a] perfectly suicidal for me to make a demonstration, now, to raise a large sum of money on our Telegraph interests. The <u>fearful delay</u> we have suffered, has caused some of Harrington's friends, <u>I know</u>, to become weak in the knees and I should simply act the part of a fool to make the least demonstration that looked like <u>hedging</u>, now.

Our Sholes Printers, (which are quite satisfactory to me) were shipped from Milwaukie some days ago, & soon we shall be all ready to move on the enemys works, and lift the biggotted fools right out of their boots. At the end of one week, one of our Perforators could do, of familiar sentences, 42 words per minute on the new Perforator. We shall get 50 words per minute, sure, from Perforator & Printer.

Little & Harrington and Edison seem to be all on the best of terms, but I cannot get one of them to touch the Closing up matter between us & them.

Why not drop into the Clarendon and have a chat with Harrington. I wish you would, and tell me the general result.

Confidentially, to you only, I will say that there is a Perforator, nearly completed, which is more simple, cheaper to make, easier to work & more rapid than Edison's.[2] I shall try to shape the interest so that it can be controlled by all the friends of Automatic Telegraphy, & not by one or two. You may depend upon it that I sleep with one eye open, all the time, & I guess you may now do the same. Truly &c

D H Craig

ALS, NNHi, Lefferts. [a]Interlined above.

1. Just prior to the date of this letter, the Automatic Telegraph Co. made preparations to begin operations. Edison, Murray, and Unger shipped a perforator to the company's main office, at 66 Broadway, on 17 July; on 20 and 21 July a room was cleaned out, the switchboard fixed, and a "carpet put down and office opened ready for work." However, the company did not open for business until 14 December 1872. Affidavit, 10 Feb. 1879, Cat. 297:310, Lab. (*TAEM* 5:996); "Automatic Telegraphy in Practical Business Operation," *Telegr.* 8 (1872–73): 556.

2. Probably the perforator designed by Frederick Grace. Memorandum of George Harrington, Cat. 297:117, Lab. (*TAEM* 5:878).

–186–

Notebook Entry:
Automatic Telegraphy[1]

[Newark,] July 31–August 4, 1871[a]

In this system I intend to use an automatic Letter puncher   in my preface[2] I mentioned that I intended to use and punch the Letters in the paper upon the stencil principle,[b] that is The letter after being punched would resemble a stencil letter,[b] the interstices of the Letters being held together by strips from outside of the paper = Thus. = ~~upon experime~~

Fig 25   B O N

upon experimenting with this and several other previously Conceived ideas, I found that the inslide of the letters or parts left after the letter was cut was very fraible or weak and that the transmitting stylus or other device would be liable to break them out in passing over them and thus destroy the legiability of the Letter at the other End of the line = The best form for punching or preparing The Letters I found was to form them of sa large number of small holes. Thus:—

Fig 26

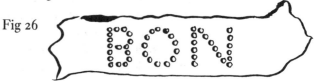

When the Letter is punched in this manner no part of the paper wLeft within a Letter after punching is materially weakened,[b] at least not weak enough to produce inconvenience in transmitting the same.

I also find that by adopting this mode and way of pm making a letter, the Mechanical difficulties which would be encountred in preparing messages by a stencil Letter would, be greatly reduced by making The Letter in the manner last described.[b] Another advantage in using the perforated Letter composed of Samall holes is that the received copy at the other End of the Line is much more legiable, and has a more handsome appearance—[3]

The copy if received from a stencil letter would resemble the following

Fig 27   BON

While that received from the Letter composed of round holes would be thus

Fig 28   BON

which will be more legiable—

ThOne of the devices with which I propose to prepare the paper with   This kind of Letter is shown in Fig 29

Fig 29

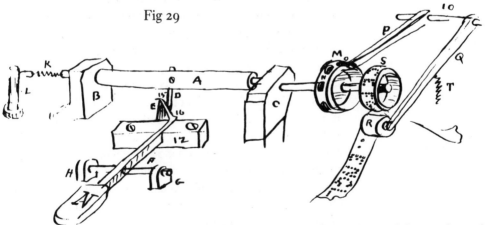

S is a dHollow Drum in which is fitted 28 letters formed Of holes.[b] The manner of making [-]each letter I will breifly explain before I give a description of the Machine[b]

A block Fig 30, of steel is drilled from end to end with holes to form the letter thus Fig 31. Then the metal around these

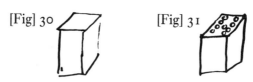

[Fig] 30   [Fig] 31

holes is filed or cut away so as to leave the edge of the hole sharp enough to cut ta hole in the paper when pressed on it. Fig 32 will show this, and Fig 33 will show one sing[l]e punch in this manner on a magnified scale

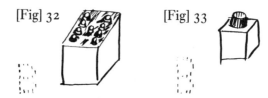

[Fig] 32   [Fig] 33

These block oeach containing a letter of the alphabet is set in the periphierry of the wheel .S. and secured firmly to it[b]

R is the roller which carries the paper and upon which the letter—is cut out.[b] This roller is formed in a peculiar manner, to withstand the constant wear of the cutting punches on it after passing through the paper.[b] Fig 34 will show the metallic portion of the roller

Fig 34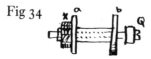

The flanges a b of this roller is filled with washes of raw hide, or cat gut, or with paper and then subjected to an immense pressure and held together by the nut x. Then turned of smoothly in a lathe. These two substances when arranged in this manner have a "peculiarity" which no other discovered substance has or is known to have,[b] that is that the Cutting edges of punches or[c] knifes or sharp tools do not wear them at all and they will stand for months of continuous cutting, and still preserve their specific peculiarity or quality[b]   Now it will be seen by referring to fig 29, that when a letter with its cutting edges is brought directly under the roller R tand the Lever Q lifted upwards with sufficient force, the cutting edges of the punches will cut through the paper on to the rawhide roller R and punch a letter—

The manner in which I bring around any desired letter & have the lever Q raise the drum R upwards to the face of the drum .S. I will not[4] explain:—

B & C are frames in which the shaft A run, upon which

shaft is two wheels S and M, the latter having teeth upon one side of^d its periphiery.^{b} on the other side is a pin o rubbing continuosly upon said periphierry. This pin is connected rigidly to the lever P which is connected rigidly with the lever Q by means of the shaft 10, so that when the lever P is raised it raises The Lever Q and drum R with it^b T is a spring to keep The pin o down ~~a~~upon the periphery of the wheel M where there are no teeth. The shaft A slides endwise, but is prevented from doing so when not operated, by the Spring K secured to the shaft and standard L^b The manner of arresting the rotation of the shaft and throwing it forward of endwise is quite simple

N is the finger key pivoted on the shaft F, which runs in the bearings H and G^b ~~and~~ the end of the key N is bent in the manner shown in the drawing, so that when The key is depressed the cam or bent end E is thrown into the path of the pin D ~~s~~on the revolving shaft A.^{b} the pin D first strikes the bent end at 15 and the shaft continuing to rotate, ~~the~~ it is cammed forward or endwise and then suddenly arrested by the straight part of the key at 16^b The moment that the shaft is moved endwise ~~the pin O on the lever P is thrown further~~ That portion of the periphery on the wheel M which has the teeth is thrown into the path of the pin O on the lever P which riding upward on one of the teeth lifts the lever Q and drum ~~d~~R and paper up to the letter which corresponds to the arm D upon the wheel .S. with sufficient force to cut it out^b Now the moment the key N is allowed to be released, The spring K pulls the shaft back and throwing the teeth on the wheel M out of the path of the pin O, and the shaft commences to revolve again.^{b} before this is done the pin O would have passed over a tooth on the wheel M and effected the cutting out of the letter & dropped down out of the way before the shaft was released. The shaft A could be provided with 28 or 30 pins set in different positions upon the shaft and directly in the path of as many keys similiar to N so that by depressed any key a certain character upon the die wheel S could be brought in the desired position. In practice it may be found necessary to dispense with the teeth ~~and~~ on the wheel M and substitue in their place undulatory "races" as shown in Fig 35   This would be a positive movement and prevent any momentum being given to the levers P & Q by the suddeness with which the teeth strikes the pin O.

Fig 35

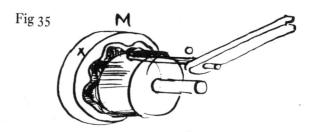

Fig 36.   [Fig] 37

 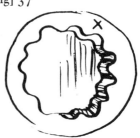

When the Wheel M is thrown forward the pin O enters one of[e] the undulatory cams at its lowest point and rides up and down one undulation ~~when~~ twhich lifting the Levers P & Q and punching the letter is thrown out of these slots by the wheel M moving back and then the pin O rides up on the smooth portion of the periphiery[b]

Fig 36 & 37 shows how these slots are made   XX being fitted inside of X and both secured together and with the shaft = The manner in which I propose to move the paper, I will now describe

~~Date of~~ written between July 31 and Aug 4 1871
Witness—   Jos T Murray[f]   T A. Edison

AXS, NjWOE, Lab., Cat. 1172:15 (*TAEM* 3:85). [a]Date taken from text, form altered. [b]"(x)" interlined above in an unknown hand. [c]Preceded by "x" in an unknown hand. [d]"side of" written in right margin. [e]"one of" interlined above. [f]Signature.

1. This entry is a continuation of Doc. 184.

2. See Doc. 184.

3. In his first patent for roman-character telegraphy (U.S. Pat. 151,209), Edison employed a perforator with a square matrix of twenty-five separate punches (various combinations of the punches producing different characters) rather than a punch wheel with a separate block for each character.

4. Edison probably meant to write "now."

Newark N.J. Aug 5, 1871

Record of ideas conceived, and experiments tried on miscellaneous Machinges and things, to which dates are appended, and such record to be used in any contest ~~regarding~~
or disputes regarding priority of ideas or invention.

T A. Edison

AXS, NjWOE, Lab., Cat. 1181:1 (*TAEM* 3:174).

1. This entry appears on the first page of the notebook; Doc. 188
begins on the second page. This notebook also contains Docs. 189, 214,
221, 226, 228, and 240.

–188–

Notebook Entry:
Chemical Recorder[1]

[Newark,] Aug 5 1871[a]

Device for recording upon ~~chemical~~ upon chemically prepared paper, by a "jump spark" or [-]Current of electricity of
high tension   Fig. 1 will give the principle of the device

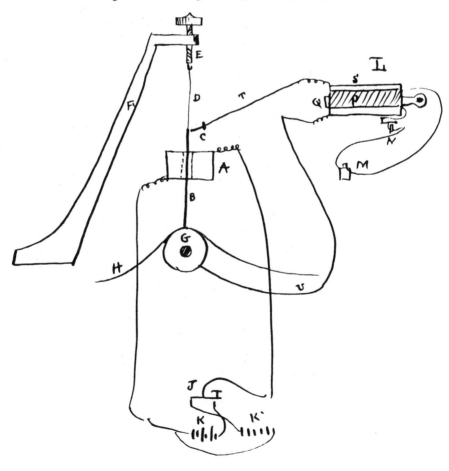

This device I do not consider fast ~~but~~ and not applicable to Automatic or fast telegraphy, but that shown in another[b] figure ~~2~~ is.—The main result which I have endeavored to obtain in this device is procuring or effecting a visible record by ~~an infer~~ a ~~a~~very feeble current. Sir Wm Thompson[2] some time since invented what is known as the Syphon Ink Recording Galvanometer,[3] which is used on the Atlantic Cable and is the first recording Apparatus which had no friction in its parts. In my device I am enabled to move the recording parts which much less Current than the Apparatus of Sir Wm Thompson and also dispense with a large amount of apparatus which[c] accompanies that apparatus.[4]

In my recorder the pen B[b] consists of an exceedingly fine steel neede of about 1½ inches in length and suspended within a helix A of very fine insulated wire, and held by a single fibre of silk to the adjusting screw E. the ~~end~~lower end of the suspended needle B after passing through the helix A nearly touches the paper H[b] ~~u~~chemicaly prepared, upon the drum G which is rotated by suitable Machinery   The helix A is connected to the line and battery K and break key J so that by opening & closing the key alternate positive & negative Currents will be sent of the line and the needle B deflected back and forth over the paper but not touching it   With the exception of having the chemically prepared paper under the end of the needle, so far it is but a Common Galvanometer.

C[c] is a point adjusted quite closely to the needle B but not touching it   This point C is connected to The secondary Coil of a powerful "Inductorum"[5] by The wire T and the Drum G to the other end by The wire U so that when the Vibrator N and primary Coil is working powerful Currents of electricty will be generated of sufficient tension to jump from the point C to the needle B thence jump through the wet paper to the drum & back to the Inductorum forming a complete circuit. this current in passing through the chemically prepared paper decomposes the substance with which it is saturated and leaving a Mark   if the paper is carried along Continuously by The drum G ~~a~~ ~~e~~and the needle B is at rest a Continuous ~~mak~~ mark will be left on the paper, the positive & negative Current passing with such inconceivable rapidity through the paper that the line is continuous   If now the needle is moved back and forth by operating the key J a Zig Zag line will be given to the right by a positive and forming a dot and to the left by a negative and forming a dash—Fig 2

Fig 2

The word "this" would be recorded[6]

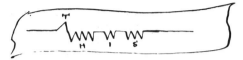

Above drawn & written Aug 5 1871
Witness. Samuel Edison[d]   T A. Edison

AXS, NjWOE, Lab., Cat. 1181:2 (*TAEM* 3:175). [a]Date taken from text. [b]Interlined above. [c]Preceded by "x" in an unknown hand. [d]Signature.

1. This entry is continued in Doc. 189.

2. Sir William Thomson (later Lord Kelvin) was professor of natural philosophy at Glasgow University and was among the premier contemporary physicists and electrical engineers. A director of the company that laid the first successful Atlantic cable in 1866, he superintended the cable's construction and developed the mirror galvanometer that enabled an operator to read feeble signals. *DSB*, s.v. "Thomson, William."

Drawing of Sir William Thomson's siphon recorder, which was used on cable telegraphs.

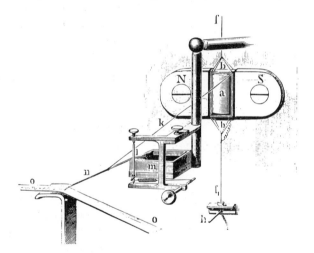

3. An article in the *Telegrapher* on 5 August 1871 described William Thomson's siphon recorder for cable telegraphy and may have prompted Edison to make this notebook entry. Thomson's device, first patented in 1867 and improved in 1871, recorded on paper the feeble signals transmitted over the Atlantic cable and was significantly easier to read than the mirror galvanometer, which had been employed previously. The siphon recorder passed the incoming signal through a coil of wire suspended between the poles of a powerful electromagnet. The resultant magnetic field caused the coil to turn one way or the other, depending on the direction of the current. A thread transmitted the motion of the

coil to a glass tube that siphoned ink from a reservoir and deposited it on a piece of paper. Coil motion recorded "dots" in one direction and "dashes" in the other. "The Syphon Recorder of Sir William Thompson," *Telegr.* 7 (1870–71): 400–401; Bright 1974, 604–16; Prescott 1877, 559–61; Maver 1892, 269–70.

4. Edison proposed using a simpler method, which he was employing in his automatic telegraphy experiments.

5. Probably a misspelling of "inductorium," a common name given to the induction coil. Shiers 1971, 83.

6. In the American telegraphic code, the letter "T" was represented by a single dash, "H" by four dots, "I" by two dots, and "S" by three dots. The zigzag shown in Edison's figure 2 is a nonsensical jumble of dots and dashes.

–189–

Notebook Entry: Chemical Recorder[1]

[Newark,] 8th August [1871]

I will mention here that the inductorum can be dispensed with and a magneto electroic machine used to generate the high tension currents.[2] I do not wish to confine myself to any particular device for making the high tension currents[a] go in a Zig Zag motion right to left & left to right and recording the jump spark upon Chemically prepared paper as it can be done in innumerable ways ªmany of which I could mention but which I consider quite unnecessary. What I will probably claim when I apply for a patent will be the General features or principle underlying the Combination to produce the effect.[3]

I will now explain a modification of this device for receiving very rapid writing. In this Case—No I will not—I was mistaken =

An Idea strikes me that sulphuretted hydrogen[4] or some other gas that will discolor certain chemical solutions Could be used in Chemical Telegraphs. Supposing a long Strip of paper was prepared with a chemical Solution Which would be discolored by a certain gas, and A fine jet of this gas was let play upon the surface of the paper, and the paper Carried forward by clock work   a continuous mark or line would be made on the paper   If the Jet was controlled ᵇso that it could be let off and on by an electro magnet placed in the main circuit, it would produce the same results, as the decomposition of the Chemical Solution by the electricity itself, and an advantage gained would be that the marks would not tail out and that the mark could be made large or small without regard to the strength of the Current which worked the magnet.

Ink recording Telegraph

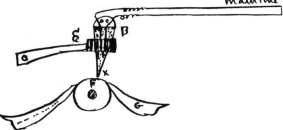

F[b] a paper carrying drum   over it suspened and held by E is a Air tight bulb B filled with ink into which is connected the main line wire D and .C. at every closure of the Current, gas[5] is generated by the current passing through the ink and decomposing the Water contained in it. this gas expands & pushes the ink out at the End of the bulb B at X through an very fine aperture, and it falls on the paper, so that as long as the Circuit is closed gas is made which continues to force the ink out through the aperture X on the paper which passing along leaves a continuous Mark   if the Current be now interrupted no More gas is generated and the ink ceases to come out of the aperture X—

Witness   Jos T Murray[c]   T A Edison

AXS, NjWOE, Lab., Cat. 1181:4 (*TAEM* 3:176). [a]Interlined above. [b]Preceded by "X" in an unknown hand. [c]Signature.

1. This entry is a continuation of Doc. 188.

2. This is a reference to Edison's proposed improvement of the Thomson siphon recorder (see Doc. 188). A magnetoelectric machine could be used in place of an induction coil to produce high potential currents of alternating polarity. The operation of magnetoelectric generators was based on Michael Faraday's discovery that the motion of a magnetic field relative to a coil of wire produced an electric current in the wire. Developed primarily as scientific instruments in the 1830s, magnetos came into use for electroplating in the 1840s and were more common in the 1870s as electric power sources for arc-light systems. King 1962c, 333–407; Maver 1892, 27.

3. No patent application has been found.

4. Hydrogen sulfide (H_2S).

5. Hydrogen and oxygen.

Device for receiving by chemical Telegraphs =

Fig 7

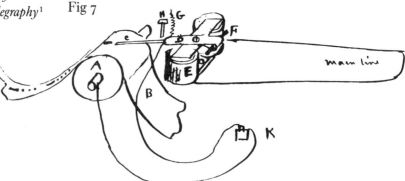

In the ordinary way of receiving on chemical paper the pen or stylus .C.[a] always remains in contact and the circuit of the main line is connected to it and the drum A, and the circuit opened and closed at the distant station leaves dots and dashes formed by action of the main line current upon the paper. But one great drawback to the practicability of receiving in this manner either by the main or Local circuit is that the dots and dash Tail out, (ie) ~~m~~continue to make a mark after the Current has ceased, the reason of this has not yet been explained[2]   The above device is to prevent this "Tailing" and is brought about by connecting the Magnet E, ~~pivot~~ in the main Circuit, and the stylus C   Drum A   chemical paper B   armature D in a local Circuit within which is the battery K. H is a stop screw for D & G a spring to draw D from the magnet   the armature Lever & pen stylus[b] holder D is pivoted at .P.

When the Circuit on the main line is closed it operates the Magnet .E, brings the stylus C down on the paper and the current of[c] Battery K decomposes the solution in the paper and leaves a mark as long as the magnet is closed   if the Magnet is now opened the ~~p~~Stylus C is lifted entirely off the paper by the action of the spring G etc thereby effectually preventing the tailing of the Character which is being received which would otherwise be the case if the pen was allowed to Stay on the paper and the Current through it simply broken.

The chemical paper might be dispensed with entirely and the record made from an inking ribbon.

If the stylus C had a point or roller at its extreme end and an inking band passed gradually by Clockwork between it and the the paper every time the magnet was actuated by the Current the roller would be brought down and pressing the ink band on the paper leaving a mark.

Or the Drum A might ~~be~~have an inking band wound around it and the pressure of the roller on the End of the stylus C would be sufficient to impress the ink upon the paper ~~u~~on the underside of the paper   Or the drum A ~~u~~might hav-~~ine~~ an inking roller to ink a slight line in the middle of its periphierry directly under the roller on the stylus .C. so that when the roller was brought down the pressure would be sufficient to make the paper take ink as long as the pressure Continued =

Witness. Jos T Murray[d]   T A. Edison.

AXS, NjWOE, Lab., Cat. 1182:6 (*TAEM* 3:53). [a]"or stylus .C." interlined above. [b]Interlined above. [c]"current of" interlined above. [d]Signature.

    1. This entry is a continuation of Doc. 180 and is continued in Doc. 191.

    2. See Chapter 5 introduction, n. 9.

–191–

Notebook Entry:
Automatic Telegraphy[1]

[Newark,] Aug 8 1871.

Fig 8

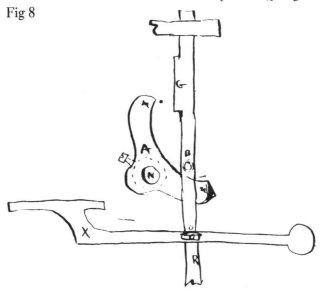

    Rock shaft for giving motion to the paper feed in my large perforator.[2]

    A is a fork peice secured to the shaft N by a screw or other means   B is a pin having a roller on it   G is a square peice upon the upright R. When the key X is depressed the upright

R is carried with it and the roller B coming in contact with the fork peice A at n. this rotates the shaft N, and at the same time the m end of the fork A comes in contact with the square G and the n end being down so far below the center that it ceases to act as a lever or means of rotating the shaft and the wheel or roller B slides down the straight part at N. It is obvious that by the two prongs being long and being arrested at two different points makes a perfect "lock" of the rock Shaft N and prevents any momentum, of that shaft from taking up the dead motion, occassioned by imperfect mechanism   There being a dead lock in this case and no loss motion to take up   Fig 9 will show the position of the fork and key when it is depressed to its full stroke and a dead lock takes place anof the Rock shaft

Fig 9

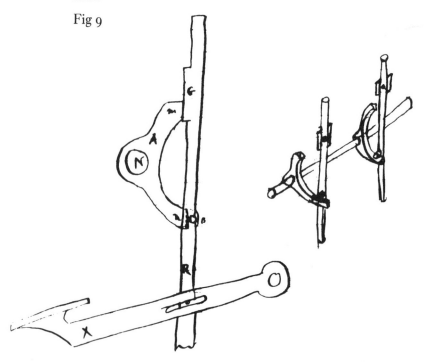

AX, NjWOE, Lab., Cat. 1182:7 (*TAEM* 3:53).

1. This document is a continuation of Doc. 190.

2. On or about 11 August, machinists at the American Telegraph Works made experimental modifications of the rock shaft for the large perforator. Edison applied for his first perforator patent on 16 August, and it included a variation in the design of the rock shaft proposed here. In the patent application, Edison did not use the square piece **G** to lock the shaft. He placed the pin **B** at varying heights on the shaft according to the distance the shaft had to move the sliding paper carrier to accom-

modate the different lengths of paper tape needed to perforate each character. This application issued as U.S. Pat. 121,601 on 5 December 1871. Cat. 30,108, Accts. (*TAEM* 20:264).

–192–

*From Charles
Williams, Jr.*

Boston, Aug 9 1871[a]

T. A. Edison

To Please send something on acct of this bill = How about that voter machine made for Roberts

C. Williams Jr

ALS, NjWOE, DF (*TAEM* 12:225). Letterhead of Charles Williams, Jr. [a]"Boston," and "187" preprinted.

–193–

*Notebook Entry:
Printing Telegraphy*

[Newark,] Aug 25 1871[a]

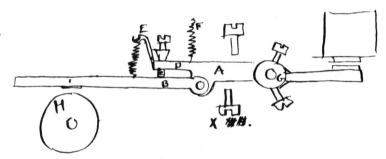

Jointed Printing Lever. The object of jointing the Printing Lever is to dispense with the device I have now on my universal Printing Telegraph Machine, which consists of an Electrical Cam and Throw over.[1] When the Type wheel is suddenly arrested and the Current through the Printing Lever Magnet[b] closed the A peice Springs forward and the arm D & Screw C ~~co~~kept in contact with the springy impression Lever B by the spring E brings that forward. at the moment that the peice A is suddenlly arrested by the screw .X. the Lever B spring forward by its momentum and impresses a Letter and fl~~y~~ies back till it Comes in contact with Screw C and when the Current is opened in the Printing Magnet all the devices fly back.

It is obvious that the printing Lever made in this manner will impress a Letter instantly & fly back out of the path of the type wheel leaving it free to start of[f].

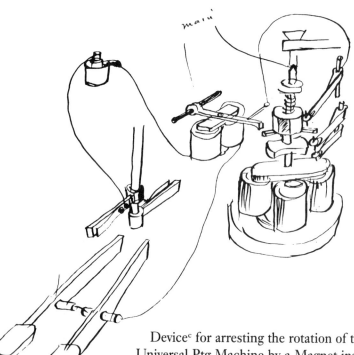

Device[c] for arresting the rotation of the break wheel on my Universal Ptg Machine by a Magnet instead of Mechanically. A Description Unnecessary[2]

All previous Written Aug 25 1871

Witness   Jos T Murrey[d]   T. A. Edison

AXS, NjWOE, Lab., Cat. 1174:4 (*TAEM* 3:9). [a]Date taken from text. [b]Interlined above. [c]"Universal Priv Line Printer" written above drawing by William Carman. [d]Signature.

1. Because Edison wanted the universal private-line printer to work quickly, the printing lever had to move immediately away from the typewheel after pressing the paper against it so that the typewheel could resume its rotation. In the original design, an arm attached to the printing lever broke the printing circuit just as the paper pressed against the typewheel, allowing a retracting spring to pull the lever back (see Doc. 165). The design sketched here resembles the printing lever of George Phelps's "financial instrument" (U.S. Pat. 110,675), used by Gold and Stock and certainly known to Edison. See Doc. 223.

2. In the original universal private-line printer design, typewheel shaft **S** was driven by a ratchet and pawl (not shown in this diagram). When a key was depressed (**K** and **K′** represent the instrument's 30 keys), it blocked arm **D** (or **D′**) and stopped the shaft. A series of levers working off the bottom of **S** stopped pulsator (breakwheel) **P** on motor assembly **N** as the typewheel halted, thereby

stopping the transmission of impulses through contact arm **C** over the main line. In this notebook entry, Edison connected one terminal of battery **B** to arms **D** and **D′**, and the other terminal through magnet **MM′** to keys **K** and **K′**. When contact between a key and **D** or **D′** completed the circuit, **MM′** charged and pulled down armature **A**, the tongue of which stopped the pulsator. **C′** and **C″** are contacts for the motor's commutator. Edison subsequently adopted this arrangement. See Doc. 208; and U.S. Pat. 131,340.

–194–

Notebook Entry:
Automatic Telegraphy[1]

[Newark,] Aug 31 1871[a]

A new system of Telegraphy, using neither dots nor dashes, but receiving the message by a Puncher or Embosser and Running it through an Automatic Translating Printing Machine I do not wish to confine myself to any particular Translating Printing Machine, as I have innumerable machines in my Mind now which I shall continue to ~~d~~illustrate & describe day by day when I have the spare time.

The Printer which I propose to use for translating from the punched or embossed paper, is as follows though I may improve it in time or adapt an entirely new one using Magnetism or dispensing with it and make the paper perform a mechanical Operation.

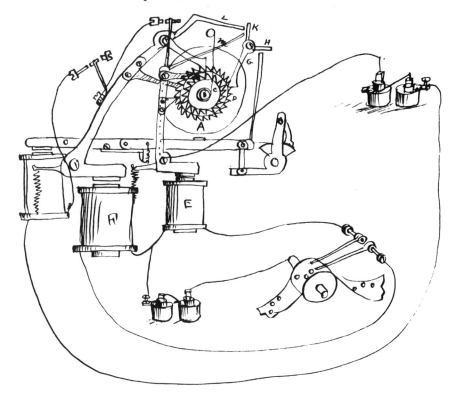

A is the type wheel[b] B its shaft[b] C is a small ratchet wheel rotated by E Magnet armature Lever pawl & detent, and D another ratchet wheel rotated by the magnet F Armature Lever click & detent.[2] The first Magnet E pushing the type wheel ahead one Letter and the magnet fF, 4 Letters. G is a pawl for preventing the type wheel from returning to Zero which it would do by the action of the rubber cord m secured to its shaft. The Printing throws this detent out of the path of the ratchet when it takes an impression[b] When this takes place the type wheel returns to Zero. This click is placed back in its original position by the first forward Movement of either ratchet Lever operated by the Magnets[b]

The Printing Lever is jointed near its fulcrum sand held to the other part of the lever by a tension, and when the Magnet attracts this Lever the forward part goes farther by its momentum than it would if the Lever was actuated Slowly and in so doing effects the impression and immediatly springs back out of the way of the type wheel and at the same moment when the Lever has reached its extreme Limit, the click G is thrown out of the ratchet and The Cord m pulls the type wheel back to Zero[b] The circuit of The type wheel rotating Magnets ~~are~~ is opened & closed by perforated or Embossed paper. The characters being[c] in two rows one row rotating ~~E~~ The type wheel 4 letters at a time and the other but one Letter at a time, the 4 Letter Magnet & devices might be dispensed with but the characters of the perforated paper would be too Lengthy. The Printing Magnet is closing when the Both[d] Escapement Levers have ceased to work and touch the back points of the printing Circuit[b] They both rest on these points for an instant, according to the length of the space between the Letter on the perforated paper[b]

To translate The Word "This" The perforated character should be thus = the holes in Line No 1 rotate the type wheel 4 Letters for each one and as T is the

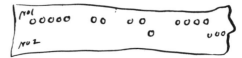

20th Letter 5 dots are used   when the space between the Letter Comes[b] this gives the Escapement Lever time enough to close both points & close the printing Lever & effect the ptg of that Letter.

The Letter H being the 8th Letter but 2 holes are necessary[b] the letter I being the 9th Letter in the Alphabet 2 holes

in Line No 1 turn the the type wheel to the 8th Character &
one hole in Line No 2 one more[c] Letter bringing it to I. S
being the 15th Letter 4 in the upper line bring it to 12th
Charctr & 3 in the Lower to the 15th which is S. 3 rows of
holes could be used but it would be hard to transmit Them
over the wire.

The Device for Running the perforated Embossed or otherwise prepared paper through may be a Drum Carried by
Magnetism Weight Spring & any other force or force storeing
device.

There are several ways by which this translating paper May
be prepared by an instrument operated by similiar paper at a
distant station.

first 1st= An Electro Magnet may have a very delicate
Valve fitted to its armature Lever and this Valve Operated a
cylinder & piston worked by Compressed air or other force or
power, and the piston of which has a punch like a belt punch,
Cutting upon a block of Leather, Raw hide or Compressed
paper and at every hole punched feed the paper ahead, or the
punching May be done while the paper is being fed Continuously by other means. The Magnet should be a polarized one
working two valves & pistons[b] one cylinder & piston punching
or Embossing holes in one Line & the other piston in another
Line, which line being determined by the polarity of the
Current. The paper might be Embossed.[c] The Lever of an
ordinary Morse Register[3] being operated by the cylinder &
Compressed air or Chemical paper might be used, the decomposed portion being of such a character that by immersing
it in another substance an action would take place to form a
salt or other matter & so raise the character in relief—which
Could be made to raised a very delicately poised Lever &
open & close ~~an~~ the electric circuit = of the Translating Printing Machine[b] or an Ink Recorder Could be used the ink of
which would be a conductor and two points closely together
& CONNECTED with either end of a battery & magnet   the
mark would ~~elo~~ connect the two points together or the ink
could be a Chemical & dipped in some solution like that
which I proposed to dip the Chemical paper.

Above written Aug 31 1871
Witness. Jos T Murray[e]   T A. Edison

AXS, NjWOE, Lab., Cat. 1172:21 (*TAEM* 3:88). [a]Date taken from text.
[b]"(x)" interlined above. [c]Preceded by "x". [d]Interlined above. [e]Signature.

1. This entry is continued in Doc. 196.

2. Edison used a similar arrangement in U.S. Patent 123,984. See also Doc. 210.

3. Edison also suggested the use of a Morse register for this purpose in a patent application draft made about the same time (Cat. 297:133, Lab. [*TAEM* 5:922]). See also Doc. 10.

UNIVERSAL STOCK PRINTER   Doc. 195

In the spring of 1871, even as Edison manufactured his cotton instruments, he began developing a new printing telegraph design—the universal stock printer. The Gold and Stock Telegraph Company received a wooden model and a prototype on 3 June and a second prototype on 14 June,[1] and in late summer Edison began manufacturing the universal machines. By the end of the year he had delivered 600 to Gold and Stock[2] and had finally executed a covering patent application.[3] Although the new ticker superficially resembled the cotton instrument, it incorporated several changes Edison had made in the earlier instrument and also displayed some novel constructions. The fundamental design became a standard in the Western Union ticker inventory for many years.

The original design of the universal stock printer refined two of Edison's earlier innovations—a mechanism that shifted the typewheels (U.S. Pat. 123,006), and the screw-thread unison (Doc. 158). Edison redesigned the shifting mechanism to

Photograph of the shifting mechanism of Edison's universal stock printer.

A Gold and Stock Telegraph Co. instruction manual includes this drawing of the shaft mountings of Edison's universal stock printer.

reduce friction. This, together with a new paper-feed mechanism, reduced the power requirement of the printing lever magnet. When a dozen instruments were on a circuit several miles long, the cumulative reduction significantly lowered battery consumption. In addition, Edison altered the unison so that the release was done by the printing lever itself, rather than by a separate set of levers actuated by the printing magnet.[4]

Edison improved the printer in two other important ways. First, he mounted the magnets so that the gap between a magnet and its armature was adjustable. Second, he changed the mounting of the shafts that carried the printing lever, the escapement lever, and the typewheels and made them laterally adjustable (cf. Doc. 136). Consequently, parts fit together far more readily and were considered completely interchangeable.[5] For several years, company instructions warned service personnel not to "attempt to file, bend or tinker with the different parts of the instrument, as they are interchangeable and accurately made, and if any part should in time get defective, supply it with a new one which will be furnished from the New York Office." By mid-decade, however, the company had decided that "it does not pay to furnish new parts for instruments, as the instruments are not all alike and consequently the parts do not always fit."[6]

Edison continued working to improve the universal stock printer after filing the patent application. Although he executed patent applications for many variations on the basic design during 1872, these features were not incorporated into the production model.[7]

THIS.IS.A.SPECIMEN.OF.THE.FIRST.INTERCHANGEABLE.PRINTING.INSTRUMENT.EVER.MADE.IN.THE.WORLD

Jan 9/72   S.S. Murray

A tape from the universal stock printer: "THIS IS A SPECIMEN OF THE FIRST INTERCHANGEABLE PRINTING INSTRUMENT EVER MADE IN THE WORLD."

1. Cat. 1183:10, 15, Accts. (*TAEM* 20:865, 867).

2. Unlike the cotton instruments, which had been made by the American Telegraph Works, these printers were manufactured by Edison and Unger (Newark Telegraph Works). Gold and Stock received roughly 640 more in 1872. By late 1874, Edison had made about 3,600. PN-71-09-06, Accts. (*TAEM* 20:974–83); 72-017, DF (*TAEM* 12:895–99); Ford 1874, 231.

3. U.S. Pat. 126,532. The patent model, now at the Edison Institute in Dearborn, Mich., was received at the Patent Office on 12 January 1872. It was universal stock printer No. 568, which had originally been delivered to Gold and Stock on 18 November (PN-71-09-06, Accts. [*TAEM* 20:978]). Edison often made patent models by modifying cotton

instruments or universal stock printers that were defective or had been returned for repair.

4. See Doc. 158 headnote.

5. Edison supplied Gold and Stock with boxes of parts for the universal stock printer. Cat. 1183:75, 77, 82, Accts. (*TAEM* 20:897, 898, 901).

6. Gold and Stock 1872, 10; idem 1876, 28. The cotton instrument's parts had little margin for adjustment. Consequently the parts of each instrument had to be individually fitted together. This is evident from the serial numbers stamped on the various parts of each machine. The early universal stock printers—at least the first 1,200 or so—had no such numbers on their parts, indicating that the parts were indeed interchangeable. Some later universal stock printers, however, such as one privately held in New York City (No. 2,041), do have numbered parts. The apparent reversion in manufacturing and repair techniques may have derived either from inadequate manufacturing standards or from the proliferation of universal stock printers of slightly varying design. For a full discussion of contemporary production techniques, see Hounshell 1984, esp. chap. 2 and app. 2.

7. U.S. Pats. 126,528, 126,529, 126,533, 131,338, 131,339, 131,341, 131,342, 131,344, and 138,869. The artifact illustrated in Doc. 195 is a production model that was used and modified; see also the descriptions of the instruments given in the company handbooks cited in n. 6.

–195–

Production Model: Printing Telegraphy

[Newark, August 1871][1]

M (22 cm dia. × 11 cm), MiDbEI(H), Acc. 29.1980.293. This universal stock printer has been modified in three ways. First, a 15-tooth ratchet wheel has replaced the original 30-tooth wheel on the typewheel shaft, allowing a doubling of the transmission speed. Edison first experimented with such a wheel in May 1872 (Cat. 1183:111, Accts. [*TAEM* 20:915]) and was making the change on his stock printers by that fall, as evidenced by the 15-tooth wheel on the model for U.S. Patent 138,869, submitted in October 1872 and now at the Edison Institute in Dearborn, Mich. (Acc. 29.1980.1335; the ratchet wheel is not covered in the patent claims; see also Doc. 338). Second, there is an adjustable spring for the printing lever in place of the original, fixed spring. Third, there is a stop to limit the upward travel of the printing lever. It is not known when or by whom the last two changes were made.

1. See headnote above. The instrument shown, No. 510, was completed and delivered to Gold and Stock on 2 November 1871. PN-71-09-06, Accts. (*TAEM* 20:978).

–196–

Notebook Entry:
Automatic Telegraphy[1]

[Newark,] Sept 4, 1871

Or[a] the ink of the ink recorder could be mixed with a sticky or gummy substance and just at the moment the recorded characters were made by the Roller a fine Spray of hard substance could be thrown upon it[b]   it adhering to the ink and raising it in relief. Sand, flour Emery, Pumice Stone, Sugar, or some hard substances of a porous nature, having an absorbant power & ground fine Could be used, or the ink might Contain an acid or chemical substance which would burn eat or rot holes in the paper.

or a lead pencil might be used instead of the ink roller and the plumbago[2] mark ~~w~~Could be made to close ~~an~~ a circuit having a very high tension Battery in it. or the end of the magnet lever might have a fine needle and prick holes in the paper and the burr which would be thrown up on the other side be made to close a circuit by means of lifting a delicately poised Lever.

or a sharp groove might be cut in the paper carrying drum ~~and~~ in which the roller having a sharp edge would just fit, and tissue or other thin paper being used When the roller came down it would enter the groove and shear the paper[b]

I[a] will now explain my Compressed air receiver

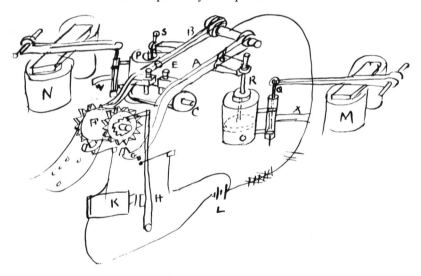

A & B are the Levers having ordinary belt punches ~~p~~Cutting on a raw hide surface or roller C[b]   D & E are the punches, F is the paper feed drum   have ~~V~~Two V shaped ratchet Wheels[b]   the End of Lever A & B having wedges which throw the paper forward after it is punched[b]   K is a magnet having a lever H which is provided with a pawl working a ratchet on the paper feed drum F. This is to feed the paper ahead between the ~~wo~~ Letters[b]   The two Levers A & B being provided with closing points so that when both Levers are at rest the[y] close the Circuit through this Extra Magnet.

But when operating they do not stop at the Contact point Long Enough to allow the magnet to act owing to the tremenduos rapidity with which the Levers act. O P are the air cylinders[b] the Valve being operated by the Main Line Magnets, one Magnet responding to Negative Currents and the other to positive Currents.

I do not wish to Confine myself to any particular form of Air Cylinder or Valve. I merely claim Punching paper or Embossing paper by Compressed air controlled by Magnetism, also postive & negative Currents for doing same[b] thing. I have been experimenting this day upon a novel idea by which the paper may be Embossed at a tremenduos rapidity & requiring very little power to Control it.

I do it by means of and Endless Chord.

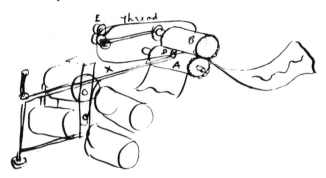

A[a] & B are two rollers rotatd by power and closely held ~~by~~ together by immoveable bearing ~~one~~ f & carried around by gear wheels. One Roller is very hard substance and the other soft. The paper passes between the rollers, and with it a Hard thread, or watch chain. This thread or chain in passing through these rollers with the paper   of Course the thread or chain is impressed its full depth in the paper and leaves an endless embossed or raised mark upon the paper. This chain or thread is Endless band like, and the Thread passes through a little hole on a head at the extreme end of the Armature Lever X at D. E is a guide roller for guiuding the chain and worked by gears or other device, from the roller A.B, so that the thread or chain will always be slack where it passes through the point on the magnet Lever. The Characters are formed by guiding the thread either one side or the other[b]   When a loop either one seide or the other is Thus made the rollers seize the thread and imme'y indent this loop in the paper. This paper after being prepared has a follower with a point, the point Running in the groove[a] and of Course giving a back & forward motion to the Lever which closes a circuit each time it passes to the right and clsg another in psg to the left.[3] Thus

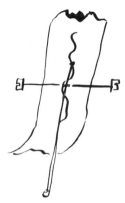

Though this may be done in many other ways[b] The Charcters being made to lift a lever or two Levers & closing the circuit that way[b]   This chain band or thread might be wound around the roller itself and a slight slack made, to allow of making characters & the thread guided by a fork [a]from the armature lever of the Electro or polarized Magnet.

or the roller could have two flanges at its end higher Than itself and innumerable little wires passing from End to End around the cylinder & free to move endwise[b]   These[a] little wires could have teeth on them two teeth in Each and when in their normal position These teeth would form two continuous wheels but if the armature lever should throw one of the little bars forward [-]One of the teeth would indent out of line & thus form a charactr. A Cam on the other side of the roller would throw the rod back in position after the impressed was taken

Witness   Jos. T. Murray[c]   T A Edison

AXS, NjWOE, Lab., Cat. 1173:24 (*TAEM* 3:90). [a]Preceded by "X" in an unknown hand. [b]Followed by "(x)" in an unknown hand. [c]Signature.

1. This entry is a continuation of Doc. 194 and is continued in Doc. 198.

2. "Plumbago" was a name given to graphite in the eighteenth century in the mistaken belief that its base was lead (plumbum). Edison frequently referred to the material by this name. Knight 1876–77, s.v. "Plumbago."

3. The creation of a zigzag line to represent dots and dashes is similar to the method used by Sir William Thomson in his siphon ink recorder. Edison experimented with this method earlier in the summer of 1871 (see Doc. 187). Here he proposed the use of a thread or chain to emboss the paper rather than a stylus or pen tip to make an ink mark.

197

*Edison and Unger
Summary Account*

[Newark,] Sept 13th 1871

T A Edison		C[redito]r
by	Cash Capital	$12360.05
	indebtness	3615.81
		$8744 24
Wm Umger		Cr
by	Cash Capital	12035.22
	indebtness	10589 82
		$1445.40

T A Edison	8744 24
Wm Unger	1445 4̶2̶0
½)	7298.8̶2̶4
	3649.42[1]

Wm Unger by note D[ebto]r to T A Edison with Inst at   per cent

D, NjWOE, Lab., Cat. 298:132 (*TAEM* 5:258). In Joseph Murray's hand.

1. An account book entry of 1 January 1872 lists a note from Unger of this amount with a term of one year (Cat. 1185:16, Lab. [*TAEM* 22:570]; see also Doc. 215). Edison gave the note back to Unger as part of the agreement that dissolved their partnership the following July (Doc. 264). The amount was stated incorrectly there. See also Cat. 1183:36, Accts. (*TAEM* 20:878).

198

*Notebook Entry:
Automatic Telegraphy*[1]

[Newark,] Sept 15 [1871]

The ink used on the Ink Recorder, or any recorder which I may adopt could be of a gummy nature, and as quick as The Record Should have been made it could be quickly Covered with gold Leaf or other atenuated metal[a] or very fine metallic dust[2] Could be mixed with The ink or thrown on after the record was made[a]

—Instead of the Thread or chain w̶for embossing after the manner described Sept 4 1871, The armature Lever Might be a tube w̶i̶t̶h̶i̶n̶ and placed at an angle and Continuiously fed with very fine Shot or round, or small substances[a]   there could be near the two rollers a stopper, and the End of The tube when in its normal position would rest against this & thus prevent the shot from Rolling down the tube. if now the armature Tube is deflected to the right the shot run out are Caught between the two rollers & the paper and they are indented in the paper, and when deflected to the left the shot roll between the rollers in another line. These indentatins Can be made to close & open an electriec circuit in Connection with the printer. the mechanism Which rotated the Indenting Rollers ~~all~~ also Carries the shot back to f Be fed through again.

At[b] the present writing I am inclined to believe that I am the originator of the idea of using perforated paper for controlling or working printing Telegraphs[3]

Witness. Jos. T. Murrey[c]   Thomas A. Edison

AXS, NjWOE, Lab., Cat. 1173:28 (*TAEM* 3:92). [a]Followed by "(x)" in an unknown hand. [b]Preceded by "(x)" in an unknown hand. [c]Signature.

   1. This entry is a continuation of Doc. 196.

   2. Metallic leaf or dust would serve as a conductor.

   3. Edison was the only American inventor working actively in both printing and automatic telegraphy at the time. The system described here, combining elements of each, was never adopted.

–199–

George Little Agreement with George Harrington

New York, September 22, 1871[a]

Whereas, George Little of Rutherford Park New Jersey, has invented and patented a system of automatic or fast telegraphy, and has a contract for the developement of said system so that the same may be adopted under the direction of George Harrington of Washington D.C. by the Automatic Telegraph Company of New York, and extended over the United States, the said Little having agreed by contract to dispose of said system and patents for certain stipulated sums of money and amounts of stock of said Automatic Company.[1]

And whereas, the said George Harrington is interested to the extent of one third in certain inventions and patent rights useful in automatic telegraphy including a copying printer[2] of Thomas A. Edison of Newark New Jersey.

Now therefore, in order to avoid all difficulties, as to the relative value of Edison's invention as compared with Little's inventions and for the further consideration of one dollar paid in hand, one to the other, the receipt of which is hereby acknowledged, it is hereby agreed by and between the said George Little and the said George Harrington that they will mutually and equally divide one with the other all such sums of money and stock as they may or shall be entitled to out of or from the payments to be made by said Automatic Company for the said inventions and patents of said Little and said Edison in so far as such inventions are applied by said Company to Automatic telegraphy and used by them in the United States.

It being understood, that the amount of interest in his own inventions and patents retained by him (the said Little) vary in none less than twenty seven one hundredths parts thereof.

In witness whereof, the said Little and the said Harrington have hereunto set their hands and affixed their seals this 22nd day of September, Eighteen hundred and seventy one in the City of New York, State of New York

Geo. Harrington[b]   George Little[b]

Sealed and delivered in presence of the words "the said Little" being interlined before execution J. W. Treadwell[3]   C. L. Vanderwater[4]

ADDENDUM[c]

[New York,] October 7, 1871[d]

The foregoing instrument applies to the following named patents, allowed cases and applications and improvements on the same and an invention of Edison of a printer for messages &c. not yet applied for.[5]

Number	Date	Inventor	Subject
108,495	Oct. 18, 1870	Geo Little	Electric Circuits
115,968	June 13, 1871	" "	Circuit Roller
96,332	Nov. 2 1869	" "	Magnetic Roller
96 333	Nov 2, 1869	" "	Automatic Telegraph
115,969	June 13, 1871	" "	Perforating Paper
91,241	June 15, 1869	" "	" "
91,240	"	" "	" "
96,331	Nov 2, 1869	" "	" "
96,330	" " "	" "	" "
	Oct. 2, 1871	" "	Automatic Transmitter
Allowed[e]	Sept. 28, "	" "	Circuit Closer
	Aug. 18 "	" "	Recording Instrument
111,112	Jan. 26, 1871	T. A. Edson	Motor Governor
114,656	May 9, 1871	" "	Transmitting Instrument
	Sept. 28, 1871	" "	Ink Recording Instrument
Allowed[e]	Aug. 18, 1871	" "	Telegraph Instrument
Application filed	Aug. 18, 1871	" "	Perforating Machinery

Signed by said Little and Harrington this 7th day of October, A.D. 1871.

Geo Harrington[b]   George Little[b]

Witnessess   J. Treadwell C. L. Vanderwater

D (transcript), NjHum. This document appears as an enclosure to "Opinion of E. N. Stoughton as to rights of Prescott to Duplex Telegraph Impts," a document related to the Quadruplex Case. [a]Place and date taken from text; form of date altered. [b]Followed by representation of a seal. [c]Addendum is a D (transcript). [d]Date taken from text, form altered. [e]A brace in the manuscript indicates that "Allowed" referred as well to the items immediately above and below this line.

1. A contract of 9 September 1869 between Little and Daniel Craig and the National Telegraph Co. (see Chapter 6 introduction, n. 8) was transferred to the Automatic Telegraph Co. in January 1871 (Doc. 141).

2. Typewriter; see Doc. 155.

3. James Treadwell was a copyist at 115 Broadway. Wilson 1872, 1156.

4. Charles Vanderwater was a student at 115 Broadway. Ibid., 1172.

5. Caveats of 10 October 1871 (DF [*TAEM* 12:628]; PS [*TAEM* Supp. III]).

–200–

From Daniel Craig

Peekskill,[a] Sept. 26/71

Dear Mr. Edison:

I want you to give me your own suggestions as to an arrangement for the introduction of your machine,[1] but as you asked me to tell you what I thought was right, I will do so, to a certain extent—but first, let me tell you what I can do:—

If you will finish three or four machines, I will have good wires put up in the city, & provide three good offices where the machines can be thoroughly tested, and if they work satisfactorily, I can raise in a single day, all the capital necessary to pay for the manufacture of all the machines that can be introduced, and to build all the lines we can find a use for.

I will undertake that you shall have <u>cash</u> and liberal pay for all the machines that can be introduced, in all the cities.

I will also guarantee my undivided and best efforts to work this business up,—will furnish all the capital that can be judiciously used, and give you <u>one half</u> of all the advantages I may obtain in connection with the business—and I firmly believe I can make your interest worth at least $5,000 per year, & probably much more.

All this I can do at once, & without serious trouble.

Of course, I <u>can</u> raise some money beyond what I have

above stated—but that will be a real trouble, & I hope you will not feel that you ought to put a very heavy load on to me. I have done all in my power to promote your interests during the past year, and expect to continue to do so for many years to come, and our appreciation of your great abilities has certainly not had a tendency to lessen the appreciation or <u>liberality</u> of Lefferts & his company.

Any kindness that you can feel justified in extending to me, will not be wasted. I claim to have more real hard work in me than any other ten men except, <u>only</u>, yourself—and I have all the ambition in the new field that you could possibly desire me to have   Truly &c

D. H. Craig

☞ Don't forget <u>several</u> samples.   .C[b]

ALS, NjWOE, DF (*TAEM* 12:462). [a]"P.O. Box <u>3237</u> = <u>New York.</u>" written above. [b]Postscript written in top margin of first page.

1. Probably a reference to Edison's proposed domestic telegraph. See Docs. 143 and 203.

–201–

From Daniel Craig

New York,[a] Sept. 27/71

Dear Mr. Edison:

I have just learnt of an entirely new movement for a strong opposition to the G.&S. Co.,[1] and I am <u>sure</u> you ought to sell your stock. There is intense hostility to the Co. & all the more since the W.U. Co. swallowed it, and any party of common sense with any decent machine can render the Co. powerless to pay another dividend.

As I have not yet got the promised specimens of your Printing (I want several short messages for interested friends) I suppose you did not get your machine[2] up, as expected, yesterday.

Give me your views, as promised yesterday, & be merciful

D H Craig

ALS, NjWOE, DF (*TAEM* 12:464). [a]"P.O. Box <u>3237</u>." written above and to the left.

1. Craig is probably referring to the work of Robert Gallaher, who had been involved with the Gold and Stock Telegraph Co. since its beginning and who left that company about this time to develop and introduce his own ticker. Early in 1872 he organized the Gallaher Gold and Stock Telegraph Co. Reid 1879, 607; "The Reporting Telegraphs," *Telegr.* 8 (1871–72): 225.

2. Typewriter; possibly the design for which Edison executed a caveat on 10 October 1871 (PS [*TAEM* Supp. III]).

From Daniel Craig

Dear Mr. Edison:

In thinking over your <u>talk</u>, today, I am moved to write you my reflections.

First, let me say, that it seems to me your programme is not calculated to advance the immediate introduction of Automatic Telegraphy—you are proposing to travel over, substantially, the same ground as Lefferts, Little and I travelled over many times between 1868 & 1870. We reached perfect Transmission at any speed up to 1000 words per minute,[1] and also the most perfect recording, both on iron (permanent) and platinum, (fleeting) paper-marks.[2] Our great wants, a year ago, were a rapid perforator and a good copying printer. I take credit to myself for having discovered in you the genious we required, and I hope you will not refuse to let me suggest to you the proper way to utilize your great inventions.

We want, to give the required confidence in our new system, the Edison perforator and printer—nothing more, and we do not need 48 hours to illustrate all we need to illustrate, after we get machines for N.Y. Phila Balto. & Washn offices.

I know, of course, that you have many most capital things to bring in to improve our operations, but I am sure we ought not to be delayed a da[y?][a] or an hour, in dem[onstr?]ating[a] what w[e can do wi?]th[a] the aid of your [perforator?][a] & printer. There [will be plent?]y[a] of time for you to [pres?]ent[a] & develop your ideas after we prove to our impatient friends that we have got a practical working system.

I <u>know</u>[b] there is the most <u>urgent necessity</u> for immediate operations by the Co., and as the Line is now all right, I hope you will bring over the perforator & printer & spend all the time needful in the office to have them well-worked, and then we can have a full test and prove all we need to prove to assure the complete sucess of our enterprize. Do, pray, use your influence to have no more [t?]ime[a] lost. If we stan[d?-------][a] our fingers no[-------]r[a] the great mom[ent? -----]ll[a] have themselves [-------][a] out to the Govern[ment boots?][a] & breeches. They are prepared to spend $10,000,000 if necessary, and Orton expects to go in to the Cabinet and will be the Engineer of the impudent fraud.[3] Do, pray urge and aid Harrington to get full tests this week, or at the earliest possible moment.

I cannot use that sample of Printing, as it is full of blunders. Will you not oblige me with several perfect & clean nice samples, soon.

Do not delay telling me what I may do about the Domestic Telegraph scheme.[4] I will do you no harm in trying.

D. H. Craig

ALS, NjWOE, DF (*TAEM* 12:465). [a]Manuscript damaged. [b]Underlined twice.

1. Craig had claimed 1,000 words per minute for the Little automatic in the 1 January 1871 issue of *Scientific American* (Craig 1871, 4). However, in tests on 11 December 1873 conducted before the postmaster general and members of Congress, the Automatic Telegraph Co. transmitted only 12,000 words in 22½ minutes, or slightly over 500 words per minute. In 1876 at the Philadelphia Centennial Exposition, the Edison automatic telegraph successfully transmitted 1,015 words in 57 seconds, a speed that was considered impressive. Edward Johnson to George Harrington, 28 Jan. 1874, DF (*TAEM* 13:52); William Thomson, "Report on Awards—Edison's American Automatic Telegraph," 30 June 1876, PPDR.

2. See Doc. 152, n. 2.

3. It is not known what plans Craig is here referring to.

4. See Doc. 143.

–203–

From Daniel Craig

New York Oct. 6/71

Friend Edison:

It is now nearly two weeks since you promised to give me your ultimatum in regard to the Domestic Telegraph invention,[1] and I have not a word from you, yet. I certainly think this is cruel. You say that the G.&S. Co. have no right to the invention and that you can arrange with me through other parties, so that you will not have any trouble with the Co.

Whatever I now do is to inure to the benefit of the Automatic Co. of which you are destined, no doubt, to be a very large stockholder.

I beg you will say something decisive at once—if you are not willing I should carry out the programme which I had supposed was settled between you and me a year ago, you have only to say so; & I shall not annoy you any farther. But say yes or no now.

Sholes has very greatly improved his printer, and it now does such work as I inclose[2] right along, and I have no doubt an expert can average 50 words per minute & change the paper besides.—

I strongly advise that you should not hold the wire a moment beyond what is necessary for you to test your Perforator & Printer. We have absolutely perfect Transmitting & Recording devices, and at the present moment it will make serious trouble with Little, if he sees an effort being made to supplant him in what he justly regards as a discovery of the means of very rapid & correct Telegraphing. We cannot afford to have any wrangling. By and by it will be in order to make improvements in the Little System—but we are now in too much haste <u>for practical work</u>, to afford even one hour for any unnecessary tests. Give us perforations & plain Print. There is glory enough in that for a whole life-time.

The W.U. Co. are <u>very anxious</u> to know our secrets—and I suggest the utmost care & prudence to keep every fact of possible value from their legions of <u>spies</u>. They begin to see that we are not the d——d fools they said we all were. Truly &c

D. H. Craig

☞ How about the gas motor?[3][a]

ALS, NjWOE, DF (*TAEM* 12:468). [a]Postscript written at top of first page and separated from rest of text by a line.

1. See Doc. 200, n. 1.
2. Not found.
3. This is the only reference to this device by either Craig or Edison; "gas motor" probably refers to an internal combustion engine, several types of which were available and in use. *Encyclopaedia Britannica*, 9th ed., s.v. "Steam-Engine: Air and Gas Engines."

–204–

Daniel Craig to Marshall Lefferts

Peekskill, Oct. 26, 1871.

Dear General:

It is a good while since I posted you upon the subject of Automatic Telegraphy, and I will now proceed to relate to you what has occasioned our long delay.

In 1869, when you relinquished your interests to myself, Little and others, you will recollect that the Perforating Machine, though performing very accurately and beautifully, could not be worked up to more than 600 to 700 words per hour. Shortly afterwards, we engaged in a long series of mechanical devices to get greater speed, and after nearly two years of hard and expensive work, we have now attained our utmost desires—having three different, and strongly patented Perforating machines, <u>which have been worked at the rate of more than 5000[a] words per hour.</u>—3,000 words per hour can be done, accurately & regularly. See sample.[1]

We have also obtained a Type-Writing Machine, which is worked by Girls at the rate of 2000 to 3000 words per hour. See sample.

Our Transmitting and Recording machines are substantially as they were first produced by Little under your direction in 1868–9, though we have in some respects considerably improved the mechanism. In the matter of speed of transmission, we have gone more than five hundred per cent above even your most sanguine hopes, and we can now Transmit, in longer circuits (we fully believe, in circuits 5 to 8 times greater) than can be worked by the Morse system—and we have actually transmitted over 300 miles of wire 1050 words per minute, with one wire and the services of two girls. You see, therefore, that whilst we used to think 12,000 words per hour over one wire, would be an enormous advantage over Morse, yet, in truth, we can do more than Sixty Thousand words per hour, if necessary, over one wire. See sample of dots & dashes.[2]

Here, I take it, are the main elements of fast & cheap Telegraphing, and as every part of the labor can be performed by Girls, I am sure you will concede to us such enormous advantages over the Morse system, that competition is entirely out of the question. If you will make an estimate even upon the basis of 2000 words per hour for the Perforators and Copyists (printers) and 30,000 for the single wire transmission, you will find that we can better afford to send 50 word messages from Chicago to New York for 25 cents, than the Morse lines can send 10 words for $1. In reality, we can, counting all things that enter into the business of Telegraphing, beat the Morse system more than 10 to 1. But I have given you the facts and you do not need my estimates or deductions from those facts.

Three years ago, you will recollect that you promised me thyou would entertain a proposition to assume the management of our new system whenever we got ready to make a vigorous demonstration looking to actual business, on a large scale. That time has now arrived. We have good men, with ample capital. We have a fully tested & complete system, capable of strangling all opposition. I see nothing more to be wished for except one thing—we need a practical man to go forward and properly systematize the introduction of our new system, and I need not tell you that of all the practical gentlemen connected with the Telegraph business of the country, you only come fully up to my ideas of the right man for our leader. I have reason to know that all of my friends and associates will gladly unite with me in a call for your services if you now feel about this matter as you did in 1868–'69. I am aware that your position has greatly changed for the better, during the past three years, and I naturally suppose that in your present comfortable quarters, you would be less likely to entertain my wishes than you would have done two or three years ago. You, I am sure will take it in no unkind spirit, when I tell you that you are not long to enjoy your present monopoly of Local telegraphing. We have an entirely new system of rapid Telegraphing, by which we can do at least twice as rapid work as you are now doing, and we can do it even more reliably than you can[b] possibly do with your last improved Edison machine. I mention this only that you may see what is coming—and you may be sure that we shall leave no stone unturned to secure and maintain all the advantages we are entitled to. I shall regret to be placed in a position, as manager of the News Department in the Automatic Telegraph Co., to oppose you, but if this should be unavoidable, I trust we shall agree to disagree, and manage our rival interests in a gentlemanly way. Of course, I know I shall beat you out of your boots, as a News Manager, and I hope you will render this disagreeable thing unnecessary by saying to me that you will now entertain an official call to assume the management of our new system.

I judge that my friends will willingly pay you $6,000 the first year, $8,000 the 2d & $10,000 thereafter, and I will give you outright ¼ of my patentee interest, and Little & others will do the same.

Your answer will, of course, be entirely private with me, but let me have it soon, as now is the time to act. Faithfully yours,

D. H. Craig

ALS, NNHi, Lefferts. [a]Underlined twice. [b]Interlined above.

1. George Little, Frank Anderson, and Edison all received patents on perforators in 1871. Since Craig apparently controlled the Anderson patent, he probably refers to it here, although the reference may also be to Little's 1869 perforator (U.S. Pat. 96,331). A sample perforation found among the Craig correspondence in the Lefferts Papers (NNHi) may be the one referred to here.

2. Not found.

–205–

*News Reporting
Telegraph Co. Circular*

Newark.[a] [October 1871?][1]
"AN AMERICAN IDEA."[b]

News Reporting Telegraph Company.[2]

This Company have already started a Private Printing Telegraph line in this City, and are locating a number of their Printing Telegraph Machines in the offices and dwellings of some of the most prominent citizens.

They are intended for giving all general news of the

Engraving of the Newark Daily Advertiser Building, where Edison and Unger established the News Reporting Telegraph Co.

world—financial, commercial, domestic and foreign—the moment such news is received in the main Telegraph Office in New York, and several hours in advance of all newspapers.[3]

The advantage to those who become subscribers we scarcely need enumerate, for hardly any event of consequence can occur in any part of the world reached by the telegraph, but which is instantly transmitted over our wires and printed out on a continuous strip of paper, in Roman Letters, hours before such news is published in the papers, and if the event should be of great consequence minute telegrams will be transmitted during the day and at a late hour at night.

The instruments are novel and very ornamental, are quite noiseless, and require very little attention.[4]

On request we locate these instruments in the Drawing Room, Library, or any part of a dwelling *Free of Charge.*

For furnishing ink, paper and the news, $3 per week is charged.

Our agent will call and exhibit and explain the instrument in a few days, and take the names of persons who desire the machine.

News Reporting Telegram Co., Daily Advertiser Building, 788 & 790 BROAD ST.,[5] NEWARK.

PD, NjWOE, Scraps., Cat. 1178 (*TAEM* 27:140). [a]Place taken from text. [b]This line and next line each followed by a centered horizontal rule.

1. Account records relating to this company cover the period 18 October–30 December 1871. Cat. 1213:100–103, Accts. (*TAEM* 20:15–16).

2. Little is known of the company. Both Edison and Unger were involved in its operations, and P. J. Boorum later recalled arranging the company's right of way and working with someone named Thompson to establish its circuits. Both Boorum and an S. C. Thompson appear in the company's account records, as does Mary Stilwell, a sixteen-year-old Newark girl who worked for one month at $6.00 per week. On 25 December 1871 she and Edison were married (see Doc. 218). Boorum to TAE, 29 Aug. 1878, DF (*TAEM* 15:1088); gas receipt for 788 Broad St., 71-014, DF (*TAEM* 12:492); Cat. 1213:100–106, Accts. (*TAEM* 20:15–18).

3. The company appears to have had an arrangement with Gold and Stock, probably for the supply of commercial news. Cat. 1213:100–104, Accts., (*TAEM* 20:15–17).

4. The instruments used by the company have not been identified.

5. Located just a few doors from the Atlantic and Pacific Telegraph Co.'s Newark office, the building at 788 Broad St. also housed the *Daily Advertiser* and the Newark office of Western Union. Edison and Unger paid a gas bill for 788 Broad St. during the period September–December 1871. The accounts indicate that the News Reporting Telegraph Co. may have had two rooms in the building, an office and a battery room. Hopkins 1873, Fourth Ward map; Cat. 1213:100–104, Accts. (*TAEM* 20:15–17).

–206–

Receipt from Ross, Cross & Dickinson

Newark, N.J., Nov. 21 1871[a]

$100⁰⁰⁄₁₀₀

Received from Thomas A. Edison the sum of One Hundred Dollars as a payment on House and Lot[1] known as No. 535 Wright St.[2] Newark N.J. sold him this day[3] for the sum of Five Thousand Five Hundred Dollar upon terms as follows: $1500 in Cash and Balance in Bond & Mortgage[4]

Ross, Cross & Dickinson Agents[5]

100[b]

D, NjWOE, DF (*TAEM* 12:226). Letterhead of Ross, Cross & Dickinson. [a]"Newark, N.J.," and "187" preprinted. [b]Written in an unknown hand at bottom of page, center.

1. This was the first house Edison bought. Only a few years old, it was one of a row of similar houses on a block in a desirable neighborhood near a trolley line. At the time Edison purchased it, he was probably boarding at 854 Broad St. in Newark. About a month later he married Mary Stilwell. He owned the house for three years. Urquhart 1913, 2:673; Hopkins 1873, Fourth Ward map; TAE to Dolly Buscho, 1 Nov. 1930, GF; Cat. 1213:151, Accts. (*TAEM* 20:19); *Holbrook's Newark City Directory* 1871, 161.

2. The seller, J. Clinton Walsh, owned the 53 and 55 Wright St. properties. Edison actually purchased number 53, but it became number 97 when it was renumbered in 1873. Walsh retained number 55. Title

search, 71-001, DF (*TAEM* 12:227–28); Hopkins 1873, Fourth Ward map; *Holbrook's Newark City Directory* 1873, 63.

3. The deed of conveyance was signed the following day, certified on 25 November, and recorded on 27 November. Deed Records, NjNECo, Z15:410–12.

4. Edison gave Walsh a $1,500 bond dated 22 November. Walsh transferred to Edison a $2,500 mortagage held by the Dime Savings Institution of Newark. Receipt from Ross, Cross & Dickinson, 17 Dec. 1872, 72-001, DF (*TAEM* 12:677); Cat. 1213:3, Accts. (*TAEM* 20:6); Cat. 1185:8–9, Accts. (*TAEM* 22:566); tax and assessment record, 71-001, DF (*TAEM* 12:229–32).

5. According to their letterhead, Charles Ross, A. L. Cross, and Jonathan Dickinson were "real estate and mercantile negotiators." Their office was at 788 Broad St., Newark—the Daily Advertiser Building—where Edison and Unger established their News Reporting Telegraph Co. (see Doc. 205).

–207–

Notebook Entry:
Printing Telegraphy

[Newark, September–November 1871[1]]

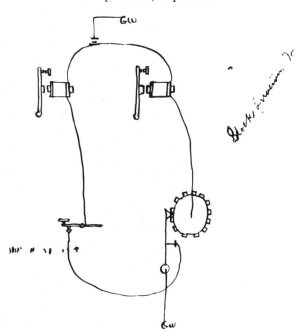

Improvement[a] in hand Transmitters, and their appliances, whereby The Second or Printing Circuit Cannot be closed by the key until The break wheel & point are on an open circuit $=$[2]

The Sketch will explain itself. This appliance may be used with one wire.[3]

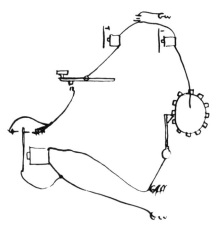

Another Device to produce the same thing—[4]

AX, NjWOE, Lab., Cat. 1174:6 (*TAEM* 3:10). [a]"Stock Quotation Trans" written by William Carman at right of drawing.

1. The preceding entry in this notebook is dated 25 August; the following one is dated 10 December.

2. In using Edison's universal stock printer at this time, it was necessary to open and then close the typewheel circuit to advance the typewheel one letter. (Edison changed this about a year later so that opening *or* closing the circuit advanced the typewheel.) Both of the arrangements sketched here guaranteed that the typewheel would be positioned correctly for printing.

3. In the diagram at right, **A** is a breakwheel; **B** is the relay for the printing circuit (not shown); **C** is the relay for the typewheel circuit (not shown); **D** is a key that closes the printing circuit; **E** is a contact point that closes when breakwheel contact **F** is between two teeth; **GW** is a ground wire; and **H** is a battery. The projecting teeth on **A** are contacts; the interstices are insulated. On all relays, it is understood that a spring pulls the armature lever away from the magnet. To operate both this device and the one shown in n. 4, an operator turned **A** to advance the typewheel to the desired letter, stopped **A** with **F** between two teeth, and closed **D** to print.

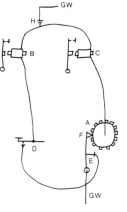

4. In this diagram, **A, B, C, D, F, GW,** and **H** are as described in the diagram in n. 3. **J** is a back point relay; it opens the printing circuit when current passes through the typewheel circuit (including magnet **K**) and closes when the typewheel circuit is open. In operation, with the breakwheel contact in between teeth and hence magnet **K** not charged, a spring would pull

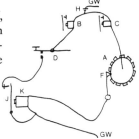

lever **J** away from the magnet and against the contact point to complete the printing circuit to **D**.

[Newark,] Dec 10 1871[a]

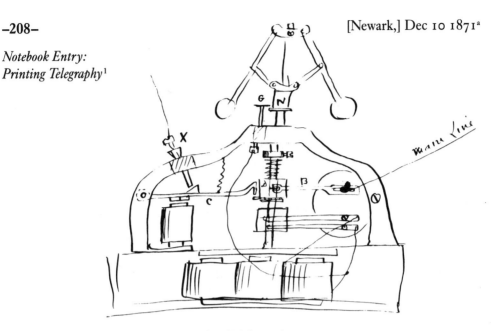

[Witness:] Jos T Murrey[b]

Electric Pulsator For Printing and other Telegraphs and adopted on my universal Printing Machine.[2]

The point X is insulated, and to it is connected the main Line   B is the break spring also Connected to the Main Line. when the armature lever C. is up against the point X the frame and break wheel D is connected to the main Line   So that when the Engine Shaft revolves the wheel is carried around and at every revolution 2 breaks are made, but if now the armature lever C is drawn down by the magnet the Main Circuit is broken and the Wheel D arrested.[3]

G is a screw insulated and having a disk on top Which disk rubs on another disk on the Governor slide when the Engine has reached a certain speed   The disk G being Connected to the Engine Magnets The current is cut off when the Governor brings the slide N up to the disk of G, or the current may be closed altogether and the magnet used to retard the motion and effect the regulation quickly or the engine Current Can be cut or broken entirely ~~or th~~

I do not wish to confine myself to using a ~~w~~disk as in G as ~~the~~ a lever might be used = or any Governor, but I prefer this Governor as it is both electrical & frictional

Neither do I wish to confine myself to using a magnet Whose lever opens the main circuit before it arrests the Break wheel thus [-]ensuring against a failure to arrest Said break wheel and prevent an extra break being made, as the magnet may be dispensed with and the same arrangement used as described in my Application for patent for my Universal Machine and the Main Line break attachment used.

I will mention here that the break wheel is covered with platina and the Little wheels of the break springs are of Copper—also that I provide a Brush rubbing Continuously on these break wheels to keep the Oxydized matter off and keep wheel clean to insure Correct breaks

Following is arrangement on the type wheel shaft Which rotates in path of the finger keys and is used for arresting type wheel and closing circuit of the arresting Magnet on the Engine, and preventing further breaks being sent.[4]

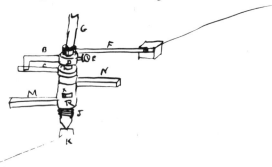

[Witness:] Jos T Murrey[b]

G shaft   D collar insulated   F continuous rubbing spring for bringing Current to D. and spring B. R is Collar with 2 arms N.M. has slot X with a stop pin in it connected to shaft to prevent ke sleeve going to far around when arrested by a key. C is a spring on collar R. J is a spiral to return sleeve to original position before arrested by key   K is a Center in which shaft G works. The Collar R being connected to other end wire   when a key is depressed and the arm say N comes in Contact with end of key while shaft is in motion the sleeve is slightly turned around an to a distance governed by the pin in slot X, bringing the point B & C in Contact and closing the circuit =

AX, NjWOE, Lab., Cat. 1174:7 (*TAEM* 3:10). [a]Both drawings carry this date; "Priv Line Printer" written by William Carman at right of first drawing. [b]Signature.

1. This entry is continued in Doc. 209.

2. Edison first planned a governor of this form as part of a transmitter

for his automatic telegraph system (see Doc. 183; and the two sketches below). He actually used it on both the universal private-line printer and the transmitter for the universal stock printer (see Doc. 211). When operators used a breakwheel for transmission on a printing telegraph circuit, they had to turn the handle "VERY REGULARLY, not by jerks, and always as nearly at the same speed as possible" (Gold and Stock 1872, 5). This was necessary because the transmitted impulses shortened as the speed of transmission increased, and a typewheel escapement magnet might not charge sufficiently with impulses that were too short. A motor-driven breakwheel (which Edison called a "pulsator") delivered

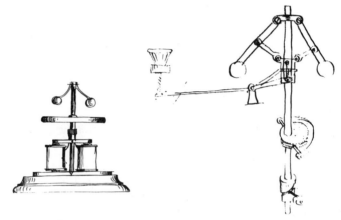

Two sketches of flyball governors from Edison's laboratory scrapbook; the one on the left is dated 15 December 1871.

regular pulses that could be made quite rapid if the main-line voltage was increased, but the speed of the motor driving the pulsator on the original universal private-line printer depended on the power supplied to it, which varied with the strength and age of the local battery. With a governor controlling the motor, the rate of transmission could be precisely set (idem 1873).

3. See Doc. 193, n. 2.
4. Ibid.

–209–

Notebook Entry: Printing Telegraphy[1]

[Newark,] Dec 12 1871

following is another device.

Fig 1 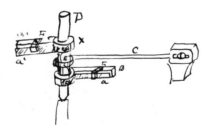 Fig 2

[Witness:] Jos T Murrey[a]

D is the shaft   X a long Collar of insulating material with 3 Raised Collars on it the middle one having a brass ring on it and a wire leading from said ring to the two ~~arr~~ Connecting springs a a'. C is a spring for carrying Current to Collar and springs a a'   B B' are arms behind these springs to as to ensure the arresting of the shaft should the springs and key fail to make a good electrical Connection. F F' are bridges to~ Connected to the arms B B' to prevent the spring from coming out too far from the arms when the proper tension is placed on them.

Fig 2 shows how the arms and springs are secured to the rubber flanges.

AX, NjWOE, Lab., Cat. 1174:9 (*TAEM* 3:11). ^aSignature.

1. This entry is a continuation of Doc. 208.

–210–

Notebook Entry:
Printing Telegraphy

[Newark,] Dec 12 1871

Escapement for Printing Telegraphs.[1]

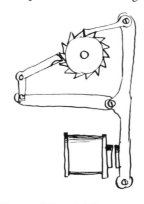

[Witness:] Jos T Murrey^a

AX, NjWOE, Lab., Cat. 1174:10 (*TAEM* 3:12). ^aSignature.

1. Cf. the design illustrated in the drawing below. Cat. 297:44, Lab. (*TAEM* 5:587).

This printing telegraph design, from one of Edison's laboratory scrapbooks, uses combinations of ratchets that are far more complex than those he used in his practical designs.

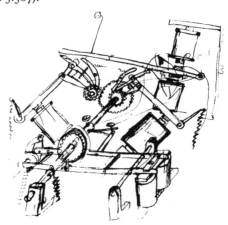

PRINTING TELEGRAPH TRANSMITTER
Doc. 211

Edison's design for a transmitter for circuits using his universal stock printer is shown in Doc. 211. When first introduced, the universal stock printer used a hand-cranked breakwheel as a transmitter. In Doc. 211 Edison was working on an adaptation of the electric motor and keyboard from his universal private-line printer. As on that device, the motor drove a pulsator (a small breakwheel). For this transmitter Edison added a gear train connecting the pulsator to a lever that swept under the keys. Depressing a key stopped the lever and, through the gears, the pulsator. Edison made the transmitter practical during the first months of 1872 at Joseph Murray's shop and

Diagram showing a hand-cranked breakwheel, from an instruction manual for Gold and Stock Telegraph Co. employees. The three circular machines at the top are universal stock printers representing three separate subscriber circuits. Below them are the relays connecting the reporting circuits to the local transmitting circuit—T marks the relays for the typewheel lines and P marks the relays for the printing lines. The hand-cranked breakwheel is in the center; to its left is a key that closes the printing circuit. The rows of connected circles at the bottom are the battery cells.

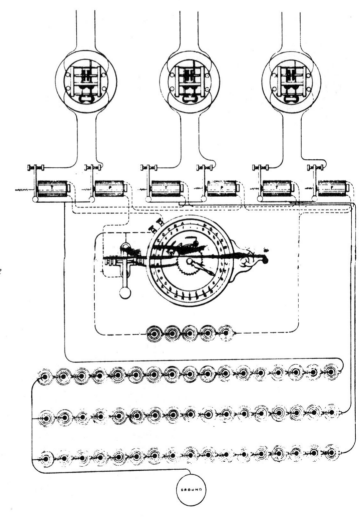

executed a patent application to cover it in June.[1] Others in the telegraph community, including Patrick Kenny and George Scott of the Gold and Stock Telegraph Co., were developing transmitters for printing telegraphs at the same time.

1. Cat. 1218:60 and Cat. 1214:4, 6, Accts. (*TAEM* 21:212, 566, 567); U.S. Pat. 131,343.

In this diagram Edison's new transmitter replaces the hand-operated breakwheel. The two wires on the transmitter's left terminals go to the typewheel and printing relays, the center right wire is the return wire for the transmitter's motor circuit, and the far right wire is the common power line. The rightmost terminal is not used; what appears to be an unused center terminal is a knob that is pushed in to start the machine.

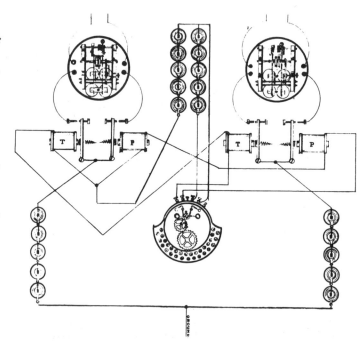

This transmitter for printing telegraphs, constructed at Edison's Ward St. shop, embodies Edison's U.S. Patent 131,343.

Patrick Kenny's printing telegraph transmitter.

George Scott's printing telegraph transmitter, described in U.S. Patent 126,336.

–211–

Technical Drawing:
Printing Telegraphy

[Newark,] Dec 15 1871

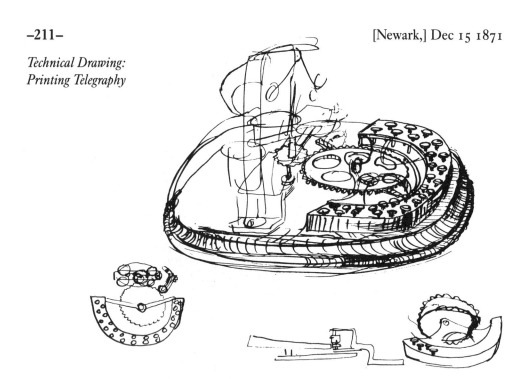

AX, NjWOE, Cat. 297:6, Lab. (*TAEM* 5:456).

–212–

Notebook Entry:
Printing Telegraphy

[Newark,] Dec 19 1871[a]

Devices for a two wheel printing Telegraph, figures and fraction wheel and Letter wheel, and preventing one from Printing while the other is. The principle is that the rim of the wheels are divided in 3 segments, and these segments have a space between them clear of about the distance of a space between the Letter these segiments are raidiated out from the Center to their full size, and then drawn in to a smaller size wheel, by the action of the printing lever. when ~~one w~~ the segiments of one wheel is swelled or radiated out to its limit the segiments of the other is drawn inward thus increasing the diameter of one and increasing that of the other

Fig 1   Fig 2

[Witness:] Jos T Murrey[b]

a b c segiments n n n studs on flange wheel with pins from segiments sliding in them to guide segiments R arms with crooked slots as fig 2 for giving outward or inward motion to segiment

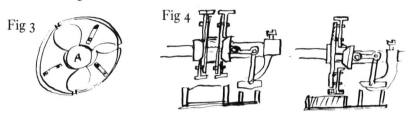

Fig 3

Fig 4

Shifting shaft

T Peice Stationary
Ratchet & evything shifts.[c]

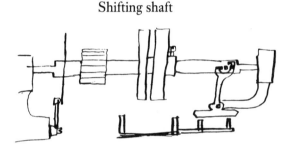

[Witness:] Jos T Murrey[b]

AX, NjWOE, Lab., Cat. 1174:11 (*TAEM* 3:12). [a]Both sets of drawings carry this date; "Stock Quotation Printer" written by William Carman at top of page. [b]Signature. [c]Followed by horizontal line drawn across the page.

–213–

Notebook Entry:
Printing Telegraphy

[Newark,] Dec 24 1871

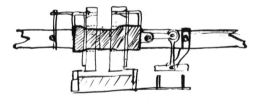

[Witness:] Jos T Murrey[a]

Loose type wheels with an incline slide sleeve so that when shifted one wheel is made rigid while The other is Loose. each wheel is provided with a flange and the Printing Lever has atwo springs (flat) on each lside So that when Ptg Lever comes up to Print the springs touch both wheels, but one being ridgid bows spring down while other being Loose the spring throws it up and keeps it away from paper =[1]

AX, NjWOE, Lab., Cat. 1174:12 (*TAEM* 3:13). ªSignature.

1. See U.S. Pat. 126,528. The patent model is housed at the Edison Institute, Dearborn, Mich. (Acc. 29.1980.1306).

–214–

Notebook Entry:
Printing Telegraphy
and Gang Saw

[Newark,] Dec 24 1871[1]

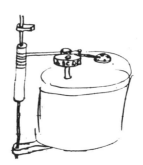

Sliding pasted reel for pasting strips of a printing Telegraph machine ein a mss form

___

To prevent sthe saws composing a gang saw from running out of Line file or cut small teeth on the side of the main cutting teeth, so that when the saw runs out the strain will make the side teeth cut and work itself back to its normal position[2] Dec 2o4 1871

___

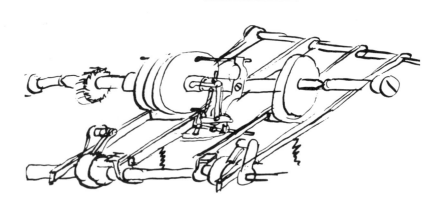

Device for printing on two strips on the Universal Stock Printer, the usual type wheel shaft being provided with an extra lever Type wheel and an independent Printing lever beneath it, and paper driver   So when you wish to stop Printing on the main strip you rotate the Type wheel around to the Letter Z where there is a pin actin. When the printing lever is brought up this pin strikes a slide bar and shoves it underneath the extra ptg lever and locks the two levers together, so that by working The main ptg lever the other one goes with it & prints characters from the extra wheel but the pad on the extra ptg lever being much higher than on the main lever it does not allow main lever to touch type wheel consequently it does not print. Now when you desire to stop ptg on the extra lever & type W, you turn around to the main dot and strike the Ptg lever and a pin from the shaft of the type wheel Throws the slide back from underneath the extra Ptg lever consequently it ceases to Print

I do not wish to confine myself to any particular way of doing this, but claim printing upon two strips of paper by Double Ptg Lever and their selection controlled by the position of the type wheel in conjunction with the power derived and the upward or downward motion of either Printing Lever Dec 24 71

I claim in a Printing Telegraph for type wheel bands of Rubber type, not hard rubber but pliable rubber type in bands and streached over a flat disk.[3] Dec 24 71

I claim fitting a band of sl silk or other fabric closely over the Type wheels of a Printing Telegh and allowing the ink roller to ink this band instead of the type direct, the object of which is to prevent clogging up the type wheel with fuz from the type wheel= Dec 24 1871
[Witness:] Jos T. Murrey[a]

AX, NjWOE, Lab., Cat. 1181:6 (*TAEM* 3:177). [a]Signature.

1. The following day Edison married Mary Stilwell.

2. Edison may have been alerted to this problem by his father, who was probably in Newark to attend his son's wedding.

3. Universal stock printers were equipped with such typewheels some years later.

An Inventive Flurry

January–June 1872

The year 1872 was a remarkably creative one for Edison. He executed thirty-nine successful patent applications, nearly twice as many as in all of his previous career. In only two other periods would he exceed his 1872 patent productivity level: the years 1880–83, when he developed his incandescent lighting system, and the year 1888, when he worked on his improved phonograph. The early part of 1872 witnessed his most intense creative work of that year. He returned from his honeymoon at the beginning of the year and during the next month and a half generated an unusually large number of notebook entries. These were concerned primarily with improvements in printing and automatic telegraphs and with his initial work on district telegraphs, a system that businesses and individuals could use to request messenger and other services. Edison also formed a new manufacturing partnership with the mechanic Joseph Murray, an association that soon grew in importance and continued for more than three years.[1]

In January and February, Edison vigorously pursued printing telegraphy, filling his notebooks with design variations for his universal stock printer and his universal private-line printer. At that time he also devoted a notebook to illustrating more than a hundred different escapement mechanisms for possible use in printing telegraphs.[2] His work on the universal private-line printer initially proved unsuccessful, but he abandoned the original design in March and had developed a new one by April. In the new design, Edison adapted his universal stock printer as the printing portion and improved the accuracy and precision of the private-line transmitter. He also fitted the new instruments with his screw-thread unison.

Left: *Experimental tape identified as the "First ink recorder message March/72 at 1800 words per minute."* Right: *Edison's experimental perforation indentified as "the first message ever perforated by us."*

In the spring Edison began to manufacture both his universal stock printer and his universal private-line printer for the Gold and Stock Telegraph Company. Officers in Gold and Stock also held an interest in the Exchange Telegraph Company of London; accordingly, when this English establishment began operations late in the spring of 1872, it employed Edison's universal stock printer on its lines.[3]

Edison's prodigious inventive activities also extended to automatic telegraphs, where he focused on perforator designs. He introduced improvements in the mechanical design and punches of his large perforator and executed a patent for a small, six-key instrument (U.S. Pat. 132,456) that could be used by local offices or by customers. Other elements of the system also received his attention. Because tailing continued to be the principal drawback to high-speed automatic telegraphy, Edison began seriously to pursue circuit designs that would prevent this problem. Also seeking to produce a reliable device for transcribing messages, he transformed his universal private-line printer into an electric typewriter and improved his ink recorder. With the latter he attained an experimental speed of 1,800 words per minute.

Both the American Telegraph Works and Edison and Unger continued to supply instruments to the various offices of the Automatic Telegraph Company, but technical problems delayed the public introduction of the system. In preparation for business, Automatic Telegraph initiated negotiations with the Southern and Atlantic Telegraph Company, hoping particularly to extend Automatic Telegraph's service to New Orleans.

In the early winter Edison also turned his attention to district telegraphy. In mid-January he made extensive notes regarding improvements, and these became the basis for a caveat he signed a couple of days later. He assigned the caveat to the American District Telegraph Company,[4] which Edward Calahan, George Field, and Elisha Andrews had created in October 1871 for the purpose of exploiting Calahan's newly invented district telegraph. Calahan's instrument allowed individuals to signal for messenger and other services from their homes and businesses.[5]

The flurry of inventive activity suggested by Edison's January and February notebook entries probably continued in the form of product development as Edison put the universal instruments into production the following spring and summer. The number of formal notebook entries declined sharply in this period, but informal and undated sketches continued to abound.[6]

1. In February, Edison and Murray established a small manufacturing shop, Murray & Co., on New Jersey Railroad Ave. In May they moved their shop to Oliver St. Murray's testimony, Quad. 71.1, p. 29; gas receipts, 72-001, 72-019, DF (*TAEM* 12:662, 919).

2. Escapement Notebook, February 1872, MiDbEI; Georges d'Infreville to TAE, 15 Dec. 1910, GF.

3. For evidence of experimental work in the shop see Cat. 1183:97–117, Accts. (*TAEM* 20:908–18). For evidence of printer production see Cat. 1183:89–140 and PN-71-09-06, Accts. (*TAEM* 20:904–30, 982–83).

4. See Docs. 226, 228, and 229; and Digest Pat. E-2:192–93.

5. Gibson and Lindsay 1962, 21–22.

6. See unbound telegraph notes and drawings, NS-Undated-005, Lab. (*TAEM* 8:207–504). Some of these may date from the second half of 1872.

–215–

Edison and Unger Summary Account

Newark Jan 1st 1872

E & U Stock a/c[1]

Tools & Machinery	$16776.76[2] cts
Finished & unfinished stock	8644.49
Raw Stock	1261.34
Bills due us not Collected	2240.86
	$28923.45
Bills we owe not Paid	6045.53
Total value ½	22877.92
	11438.96
Note from Wm Unger	3649.42[3]
T A Edison Resursece	$15,088.38

D, NjWOE, DF (*TAEM* 12:809). Written by Joseph Murray, on paper torn from a ledger.

1. "a/c" means account.

2. In another version of this yearly statement, this amount is given as $17,656.76, for a total of $29,803.45 for machinery, stock, and bills, which amount is written in the left margin of this document in another hand. Cat. 1185:378, Accts. (*TAEM* 22:704); see also ibid., 16 (*TAEM* 22:570).

3. Regarding the note from Unger, see Docs. 197 and 264.

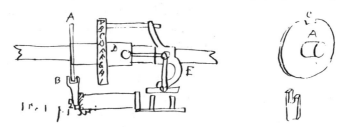

The above drawing is a sketch of my Locking device, The object of which is to place a number of my Universal Stock Printers on one circuit or circuits, and lock or unlock any one or any number from a Central Station upon that circuit.[2] I will copy here a caveat which I have previously prepared = [3]

"The object of this invention is to prevent any printing instrument among a number in the same circuit from Printing, all or any of them.

"The invention consists of an attachment to one of my Universal Stock Printers which work upon one or two wires according to the circuit devices used = [4] The attachment is throwing the lock wheel by means of the shifting devices into a forked lock peice attached to the Printing Lever, which peice acts as a lock to the type wheel and also as a block or impediment to the bringing of the Printing Lever to the face of the type wheel and making an impression. The fraction wheel[5] I have dispensed with, and use the shifting devices merely to lock and unlock the type wheels   In these lock wheels there are a slight opening .C. through which the type wheel ~~pa~~Lock arm passes at the moment of shifting. on every Machine composing a circuit The shifting devices are set in different positions so that one type wheel only of all the instruments can be shifted at a time, and unlike the ordinary mode shift all at the same time on all instruments, and also in This case I shift only when the type wheel lever is down and the magnet closed, and print when it is up   Thus preventing any of the type wheels of the several machines from shifting while transmitting a message to a particular instrument, which would be the case were I not to use a closed circuit on the Escapement Lever to do the shifting   I will mention here that after the lock wheel has been thrown into the fork, there is a slight dead motion between it and the fork which allows the Printing Lever to come up towards the face of the type wheel far enough to release the unison but not enough to effect an impression or move the paper any great distance. This fork performs two functions   The first, preventing the type wheel from jarring out or moving except in one position, and of preventing the Printing Lever from effecting an impression =

I have previously said that I dispense with the figure fraction Wheel, but this is not necessary as a second shifting devices may be attached to the ordinary Universal Stock which second shifting devices may be made to throw a disk in & out of a fork on the Printing Lever thus Locking or unlocking as is required, this second shifting device being operated when the type wheel circuit is closed whiehle the ordinary shift devices is operated as commonly or formerly when the type wheel circuit is Open.

I do not wish to confine myself to any particular device for unlocking and locking, because they would be different accordining as they are attached to the various machines now in the market. I believe at this time of writing that I am the originator of the Idea of locking & unlocking a number of step by step Magnetic Printing Machines in one or more circuits from a central Station

I am aware that Hughes[6] has an arrangement for unlocking & Locking his machines, but the machines used in the Stock reporting business, are on altogether different principles from the Hughes =

I will mention here that there are numerous ways of effect the Lock and unlocking, both electrically and mechanically One of the various modes is to have a severate magnet work a ratchet wheel, and this ratchet wheel having pins on it on upon its shaft may be thrown in the path of the Printing Lever and prevent it from Printing as long as it remains. This magnet should be polarized—or have double helix and be effected by closure of both ckts Another mode is to dispense with the shifting of the type Wheel and its shifting devices on the shaft, and place them rigid upon[a] the base attaching two pins only to the shaft thus

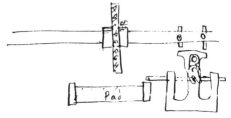

as in my shifting pad instrument already Patented[7]

A Cut in and out device might be used. This would consist of a polarized magnet working an escapement, on the shaft of which an arm might project rubbing on a periphery of metal having one insulating opening. Every instrument being provided with one of these, the insulated places being in different positions, and the magnet placed in the Printing ckt it might be made to revolve by intermitting a positive current and Cut out all the Printing magnets by reason of the arm remaining on the metallic ring and if it was desired to cut in any printing magnet the extra magnet would be rotated until the arm Came to an insulated opening when the short cut would be destroyed and the Ptg magnet would be in Ckt while all the other machines would be out   Then by using a negative Current which would not effect the directive polarized magnet, the message Could be Sent on that instrument = It is not necessary to Cut out the type wheels as they do no harm in being rotated altogether

This particular mode of cutting in and out electrically Can be varied in innumerable ways and I therefore do not wish to Confine myself to that particular way

An Extra magnet not polarized might be used the type wheel Cutting that in and that Cutting in the Ptg Lever

[Witness:] Jos T Murrey[b]   T A Edison[8]

AXS, NjWOE, Lab., Cat. 1174:13 (*TAEM* 3:13). [a]"on" interlined above. [b]Signature.

1. This entry is continued in Doc. 217.
2. See U.S. Pat. 126,533.
3. Cf. the partial caveat in Cat. 297:46, Lab. (*TAEM* 5:597).
4. The universal stock printer was often called the "two-wire printer." However, in Doc. 222 Edison sketched an "attachment to the U Stock by which . . . it can be worked on one wire."
5. That is, the wheel with numbers.
6. David Hughes.
7. That is, the cotton instrument (U.S. Pat. 113,034).
8. "53 Wright St. Newark, N.J." appeared at the end of the entry. This was the Edisons' new residence.

-217-

Notebook Entry:
Printing Telegraphy[1]

[Newark,] Jan 4 1872

Since Last writing I have made further improvements in the mechanical lock devices. As now constructed, I shift upon an opened circuit and at the Same positions on all the type wheels, and not on a closed circuit as heretofore described.[2] this makes it more practical and easier for the Transmitting Operator. The Unisons being all in different positions, it follows that if I wish to operate an instrument which says catches on the Letter B I turn around a number of times to ensure the stoppage of the Unison by the type Wheel and Stop the transmitter pointer at B and close the Printing Lever and release the type wheel   then of course the type wheel & the transmitter correspond. then I turn them both around until I come to the shifting pegs   then close the Ptg Lever   unlock the wheel and then transmit my message = In case some one of the Printing Machines did not happen to lock or got out accidentally the type wheel would not correspond with the transmitter so the message would only be a jumble of unreadable letters, but be ok on the inst which the message is intended for. I could dispense altogether with shifting and locking of the Ptg Lever, as every inst would print a jumble of letters except the inst which it is intended to receive the message, but I prefer to lock the Ptg Lever, saving paper, and, as being more seemly & convenient. I will state here that over a year ago I suggested to Genl Lefferts Should he wish to work an number of Private telegraph Printing Instruments on one circuit between a firms office & factory   Every pair of such instruments might have type wheels to Correspond but every other pair have type wheels similar but the Letters placed in different positions tSo that no firm on the same Line could read the message of another firm =

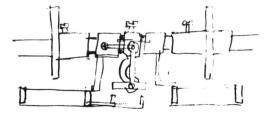

The above sketch represent a Printing instrument which has two type Wheels and Prints upon two strips of paper. This is done by one magnet working two printing Levers or a seperate magnet to each lever but both in the same circuit and locking one or the other by the shifting device and arms.[3]

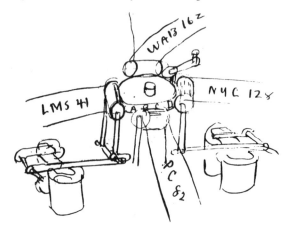

Device for Printing with a number of Printing Levers from one type wheel. a Seperate wire runs to each Printing magnet and one wire for the type wheel. This type wheel is rotated by a break wheel. 4 transmitters are used   in either is a device which ᴏthe type wheel may be arrested at any desired point by depressing a key, independent of the others in the Same Circuit. The Speed is greatly increased with this device

Witness by[a]   Jos T Murrey[b]   T A Edison

AXS, NjWOE, Lab., Cat. 1174:16 (*TAEM* 3:15). [a]"Witness by" written by Murray. [b]Signature.

1. This entry is a continuation of Doc. 216.

2. This is a reference to actions on the typewheel circuit.

3. Edison later patented two different machines employing two printing levers on one circuit (U.S. Pats. 131,338 and 131,339), but neither was of this design.

–218–

Notice in the Newark Daily Advertiser[1]

Newark, January 5, 1872[2]
EDISON—STILLWELL—Dec. 25th,[3] by the Rev. W. S. Gallaway.[4] Mr. THOMAS A. EDISON and Miss [MA]RY STILLWELL.[5]

PD (photographic transcript), *Newark Daily Advertiser,* 5 Jan. 1872, 3.

1. This notice appeared in a section headed "Married."

2. Apparently the newlyweds honeymooned in Boston. No entries in Edison's hand exist in technical or financial records from the date of the wedding to 1 January 1872, and an account entry dated 1 January lists $165.00 "By Cost Trip to Boston" (Cat. 1213:4, Accts. [*TAEM* 20:7]). About the time of this notice, the Edisons established housekeeping in their new home at 53 Wright St., Newark (see Docs. 206 and 216, n. 8). Account records show that Edison spent about $2,000 furnishing the house, providing for Mary's wardrobe, and employing servants. Prominent among the furnishings was a piano. Cat. 1213:250–52, 400, Accts. (*TAEM* 20:21–22, 25); Cat. 1185:18, Accts. (*TAEM* 22:571).

3. No official record of the marriage has been found.

4. W. Smith Galloway was the minister of the Summit, N.J., Methodist-Episcopal Church from 1870 to 1872. Prior to this he had served as minister of the Asbury Methodist Church (renamed the South Market St. Methodist-Episcopal Church), located a few blocks from the Stilwell home on Jefferson St. in Newark. *Journal of the Northern New Jersey Conference, United Methodist Church* for the years 1870–72; *Holbrook's Newark City Directory* 1869, 257; 1870, 261; and 1871, 281.

5. Sixteen-year-old Mary Stilwell (1855–1884) had worked for Edison's News Reporting Telegraph Co. in October and November 1871 (see Doc. 205, n. 2). She was the daughter of Nicholas Stilwell, a lawyer and sometime inventor, and his second wife, Margaret Crane Stilwell. Among Mary's many siblings were her older sister Alice and her younger

Edison's first wife, Mary Stilwell.

brother Charles, both of whom retained a close association with the Edisons after Mary's death. U.S. Bureau of the Census 1967c, roll 880; *Newark Daily Advertiser,* 10 Feb. 1870, 3; Edison, Mary (nee Stilwell), folder, EBC.

[Newark,] Jany 6th 1872.

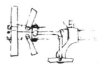

This is a double Type wheel having each wheel split in 3 or more Segiments and the inner ends pivoted on a set sleeve on the shaft. When the Shifting device is thrown one way the segiments of one type regain a complete wheel ready for Printing while the segiments of the other are thrown off at an angle thereby making it smaller than the other wheel[1]

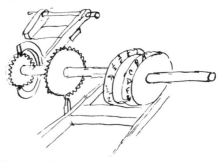

Unison Device. Extra ~~ex~~ tooth wheel with a space where there is no teeth. when the Escapement lever is up and circuit open the prong on the Ptg Lever fits the teeth in extra wheel, & no effect is produced & the Ptg takes place as usual. but if we wish to rotate the type wheel to Unison should it be out we close the Escapmt Lever circuit & bring escapmt Lever down   in so doing we bring the V teeth of second wheel in such a position that the prong on the Ptg Lever when that is brought up will shove the type wheel around half a Letter Lifting the Escapmt Lever out into half next tooth & when the printing Lever is let down the magnet working escapement which still continues closed brings escapement Lever down again and places teeth of second wheel again in a position to be thrown forward by the printing Lever, & so on until it comes to a place where there is no teeth then it is to unison. then open the Escapement Lever circuit & go ahead on an open circuit in the usual manner, until you wish to catch all the instruments at Unison, then close circuit & work Printing Lever until type wheel stops = [2]

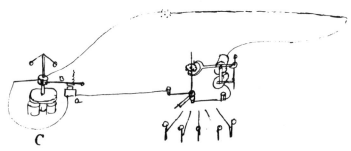

C

Arrangement for furnishing "breaks" for a whole line of Printing Instruments from a central Station, by means of my Engine break already described in my Universal Printer Patent and Caveat   This engine is of the same make as described on page 7 of this book[3] but in this case I dispense with breaking the main circuit by the points on the armature Lever of the Little arresting magnet—and I also make the armature Lever arrest the break wheel on the up stroke (ie) when the magnet is open, and not closed as employed in the usual manner. I also include this magnet in the main circuit, so that every ~~th~~ time the engine makes ~~two~~ one revolution two breaks are made and of course the arm of the arresting magnet vibrates with it but does not arrest the wheel. Now I provide the Type wheel shafts of the Several instruments with projecting arms which pass in the path of Keys   now when a Key is depressed the arm revolving around by the action of the Escapement ~~an~~ & magnet which receives its breaks from the Engine break at the Central Station. this arm is made of two springs in contact when revolving and are connected to the main circuit in such a manner that when the arm is arrested by the key the two springs are seperated and the main circuit opened. at the moment that this occurs there is not circuit on the Line it being so adjusted that it will be at the interruption of the Current on the break whe~~n~~el = Now there being no Current on the Line the armature Lever[a] of the Little arresting magnet Stays up and Catches the break wheel and arrests it, not on an open circuit but just as it commences to close. this insure a full break when the key is raised, for when it is the Circuit is instantly closed   the Little Lever is let down or drawn down and the break wheel revolves as before & so on. The ~~S~~Receiving instrument can have a vibrabting point & Local battery Connected with the Printing Lever So as to receive any message which may be sent, but I do not use any spring on the sending instrument, as I wish to dispense with a Local battery although it may be used =

I

In the instruments and Lock devices djust explained but one instrument Can be used at a time

But a shifting device might be added which would shift on a closed circuit on the Letter Z and a double unison made This shift might be made to shift and throw one unison out and the other one in   the first unison for working a Single inst and the Last to work them all. The last being set on all insts alike   the former at various places. Where it is necessary to work both the Letter wheel and Figure wheel I add an extra shifting device for Locking and unlocking =

I will state here that I Lock on the fraction dot open circuit and unlock on the fraction dot closed circuit = on my Lock Instrument =

Witnessed[b]   Samuel Edison[c]   T A Edison

AXS, NjWOE, Lab., Cat. 1174:18 (*TAEM* 3:16). [a]Interlined above. [b]Written at a slant. [c]Dotted line drawn under signature.

1. See Doc. 212; cf. U.S. Pat. 126,529.
2. Edison patented this unison device (U.S. Pat. 131,344).
3. Doc. 208.

–220–

Notebook Entry:
Printing Telegraphy

[Newark,] Jany 8 1872

I am requested by General Lefferts to invent a peculiar unison for what is known as the Callahan 3 wire instrument[1]

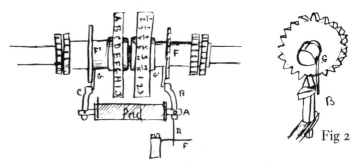

Fig 2

A is a slide peice which moves very freely through the pad or one side of it and is prevented from turning by the wire D passing between two guide wires F secured to a pillar = Upon this slide is secured two peculiar shaped click B.C. Fig 2 shows one of them to better advantage   F F′ are two stars wheels secured to the two different sleeves of the type wheels, and these wheels act the same as the unison device shown on page 18²   Some of the teeth of the star wheel being cut out at a given position = G G′ are pins secured to the ~~W~~Type wheel sleeves just back of the dots or Zero points = The purpose of these pin is toª throw the slide and its clicks ~~ei~~under one starª wheel orª other. = Supposing both wheels to stand at zero Then the slide will be in a position where one click is directly underneath the star wheel of one type wheel and the other is thrown out one side.

Now supposing that the Letter wheel was in unison But the fraction wheel was not = Now when the operator starts to rotate the Letter wheel, the pin G will come on the bevel of the click C and throw the Bar A over and bring the click directly underneath the star wheel of the Figure type wheel, which is out of Unison. (Bad drawing, shows both under). Now the first Letter printed the printing Lever ~~cu~~omes up   The click B, coming into the star teeth throws the type wheel ahead one Letter and will continue to throw it ahead at every mo~~m~~vement of the printing Lever until it comes to where there is no teeth on the star wheel then the type wheel will be at zero. Now as long as you work on the Letter wheel The slide will remain over to the figure wheel side, but if ~~not~~ now the operator brings the Letter wheel to zero, ~~or what~~ or tries to, and it is 2 Letters behind, and commences the rotation of the figure wheel. the first forward movement of the figure wheel by the aid of the pin G′ throws the slide over and the click under the star wheel of the Letter wheel and that wheel will be brought to zero by the first two upward movements of the Printing Lever   The whole thing is this   B, & clicks is a "Corrector"   when one type wheel goes around it throws [---]B & etc under the other for correcting it & vice versa

I do not wish to confine myself to a star wheel & V shaped click as I can use a ratchet wheel and ordinary click   Neither do I wish to confine myself to a sliding bar with Clicks in the Printing Lever as the star or ratchet wheels may be made to slide by the action of the rotation of the type wheel and an arm or other devices secured outside or independent of it =

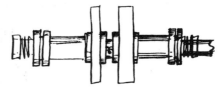

Another unison device   each sleeve being provided with two little teeth slightly Bevel and a pin on the shaft. If one

wheel gets out of unison the other wheels tooth grabs the tooth on the other wheel brings it up to the pin and they are then parted if one wheel goes ahead

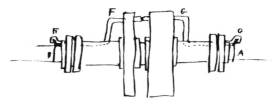

A B are flanges bent slightly at the unison point   F G are Carriers for locking the rotating wheel with the wheel to be set = one end of each has a head with a slot in it whiich works in the flanges A.B. When a type wheel is rotated around this carrier is thrown forward but is brought out of the path of the outher carrier between the type wheels by having a portion of the flanges A B bent outward just at the unison point or last one or two Letters. It is not necessary to bend the carriers up to the periphiery of the type wheels as I have shown = witnessed[b]   Samuel Edison[c]   T A Edison

AXS, NjWOE, Lab., Cat. 1174:21 (*TAEM* 3:17). [a]Interlined above. [b]Written at a slant. [c]Dotted line drawn under signature.

1. Edward Calahan's stock ticker did not have a unison when it was first designed. Telegraph inventor Henry van Hoevenbergh created one for it in 1871 (U.S. Pat. 120,133). Edison sketched another unison for this machine on 6 March 1872 (Doc. 253).

2. Doc. 219.

–221–

Notebook Entry:
Automatic Telegraphy

Perforator for Telegraphs   [Newark,] Jany 13 1872[a]

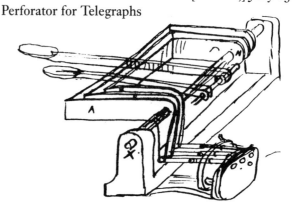

A B C are Levers   I use nine of them   they are worked Loose on the shaft X which is its fulcrum   These Levers are made of ⅛ inch sheet steel and very broad so as to give them the required strength. one is placed inside of the other, and the whole are placed immediately under a bank of keys. In practice I obtain the selection of any ~~bar~~Lever[b] by making tits or teeth on the keys   Thus

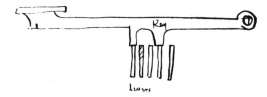

Or these teeth knobs or projections on the keys may be thus

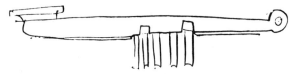

or they may be dispensed with altogether on the keys and The teeth made on the bars or Levers or notches may be Cut in the Levers   Thus

Fig 3

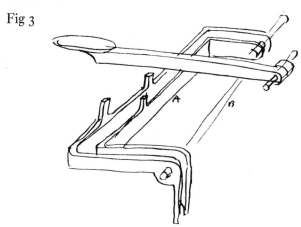

or this way

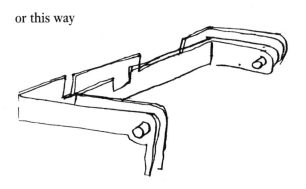

some trouble may be anticipated by the springing[1] of the Levers but this Can be obviated by Connecting a strengthning Lever as in Fig 3 from A to B   Or The Levers Can be made in this manner

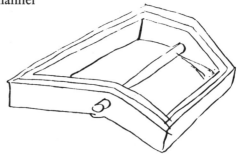

Each square swinging seperately or it Can be made in this form

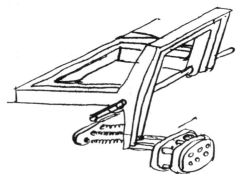

This form of Lever is probably better than the ones heretofore described owing to the strength of the Levers

The paper feed device which I intend to use in this machine is peculiar   The motive power being derived from a spring or weight or any outside power not derived from the pressure on the keys =

Theis Length of The paper feed is controlled by the Levers or by the keys.

End view

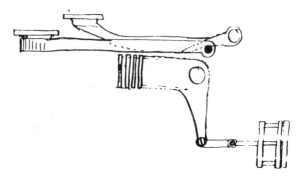

Paper Feed

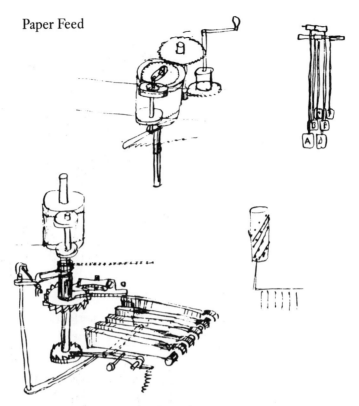

About perforator Jany 13 1872
Witnessed   Samuel Edison[c]   T A Edison   Inv

ADDENDUM[d]

[Newark, c. January 13, 1872][2]

Perforator = I do not wish to confine myself to a die and punches, for I believe they could be dispensed with and Cutting punches hollowed out, and cutting on a hard raw hide surface, be substituted with advantage. ~~when~~

Should my paper device prove to be badly arranged in an upright position, the Die may be Set flat ways instead of edgeways as heretofore Shown   Thus[e]

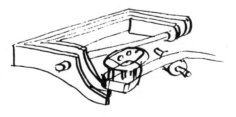

The flat bars may be dispensed with and ~~the~~ several round bars substituted   Thus

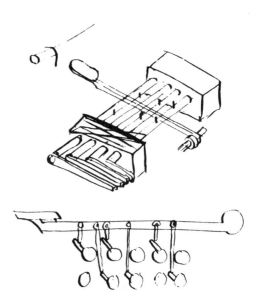

New movement for perforator

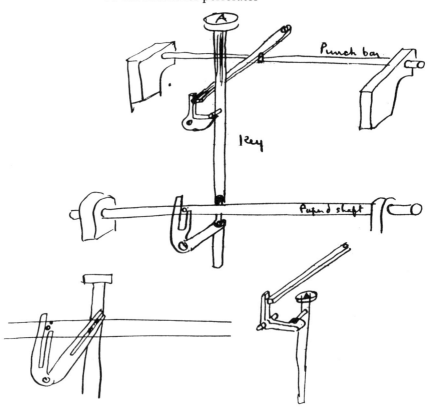

AXS, NjWOE, Lab., Cat. 1181:9 (*TAEM* 3:178). [a]Date taken from text. [b]Interlined above. [c]Signature. [d]Addendum is an AX. [e]Followed by "over" to indicate a page turn.

1. "Springing" means bending sideways under pressure.

2. This addition immediately follows Doc. 221. It was probably written soon afterward, but could have been written as late as the end of January, for the next entry in this notebook (Doc. 240) is dated 1 February 1872.

–222–

Notebook Entry: Printing Telegraphy[1]

[Newark,] Jany 13 1872 =

Universal Machine = Instead of using the two arms and Connecting springs upon the type wheel shaft and being in the path of the [-]keys   The connecting points for closing The Little Magnet in the Pulsator may be dispensed with and straight arms only used—[2] When this is done I put a wheel having 30 fine[a] points of Contact insulated between & Connected so as to revolve with the type wheel shaft. rubbing on this wheel is a spring connected to the pulsator Magnet so that by depressing a key the plain arms suddenly arrest the Type wheel and the contact spring, just comes on a contact point of the tooth wheel, and the pulsator wheel is arrested by the little magnet. The type wheel shaft being arrested just a little way from its limit   the key being raised will allow it to reach its entire limit rotating the type W Shaft a little & severing the contact points. On account of the Speed with which this wheel is rotated, the spring though touching a point every letter does it so quick that it has no effect on the little magnet of the pulsator, but only when arrested for a small period of time— It is not necessary to use a ~~w~~contact wheel with 30 points and a rubbing Spring. A single spring secured to the shaft of the Escapement lever ~~w~~might be extended out some distance and made to Close against a point just as the Escapement lever was nearly home on an open circuit

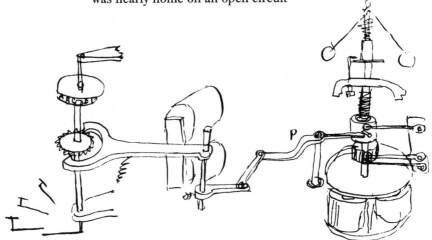

Improved method of catching or arresting the little break on the Engine[3] in my Universal Printer. heretofore I have arrested it first by a movement derived from the arresting of the pin on the type wheel shaft, and the movement of that shaft   Secondly by closing a secondary circuit by arresting a point on the type wheel and closing a magnet   the armature of this magnet arresting the break wheel. Thirdly The manner which which I describe on page 26,[4] and Lastly the above. The arresting of the break wheel in this case is done by the direct action of the Escapement lever, the movement having nothing to do with the type wheel shaft   that only being used as a means of stopping the Escapement lever at certain points by the action of the points on the shaft going in the path of the Keys and by the aid of the ratchet or Star wheel.

It will be seen by referring to the drawing that a lever is connected to the Escapement lever and that the End of this lever is in or out of the path of the ~~Esca~~ little pin on the break wheel according to the position of the Escapement lever = When the Engine is running at every half turn of the break wheel the magnet operating the Escapement lever is closed   consequently the lever P. is thrown into the path of the pin on the break wheel and would arreste it, but the break is so arranged that just as the pin is in the act of being arrested the current of the Escpmt Lever Magnet is interrupted by the break and the Escapmt lever flies back by the action of the spiral spring and throws the arresting lever P out just in time to prevent catching on the pin and the circuit is again immediately closed   the lever is thrown in then jerked out every time the type wheel advances one letter, ~~or if used on Double ckt open & close every~~ consequently the break is not interupted   b~~y~~ut if we close a key the pin on the type wheel shaft comes in Contact with it, the Escapement lever is arrested half way ~~while~~ and the lever P held in the path of the pin on the break wheel and the break wheel is arrested. now when the key is raised the [-]Escapmt lever is allowed to go its full incirsion the lever P is drawn out and the break wheel revolves as before   I do not wish to confine myself to any particular lever, but claim arresting a break wheel by a movement from the Escapement lever or any device attached to the Escpmt Magnet and holding the lever in the path of the ~~type~~ Break wheel by arresting pin or pins on the type wheel shaft by key or other devices. I Could used a seperate magnet to arrest this break wheel and place it in the main circuit and put a break point on the Escapement lever and arrest half way

Thereby opening this Extra Magnet and arresting the break wheel or put a wheel having 30 contact points on the type wheel shaft, and arrest the wheel by the keys & pins in such a position that the ~~break~~ main circuit would be open

Thus first

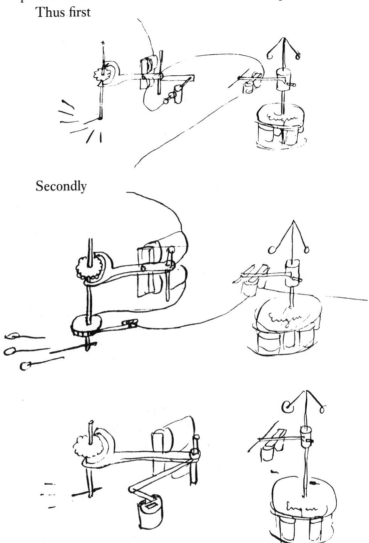

Secondly

I will mention here that the friction cloth and Spring used on the Engine shaft for friction to carry around the Little break wheel might be dispensed with, and a clutch or click used   the clutch being thrown out by the action of the arresting Lever. This click or clutch would be provided with a stop so after the click had been thrown out a short distance it would come against a stop and the wheel would be arrested. one form I here show

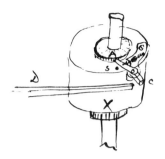

A is a smilled wheel secured to the revolving shaft   C the
click   X the Bkª wheel   D the arresting lever   S a Limiting
pin =

When the lever is not in the path of the break wheel pin the
click is drawn to the face of the Milled Wheel by a spiral or
flat spring thereby Locking the break wheel to the revolving
shaft. now if the lever D is depressed the point comes on the
click this immediatly is thrown out and the Limiting pin
S stops the click from going out further and the wheel is
stopped.

another form

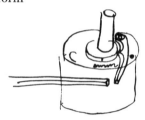

Attachment to the Universal Stock by which with the aid of
a Local it can be worked on one wire, or can be worked without a Local

Fig 1

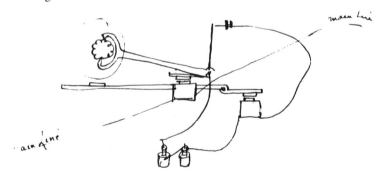

Fig 2

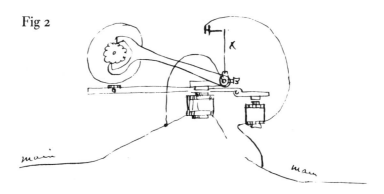

The points X are adjusted so that when the Escpmt lever has reached nearly the whole length of its play the contact point just touch. now when the Escapement is rotated rapidly the printing lever does not respond— In Fig 1 the Escpmt stops on an open ckt   in Fig 2 on a closed circuit

Witness   Samuel Edison[b]   T A Edison

AXS, NjWOE, Lab., Cat. 1174:23 (*TAEM* 3:18). [a]Interlined above. [b]Signature.

1. This entry is continued in Doc. 223.

2. For elucidation of these terms, see the schematic diagram in Doc. 193, n. 2. The "connecting points" (or "springs") here correspond to the contacts on arms **D** and **D′** in that diagram; the "Little Magnet" corresponds to **MM′**.

3. The "little break" was the pulsator, the small breakwheel that transmitted impulses in the typewheel circuit. The "engine" was the electric motor that drove the pulsator and typewheel shaft.

4. See the next three drawings in this document.

–223–

Notebook Entry:
Printing Telegraphy[1]

[Newark,] Jany 13 1872

I[a] wish it understood here that I can easily combine my Pulsator and arresting devices now used in my universal machine, to any form of Printing Telegraph (ie) attach it or to any mechanism operated by step and step motion or escapemt, Ratchet Motor, or release escpmt & weight or other power   or I can combine it in the form of a transmitter only[2] in many different ways dispensing with the type wheel Ptg lever & printing devices.

I mention here, that I can gear up directly from the engine,[3] and on the last gear wheel shaft[b] put a pin or arm or detent revolving in the path of keys, and connect this arm with a V shaped wheel working an arresting lever connected so as to arrest the break wheel on the main Shaft of the Engine, this wheel being carried around by friction, Thus making a Transmitter. The Universal as it is now operated Can by insulating the arrester & platinizing the point on the break wheel be made to Close a secondary circuit thus enabling to work instruments which use two wires. The geared transmitter I Show   thus =

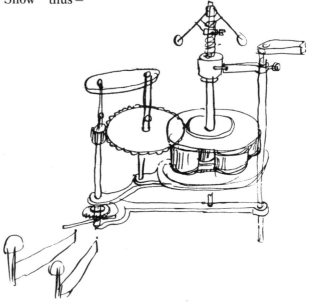

Witnessed   Samuel Edison[c]   T A Edison

AXS, NjWOE, Lab., Cat. 1174:29 (*TAEM* 3:21). [a]"Univ Priv line printer" written by William Carman at top of document. [b]Interlined above. [c]Signature.

1. This entry is a continuation of Doc. 222 and is continued in Doc. 224.

2. See Doc. 211.

3. See Doc. 251.

–224–

Notebook Entry:
Printing Telegraphy[1]

[Newark,] Jany 13 1872.

It is of course obvious that my Universal Could be Operated aupon an open & closed circuit (ie) by opening The Circuit advance the type wheel one letter and by Closing advance another letter, but I prefer to use both opening & closed to move a letter it being in my Opinion More practical =[2]

The pins on the shaft of the Universal type wheel spring a little and might be made in this way

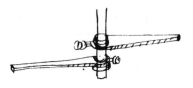

Instead of

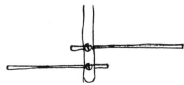

I claim a dead motion[3] or the use of a dead motion in a Star or wedge acting escapement for the purpose of giving the printing lever time to get away from The type wheel before it starts & not blurring the letter. I believing myself the first one who discoved or used this method, other parties, (Phelps[4] and Grey & Barton[5]) using devices for throwing their printing levers ~~tw~~ to the type wheel & then quickly back[6] they not knowing they are wholly unnecessary providing there is a slight dead motion in their Escapement which there generally is

I claim the use of paper in a ~~Prin~~ single wheel Machine ⅝ of an inch or more wide, when such paper is used with my paper driving devices, or in other words I have found that ½ or less paper cannot be driven or worked practically in a printing Telegraph which is operated by inexperienced persons

I do not wish to confine myself to any particular form of Electric Engine in my universal, nor even to an Electric Engine as the break wheel May be driven with gearing or any other device operated by any force   Or the revolving wheel Can be dispensed with and a Vibrating lever used, operated by a local and this self Vibrator opening & closing the Main Cirrcuit and the arrester used same as that for arresting break wheel heretofore described and this arrester made to stop The Vibrating Lever or to open its circuit by a Contact point = Or the breaks may be furnished by a clock movement or any other devices, and the arrester made to act electrically (ie) open & close[a] the Main circuit so that when arrested in a certan position it will open the Main circuit & prevent any more breaks from being sent over the wire & when released allow the breaks to be sent = In fact a point connected to the Escapement Lever of any printing Telegh Can be made to Control the devices which furnishes breaks by arresting it in certain points of its motion as well as to control the printing lever—

Witness   Samuel Edison[b]   T A Edison

AXS, NjWOE, Lab., Cat. 1174:30 (*TAEM* 3:22). ᵃ"& close" interlined above. ᵇSignature.

1. This entry is a continuation of Doc. 223.

2. See Doc. 195 headnote.

3. "Dead motion" is the initial part of the travel of the escapement lever, before the lever engages the typewheel ratchet and consequently before the typewheel moves.

4. George Phelps had developed two private-line printers (U.S. Pats. 110,675 and 126,329) for Western Union; both had jointed printing levers.

5. Elisha Gray, Enos Barton, and Western Union division superintendent Anson Stager had recently formed the firm of Gray and Barton in Chicago to manufacture telegraphic and other electrical apparatus. Later in the year they formed the Western Electric Manufacturing Co., with Western Union holding a one-third interest in the firm. Western Union also sold its manufacturing shop in Ottawa, Ill., to Western Electric, which became the largest manufacturer of Western Union's instruments. The company's products included Gray's private-line printer (U.S. Pat. 132,907), also known as the Gray and Barton printer. Although U.S. Patents 126,329 (see n. 4) and 132,907 would not issue until later in 1872, it is evident from this entry that the instruments described in those patents were already in use. Lovette 1944–45, 274; "The Story of the Western Electric Company," *Western Electric News* 1 (1912): 1–2.

6. Edison had devised such a lever himself. See Doc. 193.

Elisha Gray's private-line printer.

–225–

Notebook Entry:
Printing Telegraphy

[Newark,] ~~Dec~~Jan 15 1872

I do not wish to confine myself to any particular method of arresting the break on my Universal. I can be done this way.

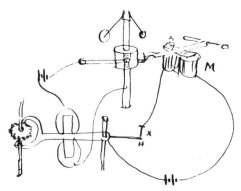

X and H are vibrating points work so quick dont allow current in arresting magnet Long enough to close lever but when key depressed and shaft arrested Escapement lever closes points X & H allows Current to circulate through M long enough to bring armature down and arrested the break wheel.

Another way

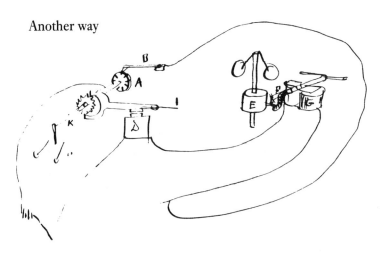

LK is the shaft awith point arrested by keys C   A a break
wheel with block points made of insulating substance. B a
spring. The pointer on the shaft is placed in such a position
that when arrested the point B will be on one of the insulated
parts of the wheel A. The insulated points always passing
under the spring B when the main circuit is broken by the
break wheel consequently the breaking of The main circuit by
the wheel A & spring B does not effect the regular pulsations.
Now the magnet G being in the main circuit every ha it vibrates its armature & arrester the same as the escapement   The arresting Lever going in & out between the pin
on the break wheel twice every revolution or as many times as
there is breaks on the wheel but does not arrest it. but if now
a Key is depressed the beak wheel A is arrested and the circuit
is broken and the arrester is thrown or rather is not drawn
down and the break wheel is arrested on a closed circuit but
the circuit remains broken at B & A   if the key is raised the
escapement advances slight distance which the key prevented
and the circuit is closed the arresterd lever is drawn down and
the instrument goes on until arrested by another key being
depressed   The wheel A & spring B might be dispensed with
and it under[1] patent break substituted by putting it on the
escapement and so connecting it that when the escapmt lever
is drawn to the magnet the circuit will be closed but when it
goes back it will be broken[2]   I Can use any devices to do this
several of these devices were described in a former caveat relating to Vibrating devices which at the time I intended to use
on my Universal instead of the Engine.
Witnessed   Jos T Murrey[a]   T A Edison

AXS, NjWOE, Lab., Cat. 1174:32 (*TAEM* 3:34). [a]Signature.

1. Probably should be "Unger's"; see Doc. 237, n. 2.

2. Edison used this idea (**I** in the drawing) in his redesigned universal private-line printer (Doc. 262).

–226–

Notebook Entry:
District Telegraphy[1]

[Newark,] Jany 15 1872

The object of this invention is to communicate different signals automatically from a number of machines on one circuit and to record & indicate such signals at a central point   The signalling machines being placed in private houses etc all a number on one circuit connected to a Central station, and are to be used for instantly summoning the police or a messenger.[2]

The invention consists in a several combinations of Mechanism to effect the object automatically. the Combinations that can be made to perform this operation are so various, &[a] unlike, and that to protect myself I shall describe a large number of ways in which the result may be obtained

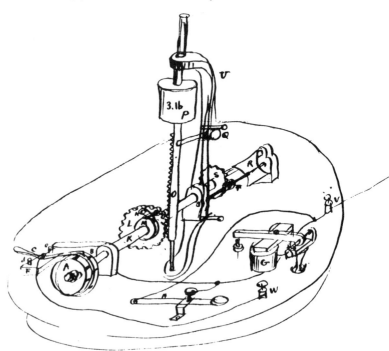

K is a shaft   S is a gear wheel   T a pinion wheel with a fan R on it   This fan being used to regulate the speed of the shaft K   L is a loose gear wheel carried around by the milled wheel M & click n by the rack .O. This rack deriving its power from the weight .P. This weight being thrown up by the finger   Q is the handle by which this weight is lifted   This handle movement is limited by two Limiting pins   It is obvious that ~~thi~~if the weight is raised to its full heighth that the shaft K will remain Stationary while it is being raised owing to the position of the click n and when it is allowed to descend it will commence to rotate the shaft K with a certain speed governed by the fan R.

On the End of this shaft K is a double break wheel A with a reversable contact point C secured to the hard rubber stud B   The point being limited by the pins d & .e. One of these wheels is provided with characters which will open & close the circuit in a manner to indicate "Police" at Central station & also the number of the house   the other wheel will indicate "Messenger" after the same manner. When it is desired to Call police the contact point is thrown over on one wheel and the weight raised, in descending it rotates the shaft K once thus giving the signal   H is a key connected on the main Line   G is a test magnet   when you are about to give a signal the key is depressed and the main circuit allowed to circulate through the magnet G which if it remains closed for a small length of time shows that no ~~si~~other signal is being sent[3]   when the Key is raised the magnet cut out and the weight raised and in descending the Signal is ~~ree~~ transmitted

I do not wish to confine myself to a fan for regulating The speed of the shaft .K. or the descent of the weight As it can be done by air thus:—

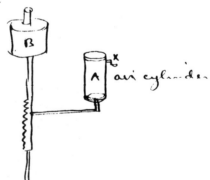

B the weight   A. a very tight air cylinder with a piston within it   the piston connecting to the rack & weight   when the weight is raised sufficient extra power is applied to force the air out through the small hole .X. but when descending the power of the weight can only force it out slowly and consequently it falls slowly

The weight might be dispensed with an a spring Substituted thus

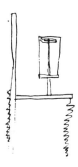

The rack break wheel and its parts might be dispensed with and the same thing accomplished in this Way   Thus:—[b]

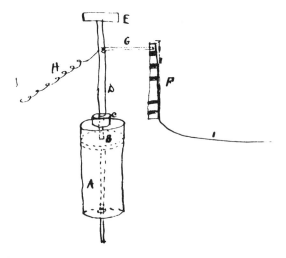

A is a cylinder with a rod D having a head B in this cylinder. E is a knob or handle & serves both the purpose of raising the Rod D and as a weight to force the air out of the cylinder as the piston head descends   G is a contact spring rubbing on either side of the Signal Bar .F. as the case may be. ~~this~~ when it is desired to give the signal upon the circuit the Rod D is raised to its full heighth and the spring G rubbing on ~~t~~either side of the signal Bar or Contact peice F makes & breaks the circuit as the rod slowly descends   as it will owing to the force necessary to ~~express~~ expel the air quick being greater than the effect of the weight of the rod & the knob .E.

The Contact may be made in this manner

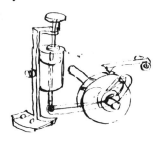

The downward movement of the piston rod moving the Contact wheel a short distant around, four or five signal being made on one revolutions, and one full signal being upon the segiment of the Circle moved—

The air cylinder may be made to regulate the speed of a shaft which may derive its power from any source. The object mainly being to dispense with as many gear wheels as possible or with them entirely. The manner of regulating is thus:—

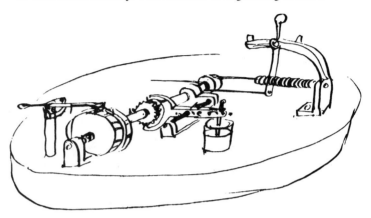

The air cylinder might be dispensed with and a spring attached, or a ball might be attached to the end of the Vibrating lever .C. The form of the teeth which gives motion to this lever can be made in a ratchet form or star, or waivng former. The ratchet wheel and Vibrating arm might be dispensed with by making it in a governor in This form

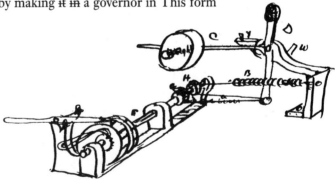

F is the shaft. G H are two is^a small pulleys   a is a cord worked by the lever D. B is a spiral from which the power is derived This spring is provided with a cord where it goes on the pulley H, it being on top and wound in opposite to that on .G.

C is a weight which is used to regulate the rotation of the shaft .F. When in its normal position the lever D is at the point W, but when a signal is to be given by the rotation of the break wheel the lever is pulled over by the hand ~~fr~~ to Y and this act ~~stores~~ ~~u~~ winds up the spring but does no turn the shaft on account of a click which prevents it going backwards. The act of moving this lever by the hand stores up the power in the weight and when ~~o~~the lever is Let free by the hands the shaft commences to turn slowly according to the distance of the weight

There is still another mode of governing the rotation of this shaft, without gears   Thus

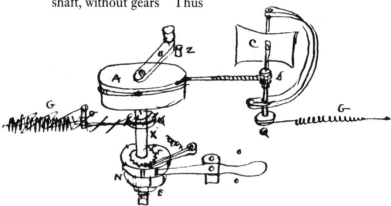

A is a drum upon which is a handle B   this is secured to the shaft X   Q is a small pulley on which the cord of the power spg G is secured   N are the two break wheels carried around by friction on the shaf X by a spring washer & cloth at .E. C is a governing fan connected with the drum A by a cord

By turning the handle B from Z to .O. The spring is wound up and when the handle is free ~~to~~ ~~m~~ the shaft commences to turn but cannot go fast as the fan an large size of the wheel A prevents it

A single gear and flat spring might be used in the following Manner

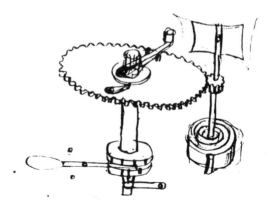

This is another form in which a governor which can is not very necessary, or slight—

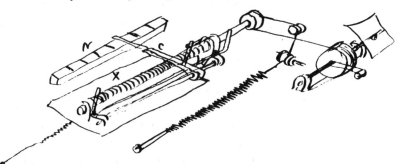

X is a screw   C a traveller   N the break peice   when the drum in wound up the shaft or screw revolves very quickly and the traveller goes over to the eother end making the breaks on N   when it gets over, it throws a little wire up anwhich throws it out of the thread and is drawn back home on this wire by a spiral spring—

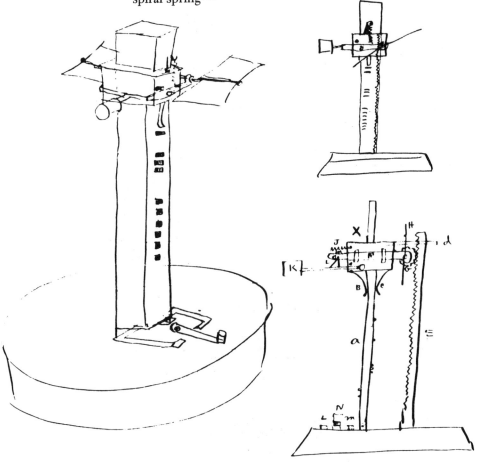

a is a rod of insulating Material with the characters sunk in on each side   B & C are the contact springs   d is a spring which rubs on the side of E and through which the circuit is conducted from E to the square X. K is a handle pivoted on X ~~and has two Limiting~~ and has a Limiting pin to prevent it from falling = G is a pinion having a fan H secured to a slide F, and is thrown out of the rack E when the handle K is lifted in the act of raising X to the top of A from the bottom and is thrown back in the rack by the spring J when the handle is released   then the block X gradually drops to the bottom with a speed governed by the size of the fans H H′   The contact spring B or C making a signal by coming in Contact with the side of a   The side which shall act being governed by the electric switch L M N   N being the button connected ~~to aX and through~~ with the Line and ~~C~~L to one side of a & M to the other

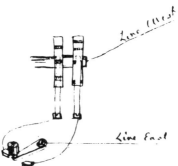

Manner of working the Double break

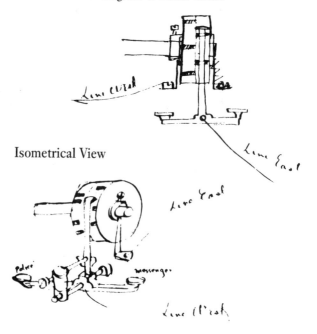

Isometrical View

About Signaling Machine
Witnessed   Samuel Edison[c]   T A Edison

AXS, NjWOE, Lab., Cat. 1181:33 (*TAEM* 3:183). [a]Interlined above.
[b]Followed by "over" to indicate a page turn. [c]Signature.

1. This entry is continued in Doc. 228.
2. Telegraph devices had been used in municipal fire alarm systems
in American cities since 1852. In 1871 Edward Calahan invented a "district and fire-alarm telegraph" to give "warning in case of fires, burglars,
or accidents" (U.S. Pat. 127,844, p. 1). Together with some Gold and
Stock Co. associates, Calahan incorporated the American District Telegraph Co. (forerunner of the present ADT Co.). The company established offices in several Brooklyn and New York City neighborhoods and
placed small, simple transmitting devices in private homes in the surrounding areas. Messenger boys stationed at the offices responded to
subscribers' signals for police, firemen, the family doctor, or messenger
service. The company grew from an initial group of 4 subscribers to over
2,000 within two years. In 1874 Edison and several others organized a
competing service, the Domestic Telegraph Co., which exploited Edison's U.S. Pat. 154,788. Reid 1879, 634–36.
3. This was a feature of Calahan's device as well.

-227-

Notebook Entry:
Automatic Telegraphy

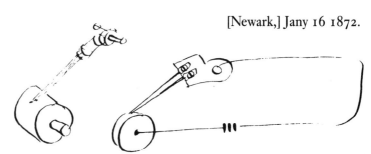

[Newark,] Jany 16 1872.

I claim two pens for receiving chemically which shall be
connected together electrically, one pen I put about the one
hundredth of an inch ahead of the other in the same line. The
object of this is to reduce the resistence of the paper and make
a double decomposition, as the resistence of the paper is reduced one half by having two points   it follows that very near
as much current will pass through each of the two[a] points
as one as twice as much current will be taken from the battery. now one pen passing over the paper will leave a mark of
a certain intensity and another pen passing over the same
mark will increase its intensity.

It is not necessary to have them one before the other for
they may be placed side by side very close so that one pen
shall make a very fine[b] mark and the other will double its size
so as to make it legiable.

Another Idea. In a iron solution (ie) prussiate of potash[1] or other solution with which an iron pen is used to decompose and in which there is considerable acid, and excess of acid may be used so that when no current is passing the acid will attack the pen and produce a mark but when the current is passed through the pen & paper it will prevent the acid from attacking it or prevent the mark whatever it may be caused by. by this means the resistence of the paper may be materially reduced on account of the possibility of using an excess of acid =

To produce this effect it would be necessary to arrange the Transmitting & Receiving Machines in this Manner

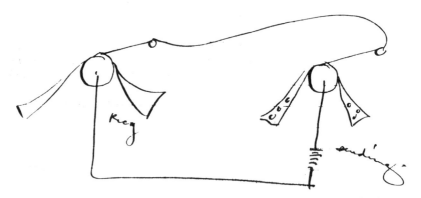

I will now illustrate and describe several methods of arranging the electrical Connections and batterys for sending and Receiving =

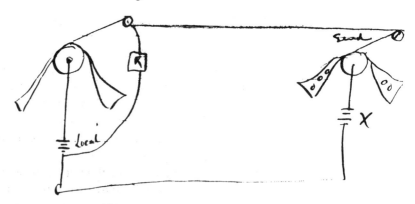

In this Case The Local is used to decompose The Chemical paper and the Main battery X to neturalize its effect, by means of the perforated paper— R is the resistence Coil for adjusting The Local so that the battery X will exactly neutralize it

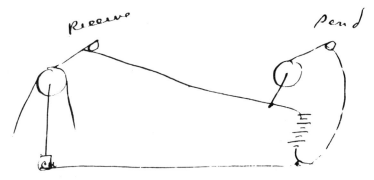

In this Case the Current or rather circuit is not broken but the battery is cut in & out by The perforated paper & pen

This Can be used with a local

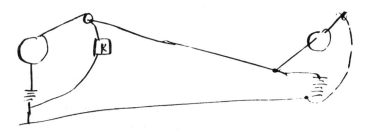

or with an induction coil   the circuit not decomposing direct but acting in the ~~see~~ primary coil induces sufficnt current in the secondary to decompose

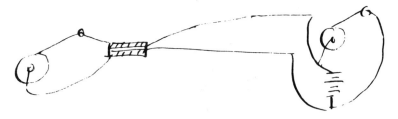

or the induction coil may be used thus

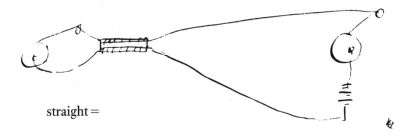

straight =

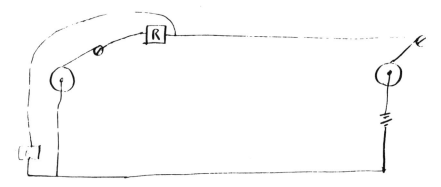

2 Resistence boxes adjusted so as to split the Current.

Local which decomposes and main Line battery which also decomposed   R coil in Local circuit to prevent the Local from making too plain a mark   The use of this Local is to electrolize the atom of chemical solution to a point where a slight increase would decompose it which is done by the Main

Poles opposite. Main Line decomposing   Local neutralizing. Exceess of main over Local produces mark   when it ceases Local cuts the mark short

fine ~~wheel~~ teeth wheel pen for receiving on chemical paper.

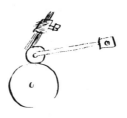

Wheel recording device on chemical paper, with a brush to keep it clean
Witnessed   Jos T Murrey[c]   T A Edison   Invt

AXS, NjWOE, Lab., Cat. 1182:9 (*TAEM* 3:54). [a]"of the two" interlined above. [b]"very fine" interlined above. [c]Signature.

1. Potassium cyanide (KCN).

–228–

Notebook Entry:
District Telegraphy[1]

[Newark, January 16, 1872?][2]

Here is quite a novel and simple ~~p~~arrangement for performing The ~~arrangemen~~ operation[3]

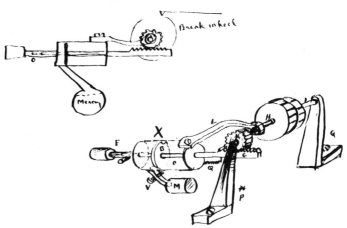

X is a cylinder   B a plunger with a rack G on it working in small gear wheel K on main shaft I which has break wheels H same as described before.[4] Q is a spiral to pull the rack back when at rest   the signals where be provided with devices for carrying only one way   L is an arm to which the cylinder X is fastened   this arm swings on the Main shaft   M is a small cylinder in which the mercury runs when the cylinder is tipped up as shown. Now if we wish to rotate the signal[a] wheels & shaft we take hold of the handle E and let the two cylinders drop by gravity so they stand upright. the moment they do so the mercury from the small cylinder runs ~~into~~

slowly into The Large cylinder, (the rapidity with which it runs being governed by a Taper screw .V.) and acting on the head of the rack plunger slowly raises it and rotates the shaft and break wheels one revolution   Now by returning the cylinder to the position shown in the Drawing the mercury runs by ready for the next signal

I obtain another slow motion of the shaft carrying the break wheels by putting a fan on it and running this fan in a tube filled with Metallic Mercury.

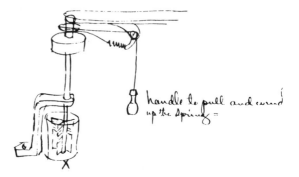

X is the tube filled with Mercury

I shall now describe the means of indicating the signals ~~sent~~ and recording the same in the Central office

As all of the breaks wheels must necessarily stop on a closed circuit   [-]to bring out the signals properly upon an ordinary Morse register a slight alteration is necessary.

The break wheel to give the signal 1 2 5 is made in this manner   one break for 1 two breaks for 2 and five breaks for 5 and then the wheel stops on a closed circuit   NThe signal to be plain should come from the Register in this form

This is done after the following manner

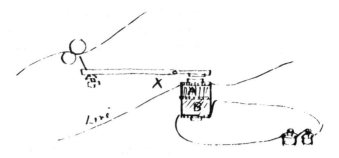

The magnet X of the register has a double set of helices, through one of[b] which ła Local passes constantly and through the other the main   Now when both are closed they neutralize each other, and no mark is made on the paper, but when the break wheel of any machine is started every time an opening ocurrs the main Line current ceases in one of the bobbins for an instant and the Local acting brings the Lever up and indents a dot. when two breaks follow quickly 2 dots & so on

The dots may be recorded in this manner   by polarizing the magnet in such a manner that when the main line current circulats through it, it neutralizes the effect of the polarization for the time being but when interrupted for an instant the polarization of the magnet is sufficient to indent a mark upon the paper

I do not wish to confine myself to an particular Machine for recording in this manner As an Ink recording Machine could be used or Chemical paper etc.

It can also be printed out. I do this in following Manner.[5]

AX, NjWOE, Lab., Cat. 1181:43 (*TAEM* 3:188). [a]Interlined above. [b]"one of" interlined above.

1. This entry is a continuation of Doc. 226.
2. This entry is clearly preliminary to Doc. 229 (17 January), which duplicates some parts of it and refines others; see also n. 1.
3. The devices illustrated here are district telegraph apparatus.
4. Doc. 226.
5. No drawing follows.

–229–

Caveat: District Telegraphy[1]

[New York,][2] January 17, 1872[a]

I, Thomas A. Edison of Newark, N.J. do hereby declare[b] the following to be a full description of Impts. in District and Alarm Telegraphs as far as perfected.[c]

The object of this invention is to communicate different signals, automatically, from a number of machines placed in one circuit and to record the same at a central point, or station. the machines are to be placed in private houses, and are to be used for summoning the police, messenger, though applicable for other purposes.

The invention consists, in a peculiar combination of mechanical devices for transmitting the signal automatically, and a Recording machine for recording these signals at the central station.

The mechanical combinations which would perform this operation are so various that to protect Myself, and to further an enterprise now being carried on, with benefit to the public, I shall describe in this caveat, as many such combinations, which I believe to be my invention.

Fig. A.[d]

In Fig. A.[e] G is the main shaft, S is the double signaling wheel or "break" having a click upon its side (.v.) Q is a ratchet wheel, secured to the shaft G, into which the carrying pawl v lays. this allows the shaft G to carry the break wheel forward but not backward. a second ratchet wheel secured to the break wheel is between Q and .S. into which the pawl R liesays, the use of which is to prevent the break wheel S from going backwards when the spring M is being wound up.

F is the contact spring pivoted on the platen B, one end resting on the break wheel S and the other end having a handle, whose object is to throw the contact spring from one side of the signaling wheel to the other, so that different signals may be made

P is a small pulley over which the cord O runs, M is the spring, which may be either spiral or flat or in fact any shape, and is used to store up the power derived from the hand in the act of moving the handle K from the dotted Lines to the upright position

T is the weight secured to an arm from K by the screw H, and is adjustable so as to allow of an increase or decrease in the speed with which the shaft G is driven.

Operation.

When a signal is to be given, the contact point F.E is thrown over to either one side or the other, according to the signal to be used. Then the handle K is moved by the hand from the dotted lines to its normal position. in thus moving it, sufficient power is stored up in the spring M to give the shaft G and break wheel one revolution which is sufficient to send the signal. It will be seen that the spring M must raise the weight T and arm and restore them to the places shown by the dotted Line and this makes the revolution of the shaft G quite slow =

The ratchet wheel on the break wheel and on the shaft might be dispensed with and the signal or break wheels secured to the shaft, and a pin put in the shaft, which would be arrested by another pin fastened to a lever, as in figure .2.

Fig .2.

and when in the act of winding up the spring M, the cord would slip over the pulley .p. and not rotate the shaft, or the pulley itself maight be provided with a ratchet and pawl, which would carry the shaft forward but not backward

I do not wish to confine myself to any particular mode of governing the speed of the main shaft as I shall explain several ways in which that object can be ~~obtained~~ attained.

Fig. B^d

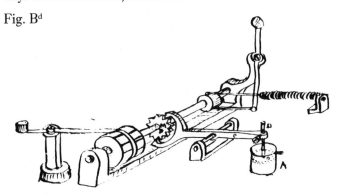

The mechanism in This machine Fig. B^f is similar to that described in fFigure A,^g with the exception of the governor, which consists of an air cylinder A within which is a plunger

B secured to the lever c by the rod D and set screw. The rod c is provided with two prongs placed in the path of the V shaped tooth wheel M secured to the main shaft. when the power is applied to rotate the main shaft ~~l~~an up and down motion is communicated to the lever c and piston or plunger .B., and as this plunger head has considerable area it retard the revolution of the main shaft. The plunger if moved slowly expels the air quite easy and takes but little power to move it ~~w~~but when the attempt is made to work it up and down rapidly, the power required is greatly increased

The air principle might be applied directly   Thus:— Fig. C.[d]

Fig. C.[d]

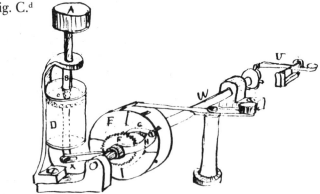

D is the air cylinder, c the head   B the plunger rod   A the handle and weight   F a fine tooth ratchet or milled wheel   H an arm moving loose on the shaft .W. and connected to B by ~~the~~ a[b] pin & slot   G is a pawl laying on the ratchet wheel   E is the break wheel, divided in four segiments each one containing the ~~s~~complete Signal. Now when the plunger Bc is raised by the hand and the shaft W released by the arm U The weight A acting by gravity forces the plunger down slowly and the break wheel is rotated a quareter of its circumference which is sufficient to give the required signal.

I do not wish to confine myself to a weight as a spring, magnet or other power may be applied to force the piston down or up as the case may be.

Neither do I wish to confine myself to that form of break as it can Be made thus   Fig. D.[d]

Fig. D.ᵈ

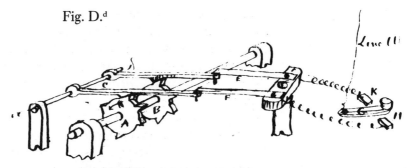

A and B being cogged wheels and raising and lowering the rods c D and making and breaking the current on the points E.F. the signal being selected by the electrical switch G and points H and K and the wires.

Fig. Eᵈ

Another form of break may be used. In Fig. Eᶜ The wheel A ~~B~~having an irregular groove corresponding to the signal to be sent and within this grove is a pin connected to an upright lever B   C is a connecting point. when the wheel A is revolved a peculiar motion is given to the upright B, which closes and opens the circuit at the point C as many times as the groove in A is thrown out of Line.

Fig. Fᵈ

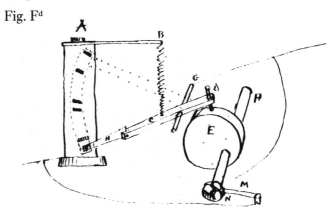

Fig. F.ᵈ is another form of break   A is a pillar of insulating substance with the signaling character sunk in this substance and all connected together with the Line. C is the contact Lever with springy point H. This lever is pivoted at .G. E is a cam wheel revolving with the shaft .F. D is a pin rubbing on the periphiery of .E. at every revolution of the wheel E and shaf F the rod c is carried down by the wheel and up by the spring B, the whole length of the pillar A   in going up the spring H rubbing on the contact points of the pillar A transmitts the signal but in coming down is prevented from opening the circuit by the wheel n and spring m which closes the circuit when coming down but allows it to break going up by a peice of insulating substance forming part of the wheel.

The pillar might have two sets of teeth set in the same line but one set connect with a strip of metal, and the other the same way, and a switch could be used to select or determine which signal is to be used

Fig. G^d

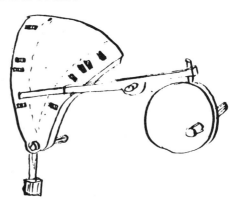

A disk might be used which would swing on a pivot and a handle used to select the signal to be transmitted as in Fig. G.^{g}

There are several modes or devices by which the signals can be se[-]lected, one of which is to shift the break wheels, so as to bring one wheel or the other under the same point.

A Another way is to have both wheels loose on the main shaft and throw a clutch from the shaft to the wheel when a signal is to be sent and arresting it at zero and throwing the clutch in an the other wheel and so on.

Or One single wheel can be used, two disks interlocking each other band seperated by a disk of hard rubber, and one spring used. either disked could be switched in or out by a switch, or changer

The machine might have a paper carrying drum, instead of the break wheel, and perf the signal or in fact a great many signals previously prepared might be selected and run through the machine.

Or if only two signals were to be used the two strips of perforated paper could be pasted on a smooth wheel and the breaks made through the paiper..

I will now describe another novel device for obtaining the slow motion necessary without the use of a number of gear wheels.

Fig. H.[d]

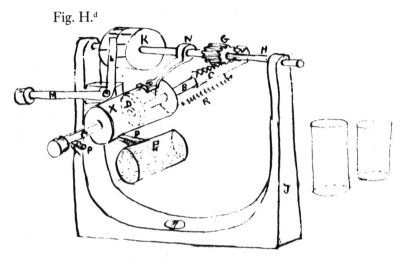

In Fig H.[e] X is a cylinder in which is a plunger D with rod B on the end of which is a rack working in the gear wheel G on the main shaft H. K are the double break wheels   LM the contact spring   The cylinder X swings on the main shaft H by the lever .N. A is a handle used for raising the two cylinders to the peg P. E is a resevoir of metallic mercury connected by a spout .F which also holds the two cylinders together. The resevoir is nearly filled with mercury and is higher up than the other cylinder. Now When the two cylinders are in athe[b] position shown the mercury runs through the pipe F into the resevoir .E and entirely out of X and the spiral spring R draws the plunger D down nearly to the mouth of the spout .F. If now it is desired to give a signal the handle A is lifted off of the peg P and the two cylinders allowed to drop by gravity. the moment this takes place The mercury runs out through F into X under the ehead D and the rack is raised slowly, and giving the main shaft H one revolution, its slowness being regulated by the size of the hole through the tube F. When the signal has been given the handle A is brought back to P by the hand and the mercury runs back into the resevoir, ready for another signal. I do not wish to confine myself to any particular devices for getting this slow motion from mercury, for this purpose, but claim the use of mercury falling by gravity from one resevoir to another and acting on a head to give motion   the flow of the mercury being regulated either by a cock in the tubes or by a greatly increased area of the piston head operated upon

Another plan to get a slow motion or rather a device for governing the speed of the main shaft is to connect several paddles or leaves of metal to the main shaft and Work them quite deeply in a small cylinder of mercury   Thus. Fig. I.ᵈ

Fig. I.ᵈ

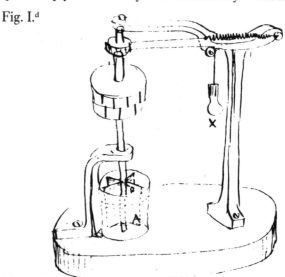

A is a sealed glass tube containing the main shaft and fan .B. and filled nearly full with mercury. When the handle X is pulled down, and the spring wound up, the main shaft will revolve very slowly by reason of the immense resistence of the mercury—

I will mention here that in some cases it will be impracticable to open and close the main circuit to transmit a signal to the central office but the signal can be sent at the same time preserve the continuity of the line intact   Thus. Fig. J.ᵈ

Fig. J.ᵈ

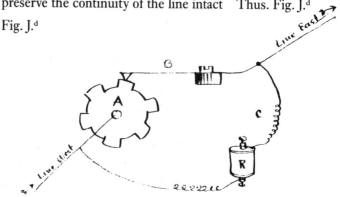

Fig. K^d

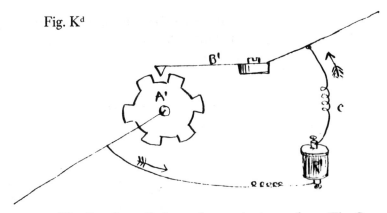

The first figure J^h shows the continuity perfect   The Second figures K^h shows the continuity perfect but the current sufficiently weakened by passing through the Resistence Bobbin to produce an effect at the Central Station. consequently when the wheel A' is rotated the strength of the current will be ~~S~~Increased & decreased as the teeth of the wheel A' passes under the contact spring B'

I shall now describe the devices by which I intend to record the signals   The first device is a common ~~m~~Morse register, as the break wheels must always stop on a closed circuit the characters would be difficult to read if received in the ordinary manner   Now to signal 1 2 5. the break wheel is constructed thus   Fig. L^d

Fig. L^d

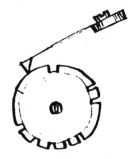

One opening for 1 two openings for two and five openings for 5. This would be received upon the register thus   fig M.^{d}

M.^{g}

Now this is quite perplexing to read and very liable to error.

When the following device is used the signal shows on the strip thus   fig. N.ᵈ

Fig. N.ᵈ

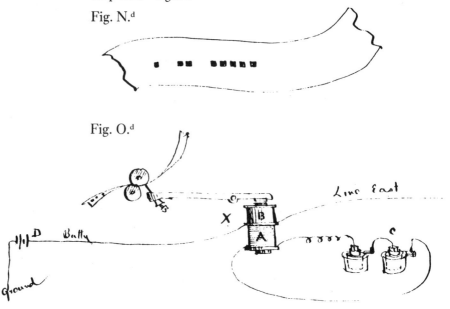

Fig. O.ᵈ

The magnet X of the register Fig. O.ᶠ has a double ~~core~~ helix or bobbins, the Local C passing through A and the main Line through B. Now when both are closed they neutralize each other and no mark is made, but when a break oceurrs in the wheel, the equilibrium is destroyed and the Local battery acting on the magnet draws the lever up and indents the paper, and when two breaks occur close to-gether two short dots are made and so on   I do not confine myself to using a local, as the extra bobbin and local may be dispensed with and a permanent magnet substituted (ie) the magnet AX polarized in such a manner that when the main circuit is closed they will neutralize each other, and no effect be produced but when a break occurs in the main line, the permanent magnetism will be sufficient to indent the paper.

An ink recorder may be used in this manner or AChemical paper. I do not wish to confine myself to any particular method of marking, but claim the manner of recording the "breaks" instead of the closure as is usual.

A relay may be used with a local   thus   fig. Pᵈ

Fig. P^d

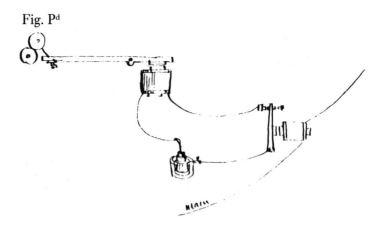

The back point of the relay closing the local through the register.

Signed by me this 17th day of Jan. AD 1872[i]

T A. Edison

Witnesses[j]   Geo. T. Pinckney[k]   Chas H Smith[k]

ADS, DNA, RG-241, Edison Caveats 37. Petition and oath omitted. [a]Date taken from text, form altered. [b]Interlined above. [c]"I . . . perfected." written by Charles Smith. [d]"Fig.", "fig", or "fig.", and letter designation written in an unknown hand. [e]"In Fig." and letter designation interlined above in an unknown hand. [f]"Fig." and letter designation interlined above in an unknown hand. [g]Written in an unknown hand. [h]Interlined above in an unknown hand. [i]"Signed . . . 1872" written by Pinckney. [j]Written by Pinckney. [k]Signature.

1. Although Daniel Craig had been pushing Edison's work on the district (domestic) telegraph in 1871 (see Docs. 143, 200, and 203), Edison assigned rights to this caveat on the day of its execution to Elisha Andrews and Horace Hotchkiss, reserving for himself the rights for "business and commercial telegraphing and electro-mechanical purposes." Andrews and Hotchkiss were officers of the American District Telegraph Co. (see Doc. 226, n. 1) and immediately signed their rights over to the company (Digest Pat. E-2:192). Lemuel Serrell, Edison's patent attorney, renewed this caveat twice (Serrell to Commissioner of Patents, 14 June 1873 and 24 Mar. 1874, Edison Caveats 37, DNA).

2. Much of the text of this caveat was taken from the laboratory notebook entries for 15 and 16? January (Docs. 226 and 228). Employees of Serrell's New York office edited the text and witnessed Edison's signature.

Notebook Entry:
Printing Telegraphy[1]

[Newark,] Jany 17, 1872.

Escapement

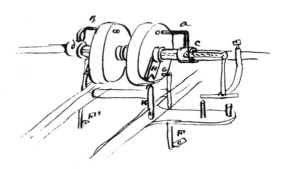

Two Loose Type wheels   clutches a B c d connected together by ~~ar~~two arms running in slots planed in the shaft
Type wheels are provided with two holes into either   the
clutch Can be thrown by the shifting devices   H is an arm
with a V cut in it   underneath on the Ptg lever is an arm
K which goes in this V in H[2]   The same is on the other
wheel   F F′ is a retarding spring   G etc is a pin from type
wheel to hold it in position when it is unlocked or not carried
around   The Operation is simple. If one wheel is locked by
shifting the other wheel will stand one letter from zero but the
arm K on the printing lever ~~thro~~ acting in H cams the type
wheel one notch to zero in the act of unlocking and the type
wheel which is thus unlocked and cammed forward is prevent
from wiggling ahead by the spring F & pin G, but when this
wheel is again locked, the force of the Magnet is plenty to
overcome the spring F and the type wheel is rotated.
Witnessed = Jos T Murrey[a]   T A Edison

AXS, NjWOE, Lab., Cat. 1174:34 (*TAEM* 3:24). [a]Signature.

1. This entry is a continuation of Doc. 225 and is continued in Doc.
232.

2. In a design submitted in a patent application on 28 May 1872 (U.S.

Pat. 131,341), Edison used a variant of the arms **K** and **H** to shift a single typewheel alternately to letters or numbers.

[Newark,] Jany 18 1872

A Chemical printing Telegraph.
This is a novel affair.

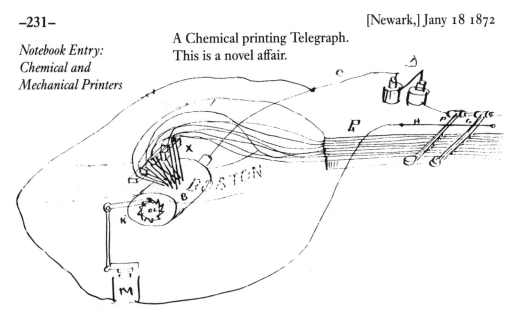

X is a bunch of flat platina wirers secured to insulating substance and Each wire insulated from its neighbor—and the Ends brought down on the platina faced drum B very close   so close that but a thickness of a peice of tissue paper seperates any of them   they make a solid mass about ⅛ of an inch square =

The chemically prepared paper passes between this brush square X and the drum RB. This drum is fed ahead after evey letter is made by the keys coming in connection with the rod H   This throwing the battery D through the magnet M as will be seen by the wires.

Now beneath the bank of keys are twenty five rods or as many rods as there are platina points in the pen square .X. Each one of these bars being connected to a platina pen forming the brush   The bottom part of the lever or keys having tits or teeth on and these teeth are so arranged that they will connect the battery D through a given number of rods necessary to form a letter to the points ⊖in X and the current will pass down these selected points Through the chemically prepared paper to the drum thence to the battery D leaving a blue or other colored mark at each Point   For instance the letter H would be thus

The first row of points would all mark & leave (a)   The second one dot of c   the 3rd another dot of c   the 4th another do & the fifth row all the marks

The End view of the square is thus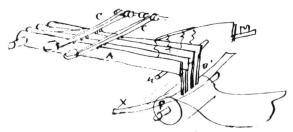

ain practice it is possible to place them much closer than this   it will be easily seen that certain points Can be selected to given an letter figure & sign in any language   Now by raising the key the battery is thrown off of the decompossing points and ~~through~~ Thrown through the magnet M which advances the Drum B ahead ready for the next letter.

I do not wish to confine myself to printing on a strip of paper, as the Brush X might be made to travel over a ~~st~~wide strip of paper side ways & then back and at evey step decompose a letter   or 40 more or less brushes migh be used and a commutator made to throw the electrical connection from one brush to another clear across the paper and after a line had been printed the paper could be advanced and the commutator made to throw the current back to the first brush & thence step by step to each brush until another line had been printed   I could dispense with the magnet ~~d~~to draw the paper and substitute any power manipulated with sutiable mechanism and controlled by the key or keys. The bars might be dispensed with and wires substituted or any arrangement by which any given key could select the required number of recording points to make its letter

I could dispense with the chemicals and record with ink in this way

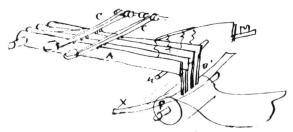

n is a stop to pevent ~~d~~ too deep an impression   X the ink band

A are the rods 25 more or less in number arranged like levers underneath the keys   these ~~e~~Levers have on their ends the ~~in~~ Printing wires are arranged together in a bunch making a ~~s~~almost solid square each fine wire being connected to a lever and working seperately from all the rest

The End of all the wires attached[a] to the levers would appear this way ☞

The word ship being printed thus,
But somewhat closer and legiable. the keys over these levers could have teeth arranged to sellect the levers or the levers could be provided with teeth which I think the most convenient.

The paper would be fed by a movement derived from the Movement of the key or controlled by such movement ~~The way I ink the Ends of~~ in this machine I used an Ink band passing ~~beneath~~ between the marking points and the drum or rather paper which passes over the drum   This Ink band is fed along by a movement derived from the key or may be driven ~~f~~by magnetism etc.

This principle of forming letters and printing them on strips of paper or other substances by the depression of keys I believe to be original with myself

I have shown the principle features here   the small devices are of no consequence being but ordinary movements well understood and easily Combined.

The ~~end~~Ordinary perforating machine of my invention now in use could be provided with 25 Bars similiar to what is called punch bar but much thinner and ~~t~~a cushion in placed of the punches & die used and the ends of these levers being provided with the printing points closely gathered and tits put one the Bars so that the depression of the key would carry forward the required number of ~~pins~~ bars to form that letter correspdg to the key[1]

[Witness:] Jos T Murrey[b]   T A Edison   Inventor Edison

AXS, NjWOE, Lab., Cat. 1182:13 (*TAEM* 3:56). [a]Followed by line indicating a page turn. [b]Signature.

1. Edison subsequently altered his original large perforator design (U.S. Pat. 121,601) to incorporate this arrangement (U.S. Pat. 151,209).

–232–

Notebook Entry:
Printing Telegraphy

[Newark,] Jany 18 1872
A Printing Machine in which the Engine carries around the type wheel, and the escapement lever on releases

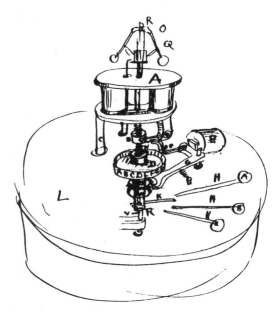

A is the Engine set up on pillars C etc   R is the shaft driven by the Engine   Q is the governor. D[1] is the break wheel arrested in the manner before described by a connecting link ff F from the escapement lever G[2]   This break wheel is arranged just the same as heretofore described[3]   D is the type wheel   N the friction[4] which carries this type wheel arrond [-]when the Engine shaft R is moving   This type wheel is provided with a long sleeve W which passes on the shaft R down through the Base nearly to where R is pivoted it being prevented from going farther pby the pin V or any devices. (The friction might be placed here)   now this sleeve has on the release toothed wheel M the vibration of the escapement G releasing the type wheel after the manner of a clock escapement.

This sleeve is provided with an arm K which passes over the path of keys and may be arrested by the depression of any key   Now when a key is depressed the Vibrating lever G is held in such a posstion that the arrester F is thrown into the path of the pin on the break wheel and arrests it and the machine at the other end is arrested by the stoppage of the break wheel and the consequent stoppage of its vibrating lever and release of type wheel

The break wheel might be Connected directly on the type wheel in this Case and the lever F dispensed with, this simplyfying it, but it would have to have 30 breaks or as many as theyr are lchatrs in type W

The Escapement might be dispensed with also and a simple magnet with vibrating arm which Vibrates in the path of the pin on the break wheel. When the key was depressed on the transmitting instrument it would arrest the break wheel and type wheel on an Open circuit, consequently the vibrating arm at the distant station would be arrested and arrested the break wheel & type wheel of that Machine and when the key was raised the sending machine type w & Bk wᵃ would immedy start off and the breaks be sent over the wire & the Magnet & its arrestedr commence to vibrate. This arrestedr is not only used to arrest the break wheel but to prevent the re'cgᵇ type wheel from going faster than the sending[5]

Witnessed   Jos T Murreyᶜ   T A Edison

AXS, NjWOE, Lab., Cat. 1174:35 (*TAEM* 3:24). ᵃ"type w & Bk w" interlined above. ᵇInterlined above. ᶜSignature.

1. Should be **B** (above typewheel **D**).
2. See Doc. 222.
3. See Doc. 208.
4. By "friction" Edison meant a spring around the shaft compressed between typewheel **D** and breakwheel **B**. There was another above the breakwheel. This arrangement rotated **B** and **D** with the shaft, while allowing the shaft to turn when **B** and **D** were halted.
5. In this arrangement, the motor ("engine") of the receiving instrument would drive that instrument's typewheel.

–233–

Notebook Entry: Printing Telegraphy

[Newark,] Jany 20 1872ᵃ

For Printing fractions, the type wheel be made thus

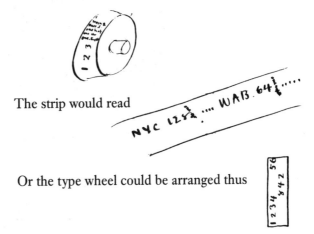

The strip would read

Or the type wheel could be arranged thus

The 8.4.2 being cut in the type wheel where the spaces should be[1] and to print them stop on a closed circuit = [2]

or it Could be made thus[3]

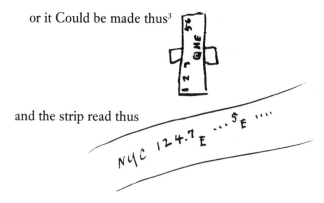

and the strip read thus

AX, NjWOE, Lab., Cat. 1174:37 (*TAEM* 3:25). [a]"Stock Quotation Printer" written by William Carman at top of document.

1. The offset numerals 8, 4, and 2 were intended to be denominators of fractions (e.g., ⅜). Edison and Pope had arranged the gold-printer typewheel this way.

2. Edison's stock printer normally printed only when the typewheel circuit was open. See Doc. 207.

3. On this wheel, **Q** = quarter (¼), **H** = half (½), and **E** = eighth (⅛).

–234–

Notebook Entry:
Mechanical Printer[1]

[Newark,] Jany 27 1872

Printing Machine.[a]

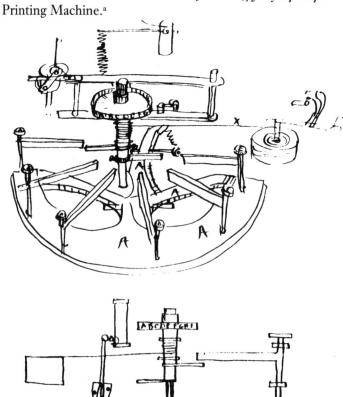

Printing Machine.

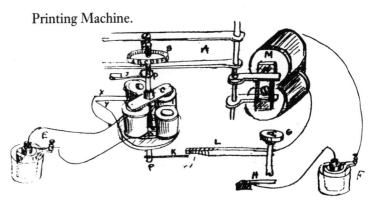

The object of this invention is to print rapidly ⊖in roman Letters upon a a continuous strip of paper.[2]

B is the type wheel   P is the shaft running between centers, & Through the bed of the Engine. C is a the engine revolving Armature. x y the engine break sp[rin]gs   E the battery   This armature C revolves with tremenduous rapidity and Carries the shaft P and type wheel B with it by friction, and when the shaft is arrested by the depression of the key A & arm L and the pin K the armature keeps right on revolving and when the key A is raised the shaft is free and the friction of G & C grabs the shaft P & Carries it forward

A represents the key of which there are 30 more or less. These keys are arranged around in a circle and have a Connection ring of which H. is a part and shows the operation   This ring is connected with one end of the battery to the printing Magnet M & thence to keys So that when any key is depressed the bottom comes in Contact with the ring represented by H and the printing magnet and the Printing is effected. TOwing to the eImmense rapidity of the revolution of the type wheel it is sure to be arrested before the key closes on the ring and the Magnet charges.

I do not wish to confine myself to this particular form as it may be necessary to adapt some devices to ensure the arrest of the type wheel before the printing magnet is closed   this may be done by putting on a break wheel aon the type wheel shaft and arresting it on a closed circuit, or the Circuit may be closed on the key A at L and the pin .K.

What I claim is rotating a type wheel and Shaft by Magnetism, and Catching it by the dKeys and Printing by Magnetism or by any other forces; and allowing the engine to go on after the type wheel has been arrested.

J is a ratchet wheel pawl which works in a ratchet wheel secured to the Shaft .P. The use of which is to prevent a rebound   when the pin K is arrrested by the key arm L of the key, the ratchet pawl[b] after the pin strikes L falls in a tooth and holds it there while the letter is being printed.

I do not wish to confine myself to this mode of ~~ree~~ stopping the rebound, as the rebound may be taken up by a roughened wheel in place of the ratchet wheel on shaft .P. and a pawl with a roughened surface connected to the printing lever which when the lever comes forward to print the pawl would give a forward motion to the type wheel shaft P and bring it to the stop L and take up the rebound thus

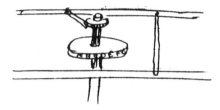

[Witness:] Jos T Murrey[c]   T A Edison

I do not wish to confine myself to magnetism for driving the type wheel shaft as a train of Gearing actuated by a spring Weight or any other force could be made to rotate the shaft, and the printing could be Effected by throwing an arm ~~into~~ from the printing lever into the path of pegs on one of the gear wheels or the motion to print could be derived from the downward motion of the key =

AXS, NjWOE, Lab., Cat. 1182:17 (*TAEM* 3:58). [a]Followed by "Typewriter for aut", written by William Carman. [b]Interlined above. [c]Signature.

1. This entry is continued in Doc. 235.

Two early experimental samples from Edison's typewriter.

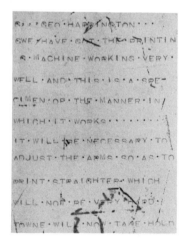

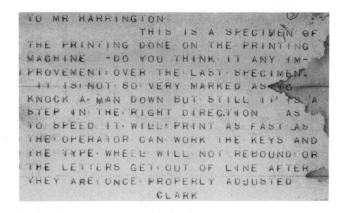

2. Here Edison has turned his universal private-line printer into a typewriter ("hand printer") by omitting the transmitting mechanism. Edison developed this machine during February and March, and on 18 April he submitted a patent application based on the second "printing machine" in this document (U.S. Pat. 133,019). Cat. 1211:42–47, Accts. (*TAEM* 21:165–67).

–235–

Notebook Entry:
Mechanical Printer[1]

[Newark,] Jany 29 1872

I have this Printer arranged in this form now

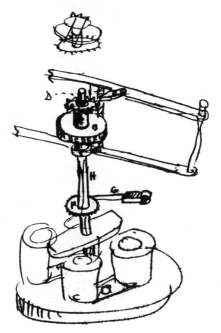

H the shaft. P & G the device previously described to prevent the rebound ᴮC a sleeve on which is secured the type wheel B, star correcting wheel a held up by the collar .X. E a detent for correcting position of type wheel. The sleeves & its devices is connected and Carried around by the shaft .H by a slot in the sleeve & a pin in the shaft shown at .D. this slot is larger than the pin and allows some dead motion so that when the shaft is suddenly arrested by a key and it should rebound so that the pawl G did not happen to fall in the right tooth the type wheel being a little loose the detent eE would bring it right when the printing lever came forward

The speed on this printer might be increased by Cutting a double Alphabet on this wheel without increasing its size and double the speed could be obtained in printing without increasing the speed of the rotation of the shaft H but the letters would have to be a little smaller.

I may find it necessary to attach the interupting break wheel to the main shaft to interrupt the printing lever current so that it cannot come forward until the type wheel has stopped. I used this on my Printing Machine the patent of which has already be applied for =[2]

T A Edison

AXS, NjWOE, Lab., Cat. 1182:20 (*TAEM* 3:60).

1. This entry is a continuation of Doc. 234.

2. U.S. Patent 133,841, for which Edison had applied on 13 November 1871.

–236–

Notebook Entry:
Chemical Printer and
Automatic Telegraphy

[Newark,] January 30, 1872.
I will now describe another printer worked chemically

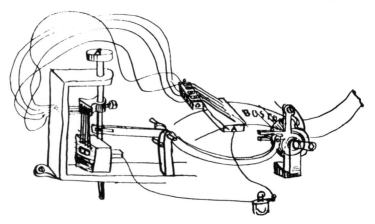

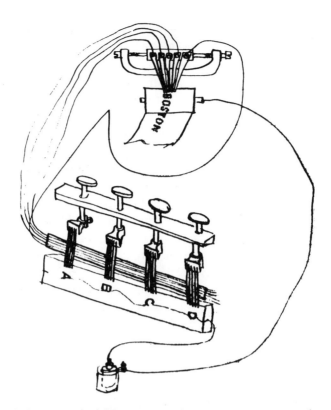

A description of this apparatus is not very necessary   there are 30 keys   Each key is provided with a comb ~~and under~~ of five contact points   underneath the keys is a long peice of hard rubber and metal letters of the Alphabet sunk in it flush with the top. ~~Each~~ there is a recording combs also of five wires   rests on chemically prepared paper rotated by a drum which is driven forward by the depression of the key   Now the first wire in every key come are all connected together   the second wire of every comb is also connected and so on   now the first wires of all the combs being connected together they are then connected to the first wire of the recording comb the second to the second   Now all the metal letters are connected together and attached to a battery and the drum is attached to the other end   So that by depressing a key its comb passes over the metal letter and ~~advanc~~ makes contact sending necessary pulsations to the recording comb to produce that particular letter after the manner of the Bonneli Type Chemical Telegraph = [1]

While the letter is being made the paper advances by the action of the key   The letters might be placed on the key and the combs be made stationary =

The recording pen might be made to travel back and forward over a sheet of Manuscript paper and Make a MSS Copy   Thus

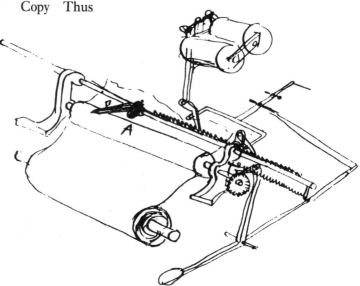

BA is the main roll   B Travelling pen   the rest is easily understood = [a] The usual devices can be used similar to that used on my Mechanical printing Machine.

This is a new movement on my perforator for throwing the punch bars forward

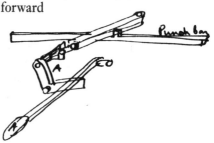

A goes forward by depressing the key until the round end of the cross bar goes over the top then the punch bar is thrown back by the action of its springs carrying the cross bar with it   Now when the belbow A comes back it throws the cross bar upward by reason of the incline on the elbow & passes under it there being enough dead motion in the brass bar to allow of this

The Cross bars have heretofore pBeen pivoted or had their fulcrum of a shoulder screw the thirty being secured on one long bar   Now I propose to dispense with drilling & taping 30 holes and using thirty screws by casting slitght tits or bosses on this bar & milling them up and putting the cross bars on them & preventing them from coming off by fastening a plate over the whole of them   thus

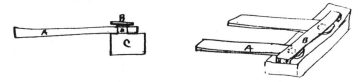

A the brass peice   D the boss or tit   C the long bar the tit & it being Cast in one peice   B is the plate which prevents these Cross peices from slipping off of the bosses

I propose to try the experiment of dispensing with a spring to each key in my perforator of which there are Thirty = by using a friction plate and have small pins on the keys and when the key rises suddenly it comes up sideways against this plate and puts a slight friction on it sufficient to hold it there until agains depressed

Now all the keys being held up this way I put a rod extending underneath the whole of these keys and connnect a couple of springs to it and when a key is depressed it carries this bar with it and when let rise the spring on the bar throws it up and the key with it.

On paper feed for the perforating Machine I propose to alter it by putting in a Swivel on the paper feed dog. Thus

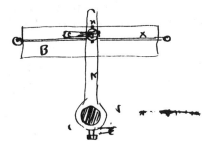

B the stage   N the swing lever connected to the rock shaft   this Lever performs an arc of a circle   n is a ~~swil~~ swivel into which the click [C] is pivoted   X is a guide pin to keep the click straight

Device to get a constant spring motion of the rock shaft of my perforator. Thus

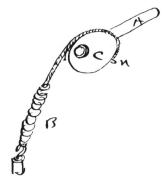

A is the rock shaft   B a spring   C a cam or eccentric shaped wheel   n is a pin to which is secured the cord connected to the spring B.

Now when the shaft is rocked the spring gets ~~col~~ closer to the fulcrum by reason of the eccentric wheel   I go to bed! [Witness:] Jos T Murrey[b]   T. A. Edison

AXS, NjWOE, Lab., Cat. 1182:22 (*TAEM* 3:63). [a]"under" interlined above. [b]Signature.

1. On the facsimile telegraph system of Gaetano Bonelli, see Doc. 46.

–237–

Notebook Entry: Printing Telegraphy

[Newark,] Jany 30 1872

On page 27 I describe a printing machine by which the type wheel returns to zero eveytime a letter is printed.[1] I will now describe ~~a~~several way in which it can be done.

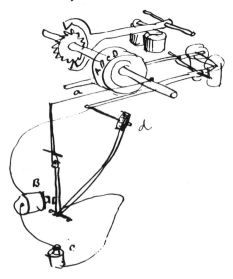

B a magnet[a]   a a cord[a]   d ungers[2] break[a]   When the type
wheel has been rotated to proper position the printing lever
come up   effects the printing   closes the circuit in the relay
and the Relay lever goes forward   tightens[b] up the Cords and
pulls the wheel & shaft back to zero   Again!

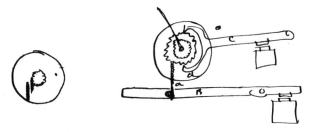

a a rack[a]   B printing lever[a]   c escapmt lever[a]   d type
wheel   when printing lever goes up rack goes into teeth cut
in shaft and throws it back to zero which is a limiting pin[a]

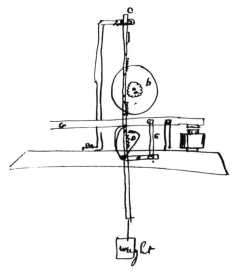

G[b] the printing lever[a]   b the type wheel[a]   a the shaft with
teeth cut in it[a]   C an intermittent rack pulled downward by a
weight and released by the printing lever moving the pallets D
by aid of bar F and link E[a]   Thisese pallet release pins on
side of .C. the action of this mechanism is easily seen.
Witnessed   Jos T Murrey[c]   Thomas A. Edison

AXS, NjWOE, Lab., Cat. 1173:29 (*TAEM* 3:92). [a]Followed by "(x)" in
an unknown hand. [b]Preceded by "x" in an unknown hand. [c]Signature.

   1. This description actually appears on page 22 of the notebook. See
Doc. 194.
   2. Nothing is known of William Unger's breakwheel.

Notebook Entry:
Printing Telegraphy

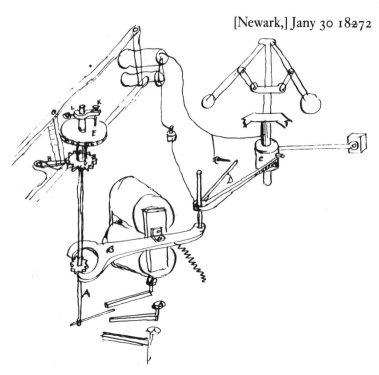

Device[a] or attachment which I use on my universal machine for printing Correctly when the type wheel is stopped by a key a Little out of Line.

In my universal printing Machine, when sending the type wheel shaft is arrested by a key on the open circuit and a Little before the escapement lever throws the type wheel to its proper position for printing. I do this to obtain a slight move-ment of the lever B after a key is raised so as to release the break wheel C by the arm .D, but in doing so the type wheel providing no devices for prevention were used would be stopped half way between a letter and when the printing lever came up would print ½ of one letter & ½ of the other Besides and objection would arise which is that it would be necessary to move the type wheel forward a half a letter before the circuit of the printing lever would be opened Now this it couldnt do Consequently the whole machine would stay Locked of course the printing lever might be thrown forward and the momentum print the letter & fly back out of the road,[1] or a device could be used so that when the type wheel was arrested the printing circuit would close for an instant print the letter & cut itself out until the next letter.[2] I have used it o such an arrangement on a previous model of the Universal but consider them all objectionable   Now in this devices now being described the key arrests the shaft A ½ way home consequently the type wheel is not in a correct position to print. now to make it print correct I use a corrector G on a sleeve F on which is type wheel F³ and guide pin K and holder L   the slot in the holder has dead motion enough to allow the type wheel & its sleeve a loss motion or play of more than a half a letter. H is the corrector detent, wedge acting & connected rigid to the printing lever. If now a key is depressed the shaft A is arrested ½ way home and the printing lever coloses   the detent H acting in the wheel G the takes up all the dead motion of the type wheel on the shaft and throws it forward to the right position for printing. Now the shaft after the key is raised can go forward & open the circuit of the printing lever without moving the type wheel   this allows the printing to fall back just in time to let the type wheel go forward   In receiving nothing is arrested   consequently the type wheel comes right every time or if not the corrector will correct

The correcting wheel and detent should be made thus

This shaped tooth allowing the detent to correct the wheel and then go ahead a little farther to allow the printing lever to print without moving the wheel.

I do not wish to confine myself to any particular devices for correcting the wheel but claim a dead motion of the type wheel upon the shaft so that a forward movement or backward movement can be given that type wheel by the action of the printing lever

Lock devices⁴

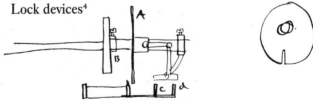

A is a disk. B the type wheel   c d the shifting pins   When you lock   rotate shaft so as to bring the Shift fork under d   close printing lever   this throws the disk over on printing lever and prevents it from coming up   to unlock advance to c & close ptg lever[b]

[Witness:] Jos T Murrey[c]

I mention here that on my universal the click on the printing lever which corrects the type wheel should be a little springy at least something should give. I have so fixed it already=

AX, NjWOE, Lab., Cat. 1174:38 (*TAEM* 3:26). ᵃ"Univ Pr line" written by William Carman above drawing. ᵇFollowed by centered horizontal line. ᶜSignature.

1. See Doc. 193.
2. See U.S. Pat. 123,005.
3. The letter F should be E.
4. This design is for the universal stock printer. The sketch at the right is of disk **A**, showing the notch through which the protrusion on the printing lever passed when the disk was shifted.

–239–

Notebook Entry:
Printing Telegraphy[1]

[Newark,] Jany ~~2~~ 31 1872

New Device on Universal

I use the correcting device,[2] and closing the Local Ptgᵃ circuit both on the vibrating spring and on The arrester, But combine another point of closure for the Printing circuit when Transmitting   Underneath all the keys is a plate   This plate is insulated and so connected to the switch which I use that when Transmitting The Local of the printing circuit cannot be closed until the arrester comes in contact with the pin on the break wheel and a key is depressed & in connection with the Insulated plate before mentioned   Now when the key is raised the circuit of the printing lever is broken before the arrester arm is allowed to release the wheel [-]so the printing lever gets away from the type wheel before it starts which is necessary as the Printing lever pressing against the Type wheel puts a large amount of friction ~~in it which~~ on the shaft and the power of the spring is not sufficient to pull the escapement lever home and rotate the type wheel without it is very strong   consequently it sometimes sticks and the break wheel is not released   by breaking and closing the Local Printing circuit in two places it allows the ptg lever to fly back before the type wheel starts   Consequently there is no friction on the shaft when this takes place

There is one disadvantage to this and that is the great number of points of contact

I have dispensed with this and use it Mechanically   thus

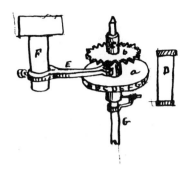

G is the shaft which is rotated by the escapement   C is the collar or sleeve on which is the type wheel a & corrector wheel b   Now this sleeves is larger than the shaft and consequently will move sidewise slightly   D is the printing [---]lever— E is a rigid arm having a round forked end around the sleeve and is just long enough to throw the whole ~~sleeve~~ dead motion or shake in the sleeve towards the printing lever.

Now when the printing lever is closed the whole pressure comes on the arm E and not on the shaft G   consequently when the escapmt acts on the shaft G to rotate it there is no friction or pressure ~~of the~~ from the Ptg lever.

[Witness:] Jos T Murrey[b]

AX, NjWOE, Lab., Cat. 1174:41 (*TAEM* 3:27). [a]Interlined above. [b]Signature.

1. This entry is a continuation of Doc. 238 and is continued in Doc. 243.

2. See Doc. 238.

–240–

Notebook Entry: Relay

[Newark,] Feby 1 1872

Relay

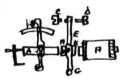

A is a permanent magnet acts on peice iron ~~o~~H on armature E   F is Line magnet   K its armature   Lever E pivoted at G plays between two points C.D. B is handle for adjusting permanent magnet closer to its armature.

The object of this invention is to dispense with a spring weight gravity etc and adjust with a peice of magnetic steel.[1]

I get around the page patent[2] in this ways not opening the circuit[3]

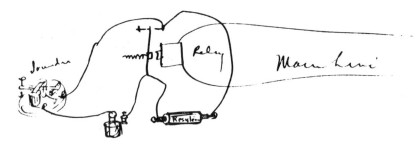

When the lever flies back the circuit is only lengthened (ie)
its resistence increased

Another way

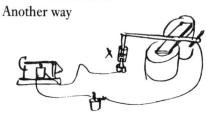

X is a tube the points being in water   when the points are
seperated the circut only passes through the water but is not
broken =

I claim increasing & decreasing the resistence of a local[a]
circuit by the action of the relay of another circuit, or mechanically increasing & decreasing athe resistence of any circuit by
resistence thrown in & out by a break or contact points to produce effects at a distance on that or other circuits[4]

[Witness:] Jos T Murrey[b]

AX, NjWOE, Lab., Cat. 1181:16 (*TAEM* 3:182). [a]Interlined above.
[b]Signature.

1. Telegraphers had to adjust relays often to compensate for changing
conditions on telegraph circuits. See Doc. 30, n. 1.

2. In 1868 Charles Page received U.S. Patent 76,654 for an
"Improvement in Induction-Coil Apparatus and in Circuit-Breakers."
Claim 14 of the patent covered the "employment of one electro-
magnetic instrument to open and close the circuit of another electro-
magnetic instrument"—that is, a telegraph relay. Western Union bought
a controlling interest in the patent in 1871 and filed infringement suits
against other companies for using relays. Post 1976, 171–81.

Edison is here working on the automatic telegraph system for which
a noninfringing patent would be important. Apparently he had Henry
Thau and David Hermann construct such a relay at their shop in New
York in June. Cat. 1213:9, Accts. (*TAEM* 20:9); Wilson 1872, 539,
1193.

3. Edison devised a similar relay in 1867. PN-69-08-08, Lab.
(*TAEM* 6:771).

4. The following drawings of magnets and armatures appear after Murray's signature:

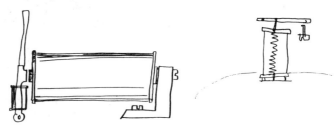

–241–

Notebook Entry:
Automatic Telegraphy

[Newark,] Feby 1 1872

Hollow[a] Cutting Punch   Manner of getting rid of the chips   thus

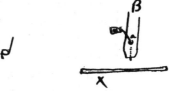

B is the punch[b]   X a raw hide platen on which the cutting punch B cuts[b]   a is a small hole through which a wire (small) runs[b] ~~sa~~this wire being stationary. When the punch moves down and cut a peice ~~th~~ from the paper   this sticks in the punch which in coming up the wire punches or shoves it out. =

I have another device by which the punch or cutter when it comes on the paper presses hard and turns around[b] thus cutting it out after the manner of a knife =

Mrs Mary Edison My wife Dearly Beloved Cannot invent worth a Damn!!

T A Edison

AXS, NjWOE, Lab., Cat. 1172:31 (*TAEM* 3:93). [a]Word preceded by "x". [b]"(x)" interlined above in an unknown hand.

–242–

Notebook Entry:
Automatic Telegraphy

[Newark,] Feby 2 1872

Perforating Apparatus

This invention consists of a set of cutting wires or punches which press on the paper and turn around at the same time by the action of the Levers worked by the keys and a steady pressure kept up by a spring on the key bar thus

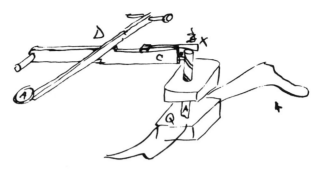

D the key   C the punch lever or key bar   A the cutting punch which has a spiral groove cut in it like a twist drill into which groove a tit from the key bar C runs   X is a spring When the key is depressed the key bar C is depressed   the spring X hitting the top of the punch A brings A down on the platen of raw hide Q & on the paper   and the downward motion of the bar C and the tit rotates the cutting punch A and cuts a round hole in the paper after the manner of a knife or rotary shear

I have only shown one punch lever and punch   I use nine   The spiral grove in the punch A might end near the top and a spring attached to the punch A to pull it down on the paper when the lever C is depressed   but the tit on this bar would lift the punch up off of the paper when at its normal position as the spring on the bar C would be much stronger than that on A.

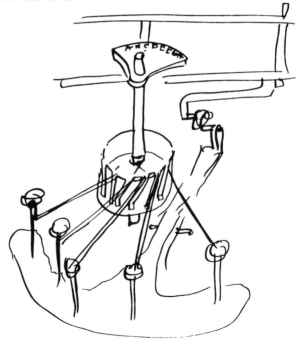

Printing Machine   X crown ~~wch~~ wheell with cam slots   depress a key   throws type wheel around in right place and lever hits ptg lvr   prints.

Described to [--]Unger

T A Edison

AXS, NjWOE, Lab., Cat. 1172:32 (*TAEM* 3:94).

–243–

Notebook Entry:
Printing Telegraphy[1]

[Newark,] Feby 6th 1872

another device

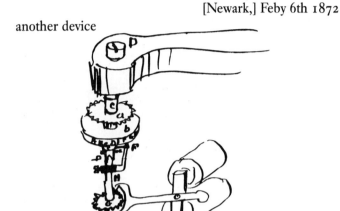

a the corrector   b type wheel   F carrier. C sleeve   D long center screw[2] on which sleeve & type w turn   E pin keep sleeve in centr screw   H shaft   G escapement   In this Case ~~shaft H don't carry aroun~~ The type wheel is not on the shaft H but on its upper center screw it being carried around by a dog or carrier F secured to the shaft H provided with a slot and into which is a pin from the type wheel sleeve C—having sufficient dead motion to ~~st~~ allow of the type wheel being corrected. the worm or unison may be on the shaft or on the sleeve   In Transmitting it is necessary to throw the unison out. This I can do by bringing the printing lever up a little toward the type wheel or by throwing the unison arm out directly— The Engine could be stopped always in a given postion by the action of the switch so as to ensure that it will start off when the current is put on which it dont always do now.

T A Edison

AXS, NjWOE, Lab., Cat. 1174:43 (*TAEM* 3:28).

1. This entry is a continuation of Doc. 239 and is continued in Doc. 244.

2. A center screw has a conically concave end into which a point fits—in this case, the end of shaft **H**. Edison had already used center screws in the universal stock printer and in the instrument for the Financial and Commercial Telegraph Co.

–244–

Notebook Entry:
Printing Telegraphy[1]

[Newark,] February 6th 1872.

I will mention here that it may not be necessary to arrest the break wheel at all in this printing telegraph   The circuit might be opened by an action derived from the pin on the type wheel shaft and the keys   a circuit opener being on these prongs or otherwise. So that when the ~~key~~prong come in Contact with a key it would opene the main circuit just at that moment and when the key was raised ~~closed~~ it   the break wheel going constantly would at all times furnish breaks but ~~can~~ they would be nullified by opening the circuit =

Unison[2]

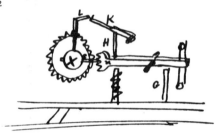

one revolution of Escapement Shaft or rather of type wheel shaft X bring pin c in tooth of rack n   lifts it one tooth   next revolution on[e] more tooth & so on until the arm H throws arm K in path of L when shaft is arrested   when printing lever comes up the arm G throws Rack & bar n down again & so on

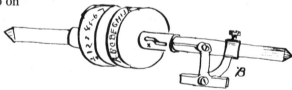

Two type wheels, fraction & letter wheel   letter wheel sets so that the letter will be opposite space on fraction wheel so that when it is printing from fraction wheel will not print its space coming over pad   by rotating type whls round right

position & closing ptg lver the T peice shifts the pin in the slot X and cams the fraction wheel ahead & right to print & same time throwing letter wheel behind so its space will come over pad

It Can be done on one wheel   letter thaen a figure letter then figure & so one   when 60 characters 30 teeth in escapement   each twooth shares two letters   So every other one being a letter letters only print now   come round to right position & shift and came wheel slightly ahead then fractions & figures will be right to print & sletter not[3]

T A Edison   Invintor

AXS, NjWOE, Lab., Cat. 1174:44 (*TAEM* 3:29).

1. This entry is a continuation of Doc. 243.

2. This unison is for the universal stock printer. Cf. U.S. Pat. 138,869.

3. Edison used this idea in two patented modifications of the universal stock printer (U.S. Pats. 131,341 and 138,869).

<table>
<tr><td>–245–</td><td>[Newark,] Feburary 87th 1872</td></tr>
</table>

–245–

Notebook Entry: Printing Telegraphy

[Newark,] Feburary 87th 1872

In my universal I have placed the Engine directly beneath the Printing Machine and underneath the base thereby Making the Machine Much More compact[1]

The break wheel on my Universal has two closures or metallic teeth and between these teeth is filled with Ivory. It is not necessary to have anything there as a limiting pin Might be used on the arm carrying the Contact rollers. [-]Still I prefer to use Some substance between these contact teeth so that the roller will not knock their edges off and also to prevent noise. Now I have found in Experimenting that glass is the best thing that I can use as the spark does not interfere with it as much as rubber ivory, bone etc and what is still better agate or garnet could substituted

There is considerable oxide formed on this Contact point and I propose to use a brush rubbing Continuously up it to keep it clean also upon the Contact roller

I will mention that the break wheel or pulsator may be made of metal Entirely and no breaks in it and the Vibrating arm may have a Contact spring upon it and this would be held on and taken off as the wheel Vibrates

It is not necessary that the pulsator should be exactly as I now use it   is may be made thus

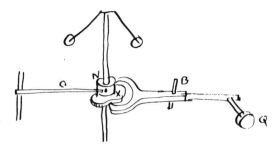

C is the arm arranged to the Escapement lever   N is the wheel carrried around by friction   X is a flange on it which has two wavey [-]teeth   B is a lever which closes & opens the main circuit on the point Q   the Vibration being had by the rotation of X and the two wavey teeth.

I have shown in former descriptions the type wheel is arrested by the key ½ way home[2]   consequently there [-]must be some device to throw the type wheel in the right position to print.

I can get rid of this device thus

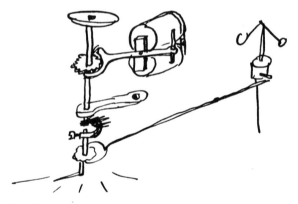

In this Case the type wheel, Escapement etc is precisely like those in ordinary printers, but an extra shaft on which is a star wheel   this star wheel giving Motion to a Lever which arrests the pulsator. Now this second lever runs under the base and is stopped by the Keys. the type wheel shaft will always be right for printing when sending and receiving, but the extra lever which is carried around by the type wheel shaft will be arrested in a position to throw the arrester arm into the path of the pin on the pulsator.

This arrangement Can be done with one shaft, the extra shaft being replaced by a sleeve

Thus

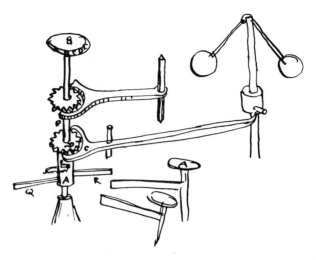

F is the Main shaft. D a wheel which gives motion to the arresting lever C   A is a sleeves with theis wheel D on also with the key arms Q and R   B is a slot having a pin in it and a spring fastened to the sleeve to pull it around against the pin   Now when a key is closed ~~one~~ the escapement ~~comes~~ Rotates[a] around ~~tuill~~ till it stops home but just as it it going home the arm on the sleeve comes against a key and the sleeve is arrested and the pulsator is caught but the Escapmt lever pulls the tyspe wheel home the dead motion in the slot allowing this. the moment the key rises the spring pulls the sleeve ~~up~~ around instantly against the pin in the slot B before the type wheel shaft starts to move.

An Extra Magnet might be used to rotate the shaft which has arms on and the Escapmt lever have arm which arrests pulsator, and The Type wheel shaft be Entirely independent and be rotated by another Magnet & Escapment lever   or a springy Connection might be attached from the Escapmt lever which rotates the type w shaft to the lever which rotates the key shaft and the Extra Magnet dispensed with.

Witnessed   T A Edison

AXS, NjWOE, Lab., Cat. 1174:46 (*TAEM* 3:30). [a]Interlined above.

1. See Doc. 250, n. 6.
2. See, for example, Doc. 238.

–246–

Notebook Entry:
Printing Telegraphy

[Newark,] Feby 10 1872

I have tried nearly all kinds of self acting breaks upon the Escapmt lever of the Universal so as to dispense with the Engine but a change of adjustment of that lever increases or decreases the speed whereas the Engine does not to any great extent. another defect in a self acting break is that it is a very hard matter to regulate the break

Escapement

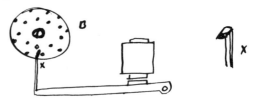

X is a dog with a cam end   B a face pin wheel. when it goes up it cams ahead by the inner row coming down by the outer row—

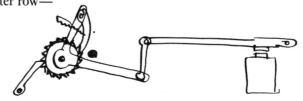

Another Escapement

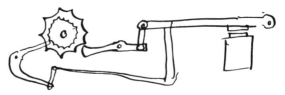

another geneva stop[1]

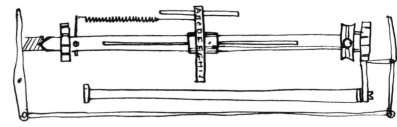

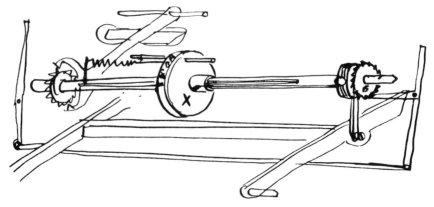

Travelling type wheel   this wheel is rotated the same way as the ordinary printing Machine by an Escapment etc

The shaft is provided with a key seat which the sleeve and type wheel travel endwise   B is a ratchet having a pulling by its side and this ratchet is rotated by a pawl from the printing lever   There is a string leading from the type wheel to this pulley which turns the corner on a little pulley wheel   the other End of the type wheel is attached a spring

Now ~~the~~every time the type wheel is printed from the upward movement of the ptg lever winds the string up the distance of one letter. when the type w reaches its full length it throws out the detention click and the spring pulls the type wheel back to its starting point and restores the detention click   an attachment may be made with the detention click to feed the paper a line ahead when it is moved

I am going to get up a machine with 26 trains of diffrential gearing   Each gear having a letter of the Alphabet and taking a hundred words and turn the gearing in such a manner that a few letters will be the result of these ~~w~~Hundred words and these letters Can be transmitted over Atlantic Cable and then another Machine in London Can be set at these Letters and rotated backwards until the 100 wds are indicated   I can either indicate or print the result[2]

Escapment wedge acting square teeth.

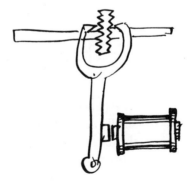

This is an old invention of mine but I record it.

Paper feed for Printing Telegraphs.

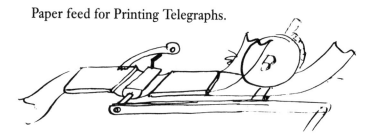

When you bow the paper the click prevents it coming back but when the bower attached to the printing lever goes up it lifts the click off and the paper straightens out by its own action thus throwing itself right for the next letter. Tried in 1869.

T A Edison

AXS, NjWOE, Lab., Cat. 1174:50 (*TAEM* 3:32).

1. A Geneva stop (also called a Maltese cross stop) was a clockwork device to prevent overwinding. The wheel with concavities on its perimeter (**B** in the accompanying diagram) was properly called a Geneva wheel (Maurice and Mayr 1980, 313). The diagram is from Brown 1896, 54.

2. Edison made no other reference to this idea.

–247–

Notebook Entry:
Printing Telegraphy

[Newark,] Feby 12 1872

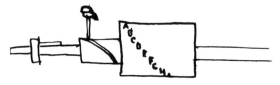

Type wheel with the letters placed as shown so as to get as small a diameter as possible. a twist draill might be used and the letters cut along the twist and be rotated so that for every movement of the ratchet a new letter would come into posision   a pin following in the groove would give the endwise motion and the ratchet would give the rotary motion   after the drill or shaft having types would reach its limit it would be thrown back instantly by lifting out the pin in the groove & using a spring

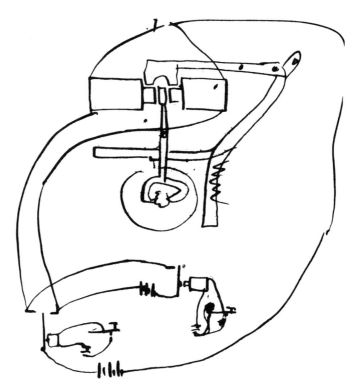

Device for a two wire instrument. close current on one wire one magnet draws escapmt armature one way close on other wire other magnet draws the other way close both mag cir- cuits and both magnets close bringing down printing lever.[1]

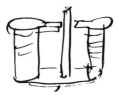

polarized relay bad iron or residual magnetism used only no steel.

or this way

Device for using two type wheels having 15 character on one and 15 on the other and a15 teeth in the escpmt and a shifter on the shaft so that in rotating it will shift on one & then the other

I claim slowing a printing lever by a piston attached to the printing lever run in a tube of mercury[2]

T A Edison

AXS, NjWOE, Lab., Cat. 1174:54 (*TAEM* 3:34).

1. Cf. Cat. 30,094, Lab. (*TAEM* 5:350).

Drawing from Edison's patent for an air-cushion printing lever.

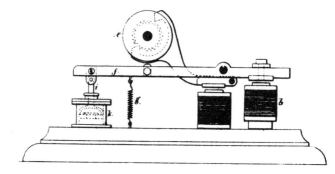

2. In U.S. Patent 128,604 Edison used a cylinder filled with air to accomplish the same purpose (see drawing above).

–248–

Notebook Entry: Printing Telegraphy

[Newark,] Feby 14 1872

any of the arrangements described in my description of an apparatus invented for Andrews described in another book[1] can be applied to the printing lever

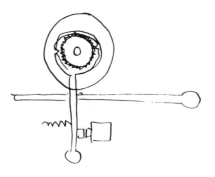

60 characters on the wheel   60 teeth in the escapmt   figure between each letter   stop on open circuit to print letters and on closed circuit to print fractions and figures

My Wife Popsy Wopsy[2] Can't Invent[a]

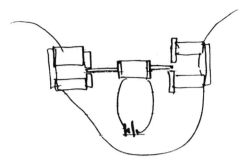

Polarized Relay tongue being polarized by a Local battery. Model constructed in 1869—In Genl Lefferts possession[3]

Escapement[4]

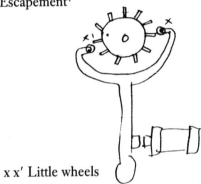

x x′ Little wheels

Escapement

Escapement motor Pin wheel.

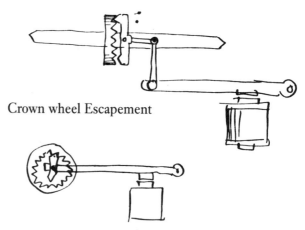

Crown wheel Escapement

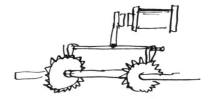

Inside star wheel escapement

Double star wheel escapement

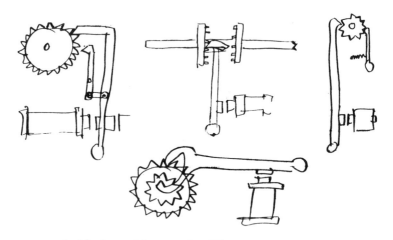

inside & outside star wheel Escapement

T A Edison

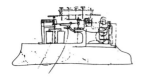

Edison's drawing of his transmitter for the English stock-reporting system.

AXS, NjWOE, Lab., Cat. 1174:56 (*TAEM* 3:35). a<u>"My . . . Invent"</u> underlined twice.

1. Although the book referred to has not been found, it is clear that Edison had been working on machines for Elisha Andrews's English stock telegraph venture since its inception the previous spring. See the accompanying sketch and, for example, Cat. 1212:96, 238, Accts. (*TAEM* 20:746, 809).

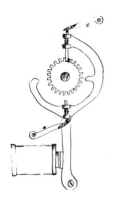

One of Edison's February 1872 notebook drawings of escapement mechanisms.

2. Robert Conot claims that the name "Popsy Wopsy" was from a contemporary music hall song. Conot 1979, 47.

3. Nothing is known of this model.

4. During February 1872 Edison filled a separate notebook with over one hundred escapement designs. EP&RI.

"BABY" PRINTING TELEGRAPH   Doc. 249

In February 1872 Edison executed a patent for his "Baby" telegraph, a compact printer with a single typewheel.[1] It is not clear whether Edison intended it for private lines, stock reporting, or some other purpose.[2] The design incorporated a polarized relay and used only one circuit wire. A single armature and magnet combination actuated both the typewheel and the printing mechanism.

The only evidence that this printer was ever used is an illustration of a printing telegraph found in a popular mechanical dictionary published in 1876–77.[3] There are no notebook entries relating to its development, and only one possible manufacturing record remains.[4]

1. U.S. Pat. 126,530. See the patent drawing reproduced here.

2. The typewheel on the patent model (Doc. 249) has thirty blank spaces, the same number of characters as on either the letter or number typewheel of the universal stock printer.

3. Knight 1876–77, s.v. "Telegraph," whence the name as well.

4. Edison paid Thau and Hermann on 8 June 1872 for two "very small printers." Cat. 1213:9, Accts. (*TAEM* 20:9).

Drawing from Edison's printing-telegraph patent, the "Baby" printer (U.S. Pat. 126,530).

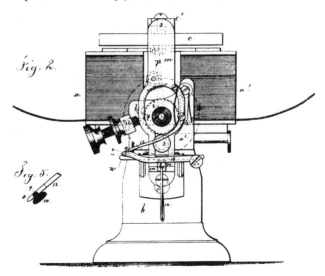

*Patent Model: Printing
Telegraphy*¹

M (8 cm × 10 cm × 12 cm), MiDbEI(H), Acc. 29.1980.1308.

1. See headnote above.

2. Edison executed the covering patent on this date. Pat. App. 126,530.

[Newark,] Feby 16 1872.

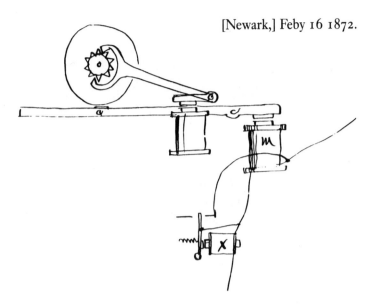

M is the printing Magnet   X is an extra magnet in same circuit. they both respond together but when the printing lever just touches the type W[heel] the relay points[1] close and cut it out and it flies back[2]

I use this device for slowly up mechanism and performing this operation

Thus

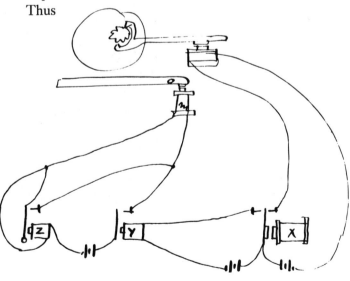

X is the main line relay　front point[3] vibrating works type wheel around　now when Type wheel in right position open main line　relay lever X flies back closes Relay Y which closes printing circuit in which is cut off relay Z[4] and printing Magnet—M—both close and the moment letter is printed Relay Z cuts current out of printing magnet and lever flies back

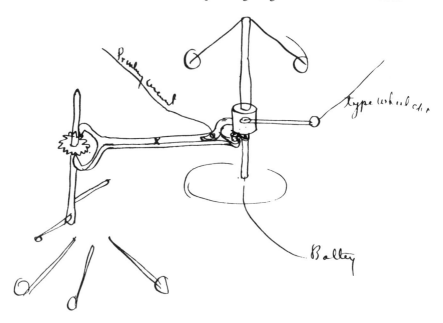

On the break wheel is a plate which has two teeth bevelled[5]　X is an arm having 2 prongs one each of which is two little friction rollers　and every revolution of the Break wheel ~~and~~ the rollers engaging with the bevelled pins an intermittent motion is given arm X which acting on the escapement wheel Rotates　for every revolution of the break wheel the ratchet is advanced two teeth there being 2 breaks on break wheel and 2 bevelled pins. one prong is only necessay by using an spring in place of the other.

by depressing a key the ratchet wheel shaft is arrested　this stops the arm X which arrests break wheel.

one of these prongs is insulated and connected to the printing wire and when the break wheel is suddenly arrested the roller comes in contact with the ~~bit~~ bevelled tooth and closes the printing circuit. I have a machine already made.

T. A. Edison

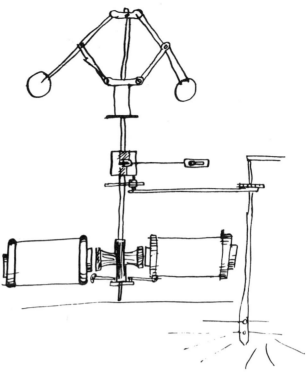

Arrangement of the Break in the universal[6]

AXS, NjWOE, Lab., Cat. 1174:60 (*TAEM* 3:37).

1. The armature of **X**.

2. This was done in an attempt to ensure that the printing lever touched the typewheel as briefly as possible. Proper operation would depend on the adjustment of the springs against which **M** and **X** pulled. The retracting spring for the printing lever is not shown.

3. The front point was the relay contact on the same side of the armature as the magnet—that is, the point touched by the armature when the magnet was charged.

4. **Z** corresponds to **X** in the first sketch in this document.

5. In this sketch Edison was again attempting to devise a mechanism to stop the breakwheel of the universal private-line printer (cf. Docs. 193, 222, and 225; and see n. 6 below). As labeled in the diagram at right the motor was at the lower end of shaft **M**; breakwheel **B** rotated with **M** by dint of a friction spring, but could be stopped without stopping **M**;

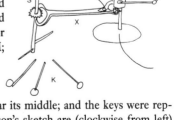

arm **X** would pivot somewhere near its middle; and the keys were represented by **K**. The labels in Edison's sketch are (clockwise from left) "Printing circuit", "type wheel ckt", and "Battery". Because Edison did not refer to a typewheel either verbally or visually, it is not clear whether this mechanism was intended to function as a receiving instrument.

6. This drawing is a clearer illustration of the idea sketched above. Note that the motor magnets are horizontal, an arrangement Edison used on his universal transmitter (U.S. Pat. 131,343). See Docs. 211 and 245.

–251–

Notebook Entry:
Printing Telegraphy[1]

[Newark,] Feby 26th [1872]

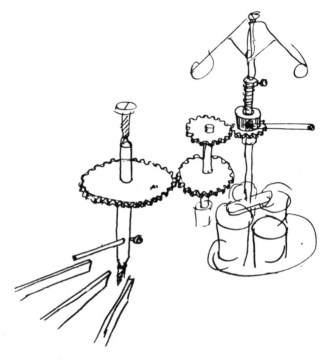

Universal break wheel working shaft that has stops on for arresting by keys with gear wheels. I have previously shown & discribed it working with a ratchet wheel for the ratchet or intermittent & substitute gears   a geneva stops wheel[2] might be used in place of the gears   having 30 Stops on the shaft wheel & one on the break or more

This gear was previously suggested to Dean[3] Feby .1. and thought before while writing description on page 29[4]

AX, NjWOE, Lab., Cat. 1147:64 (*TAEM* 3:39).

1. This entry is continued in Doc. 253.
2. See Doc. 246, n. 1.
3. Charles Dean (n.d.), a mechanic at Edison's Newark shops, was later important in the organization and management of the machine works Edison established in New York City for the manufacture of his electric lighting system. Jehl 1937–41, 676–77.
4. See Doc. 223; cf. Doc. 211.

Newark, Mch 6/72

Murry & Co. Thos A Edison.[a]   Mfrs Tel Instruments
                III NJ RR Ave.

#3700[2] Are just putting in their machiney, and will soon commence work. "E" is of the firm of E & Unger   "M" has no means of his own. "E" is said to have some little means in R[eal]. E[state].[3] & to own ½ interest in firm of "E & U"   Are v[er]y little Kn[own] here, & we cannot get information wh[ich] seems to warrant any extended Cr[edit]. 2092

D (abstract), MH-BA, RGD, N.J. 21:125. [a]"Thos Edison." interlined above; "A" interlined above that. Followed by "(Other reps 'E' see 'E & Unger' 3/238)"; the numbers "3/238" are the volume and page numbers for related credit reports.

1. R. G. Dun & Co., established in 1841 as the Mercantile Agency, was by 1871 one of two major credit-reporting firms in the United States. In the early years selected local correspondents, such as attorneys, bank cashiers, or established merchants, reported the standings of local businesses to the central office in New York. Clerks there transcribed the narrative reports into large ledgers organized by state and county, and from these ledgers they derived requested information for subscribers. Later, the agency opened local reporting offices with full-time agents. A Newark office opened in 1871. In 1933 R. G. Dun & Co. acquired its largest competitor, the Bradstreet Co., thereby becoming Dun & Bradstreet, Inc. In the early 1960s the company deposited over two thousand ledgers, covering a period from the 1840s to the 1880s, in the Baker Library of the Graduate School of Business Administration, Harvard University. Dun & Bradstreet 1974.

2. This number refers to the agent making the report. The book containing the matched codes and names has been lost. The number at the end of this document is repeated in the left margin; its meaning is not known.

3. See Doc. 206.

[Newark,] March 6th 1872

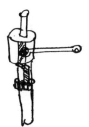

Carrier for break on the universal printer

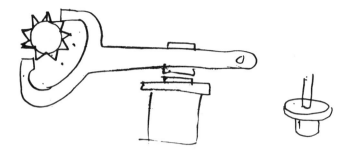

I claim Turning a type wheel by an engine (electric)[a] and releasing it step by step by an Escapement and Magnet the pulsations being derived from ~~Sam~~ a break on the shaft of the Engine which rotates the type wheel = [2]

Unison[3] for Callahans 3 wire Machine[b]

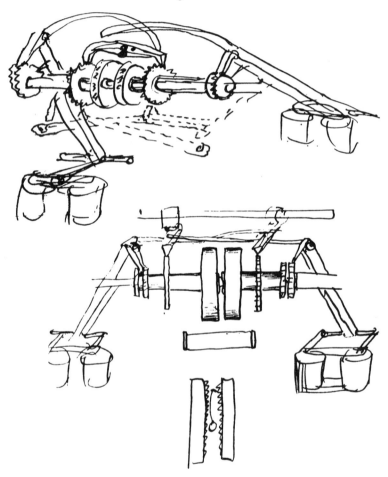

Attachment to Universal Stock Printers,[4] to prevent back lassh in Escapement wheel when they are run at High Speeds.

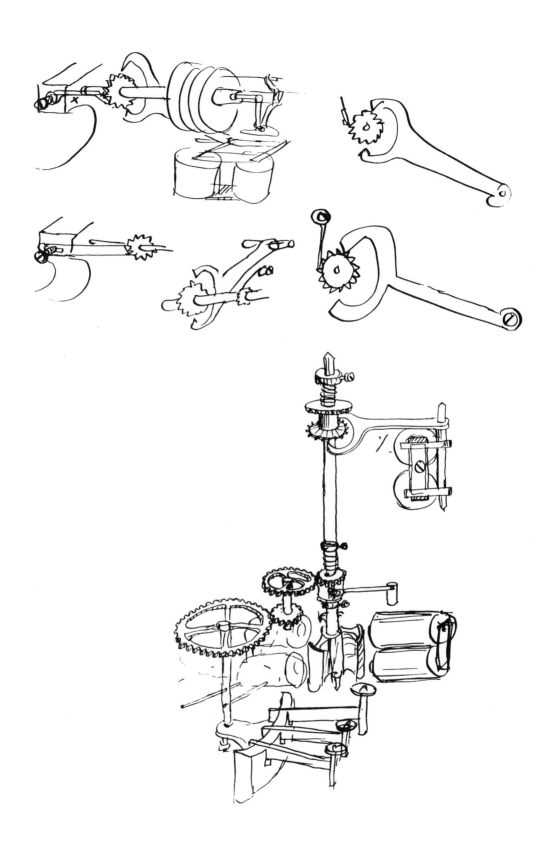

Printing Telegraph Machine[5]   The Engine eshaft carrying the Type wheel and released by the escapmnt lever & Magnt X

T A Edison   Inventor

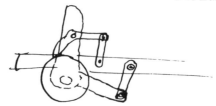

Positive Paper Feed for Printing Telegraphs[6]

AXS, NjWOE, Lab., Cat. 1174:65 (*TAEM* 3:39). [a]Written in right margin. [b]Followed by "Stock Quotn Printer", written by William Carman.

1. This entry is a continuation of Doc. 251. It is the last entry from 1872 in this notebook and is followed by four blank pages.

2. See n. 4.

3. Lefferts had requested such a unison two months earlier. See Doc. 220; see also Edison and Unger accounts for 10, 20, and 25 April 1872 (Cat. 1183:99, 101, 104, Accts. [*TAEM* 20:909–10, 912]).

4. The five sketches on p. 471 show pawls designed to hold the escapement wheel. No surviving machines exhibit such a modification.

5. In this design the motor provides the motive power for the typewheel, and the escapement acts as a true escapement, releasing the typewheel rather than driving it. The motor armature is of the form used in the universal transmitter; that is, it is three-lobed. See Doc. 211.

6. In the sketch preceding this heading, the horizontal bar is the printing lever viewed from the side and fixed to a shaft at its right end (not shown). The small circle within the large circle is a fixed shaft on which the lever for the paper feed pivots.

–254–

R. G. Dun & Co.
Credit Report

Newark, Mch 7/72

Edison and Unger.[a]   Telegraph Fixtures.   4 Ward St.[1]

3700[2]   Have been in bus some 12 Months or more. Manufacture the Gold & stock Brand Tel Instruments[3] & claim to have a monopoly. "Unger" states that they have abt 22m$[4] invested in Machinery & Stock,[5] & that none of this is borrowed, & that "E" owns some little R[eal].E[state].[6] "Edison" was formerly connected with the Am Tel Wks, Is an ingenious Mechanic & inventor. Formerly came from Ohio. It is intimated that Mr Harrington Pres of the Automatic Tel Co. 80 B'way N.Y. has claimed that "E" started his bus here in Ward

Workmen and instruments in back of Edison and Unger's Ward St. shop.

St & got his Machiney with his (H's) money.[7] We can learn very little of this firm, beyond their own statement, and without further Knowledge of them cannot advise extended Credit. "Edison" is also of "Murry & Co" a new firm just starting.[8] The Bank a/c[9] is in the name of "Wm Unger," and checks payable to the firm are first endorsed by the firm & then by "Wm Unger." Their transactions indicate a large business. 2092

D (abstract), MH-BA, RGD, N.J. 22:238. ªFollowed by "(Other reps 'Edison' see 'Murry & Co' 2/125)"; the numbers "2/125" are the volume and page numbers for related credit reports.

1. See Doc. 157, n. 2.
2. See Doc. 252, n. 2.
3. Universal stock printers.
4. "m" means thousand.
5. Cf. Doc. 215.
6. See Doc. 206.
7. Edison had begun to transfer machinery to Edison and Unger from the American Telegraph Works in the fall of 1871. Quad. 70.7, p. 705 (*TAEM* 9:722); 73-003, DF (*TAEM* 12:1001–34 passim).
8. See Doc. 252.
9. "a/c" means account.

SIX-KEY PERFORATOR   Doc. 255

Edison began working on perforators for automatic telegraphy to fulfill his 1870 agreement with Daniel Craig.[1] He designed the following perforator as an inexpensive alternative to the large machines used in main offices, intending it for local offices or customers wishing to prepare their own messages.[2]

Edison's first perforators had been large, rapid instruments with a separate key for each character. By February 1871 he was designing smaller perforators.[3] Edison apparently completed a six-key perforator about 17 May 1871.[4] However, he did not execute a patent application for this perforator until 15 March 1872, which suggests that he reworked the apparatus in the interim. The patented machine contained one key for dots and a second key for dashes. The other keys were used to space the perforations and to operate the paper feed.

Top: Edison's original design for perforations used in his large perforator (U.S. Pat. 121,601). Bottom: Altered perforations used in Edison's six-key perforator (U.S. Pat. 132,456).

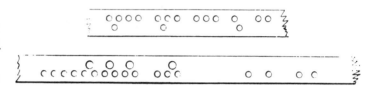

As in his earlier, large perforator, a single hole in the tape transmitted a dot. There was, however, an important change in the form of the holes that produced dashes. The large perforator punched out a dash using three holes of equal size, one above and between the other two. In his six-key perforator, Edison altered this design so that the single dot was larger and overlapped the two smaller holes. In this way, he sought to "insure a metallic contact of the brush or transmitting stylus between one of the smaller perforations and the other, thereby producing a dash-mark."[5]

1. Doc. 103.
2. U.S. Pat. 132,456, p. 1.
3. Edison worked on a three-key perforator between February and April 1871. Cat. 30,108, Accts. (*TAEM* 20:217–37 passim); 71-015, DF (*TAEM* 12:583). See also Doc. 149.
4. See Doc. 166, n. 3.
5. U.S. Pat. 132,456, p. 1.

Patent Model:
Automatic Telegraphy

M (16 cm dia. × 14 cm), MiDbEI(H), Acc. 29.1980.1606.

    1. Edison executed the covering patent application on this date. Pat. App. 132,456.

256

New Idea for Printing ~~Telegraph~~ machine[2]

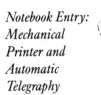

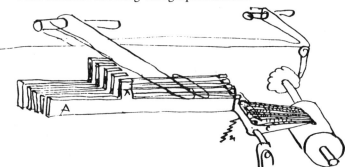

A are 7 bars ~~are~~ moving 7 levers c up and down off of the recording Drum

These seven levers are thrown down upon the drum by the forward movement of the bars A   This drum carries the paper   between the paper and the c bars ~~are~~ is[a] an ink band   The A Bars are provided with 30 rows of tits down the side of which pass the side of 30 key   Now on the side of these tits are The letters of the Alphabet in relief   The key in passing down throws forward these bars acording to these relief parts of the letters on the tits   at the same time the paper drum is being fed forward by the downward motion of the key

It will be seen that this is but a mechanical way of producing the result chemically as shown on page 22 & 23 this book[3]   communication being had with the recording pens mechanically instead of electrically.

further description unnecessary

Driving paper for a perforator

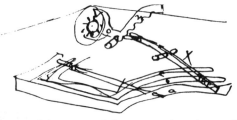

a b c some of the 9 punch levers used in the perforator   X is a lever which has a ~~rack~~ click on the end   Loose going ~~up~~ down rigid going up   n a rack connected to a release Escapment lever which releases the paper drum driven by weight, step by step   The last ~~key~~lever in the punch leves being nearer the fulcrum of X than the first will shove X further   now X is going down having a loose click does not Vibrate rack n but in going up vibrates it releasing paper drum a notch every vibration   The number of vibrations being governed by the length which X has gone down which is governed by the punch used which is nearest its fulcrum   it may be used thus

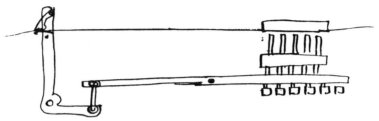

Description unnecessary

The printing device described on page 27.[4] n̶The seven recording levers might be cutting punches and cut the letter out instead of recording them then this prepared strip could be used for transmitting in a machine already described by me in another book where is use this kind of paper[5]

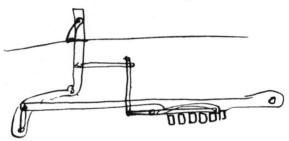

Positive Paper feed movement for perforator

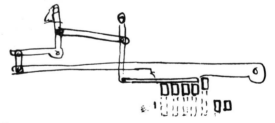

Different

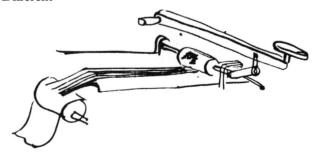

Printing machine, different shape from that described on page 27.

AX, NjWOE, Lab., Cat. 1182:27 (*TAEM* 3:63). [a]Interlined above.
  1. See Doc. 258, n. 1.
  2. Typewriter.
  3. That is, Doc. 236.
  4. That is, the first drawing in this document.
  5. Doc. 194.

*Notebook
Entry:
Automatic
Telegraphy*

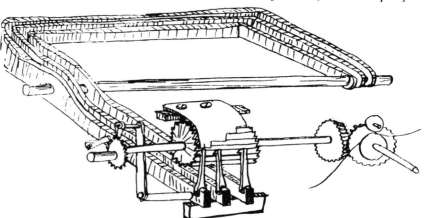

paper feed, The punch bars determining The distance
Twisting cutting punch

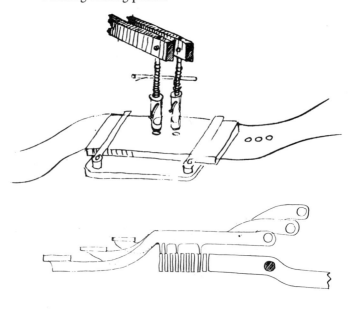

An Inventive Flurry

AX, NjWOE, Lab., Cat. 1182:31 (*TAEM* 3:65).

1. See Doc. 258, n. 1.

–258–

Notebook Entry:
Electric Engine
and Battery

[Newark, March 1872?[1]]

Electric Engine.[2]

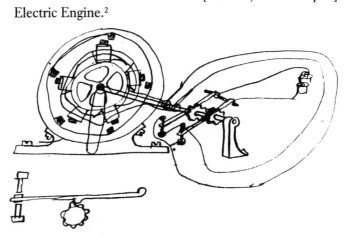

This Engine is somewhat dissimilar to that of any other heretofore made although ressembling greatly in appearance others.

The outside ring is stationary & is provided with armatures, arranged in relation to the Magnets that when all the magnets are Charged every magnet will draw itself up to the armature. the drum carrying the Magnets will then revolve the distance between one of the armatures be charged and draw themselves up again

The break consists of two toothed wheels [-] Insulated from the shaft & each other and upon which are two springs also insulated from each other ~~These~~ at the end of these springs are four platina points two above & two below so that when [-]the teeth on the wheels raise both levers up they will come in contact with both platina points and these platina points being connected to the magnets, and the two springs being connected to the battery the circuit is thrown through all the magnets Not as it is usually done but [this?][a] way

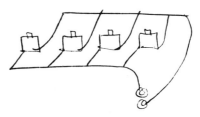

These magnets   and herein constitutes my improvement consists of very fine wire Say from No 20 to 36 or even finer When they are of finer wire each magnet is made thus

five or six Sections and each Connection or section So Connected [that?]ᵃ the current splits and divides passing through by one wire throgh the several routes provided for it & emerging to one wire again. these magnets made in this manner are connected as shown on the top sketch or may be thus

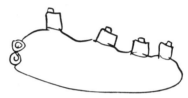

It is a law of magnetism that the greater number of convolutions of wire around the Core the greater would be the effect, taking care about the resistence.³ heretofore parties experimenting with electric engines use very large wire and thus obtain a very low resis[tance],ᵃ but have but few convolutions. I also obtain an extraordinary low resistence by providing a number of routes each having a large No of convolutions and agregately an enormous number, but they are so Connected that owing toᵇ the no of routes, combined the resistence is reduced from say 300 ohms (if the wire is connected as is ordinayly done) down to 2 ohms, for each magnet and by combining the magnets after this plan I reduce the total Resistence of the several magnets to 1 ohm for the whole and at the same time have an enormous number of Convolutions of wire.

I have discovered that if ~~100 feet of copper wire~~ if a2ᵇ spools hasv[e? ---]ᵃ or one magnet  has 6 [ohms?]ᵃ Resistence

and the current is passed through it it will lift say 2 lbs. if now we take a magnet with 24 ohms resistence each spool having 12 ohms & connect them thus

or thus

We reduce the resistence to 6 ohms but obtain twice the force or 4 lbs without any more consumption of material in the battery = [4]

another theory I take advantage of in the construction of my engine   which is that in all electric engines ~~or in a~~ now in use the current through the magnet must be broken before the magnet approaches the point where the most work is done   for if this was not done the armature passing over would be retarded owing to the sluggishness which the current or magnet discharges   I get around this by ~~prov~~ keeping my current on until up to the very center (ie) the armature is opposite the cores of the magnet then it will be observed by referri~~d~~ng to the break with its upper & lower springs that the current is reversed for an instant through these magnets & they instantly discharge   this Current is not kept in long enough to effect a recharge of the opposite polarity

Another advantage I claim is having all the magnets pulling at one time and a large momentum or fly wheel to carry it over the spaces between where ther is no~~t put~~ pull

This splitting of the magnets as described cann be applied to Other purposes

I propose to apply it to printing telegraphs   winding an ordinary Stock Printer 6 ohm 24 wire with No 30. wire[b] 24 ohm and by connecting as heretofore shown reduce to 6 ohms

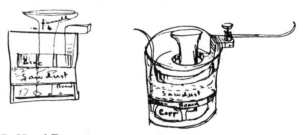

B. Vitrol Battery

I claim a board perforated with holes and sitting on leg~~a~~s an inch or so from bottom underneath which is Copper sheet.

I claim the saw dust on the perforated bo~~t~~ard or material of any kind perforated and a funnel for feeding the underneath the perforated disk with Sulphate Copper = [5]

I claim running two type wheels by two electromotors, and keeping them in unison by a governor on Each engine, and printing by a single move

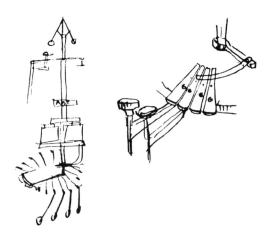

AX, NjWOE, Lab., Cat. 1182:33 (*TAEM* 3:66). ᵃObscured by ink blot. ᵇInterlined above.

1. In the original notebook, the last dated entry preceding this one is from 30 January 1871 (Doc. 236), and the sketch at the end of this notation, showing the universal private-line printer with a vertical type-wheel shaft, indicates that this was written before April 1872 (see Doc. 262 headnote).

2. On electric motors, see Doc. 165 headnote, n. 3.

3. That is, the magnetic strength of a current-carrying wire coil is directly proportional to the strength of the electric current and the number of turns of wire in the coil.

4. If the spools in both the 6-ohm and 24-ohm magnets are made with the same wire, those in the 24-ohm magnet will each have four times as many turns of wire and carry half as much current (when connected in parallel) as those in the 6-ohm magnet (connected in series). Because an electromagnet's strength is directly proportional to the product of the current and the number of turns of wire, this arrangement would result in twice as strong a magnetic field from an identical battery output (the resistance of the two circuits being equal, the battery output would be also). The limit to such an increase is the point at which the electromagnet cores are magnetically saturated, a concept Edison understood (see Doc. 40).

5. Cf. Edison's U.S. Pat. 142,999.

Drawing from Edison's battery patent (U.S. Pat. 142,999).

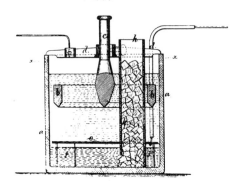

[New York?,] April 29, 1872[a]

Articles of Agreement made and entered into this Twenty-ninth day of April A.D. 1872 by and between Thomas Alva Edison of the City of Newark, State of New Jersey; and Marshall Lefferts of the City and State of New York, Witnesseth,

Whereas the said Edison has invented certain Improvements in Printing Telegraphs and the said Lefferts has agreed to act in connection with the inventions and patents on the same in Europe.

Therefore in consideration of the premises and of one dollar in hand paid each to the other the receipt whereof is hereby acknowledged the following agreement is made and entered into between the said Edison and Lefferts for themselves and their legal representatives.

First. The said Edison hereby assigns, sells, transfers and asets over or agrees so to do, unto the said Marshall Lefferts one undivided half part of the right, title and interest of every character in and to any and all Letters Patent in Great Britain that he may hereafter obtain for improvements in printing telegraphs, together with the same right in any invention of or improvement upon printing telegraphs that he has heretofore made or may hereafter make for ten years and to this end agrees to make and execute, at the expense of the said Lefferts, any and all specifications that may from time to time be required for procuring or completing Letters Patent for such inventions and for assigning or more fully conveying the rights hereby assigned or intended so to be unto the said Lefferts.

Second. The said Lefferts hereby agrees to bear the expenses of patenting such inventions in Great Britain, and also in[b] France, Belgium and such other countries as may be mutually agreed upon, but if said Lefferts unreasonably neglects or refuses to bear the expense of patenting any improvement in printing telegraphs invented by said Edison then this agreement shall not apply to the rights in that particular country or to that particular invention in that Country[c] and the said Edison shall be relieved from this agreement to that extent only.

Third. The said Edison hereby appoints the said Lefferts his true and lawful attorney to attend to any and all business in connection with the patenting, introducing, selling, transferring, or negotiating the rights in connection with such printing telegraph inventions in any and all countries of Europe where the said Lefferts may pay the expense of patenting, and agrees to execute any further or more formal power of attorney for such purposes.

Fourth. The said Lefferts hereby agrees to exercise resonable dilligence in connection with the business hereby contemplated and to well and truly account to the said Edison and pay over to him one half part of the profits that may result in Great Britain in connection with said inventions and Letters Patent, first however retaining his actual cash advances in connection with the same.

Fifth, The said Edison hereby transfers sells and sets over or agrees to assign and set over to the said Lefferts the entire right title and interest of every character in and to the said inventions and letters patent on Printing Telegraphs in France, Belgium and each and every other country on the Continent of Europe where the said Lefferts pays or causes to be paid the expenses of patenting, and agrees to execute any specifications, assignments, transfers, or other documents that may from time to time be required for carrying out the full intent and meaning hereof, or in accordance with the laws of the respective countries.

Sixth. The said Lefferts hereby agrees to exercise reasonable dilligence in endeavoring to introduce and make profit by the said inventions in the continent of Europe and to account to and pay over to the said Edison all the profits that may from time to time be realized from the rights on the continent of Europe, (first deducting his actual cash outlay) until he the said Edison shall have realized the total sum of ten thousand dollars, after which said Edison shall have no further claim in that particular.[d]

In Witness whereof the parties to these presents have hereunto set their signatures and seals.

T A. Edison[e]   Marshall Lefferts[e]
Witnesses   M. C. Lefferts[2]

DS, NjWOE, LS (*TAEM* 28:949). In unknown hand. [a]Date taken from text, form altered. [b]"and also in" interlined above in Edison's hand. [c]"in that Country" interlined above in Edison's hand. [d]5¢ Internal Revenue stamp in left margin. [e]Followed by a seal.

1. No evidence has been found to show the ways and extent to which this agreement was carried out.

2. There was only one witness, Marshall Clifford Lefferts.

–260–

R. G. Dun & Co.
Credit Report

Newark, May 21/72

Murry & Co. Thos A Edison.[a]   Mfrs Tel Instruments
111 NJ RR Ave.

#3700[1]   "M" states "that they are equal partners "E" furnishing the cash cap[ita]l 4m$[2] & he "M" off setting this by his time and mechanical skill "E" is said to own a ho[use] & lot valued @ 5m$ mtgd for 15c$[3] Is also of the firm of E and Unger. We have no means of confirming particularly above statement but judging from their machinery & stock think it must be in the main correct "M" also states that they are now about 2m$ in debt have little or no competition & ought to do a good & paying bus 2092[b]

D (abstract), MH-BA, RGD, N.J. 21:125. [a]See Doc. 252, textnote a. [b]Written in right margin.

1. See Doc. 252, n. 2.
2. "m" means thousand.
3. "c" means hundred. See Doc. 206.

-261-

*Daniel Craig to
Marshall Lefferts*

New York, ~~Jun~~May 24/72

Dr General

I am quite sure I can get a substantial pledge from the Southern Co. of the $25,000 in cash and $100,000 of their stock or some other of greater value, with the single proviso that the Fast System proves to be all right between Charleston and Washington, or New York.[1] (They now write Morse, direct & without Repeaters, in bad weather between W. & C. & in good weather between N.Y. & C—so we shall have good wires for the test.)

I wish you could get Edison's <u>own</u> interests confided to me, in such a way, <u>first</u>, that his interests and his efforts in everything pertaining to Automatic Telegraphy, <u>will go with your interests</u>, or not <u>against</u> ours, provided only that he shall be paid out of the first proceeds of anything from any telegraph Co. using any system of Automatic or Fast Telegraphy, whatever sum <u>you</u> or you and Pope[2] may agree to be fair to him, in view of what he may have produced of advantage to the fast system—and his inventions or devices being compared and estimated by similar (if any) inventions by other parties.

I also want you to give me full-control, to manage according to my best judgment, all your Automatic interests, inventions, patents and <u>thoughts</u> (so far as I can draw them out of you) upon the new system.

Pledging to you as I will and do, my best efforts to make the interests confided to me so valuable that I can pay you for the same at least $100,000 of the Stock of some respectable Telegraph Co. within the present year and at the earliest possible moment.

It is <u>important</u> for me & also for you to get Edison into such a position that he will work to promote our interests rather than the interests of our opponents, actual or possible, and I should esteem the situation very much safer if you could so manage him as to get his interests where you could fix themir account and keep him faithful to our side in any & every contingency.

I wish you could feel justified in accepting the whole Telegraph interest as collateral security against the claims of that vile dog Vail,[3] and as I know they are all moonshine, you will not be hurt in the least.

You may easily imagine that Mrs. Craig feels as though if she surrendered her splendid property at Peekskill for less than half its real value, she ought to have something sure & not hypothecated to afford her some support[4]

Do, pray, try to have this trifling point (trifling to you, but seemingly big to Mrs. C.) conceded, and then I shall hope to make her and Jim[5] reasonably well satisfied with the swap.

Harrington & Little, for some reason unknown to me, are decidedly snubbing Morris[6] & the Southern Co., which is very good for us, and now is the time for me to make a trade with them. What might be the effect of Harrington's getting down from his stilts and evincing a disposition to co-operate with the Southern Co., I do not know.

The sooner you can tell me the result of your expectreflections in regard to your interests in the Telegraph matter and the Mt Florence property the better I shall like it.

I thought you might like to see a statement of the Mt. Florence Property as enbraced in John's[7] scheme. I suppose your suggestion today applied to the Real Estate and Improvements, only, but, of course John must deliver according to the Bond—but you and I can arrange this, by letting John pay Mrs. Craig the actual (selling) value of the Personal property which inventories some twelve or fifteen thousand dollars, but, of course would not sell for that at a peremptory sale.

Of course Mrs. Craig will at my request, give to Smith[8] or whoever may be entitled to receive the conveyance, a perfect title to the property for whatever sum the Scheme can afford to pay for it. The real expectation of Finch[9] was to pay Mrs. Craig $150,000, & share with me ½ of all his profits, but I had no written agreements with Finch or other parties, and have none now with John, but expect him to do whatever may be fair, but no more.

I feel sure he can carry the scheme through & pay $150,000 for the property & all expenses, & make $10,000 to $20,000 for himself, if he can have a credit at the right time—say 1st June—for $10,000 to fee the advertising people, & I will give him this credit as soon as I get able, if ever. Truly &c

D H Craig

ALS, NNHi, Lefferts.

1. The Southern and Atlantic Telegraph Co. was negotiating with the Automatic Telegraph Co. for use of the automatic telegraph system on its lines, which extended from New York through Philadelphia, Baltimore, Washington, and Charleston to Savannah (reaching New Orleans in 1873). This was part of a complex series of negotiations between several telegraph companies, including the Atlantic and Pacific, Franklin, and Pacific and Atlantic. These negotiations were intended to produce an alliance in opposition to Western Union. Although various agreements were negotiated, the alliance was never consummated. Josiah Reiff's testimony, 1:76–113, Box 17A, *Harrington v. A&P;* Southern and Atlantic Executive Committee Minutes 1870–76, 25, 40, 42, WU; Southern and Atlantic Directors and Stockholders Minutes 1869–73, 85–86, WU; agreements between Automatic Telegraph and Southern and Atlantic, 11 Oct. 1872 and 17 Oct. 1873, WU; agreement between Automatic Telegraph and Pacific and Atlantic, 1 Mar. 1873, WU.

2. Franklin Pope.

3. Unidentified.

4. Craig planned to transfer this property to Lefferts in exchange for cash, telegraph stock, and Lefferts's interests in automatic telegraphy. Memo, Craig to Lefferts, n.d., Lefferts.

5. James Brown was to be Mrs. Craig's agent in the transaction.

6. Francis Morris was active in many telegraph ventures. At this time he and George Harrington were members of the executive committee of Southern and Atlantic. Reid 1879, 452; see also ibid., 281, 310, 370, 417–18, 426, 432, 460–61, 466, 526, 533.

7. John Lefferts, brother of Marshall Lefferts. Craig to Lefferts, Saturday 1872, Lefferts.

8. Unidentified.

9. Unidentified.

UNIVERSAL PRIVATE LINE PRINTER, 1872
Doc. 262

Sometime in mid-March 1872 Edison abandoned his attempts to make the original universal private-line printer workable, and by early April he had arrived at a new design.[1] He quickly produced several models of this design for Gold and Stock, delivering thirty-three instruments during May and June,[2] but eight of those were defective. Shipments were curtailed in late June and did not resume until August. By the end of September, Gold and Stock had received nearly three hundred,[3] and at least two had been sent to Chile. Two others were displayed at the 1872 Cincinnati Industrial Exposition, where they apparently performed poorly in competition with other private-line printers.[4]

Like the first universal private-line printer, the new machine combined a printer and a transmitter. Unable to perfect the printing mechanism on the original design, Edison used the universal stock printer as the printing portion of his new instrument. The stock-printer design remained basically unchanged, except that the new universal private-line printer used only one transmission wire and one typewheel. Only the typewheel magnet was on the main circuit.[5] The transmitter, on the other hand, retained only the motor and governor of the earlier private-line printer. By having the motor work a set of electrical contact points through a series of ratchets and gearwheels, Edison solved the problems of precision and accuracy that had plagued the first machine's pulsator (breakwheel).[6]

Edison used his screw-thread unison on the universal private-line printer. He recommended that the machines' typewheels be "allowed to run to unison before each communication so as to ensure their correct position."[7] Edison solved the problem of printing-lever rebound—that is, getting the printing lever to bounce fast enough to avoid holding back the typewheel—by mounting the fulcrum loosely, relying on the lever's upward momentum to make the impression and on gravity to pull it back out of the way of the typewheel's rotation.[8]

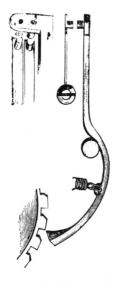

Because one of the two contact points in the universal private-line printer transmitter was slightly in advance of the other, the second point stayed clean and transmitted a clear signal.

1. See Docs. 164 and 165. The last dated notebook entry showing the vertical typewheel shaft of the original design was made on 6 March (Doc. 253). That month, Gold and Stock expenditures on the universal private-line printer rose from February's $179 ($155 labor, $24 stock) to $1,052 ($333 labor, $280 tools, $14 patterns, $425 stock). Outlays for April totaled $2,963. Cat. 1211:42–50, Accts. (*TAEM* 21:165–69).

2. Cat. 1183:101, 102, 110–13, 122ff, Accts. (*TAEM* 20:910, 911, 915–16, 921ff). Edison executed the covering patent application in April 1873 (U.S. Pat. 140,488).

3. Production figures exist only for 1872. The total number of universal private-line printers manufactured is not known. PN-72-07-12, Accts. (*TAEM* 20:986–87).

4. On Chile, see PN-72-07-12, Accts. (*TAEM* 20:986). The performance reports appeared in the *Telegrapher,* edited by James Ashley. Given Ashley's animus toward Edison, these reports must be taken guardedly. "Electricity at the Cincinnati Industrial Exposition," *Telegr.* 8 (1871–72): 457; "The Cincinnati Industrial Exposition.—The Award of Silver Medals in the Electrical Department," ibid., 482.

5. The printing magnet was on a local circuit that was routed through the typewheel escapement lever. Whenever the main circuit was open (between impulses) and the escapement lever was pulled down by its spring, the printing circuit would close. As in several of his earlier printers, Edison made the printing magnet relatively sluggish so it would not respond to the rapid impulses used to work the escapement and turn the typewheel. However, when the main circuit was opened to halt the typewheel, the printing magnet charged fully and raised the printing lever.

6. The motor drove the first gearwheel, which in turn drove the second by a ratchet and pawl. The second gearwheel drove another ratchet, which opened and closed the contact points. Depressing a key stopped the second wheel, and the contact points accordingly stopped vibrating while the motor and first gearwheel continued to spin.

7. NS-Undated-005, Lab. (*TAEM* 8:259).

8. Cf. Docs. 193 and 236.

–262–

Patent Model: Printing Telegraphy[1]

M (28 cm dia. × 15 cm), DSI-NMAH(E), Cat. 252,616. The long arm angled to the lower left would in operation have been one of two arms standing up with a roll of paper tape between them.

1. See headnote above. Edison and Unger sent this machine (serial no. 215) to the Gold and Stock Co. on 11 September 1872. However, the Smithsonian Institution received it from the U.S. Patent Office as the patent model submitted 16 May 1873. PN-72-07-12, Accts. (*TAEM* 20:987); Pat. App. 140,488.

263

*Edward Calahan to
Unknown Person*[1]

London[a] [c. June 1, 1872][2]

until they have reached the sum of £ 11 = So you see, with no freight nor packing they beat you in price. Of course they promise better workmanship—and one of them did not hesitate to pronounce ours, as d——d bad = and I immediately turned my back on him = he feels that he offended me, and writes me every few days but I take no notice of him = I have told no one about his low offer, and so long as you can fill orders promptly, and they don't get at the Officers of the Company[3]—all will go bravely on = Don't fail to send duplicate parts, especially springs—& that brass guard over the paper—Small Screws &c any scraps of parts around the shop— throw in the box—as they will come handy[b]   Andrews[4] must have told you that the 12 Insts[5] were in a very rusty condition and the straw with which they were packed was alive with New-Jersey Bed bugs—and if it was known that I had imported American Bed Bugs into Great Britain, I would be put in Newgate =[6] So look out how you compromise me—

By the way = Don't deposit that $360 with Clews.[7] Go to Duncan Sherman & Co[8] & buy ~~me~~ a draft on Bk of England, and send it to me by mail = that is the best way to do it = I hope before this letter reaches you, I will order another lot, say 50—[9] Andrews telegraphed me a few days since, that "50 Insts would go Wednesday" = I wonder if that was intended as a reply to my Message in which I asked him to "secure sixty Instruments" it came next Morning early— the letter was then on the way, with the order for sixty— Be sure and not forget the promise of "confidence" write me soon as you possibly can— Yours

E. A. Calahan

Direct letters &c. thus[c] 18 Cornhill—London, E.C.

ALS (fragment), NjWOE, DF (*TAEM* 12:893). Only the last two pages of this letter have been found. [a]Place taken from text. [b]"Small screws . . . handy" written in right margin and set off with line. [c]Followed by brace containing signature and address.

1. Although both Edison and Calahan invented for Gold and Stock and the American District Telegraph Co., there is no evidence that a close personal relationship developed. No correspondence between the two men has been located. Considering the friendly tone of this letter and the directions in the second paragraph to send money from a Wall Street bank to England, it is unlikely that Edison was the recipient. The letter was probably written to Marshall Lefferts or one of the other major stockholders in the Exchange Telegraph Co. (see n. 3), who then brought this part of the letter to show Edison in Newark, where he was manufacturing universal stock printers for the company. Edison supplied

his tickers to the company for a year or so, when manufacture began in England. The English instruments had the basic design of the universal stock printer with several details altered to increase transmission speed. Scott 1972, 16; Higgins 1877, 122–43.

2. Calahan resigned his position as superintendent of Gold and Stock at the end of March 1872 and sailed for England a month later to become the electrician for the Exchange Telegraph Co. He remained with the company for only one year. (*J. Teleg.* 5 [1871–72]: 101; "The Gold and Stock Telegraph in England," *Telegr.* 8 [1871–72]: 292; Scott 1972, 16). The second shipment of printers left Edison's shop for England on 13 June (see n. 9). These two circumstances suggest that this letter was written between mid-May and mid-June 1872.

3. The Exchange Telegraph Co. In the spring of 1872 George Field and six British men incorporated the company, became its officers, and in June began operations. "The Gold and Stock Telegraph in England," *Telegr.* 8 (1871–72): 292; Scott 1972, 15–16.

4. Elisha Andrews.

5. Sent from Edison and Unger on 24 February 1872. Cat. 1212:96 and Cat. 1183:85, Accts. (*TAEM* 20:746, 902).

6. Newgate was the prison for London and the adjacent County of Middlesex.

7. Henry Clews & Co., Bankers, 32 Wall St., New York. Wilson 1872, 207.

8. Duncan, Sherman & Co., Bankers, 11 Nassau St., New York. Wilson 1872, 319.

9. After the initial shipment of twelve, Edison and Unger next shipped twenty printers on 13 June, twenty more on 19 June, eleven on 20 July, and forty-eight on 25 July. Cat. 1212:238 and Cat. 1183:132, 137, 154, 156, Accts. (*TAEM* 20:809, 926, 928, 936–37).

-10- The Ascendancy of Manufacturing

July–December 1872

Compared to the first half of 1872, the remainder of the year witnessed an increase in Edison's devotion to manufacturing and an accompanying diminution in the number of his notebook entries and patent applications. He continued to develop new ideas for printing and automatic telegraphs, turned his attention to duplex telegraphs, and consolidated many of his business operations in the new firm Edison and Murray.

In printing telegraphy Edison concentrated on manufacturing. He received large contracts from the Gold and Stock Telegraph Company to manufacture his universal stock printer and his universal private-line printer, and he also received orders for his universal stock printer from the Exchange Telegraph Company of London, which began using his printer in June.[1]

Edison's work in automatic telegraphy consisted largely of tests conducted on the Automatic Telegraph Company line between New York and Washington and on the Southern and Atlantic Telegraph Company line from Washington to Charleston, South Carolina.[2] Because tailing remained a problem, he devoted much attention to the design of circuits to overcome this difficulty, using electromagnets, condensers, rheostats, or batteries in shunt circuits to alter the electrical condition of the line.[3] While the lines and instruments worked sufficiently for Automatic Telegraph to open for business in December,[4] the company's hopes for competing with Western Union rested on continued improvements. To this end, Josiah Reiff, secretary of Automatic Telegraph, became a party to Edison's October 1870 agreement with George Harrington and agreed to provide the inventor a salary of $2,000 per year in support of his efforts.

During the fall, Edison also began experimenting with duplex telegraphy. Western Union's adoption of Joseph Stearns's improved duplex earlier in the year probably prompted Edison's renewed interest in such systems.[5] He began discussions with Western Union president William Orton concerning the company's possible interest in other systems. No specific arrangements were made, however, until the following February.[6]

In July, Edison made a significant change in his business relations. He dissolved his partnership with William Unger and soon thereafter consolidated his manufacturing operations in the firm of Edison and Murray. This new establishment also took over the operations of Murray and Company, which Edison and Joseph Murray had formed the previous winter. The new firm located its operations in the former Edison and Unger shop on Ward Street.[7] Edison's acquisition of Unger's interest required a $10,000 mortgage and $7,100 in notes, liabilities that placed great financial pressure on the new company. Although Edison paid the $3,100 in notes due in early October, he had to refinance the remaining $4,000. At the same time, Edison and Murray delayed the payment of bills due from their manufacturing suppliers and offered notes to satisfy their creditors.

1. Production of these instruments began in the spring and continued through the summer. See Cat. 1183:89–183, PN-71-09-06, and PN-72-07-12, Accts. (*TAEM* 20:904–50, 982–83, 986–87).

2. Edison made notes regarding tests between New York and Washington. He included test strips of chemically received messages and accompanying notes that indicate he used condensers and coils for altering the condition of the line. These notes could have been made as early as the latter part of 1872 or as late as 1875. NS-Undated-005, Lab. (*TAEM* 8:217–28).

3. See U.S. Pats. 135,531, 141,772, 141,773, 141,776, and 150,848.

4. "Automatic Telegraphy in Practical Business Operation," *Telegr.* 8 (1871–72): 556.

5. Edison's interest may also have been spurred by work undertaken in the American Telegraph Works for Stearns and W. H. Mendell in late winter. The surviving records, however, do not indicate whether this was related to duplex telegraphy. 72-022, DF (*TAEM* 12:977–986).

6. Orton's testimony, Quad. 71.1, pp. 116–17 (*TAEM* 10:63).

7. Unger established a new shop in New York City (see Doc. 92, n. 1), apparently purchasing machinery and tools for it from Edison and Murray. Cat. 1183:188, 190, Accts. (*TAEM* 20:953–54).

264

Agreement with
William Unger

Newark, July 3, 1872[a]

This Agreement made and dated this third day of July one thousand eight hundred and seventy two Between Thomas A Edison of the one part And William Unger of the other part both of the City of Newark County of Essex and State of New Jersey

Witnesseth Whereas a partnership has existed and now exists between the said parties in the manufacture of telegraphic apparatus and electrical machinery, experts in electricity, application of magnetism to machinery;[1] such business being carried on in said City of Newark.

And Whereas the said parties are now desirous to dissolve such partnership.[2]

Now Therefore this Indenture Witnesseth

First. That the said parties agree with each other that the said partnership between them existing shall be and from this date, is, dissolved annulled and at an end.

Second. The said William Unger hereby bargains sells and transfers unto the said Thomas A Edison all his right, title and interest in the said partnership, the tools and machinery wherewith the said manufacture is carried on, All Contracts and choses in action[3] whatever to the said partnership belonging, excepting so far as hereinafter stated; Also the good will of said partnership and of the business aforesaid, and all property thereof whatsoever.[b]

Third   In consideration whereof the said Thomas A Edison doth hereby agree to pay and assume all outstanding debts of the said partnership and to indemnify and to save and keep harmless and indemnified the said William Unger of from and against all debts, claims, damages and demands whatsoever, which may be brought against the said partnership for and by reason of anything done in the carrying on thereof

Fourth   The said Thomas A Edison further agrees to pay to the said William Unger as part consideration herefor the sum of Twenty five hundred dollars in cash[4] and also to pay the further sum of Five Thousand dollars, for which he is to give his promissory notes to the said Unger in the sum of One[c] Thousand dollars each, payable at three four, five, six and seven months herefrom;[5] And further to deliver to said Unger his own promissory note now held by said Edison for Three Thousand Six hundred and forty two dollars and forty nine cents.[6]

Fifth. The said Thomas A Edison further agrees to pay as part consideration herefor unto the said William Unger, the

sum of Ten Thousand dollars,[7] for which said Edison agrees
to give his Bond payable on or before two years from this date
with lawful interest; such Bond to be secured by a Mortgage
of all the Stock in trade, tools, machinery, choses in action and
the property hereby conveyed[8]

In Witness Whereof the said parties have hereunto interchangeably set their hands and seals the day and year first
above written

Thomas A. Edison.[d]   William Unger[d]

Signed sealed and delivered in the presence of   Charles
Batchelor[9]

DS, NjWOE, DF (*TAEM* 12:816). In an unknown hand. [a]Place and date
taken from text; form of date altered. [b]"and all . . . whatsoever." apparently written by Edison. [c]Written by Edison. [d]Followed by a wax seal.

1. The characterization of the business was extracted from the Edison and Unger advertisement in *Holbrook's Newark City Directory* 1872,
441.

2. Specific reasons for this decision are not known, but Unger soon
set up his own electrical and machine shop in New York City. See Doc.
92, n. 1.

3. "Choses in action" are items of personal property consisting of
obligations such as patent rights and certain kinds of contracts.

4. A list of bills dated 1 March [1872?] indicates that the firm owed
Unger $2,500, which may be the source of this stipulation (72-012, DF
[*TAEM* 12:810]). On 6 August, Edison gave Unger a sixty-day note for
$2,100 (72-001, DF [*TAEM* 12:663, 665]).

5. On 3 July, Edison gave Unger five notes for $1,000 each, payable
at one-month intervals. 72-001, DF (*TAEM* 12:663, 670–72).

6. This note originated in September 1871; the actual amount was
$3,649.42, not $3,642.49. See Docs. 197 and 215.

7. Neither the $2,500 payment nor the return of Unger's note in-
volved the value of the partnership's assets. Therefore, Edison agreed to
pay Unger $15,000 for his half of the partnership, meaning that the total
value of the business was being set at $30,000. This figure agrees closely
with that in Doc. 215.

8. Neither the mortgage nor the note to Unger has been found. An
undated list of notes and bills owed by Edison and Murray indicates that
the mortgage amount was only $8,000 (72-018, DF [*TAEM* 12:911]),
but Doc. 280 (1 January 1873) shows a $10,000 obligation with an in-
terest payment of $350. A receipt from the Republic Trust Co., dated 3
January 1873, is for this interest payment (72-002, DF [*TAEM* 12:995]).

9. Charles Batchelor (1845–1910) was born in Dalston, a suburb of
London, England. He grew up in Manchester and received his early
training in textile mills. He first came to the United States in 1865 to
exhibit a textile machine. After returning to Manchester, he was employed by the J. P. Coates Co., thread manufacturers. In 1870 the company sent him to install machinery in the Clark thread mills in Newark,
N.J. By the end of October 1870 Batchelor had obtained employment
in Edison's American Telegraph Works. Later he worked for Edison and

Unger and then for Edison and Murray in the Ward St. shop. By the summer of 1873 Batchelor was assisting Edison in his experiments and he later became Edison's most trusted assistant at the Menlo Park and West Orange, N.J., laboratories. Welch 1972.

–265–

R. G. Dun & Co.
Credit Report

Newark, July 29, 72.

Murry & Co. Thos A Edison.[a]   Mfrs Tel Instruments
111 NJ RR Ave.

#3700[1]   T. A. E. states that he is a genl. ptnr. & that it will soon be consolidated with his own bus.[2] & be conducted under his own name

D (abstract), MH-BA, RGD, N.J. 21:125. [a]See Doc. 252, textnote a.

1. See Doc. 252, n. 2.
2. Edison bought out William Unger's share in Edison and Unger on 3 July 1872 (see Doc. 264), and by about October he had consolidated the Edison and Unger and Murray and Co. shops. Joseph Murray's testimony, Quad. 70.1, p. 33 (*TAEM* 10:21); Doc. 270.

–266–

R. G. Dun & Co.
Credit Report

Newark, July 29/72

Thomas A. Edison.[a]

#3700[1]   (Repts Thomas A Edison.) succ[esso]r to Edison & Unger, Recently Diss[olve]d. States that he is wor[th] abt 53m$[2] after paying off all liabs & that this includes what he owns in the firm of Murry & Co in the same bus in this city. We find that his assets are nearly all made up of his stock of Telegraph Machines on hand & the machinery he uses in bus. of manfg them   he claims to have 3 or 4m$ R[ail].R[oad]. stock[3] & 15c$[4] paid on R[eal].E[state].[5] & 8m$ due soon for instruments making on Contract. He is young & has been in bus too short a time to have accumulated a Bona fide Capl of 50 to 60m$ Especially as he Commenced on borrowed Capl a little over a year ago.[6] "E" states that he owes Unger as pr settlement.[b] at dissolution $17,500 owes on the bus abt 7m$ & owes on the bus of Murry & Co 3m$ m[a]k[in]g an indebtedness of $27,500. The detail of his assets we Cannot give beyond his statement above, except that he claims some 50m$ of mortgd stock on hand.[7] Says that he is shipping machines to England & that there is a Constant demand for them in both home & foreign trade. He has a lge amt of machinery employed, none of it expensive, keeps no Bank a/c. Is Consid talented in the way of inventive genius. Is at present Considbly spread out, being in two firms but he states that Murry & Co will soon Consolidate with him. Has so little to do with the trade here that we can get but little outside expression as to his cr[edit] or standing & his ultimate success although it might in a measure be Consid assured with Careful Management is Yet Consid problematical & cr shd be extended with a good deal of Caution   2092

D (abstract), MH-BA, RGD, N.J. 22:238–39. ᵃFollowed by "Sub-[sequent] opposite"; the second half of this report was copied on the facing page. ᵇPage change indicated here by "(Contd. opposite)"; "Thomas A. Edison Manfr Tel Instrms prev opposite. (# Contd from opposite page)" written at head of entry on continuation page.

1. See Doc. 252, n. 2.
2. "m" means thousand.
3. In February, Edison had borrowed $3,100 from Josiah Reiff to help his brother Pitt Edison with the Port Huron and Gratiot Street Railway. Quad. 70.7, p. 704 (*TAEM* 9:722).
4. "c" means hundred.
5. See Doc. 206.
6. This may be a reference to the money and machinery Edison acquired from George Harrington for the American Telegraph Works, some of which Edison transferred to Edison and Unger after he quit working at the American Telegraph Works in the fall of 1871. Quad. 70.7, p. 704 (*TAEM* 9:722); 73-003, DF (*TAEM* 12:1001–34).
7. See Doc. 264.

–267–

Lemuel Serrell to Commissioner of Patents[1]

New York, Aug 5th 1872

Sir.ᵃ

In the matter of the application No. 38, of T. A. Edison for Chemical Telegraphs[2] the present is to advise that upon consultation with him I find that he has experimented with the same considerably and his idea is that every atom of matter has a positive or negative polarity and subject to electric condition, that when "The main line current from the stylus gives to the particles an electric condition, that condition continuing, tends to prolong the mark after the pulsation ceases, the cross or counter current neutralises this electric condition or polarity and prevents attenuation of the mark, it also prevents injury to the stylus or pen by the action of the acids in the paper, because the current passing between the ~~points~~ conductors i.i. is superior to any ground currents and neutralises their action."[3]

If the portic̶l̶e̶sons included in quotation marks are substituted for the latter part of paragraph 11 after the words "<u>roller</u>

d" it is believed the action will be more clearly understood and that amendment is hereby authorised   Respectfully yours

Lemuel W. Serrell   Atty.

ALS, MdSuFAR, Pat. App. 134,867. Letterhead of Lemuel W. Serrell. [a]To this point, "New York," "1872", and "Sir." preprinted.

Drawing from Edison's
U.S. Pat. 134,867.

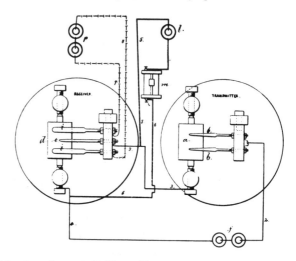

1. Mortimer Leggett. *DAB*, s.v. "Leggett, Mortimer Dormer."
2. Edison executed this application on 8 May 1872 and it issued as U.S. Patent 134,867.
3. Edison could have drawn on a number of atomic theories that were current in the nineteenth century, including those of Michael Faraday in his three-volume *Experimental Researches in Electricity*, which Edison had read. For nineteenth-century atomic theories in general, see Knight 1967 and Rocke 1984.

–268–

From Lemuel Serrell

New York, Sep. 15 1872.[a]

Dear Sir

About eighteen years ago a locomotive was fitted on the New Haven & Northampton R.R. by H Bradford[1] then of Bristol which was made thus

there was a large funnel in front of the locomotive and a pipe running along over the cars with couplings.[2]

In view of this shall the caveat be prepared for your modification?[3] Yours truly

Lemuel W. Serrell

ALS, NjWOE, DF (*TAEM* 12:682). Letterhead of Lemuel Serrell. a"New York," and "1872." preprinted

1. Unidentified.

2. This appears to be a scheme for ventilating the cars with air unpolluted by the smokestack.

3. Nothing is known of this caveat.

-269-

S. A. Woods Machine Co. to Edison and Murray

New York, Octo. 2 1872[a]

Gentlemen.—

You have assured us many times within the past four or six months, that you would soon pay us the amount due. More especially have you promised us, that during the month of September, you would surely make settlement with us,—and yet,—at this day,—"settlement" seems as far off as ever.

Our last statement,—handed to Mr. Edison by the writer shows the amount due us to be $3,053.85,—this includes the accounts of T. A. Edison, J. T. Murray & Co. & Edison & Unger.[1]

The relations between us have always been friendly.—and pleasant,—and we dislike to do anything to give you annoyance or inconvenience,—but we are the servant of the Company we represent,—and we are ordered to take measures to collect this account at once,—we must do so. Gentlemen unless our claim is paid,—at once,—we shall place the matter in the hands of our attorney,—with instructions to commence action,—immediately. This will be our last communication on the subject. Yours Very Truly.

S. A Woods Mch Co.[2] pr. Bartlett[3]

L, NjWOE, DF (*TAEM* 12:907). Letterhead of the S. A. Woods Machine Co. a"New York," and "187" preprinted.

1. The total bill consisted of $2,634.75 owed by Edison and Murray, $320.99 owed by Edison and Unger, and $99.10 owed in interest. Murray forwarded the statement for payment to "Mr. Miller" (probably Norman Miller, secretary and treasurer of the Gold and Stock Telegraph Co.), and it was paid in three installments over the next two weeks. At this time, Edison and Murray owed its creditors nearly $9,000. See Doc. 270.

2. The S. A. Woods Machine Co. manufactured wood- and iron-working machinery. Edison bought a great deal of equipment from the company, mostly lathes and lathe accessories. Cat. 1218:51, Accts. (*TAEM* 21:208).

3. Unidentified.

Edison and Murray Account

Bills against Thos. A. Edison. & J. T. Murrey.[2]

⟨Notes⟩[a] ⟨Nov 14th⟩[b]	T A Edison	J. T. Murrey	Total.
A Brinckman	418.13		418.13
Crowell & Co.	81.50	⟨paid⟩ 91.29[c]	172.89[c] ⟨Soon as possible⟩
L Bradley.	1,087.50		1,087.50
Kinney & Co.	51.65[c]		51.65[c]
Macknet & Wilson	Ɵ146.11		Ɵ146.11
Holmes Booth & Hayd[en]	303.41[c]		303.41[c] ⟨Light Draught⟩
C Walker.	13.34	173.96	187.30 ⟨Soon as we can⟩
J. Geiger	38.49		38.49
J H Thomas.	74.44		74.44
Geo. Price	672.83	⟨500.⟩[d]	672.83 ⟨is 500—[e] Labor done⟩
Hopperton	21.67		21.67
Young.		52.[c]	52.00[c]
Thropall ⟨Oct 29⟩[b]	10.55		10.55
Kirk & Co.	430.		430.00
Fried & Co.	44.82		44.82 ⟨soon⟩
Bonnell	126.82		126.82 ⟨very Sad case⟩
Gaslight Co.	81.58[c] ⟨Cut off our cock[3] soon⟩		81.58[c] ⟨Must Pay⟩
S. A. Woods[4]	336.[c]	1,634.75[c]	1,90.75[c] ⟨you can fix this⟩
W H Kirk (rent)[5]	799.98		799.98 ⟨soon as we can⟩
⟨Oct 24⟩[b]			
S. Stone & Co.	85.86		85.86
C H & J D. Harrison	43.74		43.74
Jacob Gauch	122.16		122.16 ⟨very bad case⟩
C. Batchelor		125.	125.00
Frasse & Co.	25.02	148.64	173.66 ⟨Soon as we can⟩
Lucius Pond.	135.38		135.38
C B. Smith	67.86		67.86 ⟨quick⟩

P. Frasse.	109.99 ⟨Fifty will do⟩		109.99 ⟨very bad Case⟩
A J. Davies	46.85	117.06	₂163.91
⟨Note⟩ Douglas & Sons	181.49		181.49
Cowles & Parcels	45.99		45.99
J B. Mayo.	67.50ᶜ		67.50ᶜ
S S Thorn.	78.11		78.11
Byles	6.06		6.06
Duerr & More-house	19.94		19.94
Post & Goddard	_____	93.	93.00 ⟨very Bad Case⟩
	5,774.77	2,435.80	8,210.57ᶠ
Brt. forward	5,774.77	2,435.80	
Brot over			8,210.57
Post & Kalkman		23.63	23.63
E N Wrigley		349.21	349.21 ⟨Sued put off till Jan. 6th⟩⁶
⟨rent⟩ Roberts⁷ ⟨Paid $200.⟩		375.00	375.00
			$8,958.41
	5,774.77	3,183.64	

D, NjWOE, 72-018, DF (*TAEM* 12:909). Written by Charles Batchelor; commas added. ᵃAll marginalia written by Joseph Murray except where noted. ᵇWritten between two lines. ᶜCrossed out later. ᵈWritten in another hand, possibly Edison's. ᵉ"is 500—" written in another hand, possibly Edison's. ᶠCrossed out later; followed by "over" to indicate page turn.

1. A list of notes, including those found in this document, is dated 10 August 1872 (Cat. 1183:392, Accts. [*TAEM* 20:970]). An undated but related list is found in 72-018, DF (*TAEM* 12:911). Many of these bills can be found in 72-014 and 72-019, DF (*TAEM* 12:819–75, 918–58); and in Cat. 1183, Accts. (*TAEM* 20:585–97).

2. About this time Edison and Murray consolidated their Oliver St. shop (Murray and Co.) with Edison's Ward St. shop. The new firm, located at 10–12 Ward St., was known as Edison and Murray (Joseph Murray's testimony, Quad. 70.1, p. 33 [*TAEM* 10:21]). The creditors of the two shops included craftsmen as well as manufacturers and retailers of machinery, tools, metals, leather, and hardware. Most of them were located in Newark, with a few in New York City, Connecticut, and Massachusetts. Most can be identified in *Holbrook's Newark City Directory 1872*, in Wilson 1872, or from their billheads.

3. Gas cock—that is, a valve.

4. See Doc. 269.

5. William Kirk owned the building on Ward St. in which the shop was located. This amount represents two months' rent. 71-012, DF (*TAEM* 12:480); Cat. 1221:74 (*TAEM* 21:342).

6. Edward Wrigley owned a machine shop at 19–21 New Jersey Railroad Ave., Newark, which specialized in shafting and pulleys (Ford 1874, 63). On 5 October 1872, Wrigley entered a plea in the Essex Circuit Court claiming that Edison and Murray owed him $800 for labor and material plus damages and interest. On 17 March 1873 Edison and Murray entered their plea claiming that the amount due was $445.96. The court agreed with Edison and Murray and also ordered them to pay court costs and sheriff's fees. On 27 March they paid $521.63, including interest from 17 March. Essex Circuit Court Minutes 1872–73, p. 349, NjNECo; Essex Circuit Court Judgments, 50:568–70, NjNECo; receipt from Essex Circuit Court, 73-007, DF (*TAEM* 12:1109).

7. Henry Roberts, a Newark wire manufacturer, owned the shop Edison and Murray rented at 39 Oliver St. The figure given here represents one month's rent. 72-019, DF (*TAEM* 12:946, 967).

–271–

To Norman Miller

New-York, Octo 11, 1872[a]

DrSir,

The enclosed Notes[1]

Nov. 6/72	1000.
Dec. " "	1000.
Jany " '73	1000.
Feby " "	1000
	$4000.

Have been secured by new notes as follows.[2]

Date Octo 9th 72 —[b]

at 2. 3. 4 5 6 7 8 9 mos $500 each = $4000.[c]

to order of Herman Unger,[3] and [end?] J. T. Murray[4] payb. at Public Trust Co.[5] Newark.

L (copy), NjWOE, DF (*TAEM* 12:669). Written by Norman Miller on letterhead of the Gold and Stock Telegraph Co. [a]"New-York," and "187" preprinted. [b]Followed by "Correct Copy J. T Murrey", in Murray's hand. [c]Underlined twice.

1. These were the remaining notes resulting from Edison's purchase of Unger's share in their partnership. See Doc. 264, n. 5.

2. The notes are found in DF (*TAEM* 12:672–75, 1116).

3. Herman Unger, William's brother, was at this time working as an engraver. Eventually he and his brothers established a noted jewelry firm in Newark. Johnson 1966, 12, 32–33; Rainwater 1975, 175–76.

4. The meaning of "end" here is probably "endorsed"; Murray endorsed all the notes.

5. The notes were drawn at the Republic Trust Co.

My Dear S. M.[1]

Your several letters received,[2] & I can assure you they are most heartily appreciated   I should have written you more frequently, but that when my mind is closely bent upon any one thing, I am sadly neglectful of the courtesies of Lifte— I will now try and make amends by a full Letter. First regarding your own matters—I do not at all like the position I find you are getting into, in your association with my friend Reiff. I really am filled with Pity for him in the burden he now has on his shoulders.[3] Its a big thing, you must acknowledge, to be bearing the brunt of a battle with fate, like the present one, —and if he does not act as you think he ought, there is this to remember, that he is freighted doubtless with many considerations that neither you or I—wot of.— On the other hand my sympathies and my convictions of justice, are with you most cordially— I know of course how little claim in justice Little has to the model of Instrument, in question and am ready at any & all times to assert my opinion in the matter & I assert, to you, most Emphatically that I think any attempt on his part to patent your model, ~~a~~is a downright steal[4]—& further that in the support of your claims you should receive the moral & active cooperation of our people— and thinking this, and knowing Col Reiff as I do, I believe with an unwavering belief, that he means exactly what he Says & nothing more—

I shall look for the out-come of this unfortunate contretemps—in the Success of both sides of which I am ~~n~~In a manner Interested—with no little Interest.

Interested on the ~~side~~one side you understand only as it may affect Reiff— I dont love Little or his dishonest trickery =

My Interests are not prospering to my satisfaction Either. I am in daily receipt of communications as to how to do this & how to do that, whilst the facts show that I have mastered the situation myself, & am now sending the night business of the S[outhern] & A[tlantic] Co, to Washington by automatic— while New York is not able to get a thing from me. I send to Philada with much better results, & receive from him & New York very fairly when they adhere to my plain & simple way of manipulating this thing. Edison has been trying a series of experiments—of which I heartily approve as they may result in something good—but as yet none of them evince a tendency to Improve matters much.[5]

One thing he has tried with a moderate degree of success, I have myself had in use for several days, and obtain with it the best possible results. From a very difficult working of the wire, with all Parallel wires[6] open, & all relays out, I have brought it to a practical working regardless of Parallel wires or relays—and am now doing the night business between here & Washn. (Bertram[7] is doing well)

I have Kept my own Counsel in the matter, & Edisons trial of a similar method, (though not with the same connections or as successful) will doubtless ~~secured~~ accrue ~~to~~ him the Credit. However let that go.

one thing is sure we are moving ahead slowly but surely.

(I must defer this till tomorrow)

Oct ~~2~~3oth = I was compelled to drop ~~yo~~my letter to go to work with Washington and I was not able yesterday to take it up.

Last night we Sent to Washn 22 Night Messages for S & A Co, & would have sent as many more but for the fact that Bertram had but two Copyists.[8] He will however have a full force on hand tonight So that we can send right along to him & not be compelled to wait 10 or 15 mins, after sending a batch for him to assist in Copying, & thus lose all the advantages of our system.

So far as Washington is Concerned I have things in a progressively Improving shape. But New York hangs fire, though I think when Edison gets my Big—quick pulsating Battery up[9]—we will be able to do something through ~~to~~ him =

Of the workings of Inside Matters I am utterly without advices owing to Reiff slighting me in his Correspondence, by turning me over to Mr Harrington, which is Equivalent to Keeping me in the Dark—

I will write again, soon, but you must remember that my time & mind are pretty well occupied, & that my Correspondence is necessarily pretty heavy, I write on an average 5 or 6 letters per Day—beside several sheets of a Voluminious report upon "Automatic Telegraphy" which I am preparing for Gen Palmer,[10] & which I will let you see

How Comes on the Printer[11] & Hows Barnitz[12] Yours Very Truly

E. H. Johnson[13]

I hope you will not drop off, because of my not promptly answering   E. H. J.

ALS, NjWOE, DF (*TAEM* 12:790).

1. S. M. Clark was general manager of the American Telegraph Works. New Haven Wire Co. to Clark, 8 Aug. 1871, DF (*TAEM* 12: 253); *New York Sun*, 16 Apr. 1875, Cat. 1144, Scraps. (*TAEM* 27:278).

2. Not found.

3. Perhaps the financial burden of the Automatic Telegraph Co., into which Reiff was pouring tens of thousands of dollars. See Doc. 141, n. 5.

4. Accounts of the American Telegraph Works list work done on "Clark's printer" between 2 February 1872 and 3 January 1873. Nothing is known about the device, but it was probably a typewriter to be used in transcribing messages for automatic telegraphy. Cat. 30,108, Accts. (*TAEM* 20:343–449 passim).

5. No record of these experiments survives.

6. "Parallel wires" probably refers to wires strung on the same poles. Inductive interference caused by proximate parallel wires was a particular problem for the automatic telegraph's sensitive receiver.

7. Harry Bertram was the manager of the Washington, D.C., office of the Automatic Telegraph Co. Little 1874a, 183.

8. That is, workers to transcribe the messages from the receiving tape.

9. Nothing is known of this battery.

10. William Palmer.

11. It is not known whether this is a reference to Clark's printer or to Edison's copying printer (typewriter).

12. David Barnitz was a secretary or bookkeeper at the American Telegraph Works.

13. Edward Hibberd Johnson (1846–1917) began his career as a telegraph operator in the late 1850s. In 1871 William Palmer and Josiah Reiff hired Johnson as "general supervisor of the automatic telegraph system and apparatus" for the Automatic Telegraph Co., and he continued working in that capacity with the company's successor, the Atlantic and Pacific Telegraph Co., until 1876. At that time he began working for Edison, an association that lasted many years. *NCAB* 33:475; Johnson's testimony, Quad. 70.7, pp. 737–38 (*TAEM* 9:739–40).

–273–

Patent Assignment to Gold and Stock Telegraph Co.

[New York?,] October 30, 1872[a]

"Copy"

Whereas Letters Patent have been duly granted to me Thomas, A. Edison of Newark, in the State of New Jersey, for Improvements in Printing Telegraphs, which Letters Patent bear date and are numbered as follows.

Mar 28, 1871, No 113033. = July 2, 1872, No 128604 = July 2 1872, No 128605 = July 2, 1872 No 128606 = July 2, 1872 No 128607 = July 2, 1872 No 128608 = Sep 17, 1872 No 131,334 = Sep 17, 1872 No 131,335 = Sep 17, 1872 No 131336 = Sep 17, 1872 No 131337 = Sep 17, 1872 No 131338 = Sep 17, 1872 No 131339 = Sep 17, 1872 No 131340 = Sep 17, 1872 No 131,341 = Sep 17, 1872 No 131342 = Sep 17, 1872 No 131343 = Sep 17, 1872 No 131,344.

And whereas the Gold & Stock Telegraph Company, duly incorporated and located in the City and State of New York, has purchased of me the entire right in each and all the aforesaid Letters Patent.[1]

Therefore this indenture witnesseth, that for and in consideration of the sum of one dollar to me in hand paid, the receipt whereof is hereby acknowledged I have assigned and sold, and by these presents do transfer, set over and convey unto the said Gold & Stock Telegraph Comp'y my entire right, title and interest of every character in and to each and all the aforesaid inventions and Letters Patent: the same to be held and enjoyed by the said Gold and Stock Telegraph Company for its own use and behoof and for the use and behoof of its legal representatives, to the full end and term for which said Letters Patent are granted, as fully and entirely as the same would have been held and enjoyed by me had this assigment and sale not been made.

In testimony whereof I have hereunto set my hand and affixed my seal this thirtieth day of October A.D. 1872.

Witnesses   M. Lefferts[b]   T. A. Edison[b]

D (copy), NNC, Edison Coll. This assignment is recorded in Libers Pat. D-16:209. [a]Date taken from document, form altered. [b]Followed by representation of a seal.

1. See Doc. 164.

–274–

Agreement with Josiah Reiff

[New York?,] November 5, 1872[a]

Supplemental Contract

It is hereby agreed & understood by & between Thos. A Edison of Newark New Jersey party of the first part & Josiah C Reiff of New York party of the second part, that the contract made by & between said Thos. A. Edison & Geo Harrington[1] under date of   was understood & is still held to mean that said Harrington & Edison became joint owners in the interests, inventions & improvements therein mentioned & that said Harrington became the atty acting for the joint interest of himself & the said Edison & that the net results of all such inventions & improvements then made or to be made applicable to or useful in connection with Automatic telegraphy disposed of to any Telegraph Company or other party or parties in the United States or Elsewhere were to be and are to be divided one third to said Edison or his assigns and two thirds to said Harrington & his assigns—

2nd. That in Consideration of the special attention now

being given & to be Continued in the interest of automatic telegraphy by said Edison, [-]Effort will be made at reasonable Expenditure to develope the inventions & improvements of said Edison & that patents shall be duly obtained in the United States & Foreign Countries & that every effort shall be made by said Josiah C Reiff in Connectione with said Edison & Harrington to secure the best market & most advantageous terms for the Common benefit in the proportions heretofore named as the basis of contract already Existing between said Edison & Harrington & that from any such net results if any the portion due said Edison shall not be less than Ten thousand %100 Dollars for the United States & a like amount for the foreign patents—

3rd   And it is further agreed that said Reiff will become responsible to said Edison at rate of two thousand %100 Dollars ($2,000) per annum for Income while his (Edisons) time & Efforts are given as proposed in Article 2nd preceding.

4th   And further we mutually & individually agree to labor & heartily advise and cooperate for the Common benefit of said Edison, Harrington & Reiff & the parties represented by said Harrington & Reiff.

In witness whereof, we have hereunto set our hands & seals this Fifth day of November 1872.

Josiah C. Reiff[2]

D (copy), NjWOE, DF (*TAEM* 12:794). Written by Reiff. [a]Date taken from document, form altered.

1. Doc. 155.

2. Edison did not sign this copy; however, the agreement apparently was consummated because Edison did receive $2,000 a year from Reiff for at least two years. Receipts, 73-005 and 74-005, DF (*TAEM* 12:1098, 13:57).

MULTIPLE TELEGRAPHY   (Docs. 275–279)

In the latter part of 1872 Edison began once again to design and test duplex telegraph systems. These efforts came in the wake of Joseph Stearns's successful development earlier in the year of a duplex system that employed condensers (capacitors) in the circuits. Stearns's system was adopted by Western Union and no doubt rekindled Edison's earlier interest in the subject. Edison's approach reflected two dominant themes in his telegraph inventions: the use of the polarized relay and the pursuit of many variations on his basic design themes.

Edison approached duplex telegraphy differently from most other telegraph inventors.[1] He often employed an ordinary neutral relay actuated by variations in current strength on one end of the line and a polarized relay actuated by variations in polarity of the current on the other. However, in reversing the polarity of the current in the circuit in order to operate the polarized relay, the current strength momentarily dropped to zero and briefly interrupted the neutral relay. Edison therefore sought to solve the problem of reversing the polarity while not disturbing an ordinary relay.[2] Docs. 275–79 illustrate his efforts. He sketched these designs for testing by his assistant, James Brown.[3] While Edison's work did not immediately prove successful, his progress encouraged his further efforts in multiple telegraphy.[4]

1. These differences were noted in Prescott 1877, 822, 834.
2. For some time Edison had employed both kinds of relays in the same circuit in printing telegraph designs, but they did not require that the neutral magnet be undisturbed by the reversals of polarity. See Docs. 54, 88, and 187; and U.S. Pat. 131,334.
3. Docs. 275–79 were placed in laboratory scrapbooks, probably in the late 1870s. They date from the same period and share many other characteristics. These designs probably predate Docs. 282–85 and thus were likely among those made in late 1872.
4. Thomas A. Edison, Preliminary Statements of 27 Apr. 1878 and 31 Mar. 1879, pp. 149, 151, Testimony and Exhibits in Behalf of T. A. Edison, *Nicholson v. Edison.*

–275–

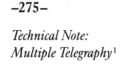

Technical Note:
Multiple Telegraphy[1]

Brown[2]   Try this[3a]

[Newark, 1872?]

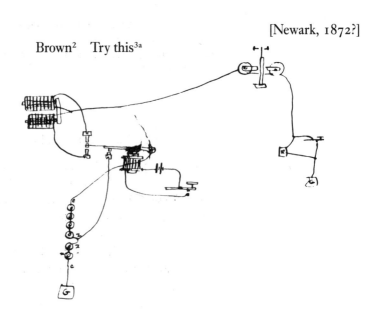

AX, NjWOE, Lab., Cat. 297:3 (*TAEM* 5:436). ᵃFollowed by "Top" here and "No 8" at lower right of the drawing, both in orange pencil or crayon, added later in another hand; a small sketch below in pencil was probably added later.

1. See headnote above.

2. James Brown assisted Edison in his shop until sometime in 1873, when he went to Great Britain to assist in the automatic telegraph experiments. He then left Edison's employ to work at the submarine telegraph cable station at Aden. See Edison's testimony, Testimony and Exhibits on Behalf of T. A. Edison, p. 12, *Nicholson v. Edison*; Cat. 297:3, 5, 21, Lab. (*TAEM* 5:433, 445, 515); and U.S. Pat. 185,622.

3. This is an example of Edison's basic design for duplex systems that employed a neutral relay on one end and a polarized relay on the other. As shown in the diagram immediately below, the components of the circuit at station 1 include a local circuit with transmitting key K^1, local battery B^3, and local relay **LR**. Ordinarily the armature arm of **LR** is in contact either with **x** or with **y** and **z**. B^1 and B^2 are the batteries for the operation of the main circuit from stations 1 and 2. Their polarity is indicated by **c** for carbon ($+$) and **z** for zinc ($-$). **NR** is the neutral relay, which is wired so that only one solenoid is in circuit at one time.

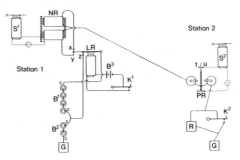

Missing from the drawing is the sounder S^1 and accompanying circuit, including the armature of **NR**, which actuates the sounder for receiving messages at station 1. Station 2 includes the transmitting key K^2, ground **G**, resistance **R**, and polarized relay **PR** with its armature at either **t** or **u**—the contact points of the switch that actuates the missing sounder (S^2) and accompanying circuit for receiving messages at station 2. This is a closed-circuit design; that is, the keys are ordinarily closed and there is current on the line when the circuit is not in use.

The polarized relay **PR** responds only to reversals of polarity of the electricity, while the neutral relay **NR** responds to changes in current strength. Key K^1 operates a transmitting switch that reverses the polarity of the current on the line and thereby actuates polarized relay **PR**, while the key K^2 can shunt resistance out of the circuit, thereby affecting current strength and neutral relay **NR**. Cf. design no. 16 in Doc. 285, and Docs. 294, 297, 300, and 315.

Technical Note:
Multiple Telegraphy[1]

[Newark, 1872?]

Brown[2]   Try it this way[3]   Keep memorandum how it work =

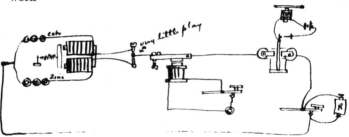

wound opposite, so that one battery will go around the wire in the same direction as the other[4]

ADDENDUM[a]

[Newark, 1872?]

I cant get the vibrating point on Sounder but what it will break circuit and when I close key it works relay in the reverse direction   the effect is when I close distant key I have to adjust very high on relay when key is open and very low when key is closed in order to work relay at all   The polarized relay works all OK

AX, NjWOE, Lab., Cat. 298:5 (*TAEM* 5:12). [a]Addendum is an AX; written in another hand, presumably Brown's.

1. See headnote, pp. 507–8.
2. James Brown.
3. This design differs from Doc. 275 in that it uses a different method of reversing the current, employing a wire loop to close the circuit instead of grounding it at each end, and showing the armature and spring for the neutral relay at the left and the sounder operated by the polarized relay at the right.
4. The coils of the neutral relay had to be wound so that the magnetic polarity of the relay's core would not change when one battery was switched into the circuit in place of the other.

Technical Note:
Multiple Telegraphy[1]

Brown[2]   Try this, one Battery[3]

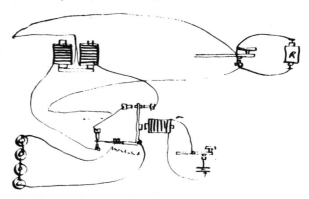

AX, NjWOE, Lab., Cat. 298:5 (*TAEM* 5:11).

    1. See headnote, pp. 507–8.

    2. James Brown.

    3. The four cells shown in this document constitute a battery, compared to the two batteries of three cells each in Doc. 276 and the opposed batteries of two and four cells in Doc. 275. Edison omitted the polarized relay of the earlier drawings in this sketch because its action was not at issue.

–278–

[Newark, 1872?]

Technical Note:
Multiple Telegraphy[1]

Brown[2]   Try this[3]

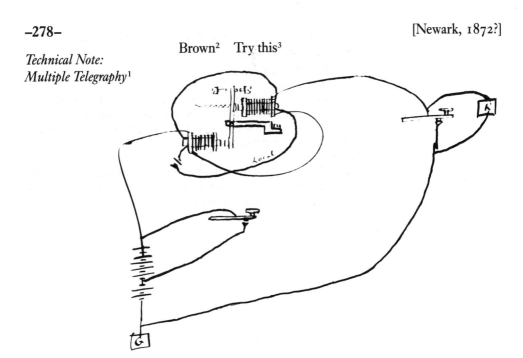

AX, NjWOE, Lab., Cat. 298:5 (*TAEM* 5:12).

1. See headnote, pp. 507–8.

2. James Brown.

3. In this sketch, each core has two coils around it, one inside the other; the inner coil is apparently in the local circuit with a battery constantly connected, while the outer coil is in the main circuit where the current is reversed and varied in strength. Edison employed this type of relay in several designs (e.g., Docs. 279, 286, 297, and 298). Cat. 1176:6, Lab. (*TAEM* 6:8); U.S. Pat. 150,846.

–279–

Technical Note:
Multiple Telegraphy[1]

[Newark, 1872?][a]

Brown[2]   try this[3]

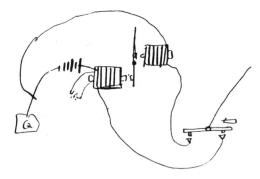

AX, NjWOE, Lab., Cat. 297:22 (*TAEM* 5:519). [a]"Top", written in orange pencil or crayon, added later to right of drawing.

1. See headnote, pp. 507–8.

2. James Brown.

3 This design involves balancing or compensating currents and does not use a polarized relay. Only one station is shown; the other end of the line would be equipped in the same way. The electromagnet directly attached to the large battery has inside it a second coil with its own local battery. The key has both front and back contacts and in practice would make contact with one as soon as it broke contact with the other. This design also has the armatures for the two electromagnets attached at different distances from the central point of the armature lever, as in Doc. 278. Cf. Docs. 297 and 302.

–11– Multiple Efforts

January–April 1873

The first four months of 1873 witnessed another eruption of inventive activity by Edison. In terms of successful patent applications and technical notes, his productivity rate equaled that of the first half of 1872, but whereas printing telegraphy spearheaded his 1872 outburst, duplex telegraphy led the early 1873 efforts. In the midst of this explosion and just a week after Edison's birthday in February, his wife, Mary, gave birth to a daughter, Marion ("Dot").[1] Despite the new family responsibilities, two months later Edison left for England to demonstrate his automatic telegraph system for the telegraph department of the British Post Office.

As 1873 opened, Edison's relationship with Western Union grew more intimate as his design of duplex and related circuits increased. Early in the year he had many diverse approaches in multiple telegraphy to offer Western Union. He continued to explore them, recording in his notes a large number of variations on those general designs while adding further approaches to those under consideration. Norman Miller, secretary and treasurer of the Gold and Stock Telegraph Company, served as an intermediary between Edison and Western Union president William Orton and helped open negotiations for support and control of Edison's inventions in this field, for it was one in which Western Union's interest was growing. By the middle of February, Edison had met with Orton and presented a selection from his designs. In several meetings they discussed potential improvements in multiple telegraphy—including diplex as well as duplex possibilities—and in instruments such as relays. They reached a verbal agreement that involved an explicit intention to cover the field and protect Western Union's patent investments. This allowed Edison to use Western Union facilities in New York to test his instruments.[2] He drafted the resultant patent applications in late March and in April. When he left for England on 23 April, he gave Miller the power of attorney to sell the duplex telegraph patents to Western Union, but the company neither purchased these inventions nor reached a formal agreement with Edison concerning his relationship to the company.

While duplex designs dominated Edison's interests at this time, he maintained his style of pursuing several lines of work simultaneously. He executed several patent applications related to earlier inventions in printing telegraphy and continued manufacturing printing telegraphs at the Ward Street Shop of Edison and Murray in Newark. He also began making notes for a proposed book on telegraphy and electricity. However, automatic telegraphy demanded increasing attention and soon opened international vistas for the prolific American inventor-entrepreneur.

Key figures in the Automatic Telegraph Company had encouraged British interest in the company's system since the time Britain had nationalized its telegraph industry and incorporated the existing system of land lines into the British Post Office. As early as 1870 Daniel Craig had corresponded with Frank Scudamore, superintendent of the telegraph department of the British Post Office, seeking to enlist his interest in George Little's automatic telegraph. Thereafter representatives of Automatic Telegraph had continued to promote the system with Scudamore.[3] Because company president George Harrington had extensive interests in Edison's inven-

Advertising card of the Automatic Telegraph Co.

AUTOMATIC TELEGRAPH CO.

CHEAP RATES & QUICK DESPATCH

BY THE

NEW SYSTEM.

LINE OPEN TO AND BETWEEN

NEW YORK, PHILADELPHIA AND WASHINGTON.

J. C. REIFF, *Sec. & Treas.*   GEO. HARRINGTON, *President.*

[OVER.]

tions, Automatic Telegraph took increasing notice of his improvements to the system, and Harrington promoted Edison's automatic inventions in Great Britain. Accordingly, Edison agreed to make a trip to England to test his automatic telegraph system. Prior to his trip, he continued to improve circuit designs[4] and also renewed his interest in a system of roman-letter automatic telegraphy after signing an agreement with Harrington and Josiah Reiff to develop such a system. Production of automatic instruments was consolidated at Edison and Murray, which acquired much of the equipment formerly used at the American Telegraph Works.[5]

1. Marion Edison birth record, Returns of Birth CE:362, Division of Archives and Records Management, Department of State, Trenton, N.J.

2. The exhibits, testimony, and arguments in the Quadruplex Case (*TAEM*, reels 9 and 10) provide the best coverage of Edison's arrangement with Western Union in connection with multiple telegraphy.

3. The promotion of the Little and Edison systems is detailed in ATF. On the British trip, see Chapter 12 introduction.

4. See U.S. Pats. 141,774, 147,312, 147,313, 147,314, and 156,843.

5. Some automatic telegraph production appears to have been shifted to Edison and Murray by November 1872, and the American Telegraph Works appears to have ceased operations at the beginning of January 1873. Cat. 1221:10–15, Accts. (*TAEM* 21:314–16); Cat. 1219, Accts. (*TAEM* 21:704–86 passim); 73-003, DF (*TAEM* 12:1000–1034); Cat. 1183:201–15, Accts. (*TAEM* 20:458–66); Cat. 30,108, Accts. (*TAEM* 20:448–49).

–280–

Edison and Murray
Balance Sheet

[Newark, January 1, 1873][1]

	[Itemized Assets]	[Total Assets]
Shafting:[2] about 3700 lb at 10c		$ 370.
Pulleys: 180 at $3		540.
Belting assorted 4201 ft		890.46
Standing Machinery:—		
2 Shiers[3]	134.	
4 Punch presses[4]	1200.	
1 Drop & 1 screwpress[5]	91.	
26 Mill Machines[6]	4525.	
3 Index[7] " ", with Vise & Centres	690.	
10 Screw Machines[8]	4617.	
4 Drill Presses	775.	
7 Engine Lathes[9]	1821.	
25 Stewart & Fitchburg lathes[10]	2120.	

66	Chucks (fitted)[11]	1005.	
1	Gould shaper[12]	295.	
53	Parker Vices[13]	304.75	
7	Slide rests[14]	284.	
4	Grindstones	100.	
2	Grinder Frames	133.	
	Polishing Machinery	250.	
88	" Wheels	440.	
3	Forges & 6 Anvils	222.	
2	Blowers[15]	43.	
		19,049.75	19,049.75

Working tools & fixtures:—[16]

	Screw Machine tools	1500.	
1	lot Gear Cutting tools	300.	
1	lot Tools cutting mills	45.	
	Lathes tools & chisels	95.60	
	Files	115.	
	Drills	94.77	
1	Set Key[17] tools	800.	
1	Set Stock Printer[18] tools	4000.	
1	Set Universal[19] " "	3000.	
	Unfinished Relay & Sounder tools	250.	

	10,200.37
	$31,050.58[a]
Carried forward	$31,050.58

Shop tools & fixtures:—

Small machinery, bench tools, office fittings, Stools, boxes for work, lacquers, & drugs, etc, etc.	2196.78	
6 Stoves & pipes	140.	
Mill Mach. pans	99.12	
Benching & drawers	298.50	
Other woodwork fitted	990.	
	$ 3724.40	$ 3724.40

Raw Stock	1922.45
Part finished Stock	6219.35
Scrap Stock	577.50
Models & Finished Insts.	2557.92
Insurance	300.00
Automatic Tel. Co. D[ebto]r.	2031.20
Exchange Tel. Co. Dr	65.18
E Wrigley Dr	7.76

		[Itemized Liabilities]	
Wm. Unger C[redito]r. By Mortgage[20]	10,000.00	
" " " By Interest on same	350.00	
" " " By Edison's notes	3,500.00	
Gold & Stock Tel Co.[21] Cr.	5,473.76	
Macknet & Wilson Cr.	50.15	
Holmes Booth & Haydens Cr.	17.03	
C. A. Dehart Cr.	3.78	
J H Thomas Cr	17.76	
Williams & Plum Cr	18.68	
Kirk & Co Cr	79.40	
Kirk, W. H. (Rent)[22] Cr	541.66	
Chas. Walker Cr	3.50	
Bell & Co Cr	5.20	
Geo. Price Cr	4.	
Hopperton Cr	5.98	
	$20,070.90	$48,455.92[a]
Carried forward	$20,070.90	$48,455.92
Jacob Gauch Cr	7.40	
Frasse & Co. Cr	8.41	
C B Smith Cr	20.	
A Hayden Cr	1.52	
Duerr & Morehouse Cr	12.75	
S. S. Thorne & Co. Cr	309.80	
D. Meeker Cr	36.64	
A J Davies Cr	4.05	
E N Wrigley[23]	399.75	
Gas	38.68	
C Batchelor (Back pay)	233.	
C Dean (" ")	407.72	
Roberts[24]	125.	
J. T Murrey (old Ledger)	625.78	
	$22,301.40	$22 301.40
[Net Worth]		$26,154.52

D, NjWOE, Accts. Cat. 1219:1 (*TAEM* 21:705). Written by Charles Batchelor. [a]End of page.

1. This inventory appears at the beginning of an 1873 account book. Two of the creditor balances in the second half of the document date it precisely: Edison and Murray received an invoice from A. J. Davis for $4.05 dated 31 December 1872, and the balance in their account with the Automatic Telegraph Co. changed from $2,031.20 on 2 January 1873 (72-019, DF [*TAEM* 12:957, 314]; Cat. 1221:10, Accts. [*TAEM*

21:314]). Several of the bills listed here were paid on 6 and 7 January 1873 (Cat. 1183:217–18, Accts. [*TAEM* 20:967–68]). Many of the creditors listed here are also found in Doc. 269. They supplied primarily machinery, tools, metals, leather, and hardware. Most were located in Newark or New York City; they can be identified in *Holbrook's Newark City Directory* 1872, in Wilson 1872, or from their billheads.

2. In a nineteenth-century machine shop, power was transmitted from a steam engine or waterwheel by means of horizontal, rotating overhead shafts that ran through the shop and were connected to the machines on the shop floor by leather belts that moved in vertical loops.

3. Shears are metal-cutting power tools. This and the definitions of machinery in the rest of the notes in this document are from Knight 1876–77.

4. A punch press punches holes in metal.

5. A drop press is a form of power hammer used for forging metal; a screw press (e.g., an early printing press) tightens by means of a large screw.

6. Milling machines shape metal the same way planing machines shape wood—the work is fastened to a bed that moves under a rotating cutting blade. Milling machines are important tools for finishing forged, stamped, or cast pieces.

7. An index is an indicator such as a dial marked with numbers; on a milling machine it indicates the position of the cutter.

8. A screw machine is a lathe especially adapted for making screws.

9. An engine lathe is a large, all-purpose lathe.

10. Stewart lathes were manufactured by the firm of James Stewart's Son, a lathe-making company in Newark. The maker of Fitchburg lathes is unknown. 70-006, DF (*TAEM* 12:165, 170); 71-014, DF (*TAEM* 12:512).

11. A chuck holds work in a lathe.

12. E. & R. J. Gould was one of Newark's leading tool and machinery manufacturers. A shaper operates like a milling machine (n. 6) except that the work is held fast and the cutter moves. This shaper came to Edison and Murray from the American Telegraph Works. Ford 1874, 73–74; 72-004, DF (*TAEM* 12:688); 72-014, DF (*TAEM* 12:873); 73-003, DF (*TAEM* 12:1001).

13. Made by C. Parker of Meriden, Conn. 70-002, DF (*TAEM* 12:45).

14. Slide rests hold cutting tools on lathes.

15. Blowers are power-driven devices that supply air to the forges.

16. Manufacturers usually had a special set of tools for each instrument they made, including jigs to guide machines; gauges to check work; cutting blades for milling machines, shapers, and lathes; and fixtures to hold each piece in place as the work was being done. The first four entries in this list are for standard tools; however, the key, stock printer, universal printer, relay, and sounder tools and fixtures were designed to shape specific individual parts and were probably useless for other work.

17. Morse telegraph key.

18. Universal stock printer.

19. Universal private-line printer.

20. See Doc. 264.

21. The source of this debt is not clear from extant records. 72-017, DF (*TAEM* 12:895); Cat. 1221:2, Accts. (*TAEM* 21:312).

22. See Doc. 270, n. 5.

23. Ibid., n. 6.

24. Ibid., n. 7.

–281–

From Lemuel Serrell

New York, Jan. 27th 1872 3[a]

Dear Sir.

In the matter of your application for patent on District Alarm Apparatus,[1] the Patent Office will declare an interference after the usual preliminary statements are filed, giving the history and date of your invention.[2] The time for filing these statements has been extended to March. 15 on account of Mr. Calahan's absence.[3]

If you will get these facts together and give them to me some time when you are at my office, I can prepare your affidavit. Yours truly

Lemuel W. Serrell pr. Walker.[4]

L, NjWOE, DF (*TAEM* 12:1184). Letterhead of Lemuel Serrell. [a]"New York," and "1872" preprinted.

1. U.S. Pat. 146,812.

2. Edison's application was rejected on 28 December 1872 as having been anticipated by Edward Calahan's U.S. Patent 129,526. On 31 December, Edison requested an interference, which was granted and ultimately decided in Calahan's favor. Edison amended his 28 December application and assigned the resulting patent to the American District Telegraph Co., which controlled the Calahan patent. Pat. App. 146,812; Digest Pat. E-2:225.

3. Calahan was in England, serving as electrician for the Exchange Telegraph Co.

4. Probably George Walker.

–282–

To Norman Miller[1]

[Newark, January 1873?][2]

I have struck a new vein in duplex telegraph; "no balance" works well enough in shop to order set made.[3] Think t'will be success.

Two messages can be sent in same direction.

In opposite directions.[4]

Way Stations can work on it and everything be made happy.

My shop is so full of non-paying work that I should like to saddle this on W.U. shop, where they are used to it.

Called to see Mr. Prescott[5] at 4 P.M. *Left.*

PL (transcript), NjWOE, Quad. 71.2, p. 13 (*TAEM* 10:228).

1. Norman Miller was at this time Edison's liaison with Western Union on multiple telegraphy. Edison testified that this note was di-

rected to Miller. Quad. 70.7, p. 269, and 71.1, pp. 248–49 (*TAEM* 9:501, 10:131); Miller to Edison, 10 Dec. 1875 (*TAEM* 13:212).

2. Edison testified that he wrote this note before his trip to England in 1873 but was uncertain about its exact date. Its content suggests a January or early February date. Edison and Orton indicated that they had spoken in the fall of 1872 on the subject of Western Union's interest in duplex telegraphy. With this note Edison sought to spark Western Union's interest. He speaks of his designs as new in type; after the middle of February he had already demonstrated such devices at Western Union.

3. In duplex telegraphy a "no balance" method represented a different approach from the common "balance" methods. In the latter, the transmitter's relay was kept neutral by electrically offsetting ("balancing") in some manner the effect of the transmitted signal so that the relay could respond to an incoming signal (see Doc. 283). One of Edison's "no balance" methods employed two different types of electrical signal on the circuit at the same time: current strength and polarity. See headnote, pp. 31–32; and Docs. 23, 28, 285, and 315.

4. This was a claim for both a duplex and a diplex, but as a choice of one or the other rather than both at once. Cf. Doc. 50.

5. Prescott was Western Union's electrical expert. See Doc. 148, n. 4.

–283–

Technical Note:
Multiple Telegraphy

[Newark, c. February 1, 1873][1]

Duplex[2] Static[a]

Impvts ~~in defence of electrical warfare~~.[b]

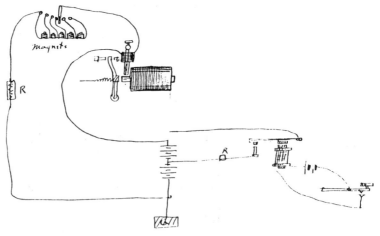

The magnets in the balancing or compensating circuit are used for the purpose of generating an after current due to the demagnetization of the iron core, which current is very powerful and by regulation made equal to the static return charge many times greater than is possible to equalize by a condenser.[3] magnets when properly placed are the sworn enemies of static currents whether dynamic or in a muscle twisting mode.[4c]

AX, NjWOE, Lab., Cat. 297:64 (*TAEM* 5:663). [a]Written in pencil, slightly above the line of the previous word; the bulk of the document is in blue ink. [b]Cancellation done in black ink. [c]"1" overwritten at bottom center of page, but nothing has been identified as a possible further page for this document.

1. This is a draft for the design "Duplex No 9" in Doc. 285. It was probably prepared around the time Edison proposed that Western Union support his work on multiple telegraphy and got his drawings ready to show Orton. See Doc. 282.

2. The design here uses the current in the circuit loop on the left to compensate for, or negate, the effect of outgoing signals on the main relay, in the center of the drawing (see headnote, pp. 31–32; and Docs. 28, 282, and 301). The outgoing line is not shown here; it would run up from the upper right end of the relay-magnet's coil in the center of the drawing. Exactly how the arrangement was to work is not certain but can be partially clarified through comparison with design 9 in Doc. 285, and Doc. 306. Cf. also Doc. 22.

Stearns's duplex telegraph with a condenser.

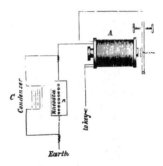

3. The problem of "static return charge" (or "kick") in overland duplex telegraphy was first coped with adequately by Joseph Stearns's innovation (U.S. Pat. 126,847), which Western Union bought and began using in 1872 (see Doc. 50, n. 3). Stearns inserted a condenser—a device to accumulate and store electric charge—into the artificial line so that it could match the electrostatic capacity of the main line and thus could counteract the effect of the kick as well as the signal on the relay, leaving the relay able to respond to only incoming signals. In Edison's design extra electromagnets were arranged to try to do the same thing.

4. The meaning of Edison's distinction is unclear.

–284–

William Orton to Norman Miller

NEW YORK,[a] *February* 6, 1873.

DEAR MILLER:

Say to Edison that I am ready to treat for his duplex, and that he may set it up in our office at any time.[1] Also that I shall be glad to consider his other propositions.[2] Very truly yours,

WILLM ORTON.[b]

PL (transcript), NjWOE, Quad. 71.2, p. 10 (*TAEM* 10:212). [a]Original on letterhead of President Orton. [b]"M" printed at bottom center of transcript.

1. See Doc. 282. The meeting led to a verbal agreement between Orton, acting for Western Union, and Edison, the nature and details of which were a subject of intense, protracted litigation (see Quad. passim [*TAEM*, reels 9 and 10]). Whatever the exact terms, Edison was given the chance to test experimental apparatus on Western Union lines at off-peak hours, working at the Western Union headquarters in New York City, and Western Union bore the expense of patent applications. In return, Western Union got at least a first chance to purchase any resulting patents.

2. The nature of these other projects may be indicated in Doc. 303. Exactly when Edison made the proposals is not known.

-285-

Technical Note:
Multiple Telegraphy

Newark. N.J. Feby. 15. 1873.

Mr. Serrell Patent Solicitor

Please file following testimony of invention in Duplex Telegraphy.[1]

The fundamental principle of Duplex Telegraphy[2] is the neautralization of the effect of the outgoing battery up[a] the receiving instrument[b]

The next important device is the destruction of the effect of the static charge of the line[3] upon the receiving instrument.[c]

The minor essentials are, simple devices for placing the battery off and on the line, and a quick and convenient compensator.[d]

I shall describe a large number of modes on which I am now experimenting preparatory to obtaining Patents.[4]

Duplex No 1[5]

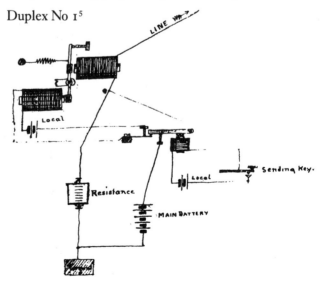

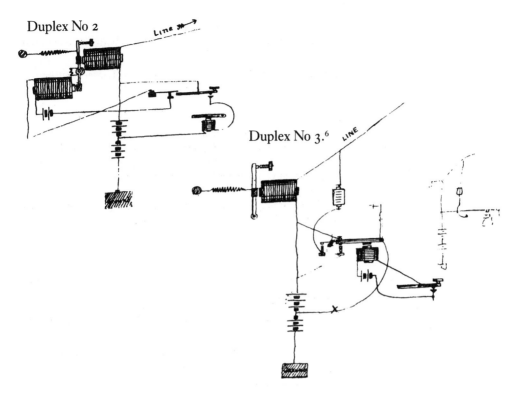

Duplex No 2

Line

Duplex No 3.[6]

LINE

An Instrument may be inserted at X with proper resistance to prevent the spark and the key made to do what the sounder now does.[7] The spring G is so adjusted the the shunt is not disconnected until the return static charge flows through it.[8]

Duplex No 4.

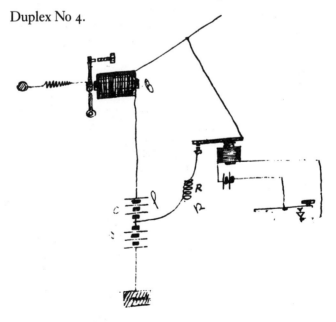

Duplex No 5.[9]

Duplex No. 6.[10]

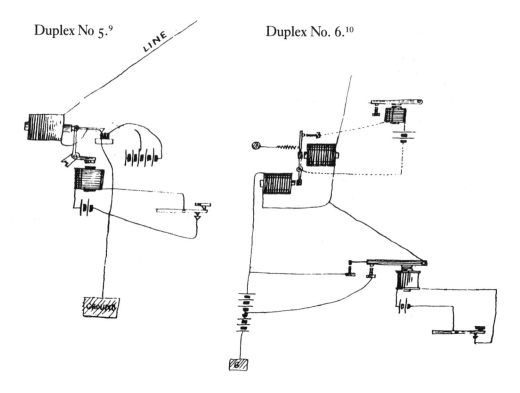

Duplex No 7[11]

Duplex No 8[12]

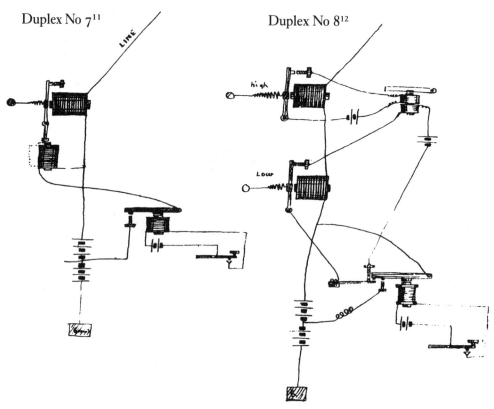

Duplex No 9ᵉ

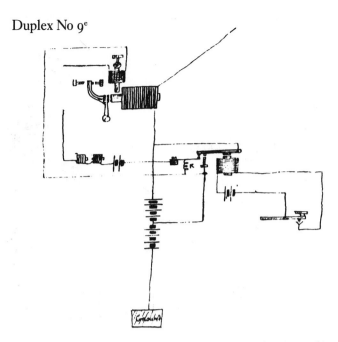

X magnets are placed in the compensating circuit so that when the main and local battery are taken off the magnts will give an inductive charge equal to the static charge of the line.[13]

Duplex No 10[14]

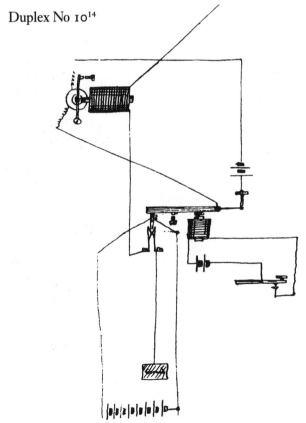

Duplex No. 11[15]

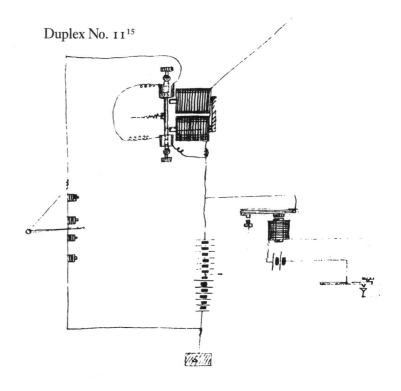

Duplex No 12[16]

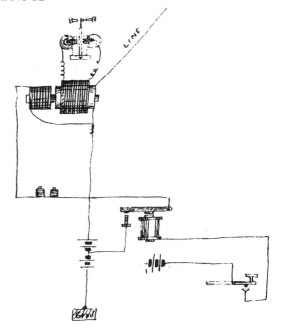

Duplex No 13[17]

Four-plex No 14
Why not.[18]

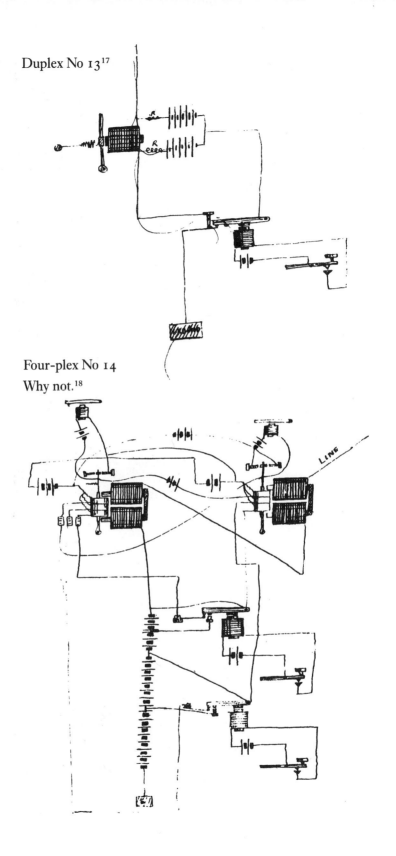

Duplex No 15[19]

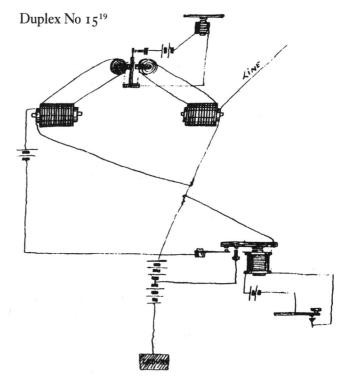

Duplex No 16.[20]

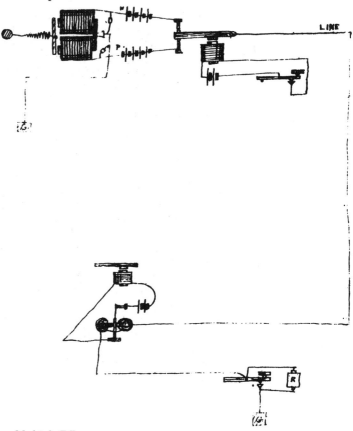

Duplex No 17[21]

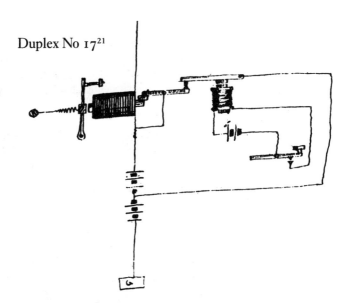

AD (photographic transcript), MdSuFR, RG-241, *Nicholson v. Edison,* Testimony and Exhibits on Behalf of T. A. Edison, pp. 222–38. Another copy is in Quad. 71.2 (*TAEM* 10:212–26). [a]The letters "on" are added here in another hand, most likely at a later date. [b]"(at the sending stations)" added here in another hand, as in textnote a. [c]"(at [-- ----]the line stations)" added here in another hand, as in textnote a. [d]"(?)" added here in another hand. [e]Dashed underline added to "9" later.

1. "Duplex" is used as a general term here, covering what would be known as diplex and quadruplex telegraphy as well (see "Four-plex No 14"). These notes were probably intended to be the basis for drafting a caveat but apparently were never sent to Serrell (Quad. 70.7, pp. 257, 328 [*TAEM* 9:495, 531]). They are related to material contained in a small notebook Edison prepared for his talks with William Orton concerning Western Union support for the development of these inventions. It is not known whether that notebook (no longer extant) contained the same number of designs as this set of notes, but the resulting trials of actual apparatus on Western Union lines involved a larger number than is found here. When applications for patents were drawn up in the wake of those tests, the patent agent involved for most of them was Munn & Co. rather than Serrell (see Docs. 302–11 and 314–16). For related drawings see also Docs. 275–79, 283, 286, 294, 297–301, and 312, and materials referred to in the notes to those documents, particularly material from Cats. 297, 30,099, and 1176, Lab. (*TAEM* 5:424, 997; and 6:1).

2. Cf. Edison on the same point in Doc. 301. See also headnote, pp. 31–32; and Doc. 28.

3. See Doc. 283.

4. The "descriptions" in most of the following cases consist of only a single drawing, and in those drawings the message-receiving aspect of the design is generally left incomplete, showing only the relay instead of the local circuit and sounder the relay would operate. In all but one instance ("Duplex No 16") only one end of the circuit is shown; actual operation would require identical equipment at the other end of the line as well.

5. Cf. "Duplex No 1" and "Duplex No 2." Possible earlier versions of these are Cat. 297:2(3), 15(1) verso, Lab. (*TAEM* 5:431, 497). Cf. Docs. 297, 302, and 304–5; and Cat. 1176:42, Lab. (*TAEM* 6:44).

6. Cf. "Duplex No 3" and "Duplex No 4." A possible earlier drawing of these designs is Cat. 298:48(2), Lab. (*TAEM* 5:97). Cf. Docs. 298, 301, 306, and 314; Cat. 297:154(2), Lab. (*TAEM* 5:995); Cat. 30,099:288, Lab. (*TAEM* 5:1066); and Cat. 1176:41, Lab. (*TAEM* 6:43).

7. The sounders in these designs were used as transmitting switches, controlled by a key rather than having the key serve directly as the switch. In this drawing the sounder is the device in the middle with a short magnet coil, as distinct from the relay, with its longer coil, shown at the upper left.

8. This refers to "kick"; see Doc. 283, n. 3. **G** is directly above **X**.

9. For an explanation of this design see the very similar one in Edison's U.S. Patent 178,223. For examples of earlier work related to this design see Doc. 13; Cat. 298:48(2), Lab. (*TAEM* 5:97); and Cat. 1176:11, Lab. (*TAEM* 6:13).

10. Cf. Docs. 305 and 312; and Edison's U.S. Pat. 178,222.

11. Cf. Cat. 297:21(2), Lab. (*TAEM* 5:515); Cat. 1176:19, 35, Lab. (*TAEM* 6:21, 37); and Edison's U.S. Pat. 131,334.

12. Edison executed a similar patent application on 22 April 1873 and later incorporated it into U.S. Patent 147,917. Related drawings are Cat. 298:71(1), 134(1), Lab. (*TAEM* 5:133, 262); and Cat. 297:3(6), 5(3), 16(2), 16(4) recto and verso, 18(1–5), 29(4), 32, 60(1), 80(1), 82(1), 85(4), 87(4) verso, 128(3), Lab. (*TAEM* 5:433, 445, 502, 505–6, 511–12, 547, 553, 639, 751, 764, 785, 796, 911).

13. Cf. the designs here numbered 10–12 and 14. See Docs. 22, 283, and 308, and the materials cited in connection with Doc. 312. Cat. 297:59(2), Lab. (*TAEM* 5:638) includes a related drawing and the following canceled line in Edison's hand: "Did you see Mr. Orton about Patents".

14. For an account of the operation of this system see Doc. 311. For related material see drawings F and G in Doc. 297; Cat. 297:5(4), 23(4), 37(1), Lab. (*TAEM* 5:443, 523, 565); and Cat. 1176:42, 51, Lab. (*TAEM* 6:44, 52).

15. See Doc. 308; Cat. 297:3(2), 5(2), 59(2), Lab. (*TAEM* 5:433, 443, 638); Cat. 1176:43, 52, Lab. (*TAEM* 6:45, 53); and U.S. Pat. 130,795.

16. For an explanation see Doc. 310 and U.S. Patent 178,221. Cf. Docs. 30, 47, 49, and 297; Cat. 298:133(3), 134(1), Lab. (*TAEM* 5:261–62); Cat. 1176:43, Lab. (*TAEM* 6:45); and Prescott 1877, 823–24.

17. This is close to the basic pattern of a bridge duplex, clear cases of which go back as far as Maron's 1863 design (Prescott 1877, 790). The term is a reference to the type of circuit known as a Wheatstone bridge (Maver 1892, 122–30). When the currents in two branches of a circuit are appropriately balanced or matched, no current will flow in a "bridge" connection between them. A bridge duplex employs this principle, putting the relay in a bridge to eliminate the effect of outgoing signals upon it while allowing it to receive incoming signals. Cf. Doc. 309; patent application "CASE No. 97," Quad. 70.6, pp. 63–64 and illustration facing 64 (*TAEM* 9:364–65); Cat. 297:1(1), 10(1), 75(3), Lab.

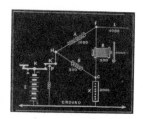

A bridge duplex design.

(*TAEM* 5:427, 468, 727); and Cat. 1176:43, 52, Lab. (*TAEM* 6:45, 53).

18. This is the earliest dated document showing Edison's work on quadruplex telegraphy and his use of diplexing as an addition, rather than an alternative, to duplexing. The label and the presence of two keys indicate that the device was designed to operate as a diplex and thus, by combination with features allowing duplexing of each of the diplex signals, to transmit and receive two different messages each way simultaneously. However, it is not clear exactly how this arrangement was supposed to operate. The possibility of putting together both same-direction and opposite-direction double-signal systems had been envisaged several times before (Prescott 1877, 829, 833–34), and Edison worked on, or at least considered, a variety of such designs in this period. Cf. the devices employed here with the designs 9–11 above and in Cat. 1176:51, Lab. (*TAEM* 6:52). Cf. also Docs. 286 and 300; Cat. 297:4(5), 37(2), 78(3) verso, 80(4) verso, 81(3), 85(3), 86(1), Lab. (*TAEM* 5:439, 570, 743, 755, 760, 784, 786); Cat. 30,099:277–78, Lab. (*TAEM* 5:1072); Cat. 1176:13, Lab. (*TAEM* 6:15); and U.S. Pat. 207,723.

19. The precise operation intended here is not clear, though it evidently involved electromagnetic induction to activate the polarized relay, which acted as the receiving apparatus. Cf. the designs numbered 6 and 12 above and the materials cited there. The idea for this design could well date back to the summer of 1872; see Cat. 297:3(7), Lab. (*TAEM* 5:436), which is written on the back of a document dating from before 30 July 1872.

20. For an account of the working of a similar design see U.S. Pat. 162,633 and Prescott 1877, 822–23. Cf. Docs. 88, 275–78, 294, 297, 300, and 315.

21. This design is based on changing the responsiveness of the relay by moving part of the "keeper"—that is, the iron connecting the cores of the coils into a single U-shaped magnet. Edison designed such a relay in the spring of 1872 (U.S. Pat. 134,868). Possible earlier versions of this design are found in Cat. 297:15(3), 16(1), Lab. (*TAEM* 5:496, 502).

–286–

Technical Note:
Multiple Telegraphy

[Newark, c. February 15, 1873][1]

To be tested[a]
No 18.[2]

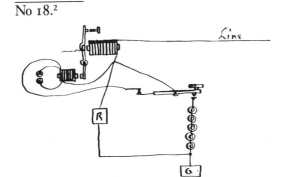

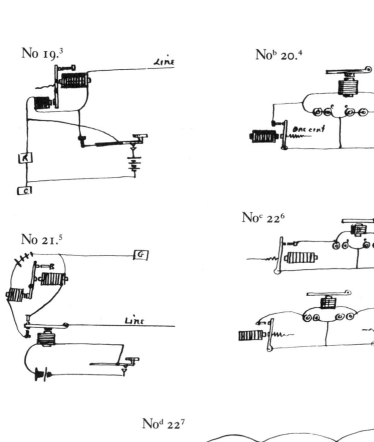

No 19.³

Line

No^b 20.⁴

One cent   other current.

No 21.⁵

G

Line

No^c 22⁶

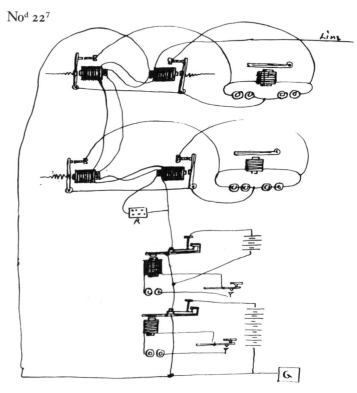

No^d 22⁷

Line

G

AX, NjWOE, Lab., Cat. 298:135, 134 (*TAEM* 5:267, 266, 264, 265); resemblances in style, paper, and ink, as well as content, link these separate scrapbook items. ᵃWritten on back of page containing drawings "No 18." and "No 19."; on the front is the letterhead of Western Union. ᵇStarts new page, also on letterhead of Western Union. ᶜFigure label smeared, possibly as a cancellation. ᵈStarts new page; letterhead cut off.

1. It is not clear whether these drawings were produced soon after Doc. 285 or were part of a prior set of drawings from which that document derives.

2. Cf. "Duplex No 1" and "Duplex No 2" in Doc. 285.

3. Cf. "Duplex No 6." in Doc. 285.

4. Cf. the incomplete and full "No 22" designs in this document.

5. Cf. Doc. 297.

6. This and the following drawing "No 22" are apparently versions of a diplex or quadruplex design (cf. Cat. 297:1[1], Lab. [*TAEM* 5:428]). This is one of many such designs from this period, some of which eventuated in U.S. Patent 207,723. See "Four-plex No 14" in Doc. 285.

7. In Doc. 292 Edison referred to testing twenty-two multiple telegraphy designs in this period, but in Doc. 303 the number he gave was 23. In any case the tested designs may well have included some not illustrated here or in Doc. 285. Many of these designs can be found with other numbers on them elsewhere in the Miscellaneous Shop and Laboratory Notebooks (*TAEM*, reel 6). Some related designs have higher numbers, indicating that more than twenty-two or twenty-three existed, but it cannot be determined whether they were tested. See Cat. 297:18, Lab. (*TAEM* 5:511–12).

–287–

From Marshall Lefferts

New York, 26. Feby 1873ᵃ

Dr Sir

You will construct for this Company 100 One hundred— "Universal Private Line Printers" to be first class in workmanship and to embrace the improvements as arranged for[1]—the whole to be made like sample Machine to be deposited in this office—

We will pay you One hundred & twenty dollars for each Machine. Yours truly

M. Lefferts   Presdt

ALS, NjWOE, DF (*TAEM* 12:1006). Letterhead of the Gold and Stock Telegraph Co. ᵃ"New York," and "187" preprinted.

1. This order, entered in an Edison and Murray daybook on 27 February, was later canceled. See Cat. 1214:29, Accts. (*TAEM* 21:578).

[New York or Newark, February 1873][1]

Mr. M.:[2]

Want order go in W.U. nights to feel the pulse of my patients.[3]

EDISON.

PL (transcript), Quad. 71.2, p. 11 (*TAEM* 10:227).

1. Edison testified that this message was sent in February 1873. Quad. 70.7, p. 258 (*TAEM* 9:496).

2. Edison's testimony identifies Miller as the recipient. Ibid.

3. Having reached an agreement to test and develop various multiple telegraphy arrangements using Western Union lines ("my patients"), Edison found he needed a pass in order to be allowed into the appropriate parts of the Western Union headquarters building in New York City after hours. He experimented at night because traffic on the lines was lightest then. Ibid.

ELECTRIC PERFORATOR Doc. 289

In early 1873 Edison patented a perforating mechanism that used electromagnets rather than muscle power to punch holes in the paper tape for automatic telegraph transmission.[1] Designed to use the same keyboard as his original, large perforator,[2] this mechanism was intended to solve two problems that characterized the earlier machine. First, by substituting simple electrical circuits for the original complicated series of cams and levers, the new design avoided the constant mechanical problems that plagued the large perforator.[3] Second, this arrangement was "less fatiguing to the hand and arm"[4] because the operator no longer had to supply the power to punch the holes. This last consideration was particularly important if the machines were to be operated by young girls, whose lower-priced labor was considered essential to reducing the costs of automatic telegraph systems.[5]

This was the only electric perforator patented by Edison and was apparently never used, perhaps because of the difficulty of developing sufficient power in the electromagnets to punch the tape cleanly.

1. Edison had filed a caveat for a similar perforator on 26 July 1871 (PS [*TAEM* Supp. III]). See also the drawings in PN-71-00-00 (*TAEM* Supp. III); and Doc. 151A.

2. U.S. Pat. 121,601.

3. An 1872 article in the *Telegrapher* (by George Harrington's brother-in-law William Barney) describes the mechanical problems of Edison's perforator. Account records from early 1872 indicate that perforator parts were shipped to telegraph offices in New York, Washington,

and Philadelphia. Barney 1872; Josiah Reiff's testimony, 1:415, Box 17A, *Harrington v. A&P*; 71-004, DF (*TAEM* 12:274–81); PN-71-09-05, Accts. (*TAEM* 20:171–86).

4. U.S. Pat. 141,775, p. 1.

5. See, for example, Docs. 144 and 148; and Craig 1872.

–289–

Patent Model:
Automatic Telegraphy[1]

M (28 cm × 17 cm × 8 cm), MiDbEI, Acc. 29.1980.1402.

1. See headnote above.

2. Edison executed the covering patent application on this date (Pat. App. 141,775). Although many other existing patent models are functioning instruments made of metal, this model is made of wood.

SIPHON RECORDER   Doc. 290

The following automatic telegraph receiver embodies primarily the ideas delineated in a notebook entry of 28 July 1871.[1] The chemical solution for moistening the paper tape was kept in a small reservoir, from which a siphon fed it through a pen point to the paper. As the paper absorbed the solution, it moved under two wire points between which the incoming signal passed. In the patent application accompanying the model shown in Doc. 290, Edison claimed that this receiver would minimize the use of chemicals by applying the solution in a thin line at the center of the paper tape and passing the receiving signal immediately through it. However, that idea had already been patented, so Edison then altered his application to claim for himself the placement of the wires that marked the paper.[2] With the two wires placed close together, just the narrow strip between them needed to be moistened. Moreover, placing the two wires on the same side of the paper, and thereby having the signal go across the surface of the paper rather than through it, avoided the intermittently high resistance that resulted when the solution failed to penetrate the paper consistently.

1. Doc. 180.
2. Commissioner of Patents to TAE, c/o Serrell, 14 Apr. 1873, and TAE, per Serrell, to Commissioner of Patents, 13 May 1873, Pat. App. 141,774.

–290–

Patent Model:
Automatic Telegraphy[1]

[Newark, March 7, 1873?[2]]

M (14 cm dia. × 10 cm), MiDbEI, Acc. 29.1980.1341.

1. See headnote above.

2. Edison executed the covering patent application on this date (Pat. App. 141,774). Like the model for Edison's electrically powered perforator (Doc. 289), this model is made of wood.

–291–

Power of Attorney to George Harrington

New York, March 14, 1873[a]

Whereas Letters Patent in Great Britain under the great seal thereof were duly granted to Henry Edward Newton[1] of London, England, as a communication from George Harrington of the City and State of New York, United States of America for "Improvements in Mechanism for perforating paper for transmitting messages" which Letters patent are dated June 10th A.D. 1872 and numbered 1,751:[2]

And Whereas Thomas Alva Edison of Newark in the State of New Jersey, United States of America has invented certain "Improvements in Circuits and instruments for chemical telegraphs"[3] for which he has made application for Letters Patent in Great Britain:

And Whereas the above recited Letters Patent numbered 1751 was obtained, and the application for Letters Patent for "Improvements in Circuits and instruments for chemical Telegraphs" as above set forth was made with the full understanding between the said Thomas Alva Edison and the said George Harrington parties hereto that the benefits and advantages in and to all of said inventions and Letters Patent already granted and applied for should be held and enjoyed mutually in the proportion of one undivided third part to the said Edison and two undivided third parts to the said Harrington,[4] and the expenses in connection with the same have been duly paid pursuant to such agreement

And Whereas it has also been understood and agreed that the said George Harrington should be entrusted entirely with all negotiations and proceedings connected with the introduction of the said inventions and improvements and the sole and exclusive right and power to negotiate and effect sales in connection with such inventions and Letters Patent:

Therefore in consideration of the premises, the said George Harrington is hereby made constituted and appointed the sole, exclusive and irrevocable attorney of the said Thomas Alva Edison to act in connection with the said inventions, improvements and Letters Patent, to attend to the introduction of the same in Great Britain, to negotiate for the sale of the whole or any portion of the right or rights in or under said Letters Patent already issued or applied for, to execute any and all contracts, assignments, deeds, transfers, licenses, or other documents and to receive any and all payments, license fees, royalties, Stock, bonds or other valuable consideration for any such rights or privileges and to give full and ample receipts for the same and to execute in the name or in behalf of the said Edison any written instrument that may properly be required for carrying out the full intent and meaning of these presents:

And the said Edison does hereby release and relinquish to the said Harrington any and all right to act in the premises and agrees that the rights and powers hereby conveyed shall be held and exercised by the said Harrington, his substitute or Attorney or his heirs or Executors as fully and entirely as he the said Edison could have held or exercised the same:

And furthermore that he the said Edison his heirs or assigns is only entitled to require of the said Harrington or his legal representatives a just and true accounting for and paying over of one third part of the net profits that may be realized from and in connection with the aforesaid inventions and Letters Patent in Great Britain.

In Witness whereof I, the said Thomas A. Edison have hereunto set my hand and affixed my seal this fourteenth day of March Anno Domini 1873, in the City of New York, State of New York, United States of America.

G̶Thomas Alva Edison   Geo Harrington
Witnesses   David G. Barnitz Gerard W. Vis.[5]

DS, NjWOE, LS (*TAEM* 28:953). Written by David Barnitz. ᵃPlace and date taken from text; form of date altered.

    1. Unidentified. On British patents, see Chapter 7 introduction, n. 2.
    2. Equivalent to U.S. Patent 121,601.
    3. British Patent 735 (1873) comprised U.S. Patents 135,531, 141,773, and 141,776 and one other circuit (fig. 14 in the specification).
    4. See Docs. 109 and 155.
    5. Unidentified.

–292–

To Norman Miller[1]

[Newark or New York, c. March 15, 1873][2]

I have tried to date with make-shift instruments, seven duplex, between New York and Boston.[3] Six of them worked charmingly. The seventh was a satisfactory failure. I have fifteen more to try.[4]

EDISON.

PL (transcript), Quad. 71.2, p. 12 (*TAEM* 10:228).

1. Edison identified Miller as the recipient. Quad. 70.7, pp. 264, 337 (*TAEM* 9:499, 535).

2. Edison reported early in April that his tests were then finished, having occupied some twenty-two nights (Doc. 303). The tests had probably begun by the end of February (Docs. 284, 285, and 288). Thus Edison probably was at the stage indicated here around the middle of March.

3. The length of the test line was double the distance between New York City and Boston. Since all the experimental apparatus was in New York, the tests were made on a "loop" circuit, connecting two lines together at the Boston office so that both ends of the circuit were in New York.

4. See Doc. 286, n. 7; and Doc. 303.

–293–

Notebook Entry: Topic List for Book[1]

[Newark, Winter 1873][2]

Duplex.
Repeater
Fac similes
Dot & Dash Chemical "Autos"
Magnetic Autos
Telegraph Inductive & discharge Currents.
Printing Instruments.
Transmitters for d[itt]o.
Perforating Machines.
Relays
Sounders
Morse Recording insts
Movements.
Applications of Magnetism
     "   " Electricity
Mechanical Electric Movements
Magneto Telegraphy
Induction Coil telegraphy
Contracts
Batteries.
Novel Connections.
Manipulations Novel.
Electromagnetic Engines[a]
Duplex.[3] Stearns—get copies patents =
    "   Farmer   "   "   "   Extension & Reissue.
    "   Siemens & Halske = Kramer,[4] get copy teleghr giving description =
          Edisons as shown in telegrapher
          Mention farmers Clockwork Double Trans
             "   Edisons Magnetical d[itt]o   same principle & result Boston Experiments

Have Double transmitter Article in Dub, Schellen
& Blavier DuMoncel Translated = [5]
Edisons No 1 as in telegraphr
Edisons No 2 Vibratory
Edisons No 3 Mechanical Equalizer
    "   No 4 Shunt with Rev battery on single relay
Reverse Current one & increase & decrease other =
Double trans Chemical
Rheos in connection
and Condenser Applications by Stearns and other
    means such as a Condenser of a duble Coil. No
    Condenser but Extra Coil on relay to take induc-
    tion also to shunt Duplex relay with water Resist
    ance =

AX, NjWOE, Lab., Cat. 1176: recto, leaf preceding page 1 (*TAEM* 6:2).
[a]End of page.

1. This is apparently a topic list for a book on telegraphic and related electrical matters which Edison was planning. See NS-74-002, Lab. (*TAEM* 7:42).

2. These notebook pages have been dated on the basis of their content and context.

3. What follows elaborates upon the subject of duplex telegraphy, the first topic in this list (see Cat. 297:103, Lab. [*TAEM* 5:855] for the beginning of a later discussion of the topic). For a similar elaboration on repeaters, see Cat. 1176:31 (*TAEM* 6:33).

4. Prescott 1877, 830–32.

5. Julius Dub, *Die Anwendung des Elektromagnetisms, mit besonderer Berücksichtigung der Telegraphie*; Heinrich Schellen, *Der Elektromagnetische Telegraph*; E. E. Blavier, *Nouveau traité de télégraphie électrique*; and Theodose Du Moncel, *Traité theorique et practique de télégraphie électrique*. Each was published in several editions.

–294–   [Newark, Winter 1873[1]]

Notebook   Duplex.[2]   wound opposite[3]
Entry:
Multiple
Telegraphy

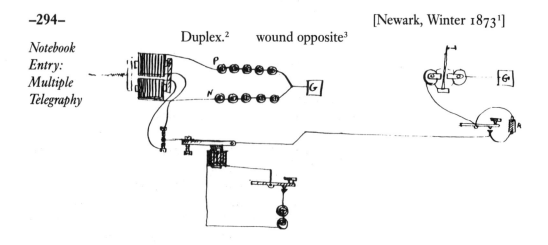

Mem[4] you cannot reverse a current through an electromagnet without opening it.[5]

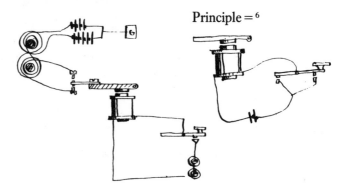

Principle = [6]

Reversing a Battery without ~~much mechanism~~ a Morse key in a circuit

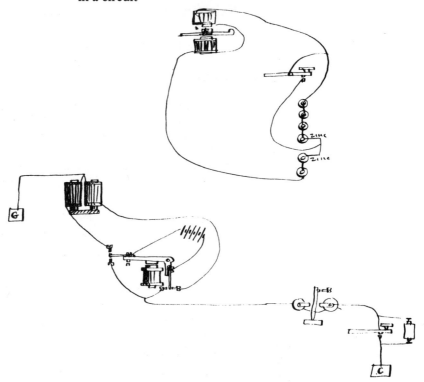

AX, NjWOE, Lab., Cat. 1176:2 (*TAEM* 6:4).

1. The date given here was determined by the content and context of the entry. Cf. Doc. 293.

2. This entry begins as a further elaboration of the topic of duplex telegraphy for Edison's projected book, as outlined in Doc. 293.

3. See Doc. 276, n. 4.

4. Edison was working on a problem here that had occupied him for some time. A solution to it was crucial for his invention of a practical quadruplex telegraph. See Docs. 275–79, 285, 286, 297, 300, and 315.

5. "Yes but you can ~~make~~ let it open if you fix your sounder right for 'Quad'" follows in Edison's hand in a different ink (probably written late in 1874).

6. Here Edison shifts from an exclusive concern with duplexes to exploring general designs for reversing currents.

–295–

Agreement with George Harrington and Josiah Reiff

[New York?,] March 31, 1873[a]

Whereas by a certain Contract and agreement bearing date the first day of October A.D. 1870, between Thomas A. Edison of the City of Newark, State of New Jersey on the one part, and George Harrington of the City of Washington, District of Columbia of the other part, wherein it was stipulated and agreed on the part of the said Edison, that for and in consideration of certain valuable considerations as therein set forth, he the said Edison should invent and develope for practical use a System of Automatic or Fast Telegraphy embracing Perforators and other instruments and devices necessary thereto, and thereafter change, modify, add to and otherwise improve said System, instruments, devices, and other matters and things connected therewith as his experience and ability as an Inventor and Electrian might suggest and allow, all of said inventions, devices, instruments and System being held as provided in and covered by said contract and agreement of October 1st A.D. 1870.

And whereas it is the desire of said Harrington that the said Edison shall, if practicable and possible, invent and develope a practical System of Fast Telegraphy, whereby intelligence or despatches may be transmitted at a rapid rate in such manner as to reproduce the intelligence or despatch so transmitted at the receiving office in Roman Characters or letters in such form and in such legible and permanent manner as will permit the delivery thereof directly to the general public or parties to whom addressed or for whom intended, without the necessity for translating, copying or transcribing as is now necessary before delivery, with the power to take drop copies at will and in Roman characters permanently imprinted at intermediate or way stations—[1]

And Whereas the said Edison is ready to undertake the discovery, invention and developement into practical use of a System producing the results as above set forth[2]   Now therefore it is hereby stipulated and agreed by said Edison, that he the said Edison shall at once proceed to invent and discover a method whereby intelligence or despatches may be transmitted over the ordinary wires now in use for telegraph purposes at a speed of Five Hundred words per minute over ordinary circuits of Three hundred miles in length, and at a rate of speed not less than Eighty words per minute over circuits of One Thousand miles in length—the intelligence or despatches so transmitted to be received at any one or more offices within the circuit, in plain, clearly defined and permanent Roman characters or letters and equally and at the same time when so desired or required at the several intermediate stations or offices as "drop Copies" all, without the necessity for further manipulation in translating, transcribing or copying, ready for delivery to the parties to whom addressed or intended.

And the said Harrington on his part or the said Harrington and J. C. Reiff his associate, of the City of New York, State of New York, stipulate and agree to advance to said Edison from time to time the aggregate sum of Three hundred dollars and to further pay to the said Edison, immediately after the satisfactory accomplishment of his undertaking[3] a further sum to a whole aggregate of Fifteen Thousand dollars, to cover the extra expenses and extra labors of said Edison in producing the System and results as above set forth; but the said sum of Fifteen Thousand dollars shall cover the cost of Three sets of instruments for terminal and way offices complete for immediate use.

And the said Edison further stipulates agrees and guaranties that the inventions, devices and System as above recited shall be practical, easily learned and manipulated, and shall be original, so as to be secured by Letters Patent in the United States and in all foreign countries, for which purpose the said Edison shall cause to be made the necessary models, drawings and descriptions when required so to do, and that all of said Patents shall be held and controlled in the manner and proportions and for the purposes as set forth in the contract and agreement first hereinbefore recited;[4] the said Harrington on his part agreeing that any of the inventions and devices of said Edison covered by Patents now held by said Harrington may under the stipulations and for the exclusive purposes herein set forth, be utilized and used as a part of said System, until the same shall be in due form assigned as herein contemplated in accordance with the contract and agreement of October first A.D. 1870 hereinbefore recited.

And the said Edison further stipulates and agrees for himself his heirs and Executors that he or they shall and will truly make and execute any and all further papers necessary or requisite in the premises, as they may be required, to complete and confirm the right, title, interest and control of said Harrington in accordance with and in fulfilment of the stipulations and purposes of the parties hereto, as hereinbefore set forth, and as pointed out in the before recited contract of October 1st A.D. 1870.

It is further stipulated and agreed by and between the parties hereto, that if the said Edison shall fail in developing practically as herein set forth, a rate of speed of Five Hundred and not less than Eighty words over circuits of the lengths hereinbefore set forth, but shall succeed at a lesser rate of speed over circuits of Three Hundred miles, then and in that case, the compensation herein provided shall be reduced as the rate of transmission falls short of the speed above set forth of Five Hundred words for circuits of Three Hundred miles, and the payment shall be as follows: If the rate of speed for circuits of Three Hundred miles shall be, but not exceed Four Hundred words per minute the compensation shall be ($12000) Twelve Thousand dollars— If the rate of speed shall be, but not exceed Three hundred words per minute, the compensation shall be^b ($10,000) Ten Thousand dollars— If the rate of speed shall be, but not exceed Two hundred words per minute the compensation shall be ($7,000.) Seven Thousand dollars. If the rate of speed shall be, but not exceed one hundred and fifty words per minute, then the compensation shall be ($5000.) Five Thousand dollars. But if the rate of speed shall not attain this latter recited speed of One hundred and fifty words per minute, then no compensation shall be claimed or allowed for extra services and labor—provided That for circuits of One Thousand miles, the minimum rate of speed shall not be less than Eighty words per minute.

It is further stipulated and agreed that the foregoing contract and agreement shall be held confidentially, that is to say, the proceedings of Edison and his purposes shall not be communicated to any one whatsoever and that neither the said Harrington, nor the said Reiff shall offer for sale the said inventions until all others now completed and in practical use are fully disposed of.

In Witness whereof the said Thomas A. Edison and said George Harrington and the said Associate of said Harrington J. C. Reiff have hereunto set their hands, and affixed their

seals this Thirty First day of March Anno Domini eighteen
hundred and seventy three.

Thomas Alva Edison[c]   Geo Harrington[c]

                                                 Josiah C Reiff[c]

In presence of   David G. Barnitz Eugene H. Gibson.[5]

DS (copy), NjWOE, LS (*TAEM* 28:956). Written by David Barnitz.
[a]Date taken from text, form altered. [b]"shall be" interlined above.
[c]Followed by representation of a seal.

1. Two major obstacles to the success of automatic telegraphy were
the system's inability to drop copies at intermediate stations and the time
required to transcribe received messages. By fulfilling the agreement
outlined here, Edison would overcome both difficulties.

2. Edison had been working on a roman-character system intermittently since the summer of 1871. See Docs. 184, 186, and 194.

3. Edison did not succeed. He tackled the problems again in 1875 at
the behest of Jay Gould.

4. Edison received U.S. Patents 151,209, 172,305, and 173,718 for
roman-character telegraphs. He assigned all three jointly to Harrington
and himself.

5. Unidentified.

–296–

To Norman Miller

[Newark or New York, March 1873][1]

Mr Miller

You get me a Phelps Relay 125 Ohms & Phelps key   I will
fix rest[a]

AL (photographic transcript), NjWOE, Quad. 71.2, facing p. 14 (*TAEM*
10:229). [a]"Borrowed of W U Co. Apl 1/73 by NCM order of V. P.
Mumford." written in an unknown hand at lower right.

1. Miller left this note with Phelps at the Western Union shop when
he borrowed the specified equipment on 1 April 1873, so Edison must
have written and sent it sometime earlier, probably close to the end of
March. (Testimony of George Phelps, Jr., Quad. 71.1, pp. 314–16
[*TAEM* 10:164]). The equipment apparently was never returned to
Phelps's shop.

MULTIPLE TELEGRAPHY DESIGNS   Doc. 297

The devices and arrangements in Doc. 297 reveal aspects of
Edison's work on duplex and related telegraph designs during
the months after he began testing some of his apparatus on
Western Union's lines and before selected designs were specified in patent application drawings and models.[1] In these

drawings, Edison was refining or modifying some prior designs, such as the current-reversal mechanisms sketched late in 1872, while continuing to envisage and analyze additional possibilities, such as the last two drawings in Doc. 297. His considerations were related to his proposed book as well as to his inventive work for Western Union.[2]

1. The drawings appear on four consecutive pages of a notebook. The pages immediately before and after are devoted to different subjects. Related sets of duplex designs from this period (bounded by Docs. 275 and 315) not published in this edition include: Cat. 298:133(3), 134(1), Lab. (*TAEM* 5:261–62); Cat. 297:13(2), 135–40, 146(3), 152–53, Lab. (*TAEM* 5:489, 929–51, 968, 989–93); Cat. 30,099:277–90, Lab. (*TAEM* 5:1065–74); and Cat. 1176:41–43, 46–47, Lab. (*TAEM* 6:43–45, 47–48).

2. This entry is in the notebook that began with plans for Edison's projected book (Docs. 293–94). See also Doc. 303.

–297–

Notebook Entry:
Multiple Telegraphy
and Miscellaneous[1]

[Newark, March 1873?[2]]

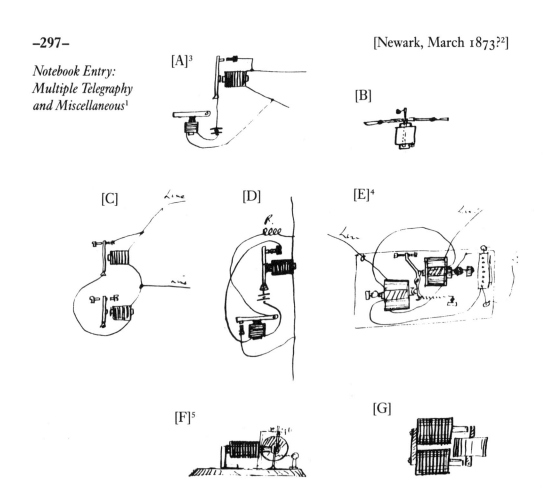

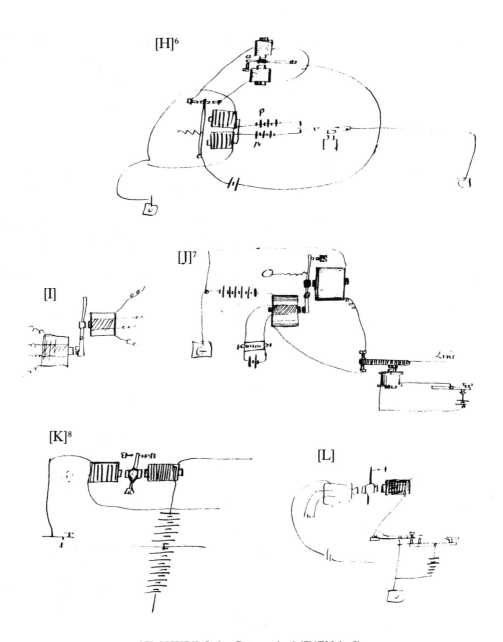

AX, NjWOE, Lab., Cat. 1176:36 (*TAEM* 6:38).

1. See headnote above.

2. The date has been estimated in light of the physical, stylistic, and topical relations among these pages and other materials from this period.

3. The designs here designated A, C, and D are related, the second being a simpler, and the third a more complex, modification of the first. Their purpose is not clear, nor is that of the device sketched in B. However, A, C, and D are clearly devices with sequential effects; one electromagnet causes another to act, and that one in turn affects the status of the first one.

4. Designs E and I are related to Doc. 278. Cf. also Cat. 1176:6, Lab. (*TAEM* 6:8); and U.S. Patent 150,846.

5. In regard to this and drawing G, see designs 10 and 14 in Doc. 285; Cat. 1176:42, 51, Lab. (*TAEM* 6:44, 52); and Doc. 311.

6. Cf. Docs. 275–77; design 16 in Doc. 285; and Docs. 294 and 315.

7. Cf. Doc. 279; "No 21" in Doc. 286; and drawing L of this document.

8. Regarding this and the last design, cf. Doc. 28; Cat. 298:48, Lab. (*TAEM* 5:97–98); Cat. 297:1(1), 4(6), 31(4–5), 34(2), 138(4), 139(3), 140(2), Lab. (*TAEM* 5:427, 438, 551–52, 557, 942, 945, 949); and Cat. 1176:5, 17, 41, Lab. (*TAEM* 6:7, 19, 43).

–298–

Technical Note: Automatic and Multiple Telegraphy[1]

[Newark, March 1873?[2]]

Try this[3]

Duplex Static Current.[4]

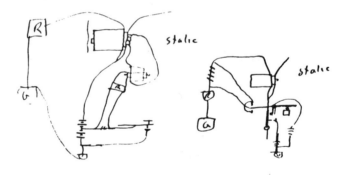

no g[u]es[s] not   Try this[5]

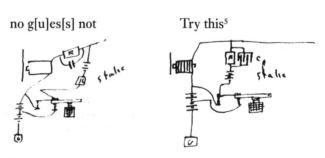

AX, NjWOE, Lab., NS-Undated-005 (*TAEM* 8:331).

1. In these undated sketches dealing with both automatic and duplex telegraphy, Edison was trying to find superior ways to cope with the problems of static discharge and induction currents on telegraph lines, whether overhead wires or submarine cables. The last two sketches illustrate the use of condensers in duplex telegraphy, a technique introduced by Joseph Stearns; the first sketch illustrates the use of a condenser in a cable system, in which context they had been commonplace for many years (Doc. 283). Possibly earlier related designs are found in Cat. 297:26, 59, Lab. (*TAEM* 5:532, 638). The other two arrangements here involve coils inside coils, creating intensified inductive effects. Edison considered a multitude of such designs in combination with several of his other approaches (cf. Docs. 278, 279, 297, and 312; Cat. 297:78, Lab. [*TAEM* 5:743]; and Cat. 1176:47, Lab. [*TAEM* 6:48]). These experiments among other things led to the system described in Edison's U.S. Pat. 147,311.

2. This date was determined through links in style and content with other materials.

3. This drawing is similar to the designs found in Docs. 299 and 317, in which Edison employed a condenser in his artificial line system for placing the receiving instrument of an automatic telegraph at a neutral point in order to overcome static discharge.

4. See n. 1 above; similar designs in NS-Undated-005 (*TAEM* 8:329–32, 412); Cat. 297:26(1), 34(3), 75(2), 80(1), 81(1), 146(3), Lab. (*TAEM* 5:532, 556, 726, 751, 757, 968); and the possibly later Cat. 297:81(1) verso, 81(5), 89(1), Lab. (*TAEM* 5:756, 762, 803). Cf. also U.S. Patents 178,222 and 180,858 in connection with the upper left-hand sketch, and for the sketch at the upper right, U.S. Patent 150,846 as well as the material cited in Doc. 297, n. 8.

5. See n. 1 above; and cf. "Duplex No. 1" and "Duplex No 2" in Doc. 285, and Doc. 314.

–299–

Notebook Entry:
Automatic Telegraphy

[Newark, March 1873?][1]

Cables
Edisons system of Cables working by Centers of resistances and static accumulation[2]

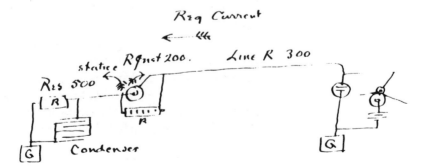

This system gives perfect work between NYork and Washington at 1600 words per minute[3]

Experiments to be tried to lessen The resistance of the artificial line—

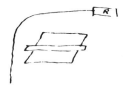

Still later 3180 words per minute areadable =

AX, NjWOE, Cat. 1176:45, Lab. (*TAEM* 6:46).

1. Edison executed his U.S. patent application on this method of working cables and long land lines on 23 April 1873 (Doc. 317). Earlier he had executed a British patent application on the same system, and it was filed on 25 April 1873 (Brit. Pat. 1,508).

2. In his patents for this system (n. 1), Edison claimed that he could neutralize the effects of the static charge on a line by placing the receiving instrument at "a point of no electric tension or zero, as regards the static charge." He did this by balancing the line capacitance and the combined resistances of receiving instrument and line with resistance and capacitance on an artificial line at the receiving end (as shown in the first drawing). Adjustable condensers allowed the static charge of this artificial line to be increased or decreased in order to maintain a balance.

3. Edison probably tried this system on the New York–Washington line of the Automatic Telegraph Co.

MULTIPLE TELEGRAPHY CIRCUITS   Doc. 300

In the unfinished circuit designs in Doc. 300 Edison sought an arrangement of devices that would increase and decrease current strength and reverse current direction. Such a design would permit diplex and quadruplex telegraphy, rather than just the duplex system.[1] Drawing A is similar to the type of design with which Edison later achieved success in this field.[2] However, the problem of current reversals breaking up the other signals was not dealt with here, and Edison knew that without a solution for that problem the system remained impractical.[3] In sketches B and C he apparently considered alternative arrangements for the terminals, but their unfinished character leaves their intended operation unknown.

1. To analyze designs of this type, as well as others for diplex and quadruplex telegraphy working on different principles, cf. Docs. 275–78, 285–86, 294, 297, 315, and the materials cited there.

2. Prescott 1877, 832–42.

3. Edison marked many other circuit designs "OK" or otherwise indicated they were satisfactory or worth further work, but those in this entry were neither so labeled nor even completed. Cf. Cat. 297:154(2), Lab. (*TAEM* 5:995); and Cat. 30,099:288, Lab. (*TAEM* 5:1066); see also Doc. 294.

–300–

Notebook Entry:
Multiple Telegraphy[1]

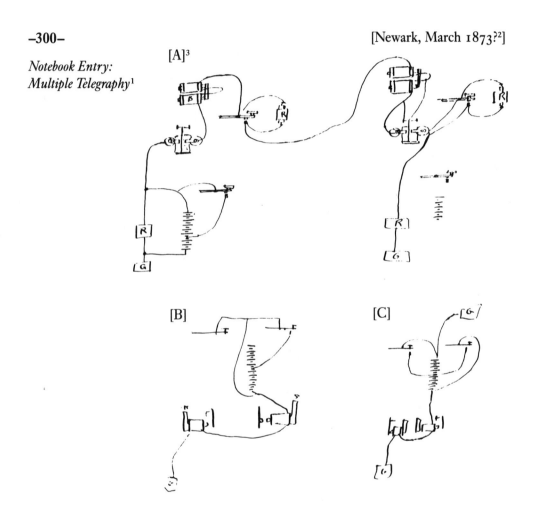

AX, NjWOE, Lab., Cat. 30,099:282 (*TAEM* 5:1069).

1. See headnote above.

2. The date has been estimated through the document's physical relationship to other items in the notebook and the correspondence of its content and style to other documents of this period.

3. In this drawing, **B** is apparently a neutral (ordinary) relay. Immediately below **B** is a polarized relay. Each **R** is a resistance, probably adjustable, and each **G** is a ground connection.

Telegraphing.

The object of this invention is to transmit two[a] despatches in ~~t~~oposite and the same direction over one wire at the same time[b]

The invention consists in ~~m~~the m~~a~~ode of rendering nil the effect of the outgoing current upon the receiving magnet[3]

Fig .1.[4] represents one device[c]

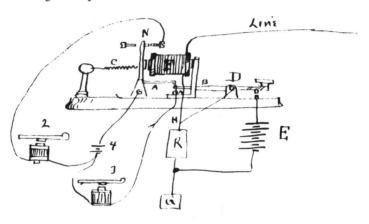

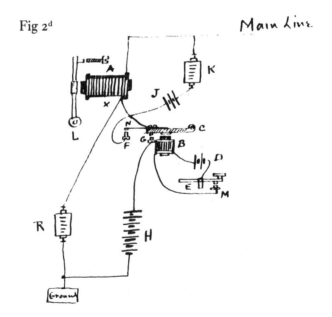

Fig 2[d]

A is the relay   B is the sounder   E the key   D a local battery   R & K resistances   J a neutralizing battery   H the main battery   When the Key E is worked the sounder works with it so when it is closed the Lever C is brought down ~~thro~~

putting the battery on the line ~~o~~through the point G a portion of the current from the battery going to the ground through R at^e the instant the [l]ever^f C closes on the point G the points N & F close and ~~Shunt the~~ throw an opposing Battery J on the relay, which in its effects is equal to the battery H. but in opposite character.^g Consequently the effect is neutralized any variation between the two batters being Compensated for by adjusting the Rheostat .K. to make the effect of the battery J weak or strong.⁵

[Fig? 3?]^h Showsⁱ

ADf (fragment), NjWOE, Lab., Cat. 297:4, 3, 14 (*TAEM* 5:440, 434, 492). These pages were placed separately in a scrapbook at a later date, but since there are physical indications that they were fastened together earlier and they are linked in content, they are assembled here as parts of one document, some other parts of which are missing. ^aInterlined above. ^bLined through with orange pencil or crayon at a later time. ^c"Top" written in orange pencil or crayon at top right of the first drawing; the sketch to the right of "E" is in pencil; the main drawing and text are in blue ink. Two pages following the first drawing are missing. ^dThis drawing begins a page, with "4" written at top center. "Top" appears at top left and "No 7" at bottom right of the drawing, both in orange pencil or crayon. ^eThis starts a new page, with "5" written at top center. Obverse has another, much less detailed and less finished sketch on it. See n. 3. ^fInitial letter obscure. ^g"but . . . character" interlined above. ^hAlmost entirely missing at torn edge of paper. ⁱLower edges of letters torn off; remainder of line and document completely lost.

1. This document is unlabeled but it has the phrasing and organizational character of a caveat rather than a patent application as it deals with quite different devices or arrangements.

2. This document clearly predates the drawing up of individual patent applications such as Docs. 304–5, 308–11, and 314–15. It is possible that this draft dates from one or two months earlier, about the time of Doc. 283.

3. This is the general point of duplex telegraphy. See headnote, pp. 31–32; and Docs. 28 and 285.

4. In this design, **A** is a spring contact fastened to the pivoted armature **N**. Pressing key **D** operates sounder 3 and sends a signal out on the line through magnet **F** by cutting in battery **E**, but the pressure on **A** prevents **N** from responding to the pull of **F**. Unopposed incoming signals charge **F** and attract **N**, thereby activating sounder 2. When incoming and outgoing signals are simultaneous, the combined current charges **F** sufficiently to overcome spring **A**, operating both sounders 2 and 3. Many related designs survive. See, for example, Cat. 297:1(1), 4(4), 15(1), 75(2), Lab. (*TAEM* 5:427, 441, 496, 726); and, in particular, Cat. 30,099:288, Lab. (*TAEM* 5:1066).

5. Cf. Doc. 285, designs 3 and 4; Doc. 314; Cat. 30,099:288, Lab. (*TAEM* 5:1066); Cat. 1176:41–43, Lab. (*TAEM* 6:43–45); and Cat. 297:154(2), Lab. (*TAEM* 5:995).

Draft to Munn & Co.

Munn & Co²
Please mention in case no. 1. a slight alteration³   Thus.ᵃ

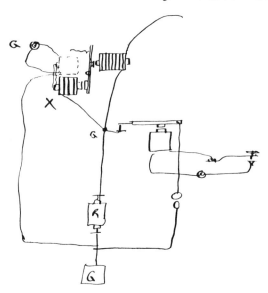

The compensating magnet .X. is split   a local current constantly circulating in one helix. the other coil is inserted into an artificial line, the circuit of which is opposed to the local G.—

Alsoᵇ mention that an extra front point may be used instead of back point and when the main circuit was closed the local ~~spe~~ magnet would be cut off   also that

AD (fragment), NjWOE, Lab., Cat. 297:30 (*TAEM* 5:550). Letterhead of the Gold and Stock Telegraph Co. ᵃ"Top", written in orange pencil or crayon above the drawing, was probably added later. ᵇStarting with this word, the text runs vertically along the right margin of the drawing.

1. The date of this entry has been estimated on the basis of several considerations. The note deals with the modification of a patent application that was being prepared by the patent solicitors. Doc. 303 indicates that this was in process by 4 April 1873, and Doc. 304 shows that it was done by 9 April 1873. The existence of early notes and draft stages of such applications means that a fair amount of time was involved in the preparation. See unsigned technical notes in 73-009, DF (*TAEM* 12:1187–97).

2. Munn & Co. was a patent agency Edison was dealing with at this time; it was also the publisher of *Scientific American.* See Docs. 303 and 316.

3. Presumably this is a reference to what became Doc. 304. But cf. "Duplex No 1" and "Duplex No 2" in Doc. 285; "No 18", Doc. 286; Cat. 30,099:276, Lab. (*TAEM* 5:1074); and Cat. 297:88(4) verso, Lab. (*TAEM* 5:802).

To Norman Miller

[Newark?,] *April* 4, 1873.

Mr. MILLER:

Please inform Mr. Orton that I have accomplished all I agreed to with one exception, and am now ready to exhibit and close the thing up.[1]

DUPLEX.

I experimented twenty-two nights; tried twenty-three duplex systems;[2] nine were failures, four partial success, and ten were all right; one or two of the latter worked rather bad, but the principle is good, and if they were to be used could be improved in detail; eight were good, one of which requires no special instruments; a single wire run in a peculiar manner in a Morse set of instruments transforms them into a duplex. Ten models for ten different duplex have been delivered to Munn & Co., patent solicitors.[3]

"WORKING PLAYED OUT WIRES."

Several experiments made on Washington wires after heavy rain. With the ordinary relay the signals came hard; with attachment to ordinary relay the signals came strong, sharp and clear; two models of this have been delivered to L. M. Serrell, patent solicitor.[4] One attachment is being made for exhibition. I will be ready in five days or sooner.

"WORKING LONG CIRCUITS."

Twenty-four hundred miles were worked by me at different times without repeater, but so far I do not think the devices I have are of any practical value. I shall not give it up.[5] The patents will be allowed in about three months.[6] In the meantime if I run across another duplex I will take steps to confine it in the Patent Office immediately, so that duplex shall be a patent intricacy, and the intricacy owned by the W.U.[7]

Please ask Mr. Orton what I shall do next. Yours,

EDISON.

P.S.—I have full records of all experiments to the minutest detail, with dates.[8] I also go back on duplex prior to Stearns'.[9]

PL (transcript), NjWOE, Quad. 71.2, p. 11 (*TAEM* 10:227).

1. The terms of Edison's agreement with Orton, stemming from their February 1873 negotiations, later became one focus of the extensive litigation known as the Quadruplex Case (Quad. [*TAEM*, reels 9 and 10]). See also Docs. 284, 288, 292, and 316.

2. Cf. Docs. 285, 286, 292, and 316. What constituted a distinct system, as opposed to a mere variation within a system, is not clear.

3. Eight of these designs are found in Docs. 304–5, 308–11, and 314–15. No extant design fits Edison's claim of making a duplex by running one wire a special way in an ordinary telegraph apparatus. The

closest designs to this are Doc. 314, nos. 3 and 4 in Doc. 285, and especially "Fig. 2" in Doc. 301.

4. This is probably a reference to a design covered in U.S. Patent 150,846.

5. This is probably a reference to the design in U.S. Patent 147,311.

6. The earliest related patent was issued 10 February 1874 (U.S. Pat. 147,311), and many applications were never approved. See headnote below.

7. During this same period Edison applied for a patent on a duplex design using electrochemical telegraph receivers; he assigned it to Harrington and himself. He also applied for patents for at least two other duplex systems through Lemuel Serrell. See U.S. Pats. 156,843 and 147,917; and Doc. 316. On one drawing of related circuits Edison wrote "DUPLEX. = W.U." Cat. 297:135(3), Lab. (*TAEM* 5:930).

8. No such dated, detailed records remain; however, see Cat. 297:146(3), Lab. (*TAEM* 5:968).

9. This is probably a reference to Edison's early experiments with duplex telegraphy, which predated Stearns's introduction of duplex instruments on the lines of the Franklin Telegraph Co. Edison testified that he began such experiments in 1865. See, for example, Edison's testimony, pp. 22–23, *Nicholson v. Edison.*

CASES A–H: PATENT APPLICATIONS FOR DUPLEX TELEGRAPHY SYSTEMS
Docs. 304–305, 308–311, and 314–315

During the fourteen days preceding Edison's April 1873 departure for England, he executed through Munn & Co. a series of eight patent applications for systems of duplex telegraphy.[1] These applications came to be known as Cases A–H owing to the designations apparently assigned to them by Munn & Co. They constitute a further development of Edison's ideas on duplex telegraphy derived both from his many tests on Western Union lines and from further thought about still other designs or variations. All are variations of earlier designs; some are elaborations or refinements, and others are major modifications.[2] The last of this series—Case H (Doc. 315)—became the focus of much further work.

Edison signed these applications, but he later said they were poorly done and that he found them confusing and partly unintelligible.[3] He had provided models and drawings of the designs, as well as some notes, drafts, and comments about them, but he left the actual wording of the applications to the firm of patent agents.[4] All of these applications were initially rejected; only for Case H did Edison eventually receive a patent, and that only after considerable amendment of the description.[5]

1. The first of these applications was executed on 9 April 1873, the last on 22 April 1873, and Edison left for England the next day. These systems were all strictly for duplexing, though the designs and tests leading to them had also involved diplex and quadruplex attempts. The original patent application file survives only for Case H; for the rest there are only typed copies (photographic copies of the drawings) of the files provided by the Patent Office to Edison in 1907 for use as evidence in a legal proceeding (see folder introduction, *TAEM* 8:548). Though these copies were certified to be "true," some apparently contain errors of transcription. See also Docs. 302, 303, and 316 regarding the preparation and filing of the applications. During this period Edison executed several other patent applications that he filed through his regular patent agent, Lemuel Serrell. Some of those covered duplex designs as well (see Doc. 316).

2. Cf. Docs. 275–79, 283, 285–86, 294, and 297–302. Consider in particular the designs numbered 1–4, 10–13, and 16 in Doc. 285.

3. Edison's testimony, Quad. 70.7, p. 344 (*TAEM* 9:539).

4. Ibid.; Docs. 301–3; and unsigned technical notes, 73-009, DF (*TAEM* 12:1187–97).

5. Some designs, such as that described in Doc. 310, were fully or partly incorporated in later, separately filed, Edison patents. See also Doc. 316, n. 2.

–304–

Patent Application:
Multiple Telegraphy[1]

New York, April 9, 1873[a]

Specification—describing a new and Improved Duplex Telegraph Apparatus: invented by Thomas A. Edison of Newark, in the County of Essex, and State of New Jersey—

My invention relates to apparatus for simultaneous transmission of two dispatches or signals from opposite ends over the same line wire; and consists in combination with opposing relay and local magnets, of a device which by mechanical means, prevents the lever vibrating between said magnets from responding to the signals transmitted from the home station, but does not prevent the same from responding to the signals from the distant station—[2]

In the accompanying drawing,—

Figure 1, represents a plan view of my apparatus for double transmission and

Figure 2. a modification of the same, showing the arrangement of an opposing & local battery for neutralizing each other, whereby it is not necessary to break the circuit.

Similar letters of reference indicate corresponding parts,—

Case A[b]

Fig: 1.

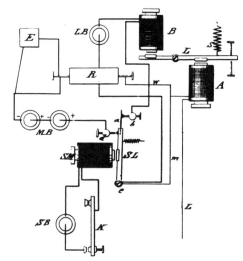

Fig: 2.

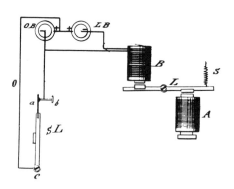

Inventor:[c] T. A. Edison Per[c] Munn & Co. Attorneys.[c]
Witnesses:[c] Chas. Nida.[3] [P.?][d] Sedgwick[4e]

In the drawing A, is the receiving magnet. B. is an opposing magnet, operated by a local battery, L.B. placed in circuit with the sounder lever SL, by spring extension a, and contact b.

L. is the armature lever of the receiving magnet A, and local magnet B, pivoted so as to vibrate between them.

S, is its spring, the tension of which is greater than the power of each magnet singly, but of less tension than the combination of either magnet with the current of the line.—

SM is the sounder Magnet, operated by the sounder battery SB, and key R,[5] d, the contact stop connecting main battery MB, by sounder lever SL, and wire m, to the line wire, R, A[6] rheostat, placed between relay A, and the earth—plate E, and transmitting part of the force of the main battery to the earth, sending the other part of the current to the distant station without overcoming the tension of spring S—

The local circuit is closed, when the home station is not transmitting, the local magnet being not strong enough to attract the lever L, on account of the resistance of spring S. If however the distant station is sending, the combined power of the line current and local current overcomes the resistance of the spring S. and the relay responds to the signals of the distant station.

When, however, the local circuit is opened, by the depression of key R, and attraction of the sounder lever SL, to its magnet, the sounder lever breaks contact with b̲, and closes the circuit with the main battery at contact d̲, the current passes through the relay A, to the line without moving the lever L, not being strong enough to overcome the resistance of spring S—Thus the signals of the home station are transmitted through the receiving magnet, which is always in circuit with the distant station, without responding to them—If however, a current be transmitted from the distant station simultaneously with the closing of the main circuit at the home station, the power of the electro magnet A, is increased, a greater power is exerted on the vibrating lever L, namely: the combination of the power of both main batteries and the resistance of the spring S, is overcome, so that lever l^7 moves, transmitting the signals of the distant station:

The relay at the other end of the line responds in similar manner to the signals of the home station, transmitting thereby simultaneously the signals from either station at the same time.

By connecting the local battery LB, with an opposing battery OB, in the same circuit they will neutralize each other, but preserve at the same time the continuity of the circuit.—

The magnet B, discharges itself within its own circuit for the purpose of being rendered more sluggish, avoiding thereby the danger of attracting lever L, and creating a confusion of signals.—

Having thus described my invention—

I claim as new and desire to secure by Letters Patent.

1st The armature lever L, placed between the receiving and local magnets A and B, having strong spring S, to be vibrated by the joint action of either with the line current, when singly they are too weak to change its position substantially as described.—

2d The sounder lever S,L, having spring extension a̲ or equivalent, in combination with the contact stops b̲, d̲, to constitute a joint conductor for the current of the main or local battery, as described.—

3rd   The combination of the receiving magnet A, with the main battery MB, sounder lever SL, and rheostat R, to regulate outgoing current of main battery and establishing earth circuit, substantially as and for the purpose described.—

Thomas. A. Edison

Witnesses: Paul Goepel.[8] T. B. Mosher[9]

TD (transcript) and D (photographic transcript), NjWOE, PS, Abandoned Patent Applications (*TAEM* 8:552, 582). Petition and oath omitted. This copy of the original file (certified to be "true"), including the wrapper and related correspondence, then still in the U.S. Patent Office records, was provided to Edison in 1907 for use as evidence (see *TAEM* 8:548); the original no longer exists. [a]Place and date taken from oath and petition. [b]"Case A" written in an unknown hand. [c]Preprinted. [d]Possibly "C." [e]Unclear term or name follows in lower left corner, and "Rej Apl 3oh 1873" is written in bottom margin, with "'June 5h'" under that; drawings and signatures are on a separate page.

1. See headnote above.
2. This is Case A; however, this initial characterization matches neither the rest of the description nor the drawings. Rather, it apparently relates to the first design shown in Doc. 301. For the design shown and described here, cf. "Duplex No 1" and "Duplex No 2" in Doc. 285 and the materials cited there; see also Doc. 302.
3. Unidentified.
4. Unidentified.
5. Should be "K."
6. Not a label; should be "a."
7. Should be "L."
8. Unidentified.
9. Unidentified.

–305–

Patent Application:
Multiple Telegraphy[1]

New York, April 9, 1873[a]

Specification[b] describing a new and Improved Duplex Telegraph Apparatus invented by Thomas A. Edison of Newark, in the County of Essex, and State of New Jersey.

My invention has for its object the simultaneous transmission of two signals over the same wire, but in opposite directions and consists of the combination of two relays with their armature lever, pivoted between them and placed at different distances from the same, so as to prevent the relays to respond to the signals transmitted from the home station without being prevented from receiving the signals of the distant station.[2]

Case B^c

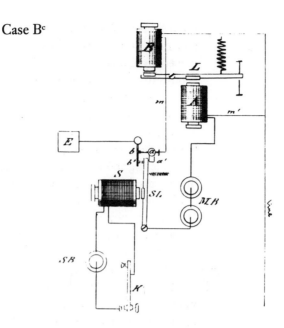

Inventor:^d T. A. Edison Per^d Munn & Co. Attorneys.^d
Witnesses:^d Chas. Nida.[3] [P.?]^e Sedgwick[4f]

The accompanying drawing represents a plan view of my improved apparatus for double transmission, in which A. and B. are the receiving magnets. L. their armature lever, pivoted between them and adjusted at greater distance from magnet A. than from magnet B. requiring therefor an increased amount of power to attract the same to A. instead to magnet B. The magnet B. is connected by wire m, with contact stop a, and by spring contact z,[5] to the earth. An insulated extension a', of contact, a, serves also as back stop for sounder lever SL. which acts, on closing to the sounder magnet S. on a second contact z', of spring contact z, disconnecting the stops a b, and throwing relay B, out of circuit. Magnet A, is connected by wire m' with the line and over the main batter M.B. to sounder lever S.L. operated by means of key k, sounder battery S.B. and magnet S.

By closing the sounder lever S.L. the same acts on contact z' separating contacts a.z. and throwing magnet B. out of circuit. The current of the main battery MB. passes through magnet A. to the line. Magnet A. is placed at such a distance from lever L. that the power of the main batter M.B. is insufficient to attract the same, the signals are therefore transmitted to the distant station without being responded to by the home station. If however the sounder lever SL. be open, the main battery M.B. and magnet A, are thrown out of circuit and magnet B. being adjusted much closer to lever L. and placed by contacts a.z. in circuit, attracts lever L. and responds to the signals of the distant station. The line current is then conducted by contacts a.z. to the earth B.[6] When both stations are transmitting, so that sounder lever S.L. is closed at each station, the combined strength of the main batteries both of the home and distant station is passed through magnet A. attracting the armature lever L. and responding thereby to the signals from the distant station.

Having thus described my invention—

I claim as new and desire to secure by Letters Patent.

1st. The pivoted armature lever L. in construction[7] with the magnets A. and B. adjusted between them as set forth and operated as described.

2d   The sounder lever SL. in connection with spring contacts z.z′. and contacts a.a′. to throw either magnet A.B out of circuit, substantially as set forth.

Thomas A. Edison

Witnesses: Paul Goepel[8] T. B. Mosher[9]

TD (transcript) and D (photographic transcript), NjWOE, PS, Abandoned Patent Applications (*TAEM* 8:586, 603). See textnote for Doc. 304. [a]Place and date taken from oath and petition. [b]"Case 'B'" written in upper right margin. [c]"Case B" written in an unknown hand. [d]Preprinted. [e]Possibly "C." [f]Unclear term or name follows in lower left corner, and "Rej Apl 30h 73" is written in bottom margin; drawing and signatures are on a separate sheet.

1. See headnote, p. 556.
2. This is case B. Cf. "Duplex No 1" and "Duplex No 2" in Doc. 285 and material cited there.
3. Unidentified.
4. Unidentified.
5. Throughout the document "z" and "z′" should be "b" and "b′."
6. Should be "E."
7. Probably means "connection."
8. Unidentified.
9. Unidentified.

DUPLEX DESIGNS WITH ADJUSTABLE RESISTANCE COILS   Docs. 306 and 307

Edison drew several closely related duplex designs that employed adjustable resistance coils to make a relay insensitive to outgoing signals but responsive to incoming ones, which was the fundamental goal of a duplex circuit.[1] Edison applied

Right: *Drawing from Edison's patent of a relay with a shunt containing a set of coils for varying resistance (U.S. Pat. 160,405).*

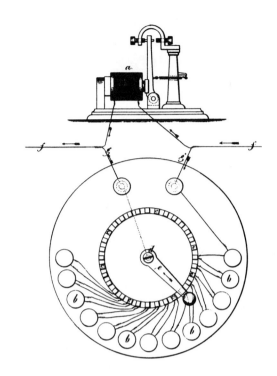

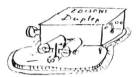

Above: *Sketch of a relay shunted by adjustable resistance coils (i.e., a rheostat) for use in one of Edison's duplex telegraph designs, comparable to Docs. 306 and 307.*

for a patent on such an adjustable relay at the same time that he was applying for many duplex patents in April 1873; the relay patent was granted as U.S. Patent 160,405. He also appears to have used a similar design in his work on automatic telegraphy (see Docs. 324 and 336).

1. Edison drew these designs on the back of Western Union forms and other scraps of paper. Cf. the accompanying sketch; Cat. 297:31(2), 33(1, 3, 5), 34(1) recto and verso, Lab. (*TAEM* 5:551, 554–55, 557–58); and Doc. 285, designs 3 and 4. Similarities of style, ink, and paper, combined with considerations of content and context, indicate that these designs all date from about the middle of April 1873.

–306–

Technical Drawing: Multiple Telegraphy[1]

[Newark or New York, c. April 15, 1873]

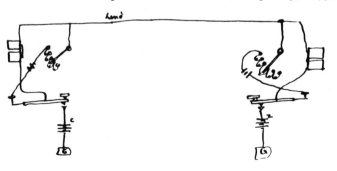

AX, NjWOE, Lab., Cat. 297:33 (*TAEM* 5:555). "Top" added at top of drawing in another hand.

1. See headnote above.

–307–

Technical Drawing:
Multiple Telegraphy[1]

[Newark or New York, c. April 15, 1873[2]]

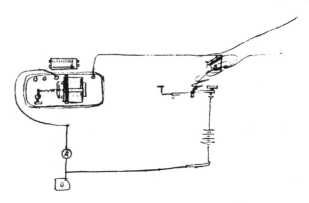

AX, NjWOE, Lab., Cat. 297:33 (*TAEM* 5:554).

1. See headnote, pp. 562–63.

2. This sketch is drawn on the back of a Western Union New York office form that is dated 15 April 1873.

–308–

Patent Application:
Multiple Telegraphy[1]

New York, April 16, 1873[a]

Specification[b] describing a new and Improved Duplex Telegraph Apparatus invented by Thomas A. Edison of Newark in the County of Essex and State of New Jersey.

This invention relates to apparatus for simultaneous transmission of two dispatches or signals over the same line wire in opposite directions and consists in the neutralizing of the effect of the out going current on the receiving instrument by an adjustable opposing magnet, operated by a local battery so that the relay is prevented from responding to the signals of the home station. It also consists in the arrangement of an induction magnet in connection with the local battery and the main line, for neutralizing the static current of the line.[2]

In the accompanying drawing

Figure 1—represents a plan view of my improved apparatus for double transmission, and

Figure 2—is a side elevation partly in section, of the relay and the opposing local magnet.

Similar letters of reference indicate corresponding parts.

Case D^c

Fig. 1.

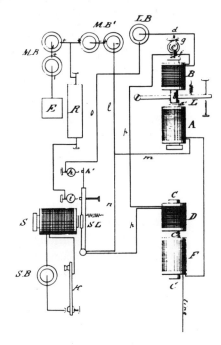

Fig. 2.

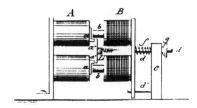

Inventor:^d T. A. Edison Per^d Munn & Co. Attorneys.^d Witnesses:^d Chas. Nida.³ [P.?]^e Sedgwick^{4f}

A represents the receiving magnet, the cores a of which are provided with projecting pieces a′ placed sideways of the axis of the cores towards each other, requiring a smaller armature, of lever L. The local magnet B having projecting cores b, is placed against the magnet A, with opposing poles, the N and S poles of the local magnet respectively. The lever L, with its armature is working between the forward projecting cores of the magnet B. The local magnet B is connected by guide rods d and d′ with pillar e and made adjustable towards magnet A by means of spiral spring f and thumb screw g. The magnet B may thereby be adjusted so as to exactly neutralize the effect of the out going current of the main battery on lever L. The local magnet B is operated by the local battery L.B. placed by contact h in connection with spring contact h′ of sounder lever SL which is again operated in the usual manner by sounder battery LB, Key K and magnet S. The local battery may be dispensed with and a secondary current sent from the main battery be made in the usual manner. The main battery connects by wire m to the relay A and the line and by wire n to the sounder lever which is placed by contact stop i through rheostat R in a shunt circuit with the carbon poles of the main and neutralizing batteries MB and MB′, the latter being connected to the earth E. C is an induction magnet, having one coil D in the same local circuit, the other coil F in circuit with the main line.

On operating the apparatus, the out going current of the main battery MB, on closing sounder lever SL, is divided, one portion passing through relay A to the line, the other through contacts i and rheostat R to battery MB and the earth. Rheostat R is of slight resistance, to prevent too much spark on contact, point i. The local circuit is connected at the same time by contacts h h′ and thereby the effect of the outgoing current upon the cores a, of the relay neutralized by the opposite polarities of the cores b of the local magnet B. The current passes therefore to the distant station, without effecting the lever L, as the sounder lever closes the local circuit, at the moment the main battery is thrown on the line.

When the sounder lever SL is open the armature lever L responds to the current of the distant station, as no opposing polarity of the local magnet prevents its attraction.

When the sounder levers are closed at both stations, the current of the main battery at the home station is neutralized by the local magnet, but the current of the distant station, being of equal polarity with it, attracts the lever L and responds thereby to the signals of the distant station.

The static current of the line is neutralized at the moment of closing and opening the sounder lever SL. When the sounder lever is closed the induction coil D of the local battery is acting upon the iron core C, which induces a momentary current into F and upon the line opposite to and of equal duration with the static current thereby neutralizing each other. At the moment of opening the sounder lever, the static current is at opposite polarity and is then neutralized by the induction current of coil F, which is also in opposite direction. The amount of induction electricity can be regulated by means of a rheostat shunt around the magnet D.

Having thus described my Invention—what I claim as new and desire to secure by Letters Patent is

First—In apparatus for double transmission, the receiving instrument A, having inside projecting cores a in combination with the opposing magnet B having projecting cores b, adjustable towards the former and operated substantially as for the purpose described.

Second—The induction magnet C, having coils D and F placed in the local battery LB and the main line, to neutralize the static current on the same, substantially as and for the purpose described.

Thomas A Edison

Witnesses   Paul Goepel.[5] Alex F Roberts[6]

TD (transcript) and D (photographic transcript), NjWOE, PS, Abandoned Patent Applications (*TAEM* 8:634, 656). See textnote for Doc. 304. [a]Place and date taken from oath and petition. [b]"—Case D—" written at top center. [c]"Case D" written in an unknown hand. [d]Preprinted. [e]Possibly "C." [f]Unclear term or name follows in lower left corner, and "Rej Apl 3oh 73." is written in bottom margin; drawing and signatures are on a separate sheet.

1. See headnote, p. 556.
2. This is Case D. Cf. Docs. 285 and 312, and U.S. Pat. 180,858. Cf. also Stearns's patent of 18 March 1873 (U.S. Pat. 136,873), which used electromagnets instead of condensers to compensate for the "kick."
3. Unidentified.
4. Unidentified.
5. Unidentified.
6. Unidentified.

–309–

Patent Application: Multiple Telegraphy[1]

New York, April 16, 1873[a]

Specification[b]—describing a new and Improved Duplex Telegraph Apparatus: invented by Thomas A. Edison of Newark, in the County of Essex, and State of New Jersey.—

My invention relates to apparatus for transmitting dispatches or signals simultaneously over the same line wire in opposite directions and consists of the neutralization of the effect of the out going current by the main batteries themselves, which are connected with same poles to both sides of the relay, the other poles being connected to the sounder lever and thence to the earth.[2]

Case E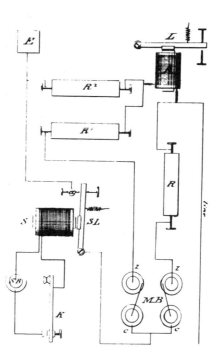

Inventor:[d] T. A. Edison Per[d] Munn & Co. Attorneys.[d]
Witnesses:[d] Chas. Nida[3] [P.?][e] Sedgwick[4f]

The accompanying drawing represents a plan view of my improved apparatus for double transmission in which

A, represents the receiving instrument or relay, L, its armature lever and M,B, two main batteries of equal strength, arranged at both sides of the relay in such a manner that the zink pole of one main battery is connected to one side of the relay, and the zink pole of the other main battery to the other side of the same. The carbon poles of both main batteries M,B, are connected to the sounder lever SL, which is operated in the usual manner, by its key K, sounder battery S,B, and magnet S, Both zink poles of the main batteries are connected to the relay A, through rheostats R, and R'[5] adjusted with slight resistance to prevent shunting or short circuiting of the relay A. A third rheostat R[2], is placed between the relay A, and the earth plate E—The contact stop a, of the sounder lever SL, is also connected with the earth.

On closing the sounder lever S,L by the depression of key K, the carbon current is rounded through contact a, to the earth, the zink currents of both main batteries, pass through the relay A, to the line, but acting against each other, neutralize their effect on the relay A. The circuit with the main battery at the distant station is thereby closed, and the signals are transmitted over the line, without affecting the armature of the relay. When the sounder lever is open, the signals from the distant station pass through the relay A, and the rheostat R^2, to the earth. The lever L, responds to them as the two main batteries being in a shunt around the relay and opposing each other, produce no effect on the same.—

When, however, both sounder levers are simultaneously closed at the home and distant station, the relay R,[6] responds in similar manner to the current from the distant station, and the relay of the distant station to the outgoing zink current of the home batteries so that thereby two signals are transmitted at the same time, one from either station.—

Having thus described my invention.

I claim as new and desire to secure by Letters Patent—

1st   The receiving relay, in duplex telegraph apparatus, combined with the same poles of two main batteries, to neutralize effect of outgoing current on the relay, substantially as set forth.—

2nd   The rheostats R R', placed within the circuits of the opposing main batteries for the purpose described.—

Thomas A. Edison

Witnesses   Paul Goepel.[7] Alex F. Roberts[8]

TD (transcript) and D (photographic transcript), NjWOE, PS, Abandoned Patent Applications (*TAEM* 8:660, 677). See textnote for Doc. 304. [a]Place and date taken from oath and petition. [b]"Case 'E'" written at top center. [c]"Case E" written in an unknown hand; the drawing was later written over in pencil, probably at the U.S. Patent Office. [d]Preprinted. [e]Possibly "C." [f]Unclear name or term follows in lower left corner, and "Letter May 2d 1873." is written in bottom margin; drawing and signatures are on a separate sheet.

1. See headnote, p. 556.
2. This invention is closely related to "Duplex No 13" in Doc. 285. The design is essentially a form of bridge duplex, though neither Edison nor the patent examiner gave any indication of recognizing that. Since bridge duplexes were an old form, the general mode of operation of this one should have been readily apparent and only its particular variations could be patented. The character of the circuit can be seen through comparison of the drawing in the document with the diagram shown here and with the illustration of a bridge duplex on page 530. All the components and connections shown in the drawing in the document are reproduced and similarly labeled in the

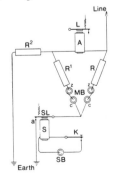

diagram, but some have been partially rotated to enhance clarity. To facilitate comparisons the two wires connecting to the ground or earth (at **E** in the document) are shown separately in the diagram. Comparison of this diagram with the Wheatstone bridge shows the close similarity of their general designs. The main differences are Edison's use of a local sounder circuit (**K, SB, S, SL**, and **a** in the diagram) instead of just a key to send the signals, and his use of two equal batteries in the branches of the main circuit instead of a single larger battery located between the earth and the divergence of the branches (i.e., between **Earth** and **a** in the diagram). Joseph Stearns had been granted a patent (U.S. Pat. 132,932) on another form of bridge duplex on 12 November 1872 during the period when Edison renewed his efforts in this field. The present application (Case E) was rejected on 2 May 1873 and again, after an amendment prepared by Munn. & Co. had been submitted, on 7 June 1873, without reference to its generic character as a bridge duplex design. Commissioner of Patents to TAE, PS (*TAEM* 8:665, 671).

3. Unidentified.
4. Unidentified.
5. Probably means "R^1."
6. Should be "A."
7. Unidentified.
8. Unidentified.

–310–

Patent Application:
Multiple Telegraphy[1]

New York, April 16, 1873[a]

Specification[b] describing a new and Improved Duplex Telegraph Apparatus invented by Thomas A. Edison, of Newark, in the County of Essex and State of New Jersey.

This invention relates to apparatus for the simultaneous transmission of two dispatches or signals from opposite ends over the same line wire and consists in the working of the receiving instrument by induction currents, generated in a secondary helix by the incoming current, when the outgoing current is neutralized by helices wound in opposite directions on the same magnet, so that the receiving instrument responds to the signals of the distant station without responding to the signals of the home station.— By means of an electro magnet placed between the battery and the induction coils the effect of the discharge of a static current is neutralized by the charge and discharge upon its iron core.[2]

In the accompanying drawing—

Figure 1. represents a plan view of my improved apparatus for duplex transmission worked by induction currents and

Figure 2. a detail side elevation of the secondary helix placed at right angles to the primary coils.[3]

Similar letters of reference indicate corresponding parts

Case F^c

Fig: 1.

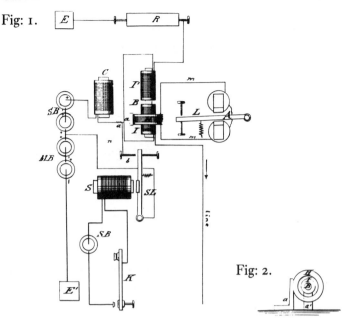

Fig: 2.

Inventor:^{d} T. A. Edison Per^d Munn & Co. Attorneys.^{d}
Witnesses:^{d} Chas. Nida.^{4} [P.?]^e Sedgwick^{5f}

A, in the drawing represents the receiving instrument,
being a polarized relay or other electro-magnet of the usual
form. It is connected by wires m. with the secondary helix H,
supported by a standard a and placed in a plane, vertical to
the axis of the magnet B. Two coils I, and I' are wound upon
the magnet B in opposite directions, helix I connecting with
the line and the battery MB, helix I' with the battery M,B,
and a rheostat R to the earth E.— These helices may also be
wound in opposite directions in one coil instead of being
separated. The secondary helix H embraces vertically the helix I, the core B of which is supported by standards a'.— C is
an electro-magnet placed between battery M,B, and the magnet B and connected by wire n to contact stop b of the sounder
lever S,L, which is again connected to the zinc poles of the
two opposing batteries M,B and M,B' and the latter with the
earth plate E'.

The sounder lever S,L, is operated in the usual manner by
sounder battery L,B, Key R and sounder magnet S.

The working of this apparatus for double transmission is
based on the principle of galvanic induction, according to
which by each closing and opening of the battery momentary

induction currents of opposite directions are produced. When therefore, a current from the distant station passes over the main line into helix I of core B, a momentary current in opposite direction to the same is induced in helix H, which throws the lever or tongue L of the polarized relay A over to one side.—

When the battery at the distant station is disconnected, the induction current in the secondary helix, though of equal direction with it, is in opposite direction to the first induction current and throws therefore tongue L back to its former place. The relay A responds in this manner to the signals from the distant station by the opposite induction currents of helix H.

On transmitting signals from the home station to the distant station, the current divides, before entering into coils I and I'.— They being wound in opposite directions on magnet B, exercise no influence upon it and induce consequently no currents in the secondary helix H. One part of the main current passes therefore to the distant station, the other part through rheostat R to the earth. The outgoing current exercises no effect on relay A, the relay at the distant station responding to the same.—

When both stations are transmitting at the same time, the current from the distant station produces induction currents in the secondary helix H so that the relay A responds to the signals of the distant station. The outgoing current being neutralized in the manner described, operating the relay at the distant station.— The conditions of double transmission are therefore fulfilled and the respective relays responding simultaneaously to the currents from the other stations.—

The object of the electro magnet C is to neutralize the static current and to prevent a spark at contact b̲. The magnet C could also be placed on the compensating circuit, but would not be so effective.—

The disturbing effects of a discharge of a steady current on the magnet B and its helices I, amd I' are neutralized by the charge and discharge of the current upon its iron core.—

Having thus described my invention, what I claim as new and desire to secure by Letters Patent is

First. In apparatus for double transmission, a polarized relay A, placed in a secondary circuit to be worked by induction currents, generated by primary and secondary helices I, and H, as described.

Second.— The secondary helix H in connection with magnet B, having opposite helices I and I′ to neutralize effect of outgoing current, substantially as described.—

Third.— The electro magnet C, arranged as set forth, to destroy the static discharge, as described.—

Thomas A Edison

Witnesses   Paul Goepel[6] Frank Blockley[7]

TD (transcript) and D (photographic transcript), NjWOE, PS, Abandoned Patent Applications (*TAEM* 8:681, 708). See textnote for Doc. 304. [a]Place and date taken from oath and petition. [b]"—Case F—" written at top center. [c]"Case F" written in an unknown hand; the drawing was later written over in pencil in two places, probably by U.S. Patent Office personnel. [d]Preprinted. [e]Possibly "C." [f]Unclear name or term follows in lower left corner, and "Rej May 2d 73" is written in bottom margin; drawing and signatures are on a separate sheet.

1. See headnote, p. 556.
2. This is Case F. The design is closely based on "Duplex No 12" in Doc. 285; it is also very similar to that of U.S. Patent 178,221, where the general operation is described far more clearly than is the case here. The idea involved is to combine the general pattern of a differential duplex with an induction coil (see Doc. 30, n. 2). The basic arrangement here is that of a differential duplex, but rather than having an armature affected by the two opposing primary coils (I and I′), a concentric secondary coil (H) is affected by the two primary coils through electromagnetic induction. That secondary coil would in turn operate a polarized relay. This may well be related to some of Edison's very early inventive efforts. Cf. Docs. 30, 47, and 49.
3. The drawings show that the coils are concentric rather than at right angles to each other (cf. Doc. 311). The labels of the upper batteries at the left of the drawing do not match the following description; probably the upper one should be **MB** and the lower one **MB′**.
4. Unidentified.
5. Unidentified.
6. Unidentified.
7. Unidentified.

–311–

Patent Application: Multiple Telegraphy[1]

New York, April 16, 1873[a]

Specification[b] describing a new and Improved Duplex Telegraph Apparatus by Thomas A. Edison, of Newark, in the County of Essex, and State of New Jersey.

This invention relates to apparatus for simultaneous transmission of dispatches or signals over the same line wire in opposite directions and consists in encircling the armature of the receiving instrument by a double coil and sending an equal current in opposite direction to the outgoing current through the same, so that the effect of this current is rendered nugatory on the receiving instrument. It consists further in placing an electro-magnet in the circuit of the main battery, for generating induction currents, and neutralizing the effect of the static current on the receiving instrument. By spring connection of the sounder lever, the main battery is inserted on closing, and the continuity of the circuit preserved, on opening the same.[2]

In the accompanying drawing.—

Figure 1, represents a plan view of my apparatus for double transmission, and—

Figure 2, a detail side elevation of the receiving instrument with the double helix, encircling its armature.

Similar letters of reference indicate corresponding parts
Case G[c]

Fig: 1.

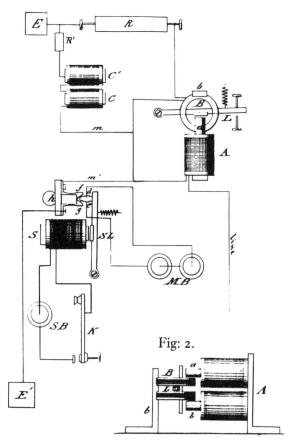

Fig: 2.

Inventor:[d] T. A. Edison Per[d] Munn & Co. Attorneys.[d]
Witnesses:[d] Chas. Nida.[3] [P.?][e] Sedgwick[4f]

A, in the drawing is the receiving relay, L its armature lever. B the double helix encircling horizontally the armature above and below the lever L and between the extended cores a, of the relay A. A vertical standard b, supports the helix B.

The current of the main battery is divided, passing equally but in opposite directions through relay, A and helix B. A regulating rheostat R connects the helix B to the earth at E. A third circuit branches off by wire m, from the main current and passes through the coils of the magnets C and C' to the earth. M,B, is the main battery connecting with both poles to a wedge shaped double contact d, e, of sounder lever S,L. Contacts d, and e, are insulated from each other, but act, on closing the sounder lever to its magnet S, respectively on two spring contacts f, g, also of wedge shape, which are supported by standard h,. When the sounder lever S,L is open, the edges of spring contacts f, g, touch each other, and being connected to the earth at E', preserve the continuity of the circuit. The wedge contacts d, e, by separating contacts f, g, insert thereby main battery M,B, into the circuits.

Wire m' conducts the current of the main battery from spring contact f, to the relay A, helix B and magnets C, C', on the closing of sounder lever S,L. The latter is operated in the usual manner by sounder battery S,B, key R and magnet S.

When the distant station is sending, the armature of the receiving relay A, responds to the signals, the current passing over wire m', and spring contacts f, g, to the earth. When however the home station is transmitting the sounder lever S,L, separates by its wedge contacts d, e., the spring contacts f, g, throwing thereby the main battery M,B, into the circuit. The current passes equally through the relay A to the line, and through helix B, in opposite direction to the earth, preventing the action of the armature, by balancing the magnetic current of relay A. The relay at the distant station responds therefore to the signals of the home station, the effect of the outgoing current on the relay of home station being neutralized. By placing the coil B in this manner around the armature the generation of induction currents may be prevented in the relay itself, when one helix is enclosed within the other.

When both stations are transmitting at the same time, relay A responds to the signals of the distant station, as the outgoing current is neutralized in the manner described. The relay of the distant station responds to the signals of the home station, transmitting the dispatches simultaneously over the line.

The magnets C, C', form a third circuit of the main battery M,B, and generate by their charge and discharge induction currents equal to the static currents of the line.

These induction currents act on the double helix B in opposite directions as the static currents on the relay A, and neutralize therefore their effect on the same. The regularity of the working of the relay and helix are thereby secured and confusion of signals effectively prevented.

Having thus described my invention.

What I claim as new and desire to secure by Letters Patent, is—

First. The armature of the receiving instrument Λ encircled by double helix B, placed between the extended cores a, of the relay A, substantially as set forth.

Second.— The sounder lever S,L, having insulated wedge contacts d, e, in combination with spring contacts e, f, to insert main battery and preserve continuity of circuit, substantially as described.

Third. The induction coil or magnets placed within, a compensating circuit for neutralizing the effect of the static current, substantially as shown and described, and for the purpose set forth.

Thomas A Edison

Witnesses   Paul Goepel.[5] Alex F. Roberts[6]

TD (transcript) and D (photographic transcript), NjWOE, PS, Abandoned Patent Applications (*TAEM* 8:712, 737). See textnote for Doc. 304. [a]Place and date taken from oath and petition. [b]"Case G" written at top center. [c]"Case G" written in an unknown hand. [d]Preprinted. [e]Possibly "C." [f]Unclear term or name follows in lower left corner, and "Rej May 3d 1873." is written in bottom margin; drawing and signatures are on a separate sheet.

1. See headnote, p. 556.
2. This is Case G. The design is based on "Duplex No 10" in Doc. 285 and on Doc. 297; cf. the materials cited in connection with those documents.
3. Unidentified.
4. Unidentified.
5. Unidentified.
6. Unidentified.

–312–

Notebook Entry:
Multiple Telegraphy

[Newark,] April 17 1873[1]

[A][2]

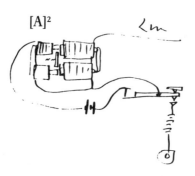

[B][3]

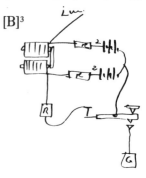

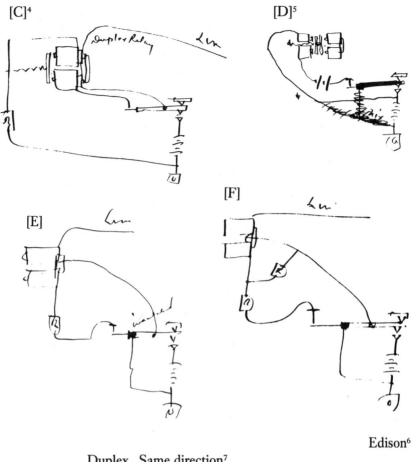

[C][4]   [D][5]

[E]   [F]

Edison[6]

Duplex   Same direction[7]

[G]

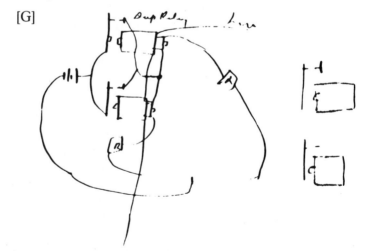

AXS, NjWOE, Lab., Cat. 1176:52 (*TAEM* 6:53).

1. This date, with signature, labels design F; drawings A through E
could date from somewhat earlier. However, all the drawings here share

a common ink, style, and topic; the preceding page is on a different (related) topic and is so headed; and the following page is blank. Thus the date probably applies to all the designs.

2. This is a variation on the design in Doc. 308 ("Case D") and is related as well to designs 9 and 11 in Doc. 285 and to U.S. Patent 130,795. See also Cat. 297:15(4), 16(4), 59(2), 86(3), Lab. (*TAEM* 5:498, 505, 638, 789).

3. This is clearly related to the design in Doc. 309.

4. This is very similar to the basic (pre-1872) differential duplex design of Frischen and Siemens-Halske. However, as drawn, this will not work as a duplex because each time the key is opened it breaks up any incoming signal. Some additional ground connection is needed.

5. Cf. Cat. 297:3(2), Lab. (*TAEM* 5:433); and see n. 2 above.

6. Though not immediately apparent, this design and the preceding drawing are closely related to "Duplex No 6." in Doc. 285. A more direct indication of their operation can be found through comparison with Edison's U.S. Patent 178,222. Similar sketches can be found in "No 9" in Cat. 298:134(1), Lab. (*TAEM* 5:262); Cat. 297:2(4), 13(2), 16(3), 57(2), 59(2), Lab. (*TAEM* 5:430, 489, 503, 633, 638); and Cat. 30,099:284, Lab. (*TAEM* 5:1068).

7. The incomplete nature of this drawing makes it uncertain how this mechanism is to work, but the label indicates that this is a diplex design. See also Cat. 297:18(2,4,5), Lab. (*TAEM* 5:511–12); Cat. 1176:55, Lab. (*TAEM* 6:55); designs 8 and 14 in Doc. 285; "No 22" in Doc. 286; drawing L in Doc. 297; and Doc. 300.

–313–

To Marshall Lefferts

[Newark,] April 21, 1873

General

Mr Bergman[1] was contracted with[2] to build a Universal Model for 125   This model will be finished and delivered to you probably by wednesday P.M. I put the price low and out by peice work so as to prevent shirking work   Bergman has altered this model several times at my suggestion   some of the alterations were very sweeping, and were made on account of a doubt

If he only receives 125 he will have lost money and as he has worked concientiously and hard I think that 200 is but fair and at that it is a very cheap model which on the old plan of day work would cost double   I think you will find that This machine is about perfect   It is very convenient.

The results are quick with the machinery working slow

I can do no better   I consider it <u>the private line machine</u> and my experience covers over 63 different printing telegraphs of my own and a constant work of 5 years 17 hours a day = in a diversity of mechanisms as complicated as 40 bushels of spiders webs

I believe it would be to the interest of the Co[3] to give the order to me to build more than 100, but I do not know if it would be to my interest   Yours Resp.

T A Edison

ALS, NjWOE, DF (*TAEM* 12:1110).

1. Sigmund Bergmann (1851–1927), born in Thuringia (later incorporated as part of Germany), came to the United States as a trained machinist in 1869 and joined Edison's Ward St., Newark, shop in 1870. Bergmann possessed unusual mechanical skill and quickly assumed an important role as an Edison associate. He was also a keen businessman and in 1876 established his own shop in New York City. There he manufactured Edison's phonographs, telephone equipment, and other inventions, expanding his quarters steadily. When Edison's electric light work succeeded, Bergmann became the principal manufacturer of accessories for the system. He left the United States in the mid-1890s to devote himself to manufacturing electrical equipment in Germany. The Bergmann Electrical Works, one of Germany's largest, later merged with the Siemens interests. Pioneers Bio.; *New York Times*, 8 July 1927, 19; "Sigmund Bergmann Dies," *Elec. W.* 90 (1927): 80.

2. Payroll records indicate that Bergmann did not work at Edison's shop from the fall of 1872 until the end of June 1873. Murray's letter of 12 June 1873 (Doc. 338) suggests that Bergmann was working at Henry Thau's shop in New York City. Bergmann and Thau became partners for a brief period in late 1874 or early 1875. PN-72-08-16, Accts. (*TAEM* 22:219–78); Cat. 1218:204, Accts. (*TAEM* 21:285); Thau's testimony, pp. 61–63, *Wiley v. Field.*

3. The Gold and Stock Telegraph Co.; see Doc. 287.

–314–

Patent Application:
Multiple Telegraphy[1]

New York, April 22, 1873[a]

Specification[b] describing a new and improved "Duplex Telegraph Apps" invented by Thomas A Edison of Newark, in the County of Essex and State of New Jersey

This invention relates to apparatus for simultaneous transmission of two dispatches or signals from opposite ends over the same line wire and consists in placing a shunt circuit around the relay, which in connection with an equating battery and adjustable rheostat neutralizes the effect of the main battery on the receiving instrument, preventing it thereby, to respond to the signals transmitted from the home station without preventing it to respond to the signals from the distant station.—[2]

Case C[c]

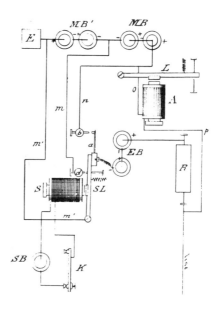

Inventor:[d] T. A. Edison Per[d] Munn & Co. Attorneys.[d]
Witnesses:[d] Chas. Nida.[3] [P.?][e] Sedgwick[4f]

The accompanying drawing represents a plan view of my improved apparatus for double transmission in which, A is the receiving relay   L its armature lever, M,B, and MB' two main batteries of equal strength, but with opposing poles, the latter being connected to the earth at E. The equating Battery E,B, is placed in a shunt around the relay A with its current in opposite direction to that of the main battery MB, neutralizing thereby the effect of the same in relay A at the same moment, when the battery MB is put on the line. The resistance of the shunt and the consequent increase and decrease of the equating battery EB is obtained by the adjustable Rheostat R, placed between the battery EB and the relay A. The other pole of the equating battery EB is connected to the spring contact a of sounder lever SL, which is operated in the usual manner in duplex instruments by sounder battery SB, key R and sounder magnet S. The sounder lever SL is placed by means of wires m m' and contact stop d in circuit with the opposing battery MB', its insulated spring contact a, connecting the shunt circuit over contact stop b and wire n.

When the home station is not sending, so that the sounding lever is open, the line current passes through the relay, attracting lever L and thence to the earth at E. When however the sounder lever SL is closed, three different circuits are formed by spring contacts a, b, and contact d, viz: the circuit of the opposing main battery M,B', over wire M, sounder lever SL, contact d and wire m, the shunt circuit through rheostat R and relay A and the main circuit from battery M,B, through relay A and line to the distant station. In the circuit of the main battery MB' a slight resistance may be thrown in to prevent spark at contact d. The outgoing main current is rendered nugatory in its effect on the relay, by the neutralizing influence of the opposing current of the shunt battery. One part of the main current passes around the relay over the shunt to the line, and transmits thereby the signals to the distant station, the relay A being prevented to respond to them.

When, however, both stations are transmitting signals at the same time, the current from the distant station operates the relay A, the outgoing current, being neutralized in its effect thereon, working in similar manner the receiving instrument at the distant station.

Having thus described my invention

I claim as new and desire to secure by Letters Patent

First, In apparatus for double transmission, the combination of the receiving instrument with a shunt circuit and equating battery, to neutralize effect of outgoing current substantially as described.

Second, the sounder lever SL having spring contact a in connection with contacts d and b, to close circuit of opposing main battery MB' and shunt circuit, substantially as and for the purpose described

Thomas A. Edison

Witnesses   T. B. Mosher[5] Alex F. Roberts[6]

TD (transcript) and D (photographic transcript), NjWOE, PS, Abandoned Patent Applications (*TAEM* 8:607, 630). See textnote for Doc. 304. [a]Place and date taken from oath and petition. [b]"Case. C." written at top center. [c]"Case C" written in an unknown hand; the drawing was later written over in pencil, probably at the U.S. Patent Office. [d]Preprinted. [e]Possibly "C." [f]Unclear term or name follows in lower left corner, and "Rej Apl 3oh 73" is written in bottom margin; drawing and signatures are on a separate sheet.

1. See headnote, p. 556.
2. This is Case C. The design is closely related to "Duplex No 3." and "Duplex No 4." in Doc. 285 and to the materials cited there; see also U.S. Patent 156,843.
3. Unidentified.
4. Unidentified.
5. Unidentified.
6. Unidentified.

Patent Application:
Multiple Telegraphy[1]

Specification[b] describing a new and useful improvement in Duplex Telegraph App's—invented by Thomas. A. Edison of the City of Newark in the County of Essex and State of New Jersey.

The invention has for its object the simultaneous transmission of two dispatches or signals over the same line wire from opposite directions and consists in working the relays at the distant stations by means of reversals of the current at the home station, while transmitting the signals from the distant station by the increase and decrease of the strength of the current of the line.[2]

The accompanying drawing[c] represents a plan view of my improved apparatus for double transmission, showing connection of home station with distant station.

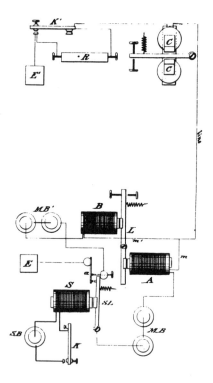

Inventor:[d] T. A. Edison Per[d] Munn & Co. Attorneys.[d]
Witnesses:[d] Chas Nida.[3] [P.?][e] Sedgwick[4]

A and B are electro magnets of equal strength and placed at equal distances from the armature lever L pivoted between them. Both magnets are arranged with seperate batteries battery MB being connected with magnet A and with its positive pole to the line, battery .M,B′, with magnet B, being negative to the line.

Batteries MB and M,B′, are alternately placed into circuit by the action of sounder lever S.L on double spring contact a and then conducted to the earth.

The ~~sounder~~ key[f] lever S,L is operated as usual by its battery S,B, key R[5] and magnet S.

Magnets A and B communicate over wire mm′ and the line wire with the distant station, where C is a polarized relay, to be operated by positive and negative currents, K′ a Morse key and R a Rheostat connected to key R′ and the earth at E′. The object of the key at K′ and rheostat R is, to increase and decrease the current upon the line, so as to affect the lever of the relays A,B. This increase and decrease of the current does not affect the polarized relay C, so that signals may be transmitted by the positive and negative currents of the batteries at the home station, which operate the polarized relay, while at the same time signals may be sent to the home station which are caused by the depression of key K′ and consequent effect on the relays A,B.

On the closing the sounder lever S,L, relay B and its battery M,B′, are thrown out of circuit and relay A and battery M,B with its positive ~~polaritye~~ to the line[g] placed ~~into~~ the circuit. In like manner when the ~~sounder~~ key[f] lever SL is open, ~~relay~~ magnet[f] A is thrown out of circuit, and ~~relay~~ magnet[f] B with its negative ~~polaritye~~ to line,[h] thrown into the circuit. The armature lever remains thereby constantly attracted to the electro-magnets A,B, as the instantaneous transfer of polarity permits no seperation of the same. If both positive and negative currents were passed through one magnet only, a charge and discharge would be produced with the change of polarities, and the armature would be attracted and repulsed.[6]

The effect on the line is the same, whether a current of one polarity or the other is sent, but as each magnet receives a current, of the same polarity, reversal of the current takes place on the line without affecting the relays A,B. The polarized relay is self adjustable and follows the positive and negative currents, whether the tension of the batteries is suddenly increased or decreased.

The polarized relay can be placed at a number of stations on the line and each will be able to receive the signal from the transmitting station.

The simultaneous depression of the keys ~~R~~K[i] and ~~R′~~K′[i] at both stations produces the responding of the polarized relay, at the distant station to the signals of the home station by reversal of the current, while the relays at the home station respond to the signals of the distant station by the decrease and increase of the strength of the current on armature lever L̲

Having thus described my invention

I claim as ncw and dcsirc to sccurc by Lcttcrs Patcnt.

First. The armature lever L̲, pivoted between relays A̲,B̲ to b[e][i] operated by key R̲′ and rheo[stat][j] R̲ from the distant station [by][j] the increase and decrease [of][j] the current, substantially as described.

Second. The sounder lever S̲L̲ in combination[k] with double spring contact a̲, to throw alternately the battersies M̲B̲ and M̲B̲′ with reversed polarities on the line for working polarized relays at distant stations, substantially as described.

Thomas. A. Edison

Witnesses— Paul Goepel.[7] T. B. Mosher[8]

DS and PD, MdSuFR, RG-241, Pat. App. 162,633. Petition and oath omitted. In hand of Paul Goepel. [a]Place and date taken from oath and petition. [b]"Case H" written in top margin. [c]Drawing and accompanying signatures from printed patent. [d]Preprinted. [e]Possibly "C." [f]Interlined above in another hand, possibly Edison's. [g]"e to the line" overwritten and interlined above in another hand, possibly Edison's. [h]"e to line" overwritten and interlined above in another hand, possibly Edison's. [i]"K" and "K′" written in left margin. [j]Paper damaged. [k]"in combination" interlined above in another hand, possibly Edison's.

Edison submitted this patent model with his patent application for a duplex telegraph (U.S. Pat. 162,633).

1. See headnote, p. 556. After initial rejection, a significantly amended version of this application (Case H) was approved in 1875 as U.S. Patent 162,633.

2. This design is a reformulation of the group of earlier designs represented by, for example, "Duplex No 16." in Doc. 285. It used changes of both current strength and current direction, each acting indepen-

dently of the other. While this design solved the problem of combining these changes for duplex transmission, it did not allow diplex, and thus quadruplex, transmission. Edison later testified that he thought up a modification to allow diplex working during his subsequent trip to England (Edison's testimony, Testimony and Exhibits on behalf of T. A. Edison, pp. 24–25, *Nicholson v. Edison*). The patent resulting from Case H, as later amended, became a point of contention in the extensive legal battles over Edison's quadruplex. Quad. passim (*TAEM*, reels 9 and 10).

3. Unidentified.

4. Unidentified.

5. The keys designated **K** and **K'** in the drawing are sometimes miscopied in this description as **R** or **R'**.

6. Cf., for example, Docs. 276, 285, and 294.

7. Unidentified.

8. Unidentified.

–316–

Power of Attorney to Norman Miller

NEW YORK, *April* 23, 1873.[1]

I hereby appoint Norman C. Miller my attorney, sole and exclusive, to arrange, sell, bargain, transfer, convey for any sum he may see fit all my right, title and interest of every conceivable description in eight duplex telgh. patents, obtained by Munn & Co.,[2] and three duplex telgh. and two compensating relay patents, obtained by L. Serrell,[3] to any corporation which shall go by the name of the Western Union Telegraph Co.[4]

Witness: J. C. MASEA.[5]   THOMAS A. EDISON.[a]

PD (transcript), NjWOE, Quad., 71.2, p. 12 (*TAEM* 10:228). [a]Followed by "M." at bottom center.

1. Edison executed several patent applications the same day (see Doc. 317), resulting in U.S. Patents 140,488, 147,311, 147,313, 147,917, 150,846, and 160,405.

2. See headnote, p. 556. Doc. 302 indicates that Edison had intended for Munn & Co. to prepare at least ten patent applications; no reason is known for the change in number.

3. The identity of these applications is uncertain. U.S. Patent 150,846 almost certainly encompasses one of the relay designs and either U.S. Patent 141,777 or 160,405 probably represents the other. One of the duplex designs is almost certainly that of U.S. Patent 147,917, but the other two duplex designs have not been identified. See Cat. 1176:51, Lab. (*TAEM* 6:52) for drawings of compensating relays.

4. No sale resulted and at some point this arrangement ended. In the Quadruplex Case George Harrington and others contended that Western Union's failure to buy these patents ended Edison's agreement with that company, but Western Union regarded this as a mere interlude in a continuing relationship. See the various lawyers' arguments in Quad. 73.1–15 (*TAEM* 10:346–772).

5. Unidentified.

case 82.

1[b] Specification describing an Improvement in Electric Telegraphs, invented by Thomas A. Edison of Newark in the County of Essex and State of New Jersey.

2 In cables and long telegraph lines, there is a limit to the speed with which perfect signals can be transmitted and received, whether the receiving instrument consists of an electromagnet, a galvanometer, a relay or a chemical telegraph instrument; this limit in speed arises from the fact that the moment the line or cable is charged by the battery being connected, a static charge is instantly set up which is in an opposite direction to the dynamic charge, and the tendency is to defer the reception of the signal at the distant station,[1] and at the moment of breaking the battery connection, the static charge disperses by dividing at the center of resistance and going in both directions one part going to the ground at the transmitting station in a direction opposed to the battery, and the other part going towards the receiving instrument in the same direction as the previous current from the battery.

3 This electrical condition is of sufficient duration to render the signals unintelligable at the receiving instrument after a certain speed is attained.

4 The time of discharge is directly proportioned to the resistance at the points of discharge at the ends of the line, and the result is, that the speed of the instruments is limited to the speed with which the line will free itself through the channels aforesaid.

5 My invention relates to the discovery of a method of neutralizing the effects of the static charge in any length of line or cable by balancing the electric forces and the discovery of a point of no electric tension or zero, as regards the static charge, so that the receiving instrument when located at that point will be operated by the rise of tension produced by a pulsation that is connected at such receiving instrument and made as instantly and definitely operative as the pulsation given at the transmitting station.[2]

6 I obtain this point of no tension by forming at the receiving end an artificial line having an equal or nearly equal resistance and electro static capacity, or capacity for producing static charges, as that of the cable or land line, and connect this with the line or cable, and place between the cable and the artificial line the receiving instrument, which hence is in the center of resistance and static accumulation. When this

balance is obtained the signals are received perfect and the rapidity is governed only by the strength of the battery.

7 The artificial line is made with an adjustable rheostat: liquid in a tube is preferable.

8 I connect, between the receiving instrument and the earth, one or more condensers or other accumulators of static electricity, which are made adjustable by having them in sections and bringing one or more sections in or out by a switch so as to increase or decrease the static charge from the artificial cable: it may also be done by placing a very high adjustable resistance coil between one leaf of the condenser and the artificial line.

9 I maintain a very low resistance between the line and the ground at the transmitting station so as to discharge the static current at this end as rapidly as possible.

10 The mode which I prefer, is to keep my transmitting battery in circuit at all times and include in the same circuit another battery of equal power with opposite poles so that when both are in there is no current generated and the resistance of the wire to earth is no more than the resistance of the battery.

11 The transmission of a pulsation is made when the circuit is closed through the perforation in the paper or otherwise, so as to short circuit or shunt the neutralizing battery and send a current upon the line.

12 The current at the receiving paper is shunted through a resistance so as to preserve a constant and equal resistance, which the chemical receiving paper does not give, owing to being more damp in one place than another.

13 In balancing the resistance and static current, the resistance of the instrument is to be added to the line and the resistance of the two equalized by the same amount of resistance in the artificial cable or line.

14 If the receiving instrument is out of the center of resistance towards the line, the pulsations will be weakened by the static charge acting against the pulsation, but if the instrument is towards the artificial cable on the other side of the zero point, the signals or characters will be slightly prolonged owing to the static charge discharging in the same direction as the current. It is at this point that I prefer to place the instrument because by placing an electro magnet in the shunt of the receiving instrument I obtain enough counter discharge from that magnet to cut off this prolongation locally and this discharge from the magnet will not interfere with the line but

has only a local effect on the receiving instrument to prevent tailing on the chemical paper.

Fig. 1.[b]

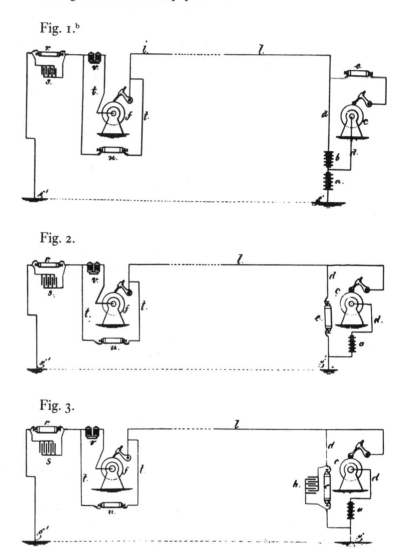

Fig. 2.

Fig. 3.

Inventor   Thos A. Edison, per Lemuel W. Serrell atty.
Witnesses   Chas H Smith, Harold Serrell

15   In the drawing, Figure 1. illustrates my invention in the form that I find most generally available. The batteries a, and b, are connected in opposite positions, the positive poles being towards each other and the negative poles connected to the ground g, and line l, respectively.

16   The transmitting instrument c, is in a circuit d, to the

battery b, in which circuit d, is a resistance e. When the circuit d, is broken the two batteries a, b, neutralize each other and there is no charge sent on the line, but when the circuit d, is closed through a perforation in the paper, or by a key, or otherwise, the battery b, is short circuited and the battery a, being unbalanced sends a pulsation on the line. The artificial line between the receiving instrument f, and the earth g', is made by introducing a resistance or rheostat at r, preferably a tube containing liquid with adjustable points: this rheostat is made to balance or equal, or nearly so the resistance of the line l, and the instrument f, and the condenser s, or other accumulator of static electricity is of a capacity to about equal that of the line, hence the receiving instrument will occupy a zero or neutral point in regard to the static charge from which the static charges will discharge both ways to g, and g'.

17 The condenser or accumulator s, should be in sections to bring in a greater or less number of sections by switches.

18 When the receiving instrument is chemical, the paper is[d] preferably prepared by dipping it in a solution of at least one pound of iodide of potassium in one gallon of water, to which is added a small quantity of flour: this paper cannot be maintained at uniform moisture hence its resistance to the passage of electricity varies; this is compensated for by the shunt circuit t, in which is a resistance u, sufficient to direct the necessary amount of electricity to the paper to make the mark and allow the remainder to pass to the artificial line, thus the varying condition of the paper does not change the resistance of the line.

19 I have discovered that when an electro magnet is energized and the circuit broken, a pulsation is set up in the opposite direction to that passing into such magnets;[3] I avail of this to prevent tailing upon the chemical paper and at v, I have shown an electro magnet for this purpose. It will be seen that this electro magnet will discharge itself within a short local circuit containing the receiving instrument, and that the reactionary current therefrom moving in the opposite direction to the main current frees the receiving instrument from the tailing caused by the discharge of static electricity, and this magnet v, may be employed in many places to effect the object before named, even when there is not an artificial line.

20 In some instances with very long lines, there may be intermediate artificial lines arranged as aforesaid or reactionary magnets with branch circuits to the earth, to either receive drop copies in such branch circuits or to free the line of static

electricity and aid in obtaining the signals perfectly at the last receiving station with the greatest rapidity.

21 Figures 2, and 3, represent the same parts as before described but in Figure 2, only a single battery is shown and the resistance e, is between the line and the earth to regulate the proportion of electricity sent over the line, by adjusting such rheostat to prevent too great return to the battery through such rheostat.[4]

22 In Figure 3, a condenser h, is introduced in addition to this rheostat that it may react between the pulsations of electricity on the main line to aid in clearing such line of the static charge.[c]

I claim as my invention,—

An artificial line between the receiving instrument and the earth to balance the resistance and static charge or nearly so at both sides of the receiving instrument substantially as set forth—

Signed by me this 23rd day of April A.D. 1873.

Thos A Edison

Witnesses   Geo. D. Walker. Geo. T. Pinckney ~~Chas H Smith~~

DS and D (photographic transcript), MdSuFR, RG-241, Pat. App. 147,311. Oath omitted. [a]Place taken from oath, date taken from text of application; form of date altered. [b]Section numbers written in margin in another hand. [c]Drawings and accompanying signatures from printed patent. [d]Interlined above. [e]To this point, written by Walker; remainder of document, except signatures, written by Smith.

1. In an undated fragment, probably written prior to this patent application, Edison described the effect of the static charge in the following manner: "The transmission of waves of electricity are instantaneous no matter what is the length of the cable. The retardation noticed by electricians in cables is ~~due to the~~ not properly retardation but the leyden jar charge sending its current against the charging current. The same as an electromagnet placed in an electric current   The first part impulse will be weakened by the counter charge against the magnetizing current." Cat. 297:12(2), Lab. (*TAEM* 5:478).

2. See Doc. 299.

3. In his British patent for this invention (Brit. Pat. 1,508 [1873]) Edison added a second claim regarding this "discovery." He called the opposite pulsation set up by the energized magnet a "reactionary discharge."

4. The British patent (1,508) reads the same to this point. The description of the third figure is slightly different, and a fourth figure and its description were added. These changes related to Edison's accounting for the "reactionary discharge" set up by the electromagnet.

The English Venture

May–June 1873

Edison departed for England on 23 April and returned to the United States on 25 June. This two-month trip was his first overseas and it provided him with an international perspective at an early point in his career. The purpose of the trip was to demonstrate the automatic telegraph system of the Automatic Telegraph Company for the telegraph department of the British Post Office and to quicken British interest in the system. Edison achieved this goal. In addition he had an opportunity to think further about the application of his "balancing the line" approach to automatic telegraphy and to experience the frustration of working with a coiled undersea telegraph cable.

As a representative of the Automatic Telegraph Company, Daniel Craig had sought unsuccessfully to have George Little's automatic system tested on English lines since early 1870.[1] The English were already well acquainted with automatic telegraphy, having used Charles Wheatstone's system since the late 1860s,[2] and both chief telegraph engineer Richard Culley and superintendent Frank Scudamore were wary of Daniel Craig's extravagant claims of having successfully transmitted from 1,000 to 1,500 words per minute. George Harrington then began to ease Craig out of Automatic Telegraph, which held both Edison's and Little's patents, and in the first months of 1873 company agent George Gouraud met with Scudamore in London. As a result, Culley and Scudamore agreed to a demonstration of the American system, but they remained skeptical about its novelty and performance. Indeed, Culley believed that changing to what was then known as the Little system would be impractical even if it could perform as claimed, because the British Post Office did not even use the slower, Wheatstone system to full capacity.[3] Two American operators were to work the equipment for the test. If they achieved a minimum of 500 words per minute over a 300-mile wire, the telegraph department of the British Post Office would shoulder the expenses of the demonstration and consider adoption of the system.

On 23 April, Edison and Jack Wright, a co-worker from the Automatic Telegraph Company, sailed for England, arriving in early May with transmitters, receivers, a supply of chemically treated paper, and strips of perforated tape prepared in America (they brought no perforators). Edison went to the postal service office on Telegraph Street in London, and Wright departed for Liverpool. Wright's two combination transmitter-receivers were damaged in transit when porters dropped them, but one was salvaged.[4] For about two weeks, Edison and Wright tested the instruments and adjusted the circuits in preparation for the demonstration. They found that the underground wires at each end of the line distorted the signals, so they moved to the outskirts of the cities where they could connect directly to overhead lines. The trials took place on Friday, Monday, and Tuesday, 23, 26, and 27 May 1873. The original agreement required eight transmissions of 1,000 words on each of three days at an average rate of 500 words per minute; the actual number of transmissions was successively six, eight, and nine. The system, which became known as Edison's instead of Little's, performed as stipulated. It tested with a low of 437 and a high of 572 words per minute.[5]

After completing these trials, Edison stayed in England for another two weeks, during which he experimented with the automatic telegraph system on a coiled 2,200-mile cable that was in storage at Greenwich.[6] His tests on this cable failed, however, and he had difficulty understanding why. In mid-June he sailed for New York, arriving home on 25 June.[7] Although Edison's automatic telegraph had passed the British Post Office's test, the British did not adopt it. Instead, they hoped to devise improvements in the Bain system and circumvent Edison's patents.

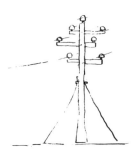

Edison was impressed by the quality of British telegraph lines. Next to this sketch he wrote: "The wires from London to Liverpool are the best strung wires I ever saw."

1. Except as noted, all correspondence, memoranda, and reports relating to the British tests of the American automatic telegraph system are in ATF.

2. Bain's automatic system also had been used earlier. Kieve 1973, 82.

3. Richard Culley to Frank Scudamore, 12 May 1873. The automatic telegraph of Charles Wheatstone worked as fast as 120 words per min-

ute, depending on the length and wiring (undersea, underground, or aboveground) of the circuit. Prescott 1877, 702–11.

4. Jack Wright to TAE, 14 May 1873, Cat. 299, Lab. (*TAEM* 6:133).

5. Test results are in six tables in ATF.

6. See headnote, p. 604.

7. Joseph Murray's testimony, Quad. 70.9, p. 99 (*TAEM* 9:813).

–318–

J. W. Eames to Henry Fischer[1]

[London,] 28.5.73

The Controller—

The trial of Mr. Edisons Automatic instrument commenced on Friday the 23rd inst at 2 pm—

It was found practically impossible to carry out the trial strictly according to the instructions laid down,[2] owing to the limited supply of prepared paper, and from the fact that a large quantity had been consumed in the preliminary experiments—[3]

Instead therefore of a column showing the time occupied in sending a thousand words, a column was made with the heading "No of words received" and the time occupied in receiving a given number was entered under the heading "Total[a] time"—

During the first day of the trial no adjustments were made after starting at 2.8 pm.[4] The speed attained each half hour exceeded 500 words per minute, with the exception of one instance when 455 words only were received— This reduced speed was however evidently caused by the sender at Liverpool[5] miscalculating the number of revolutions made per minute by his sending gear and not by any defect of the wire or apparatus.

The marks in each trial were good but manifested a tendency to run together— this tendency was however much less marked when the receiving band was made to travel faster—

On the second days trial the speed was in three instances[a] below 500 owing as on the first day to the sender not properly computing the rate of speed at which he was propelling the punched band—

~~During~~ Between[b] the fifth and sixth trial the wire was used three minutes by Mr Edison in obtaining a score or two of words from L'pool, which resulted in his making a readjustment occupying but a second— Two other adjustments were made [--]between the two following half hourly trials but they did not occupy more than a fraction of a second each—

The stock of prepared paper having run out on the previous

days trial, a fresh supply had been made, but this paper[a] was very wet, and unsuited for immediate use—

It is believed that the adjustments mentioned were made solely with the view of shunting part of the current[6] in order to remedy this evil as far as was possible—

The average speed maintained through the day was 501 words per minute—

The third and last days trial was commenced by the two first half hourly results indicating a speed of 467 and 482 words per minute— The three following results shewed that the rate had exceeded 500 words per minute— The character of the marks in each of these five trials could not be said to be good, but they were readable—

The wire was changed before[a] the sixth trial, and a marked improvement took place, the signals being very good and clear— With one exception the marks were recorded at the rate of 500 words per minute, and that exception was caused by the difficulty of precisely fixing the actual speed of transmission—

It appears to me that the sole advantage of this system over the Wheatstone is the established fact of its much higher speed—[7] It has the disadvantage that the prepared paper being damp is easily torn by the accidental pressure of the fingers or by other untoward means— The marks also are not durable, vanishing from the paper in a few hours[a] and liable by contact to become transferred to other portions of the same band—[8] Should it however become advisable to work at a higher rate of speed in preference to increasing the number of wires I should consider these objections of but minor importance— The punched paper could be preserved for record, and the chemically prepared paper thrown aside—

It is perhaps right to point out to you that by this system it would be necessary to maintain a staff of writers of sufficient strength always in readiness to deal with the greatest amount of work likely to be sent through at any one time—

An accumulation of slip for[a] even a few minutes whilst additional hands were being brought up to meet the momentary pressure, would be—for the reasons previously mentioned—fatal to the work—

Bearing in mind these conditions I can see no reason why with a proportionate increase of staff over the Wheatstone complement there should not be proportionally increased results—

I think that I should not be under = estimating the proper

working speed on our L'pool wires to be 400 words per minute, at which rate the marks have invariably been perfect—

I might mention that no adjustment appears to be necessary beyond that of the shunt—and that the instrument can readily be worked[a] by an ordinary good Morse clerk, and by care being taken that the prepared paper is neither too wet nor too dry—

J. W Eames[9]

ALS, UKLPO, ATF, Item 73. On paper headed "TELEGRAPHS." in upper left corner under embossed seal. "Enclosure to Mr. Fischer's report of 30 May 1873—Report (with 4 enclosures) from Mr Eames to the Controller of the Central Telegraph Office with ene" written in an unknown hand at top of first page. [a]Repeated at end of one page and beginning of next. [b]Interlined above.

1. Henry Fischer was controller (i.e., financial officer) of the central office of the British telegraph service in London.

2. On 22 May, Culley wrote out the "Conditions of the final trial of the Little Automatic System" (Item 73, ATF). The test was supposed to determine the time needed to transmit 1,000 words; instead, the receivers recorded the number of words sent in two minutes.

3. The paper for receiving that Edison brought with him. Compare the explanation given by David Lumsden in Doc. 319.

4. That is, 2:08 P.M.

5. The sender was Jack Wright (n.d.), who had come from America with Edison for the tests. Wright, a telegrapher since 1862, had been night manager of the Western Union office in Boston and had roomed with Edison when Edison worked there in 1868–69. He had come to work for the Automatic Telegraph Co. in New York in 1872. Cat. 299, Lab. (*TAEM* 6:133); Brief for Field, p. 61, *Wiley v. Field.*

6. Edison put shunts around both the receiver and the transmitter; see the diagram in Doc. 319.

7. See Chapter 12 introduction, n. 3.

8. By the end of 1873, Edison had developed a chemical solution that made permanent marks on the paper. Sir James Anderson to Scudamore, 5 Jan. 1874, Item 67, ATF.

9. J. W. Eames was a staff member in the central office of the British telegraph service assigned to monitor the tests. Henry Fischer to the Secretary, 20 May 1873, Item 67, ATF.

–319–

David Lumsden to
Richard Culley

[London,] 28 May 1873[a]

The Engineer in Chief

In reporting upon the trial of Edisons modification of the Bains ⟨or rather Littles⟩[b] Automatic System of Telegraphing I found that it was impossible literally to carry out the form suggested in the instructions— the paper ran out at such a speed that it would have been impossible to note with any certainty the point at which marks commenced or ceased—[1]

The method adopted therefore was to run a a number of words through from an endless band noting by a chronograph the time taken in receiving a given length of slip, the number of words on which were counted, the time taken up by adjustments,[2] if any, being likewise noted—

The facts during the 3 days trial were as follows—

1st Day— we used some rolls of Chemical paper brought over by Mr Edison from America which were of sufficient length to record over 1100 words—the actual number of words received in 2 minutes being noted   with one exception the numbers were in excess of 500 words per minute   the exceptional[c] case being 455 words per minute— this however was simply due not to any defect but to the fact that as both Transmitter and Receiver are turned by hand, the transmitting clerk at L'pool had guessed the time badly.

The Battery used was Groves Carbon & Zinc as in America—60 cells—[3]

No adjustment beyond that necessary at the first Start was required throughout the day—

The marks throughout the day were quite legible and readable, although there was a slight tendency in the dots to run together— this was partly due to the receiver being turned rather slowly in order to economise paper   a spurt being given now and then to [-]shew the effect of increasing the speed of the paper— When the receiver was turned at the same speed or in excess of the transmitter the marks came out quite distinctly—

At the finish I had a slip run through at about 700 and also one at 400 words per minute—the former although not what could be called good was readable the tendency of the marks to run together being much greater than at 500—at a speed of 400 the marks were perfect

2nd[c] Day— It was found that the Battery at Lpool had either been inadvertently left on Short circuit—or allowed to run down and the acid had to be replenished—8 half hourly trials were made during the day—

All the old paper having been used on Friday some fresh paper was prepared on the Saturday bwith a pencil[4] by hand   this paper was much too wet for giving the best results—

We commenced with the same adjustment as on Friday—the marks were the same as on that day except that they were inclined to drag a little more owing to the wet state of the paper (this is an old fault in the Bains system and gave rise to much trouble).[d] ⟨It is practically difficult to secure paper exactly of the proper degree of dampness.⟩[b]

At the 5th trial 3 mins was taken up in altering the compensation[5] to suit the paper—the marks afterwards coming out very fairly—

during the 7h trial the slide was moved once while receiving—[6]

On 3 of the trials today the speed was under 500 words per minute—due as the Lpool clerk not turning fast enough, his object of course being to keep as near to the 500 words as possible— The Rolls of paper used today being wound loosely by hand would not record more than from 250 to 500 words at each trial, the variation in the total number of words received being due partly to this fact and partly to the paper breaking before the rolls were run out, from their being too wet— the best results as is well known are obtained when the paper is just sufficiently damp to record signals— The average speed obtained during the day was 514 words per minute—at the finish I again had slips put through at

800 words marks rather light and just readable

400 words marks perfect—

100 words <u>without</u> compensation a a continuous line was received and had to reduce speed to about 30 words per minute to obtain readable signals—and even at this speed the static discharge from the wire was very evident—

On the 3d day Weather very wet and slight Contact[7] on all wires, sufficient to[c] shew on the sensitive paper employed— on the first two trials the clerk at Lpool did not turn fast enough, and I had an extra slip put through to see what marks over the 500 words per minute were like. they were readable but not good and would not do for general traffic on such an inclement day as this. it might have been got over by additional Battery power but there being only 60[c] cells at Lpool this could not be done— The wire used was via the canal[8] and shewed a good deal of leakage and the shunts employed 250 ohms at Lpool and about 60 at TS[9] the line being over 4000 allowed very little current fto do the work. This will perhaps be best shewn by a rough sketch of the connections—

The magnets in the shunts forming the compensation to counteract the static[c] discharge from the line— I had a Rly wire[10] joined up for the 2 pm and subsequent trials—after which the marks at 500 words were received quite clear and distinct—the whole of the trials were made on a No 8 wire,[11] but at the close I tried the effect of substituting a No 4 wire,[12] the result being that the received signals at 500 words were as anticipated much more clear and heavy, and readable although inclined to run[13] at 740 words—

The average results of the 3 days trial may be summed up as follows

    1st day 534 words per minute
    2nd day 514   "   "
    3d day 513   "   "

The signals at these speeds were quite legible and readable taken simply as an experimental trial, but in practice I consider that 400 words would be the maximum that could be obtained on a good wire between TS and L'pool—that is, on a wire free from contact—[f] The static inductive effect can be compensated for while contact cannot

The apparatus is simply a Bains there[c] being nothing new in the application to it of the Automatic principle or in the Chemicals employed, but it has the important addition of the Compensating Magnets. this as applied to the Bains System is new and it is this addition that enables these high speeds to be maintained[14]

Mr Edison informs me that he has tried all the methods of compensating by condensers and Batteries but none of them give results equal to the magnets.

The chemicals used being a solution of Iodide of potassium and starch the marks only last for a few hours—

D Lumsden[15]

ALS, UKLPO, ATF, Item 75. On paper headed "TELEGRAPHS." in upper left corner under embossed seal. "Enclosure to Mr. Culleys report to Mr Scudamore of 6 June 1876   Report from Mr Lumsden to the

Engineer in Chief" written in an unknown hand at top of page. [a]Written at end of document in same hand as note at top of page. [b]Marginalia written in same unknown hand. [c]Repeated at end of one page and beginning of next. [d]"old fault . . . trouble)." underlined in an unknown hand. [e]"have . . . 60" underlined in an unknown hand. [f]"a good wire . . . contact—" underlined in an unknown hand.

1. See Doc. 318, n. 2.

2. "Adjustments" refers to the resistance Edison placed in shunts around the instruments to counteract the self-induction of the line. See the sketch accompanying this document.

3. The modified Grove cell—also called a Bunsen cell—in use on American lines produced about 1.9 volts, compared with the approximately 1.1 volts of the cells used on British lines. Because the Grove cell also possessed a lower internal resistance than its British counterpart, it put out a considerably higher current. At some point in his experiments, Edison realized that reception was improved by increasing the current on the line. Culley later mentioned his fear that the heating effect of that current would set the covering of insulated lines afire "if incautiously applied." Atkinson 1910, 868–71; King 1962a, 241–43; Culley to Scudamore, 22 Dec. 1873, Item 102, ATF; Doc. 337. Compare the story that Dyer and Martin attribute to Edison about the battery used in this test (1910, 1:150).

4. A fine brush.

5. See n. 2.

6. That is, the resistance in the shunt was changed.

7. Leakage of current to ground.

8. Britain's extensive canal system connected Liverpool and London via several routes. One source identifies the canal referred to here as the Bridgewater Canal, part of which ran south from near Liverpool. Dyer and Martin 1910, 1:150.

9. Telegraph St. in London, the location of the central office of the telegraph service.

10. A wire running alongside a railway.

11. In the Birmingham gauge, commonly used in England, no. 8 wire was 0.165 inch in diameter.

12. No. 4 wire was 0.238 inch in diameter, Birmingham gauge.

13. Smear together.

14. In a 6 June 1873 note to Frank Scudamore, Culley wrote, "The invention which has been tried . . . is quite new. It is distinctly different to the processes which have been previously brought before my notice." Item 74, ATF.

15. David Lumsden, an experienced telegraph operator, was submarine (i.e., cable) superintendent and electrician for the British postal service during the early 1870s. Baker 1976, 110–11.

AUTOMATIC TELEGRAPH CIRCUITS Doc. 320

The following document is from a notebook Edison used while in England conducting tests of his automatic telegraph system in May and June of 1873.[1] Only a few of the entries in the notebook relate directly to these tests, however.[2] Instead, Edison recorded ideas for improvements of the automatic telegraph system, particularly the design of circuits that had occupied him in the six months prior to his trip to England.[3] Doc. 320 illustrates how he varied a basic design, seeking new ways of achieving a goal. In this case he was arranging combinations of resistance, capacitance, and induction, working from the "balancing" principle explicated in the patent application he executed in New York City the same day he sailed for England.[4]

1. PN-73-00-00.2, Lab (*TAEM* 6:872–906).
2. One drawing, however, appears to depict an automatic instrument:

Edison drew this sketch of an automatic telegraph instrument while in England conducting tests of his system for the British Post Office.

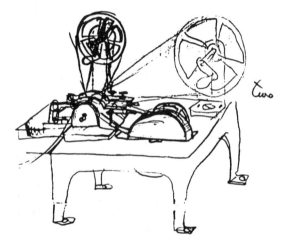

3. In this six-month period Edison applied for and was issued eight patents on such circuit designs: U.S. Pats. 135,531, 141,772, 141,773, 141,776, 147,311, 147,313, 147,314, and 150,848.
4. Doc. 317.

–320–

Notebook Entry: Automatic Telegraphy[1]

[London, May 1873]

[A][2]

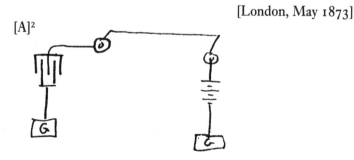

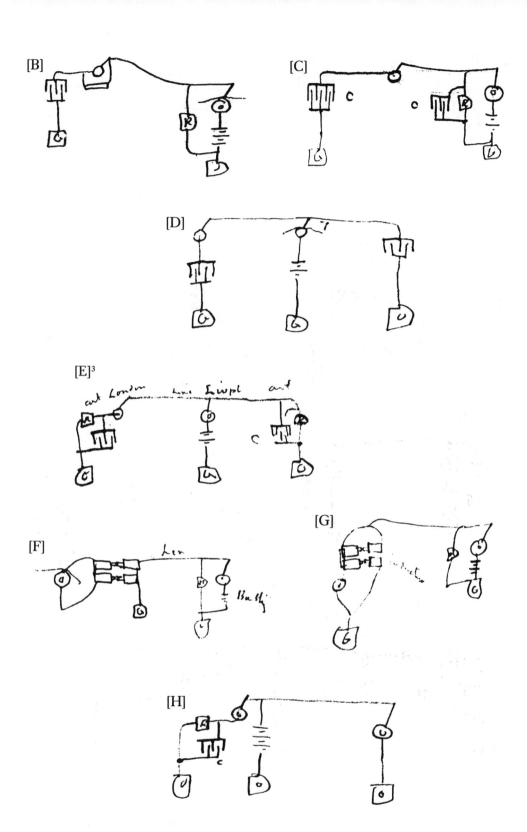

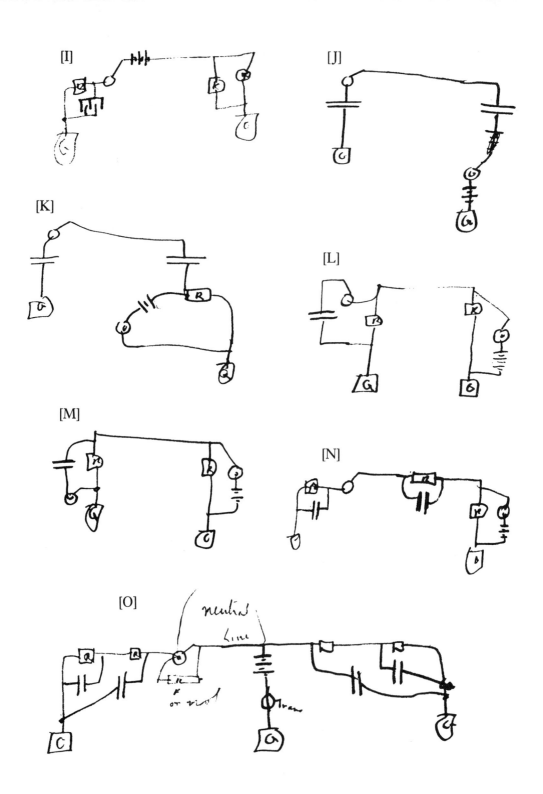

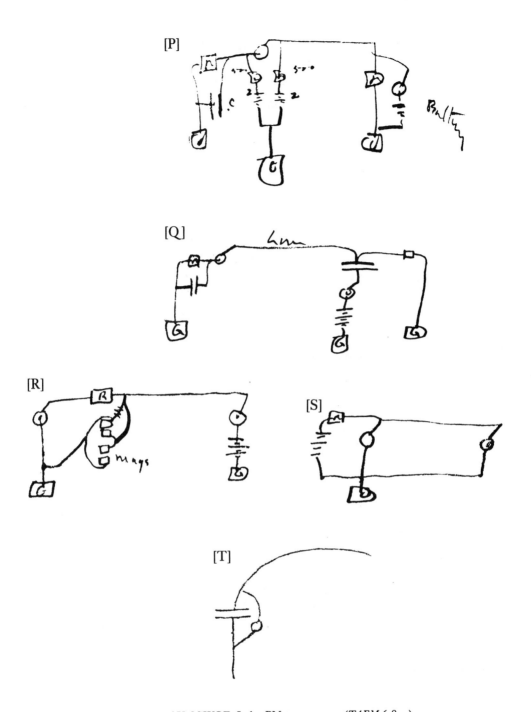

AX, NjWOE, Lab., PN-73-00-00.2 (*TAEM* 6:874).

1. See headnote above.

2. The circles in these diagrams represent automatic telegraph transmitters and receivers.

3. The label "art" means "artificial line."

GREENWICH CABLE TELEGRAPH POCKET NOTEBOOK   Docs. 321–336

Edison used this notebook while experimenting with his automatic telegraph on a cable at the Greenwich works of the Telegraph Construction and Maintenance Company.[1] These experiments took place in early June 1873, following his demonstration of his automatic telegraph system for the British Post Office. Edison obtained access to a coiled cable awaiting installation on the Brazilian line. He was unaware that the distortion typical on a long undersea cable would be severely augmented by the coiling. Edison later said that the first dot he transmitted came out on the chemical paper as a twenty-seven foot mark. The best transmission speed he could achieve in two weeks of experimenting was about two words a minute, while contemporary transatlantic cables transmitted from ten to seventeen words a minute.[2]

1. PN-73-00-00.1, Lab. (*TAEM* 6:820–71). An undated, two-page fragment labeled "Greenwich," which was probably created by Edison during or immediately following his trip to England, shows several circuit diagrams for working a cable. Cat. 297:75(4), 76(1), Lab. (*TAEM* 5:728, 731).
2. Dyer and Martin 1910, 1:151–52; Smith 1974, 317–18; Bright 1974, 105, 107, 556.

–321–

Notebook Entry:
Automatic and Cable
Telegraphy[1]

[London, c. June 10, 1873]

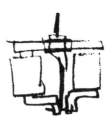 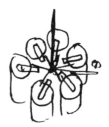

Resistance pretty dry paper   1 Cup throgh Mirror Gal-[vanometer] Thomson[2] 500 000 ohms   with 5 cells 100 000 showing proportionate to cells applied

Thomsons Gal will give quick deflectin throgh 4 millions ohms

Mine[3] gives dot in 10 seconds throgh 1 millins ~~mi~~ ohms ~~+~~
~~cup~~ & & mirror & it own R of 500 000   with 5 cups give mark
right along

hence to make paper very sensitive ~~p~~ in high resistance circuit add at rec[eivin]g inst enough cups to just make a scarcely
perceptable mark & work upon the batty at other end thus

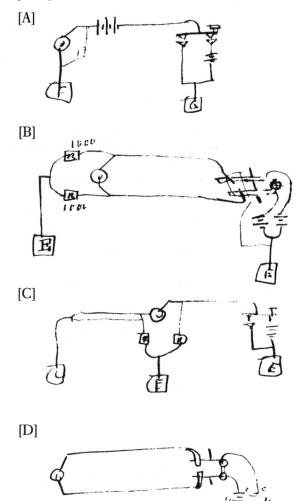

705 miles cable

AX, NjWOE, Lab., PN-73-00-00.1 (*TAEM* 6:824).

1. See headnote above.
2. Thomson's mirror galvanometer was used for receiving signals on
cable telegraphs.
3. Edison's automatic receiver.

Notebook Entry:
Automatic and Cable
Telegraphy[1]

6000 ohms

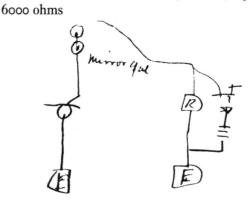

No tailing at 20 words per minute but could see that at higher there would be Especially if the Circuit or Resistance for discharging was increased

Lester[2] says 1200 ohm Mirrors[3] used for receiving insts low R

The internal resistance of 300 those Gutta percha batteries[4] is 37000 ohms or 123½ ohms per cell   oh god

Length sections wire[5]

Sect.	Knots		
1 & 7—	154.644		1 & 2
2 & 3—	187.071		3 & 4
5—	130.662		5 & 6
6 & 7—	140.596		7 & 8
	611		

Lester says he & Wy Smith[6] watched Earth Currents[7] all one night at Valentia[8]   Changed from P[ositive] to N[egative] Sometimes 2½ minute ~~he~~ others 15 minutes   he measured potential   one time twas 50 cells =

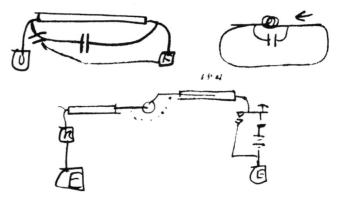

Joints never test as well as regular core   impossible make perfect joint   Wetherell[9] says guess thats reason wont allow more battery—chemical action =

AX, NjWOE, Lab., PN-73-00-00.1 (*TAEM* 6:828).

1. See headnote, p. 604.

2. Probably James Laister, who was a London member of the Society of Telegraph Engineers. "Supplemental List of Members to April 9, 1873," *J. Soc. Teleg. Eng.* 1 (1873): 4.

3. Mirror galvanometers.

4. This is probably a reference to the Menotti battery, whose flat, circular anode was soldered to a gutta-percha-covered wire. It was used primarily for testing and for long cable lines. Prescott 1877, 60–61; and Ternant 1881, 246–47.

5. The numbers here are the same as those indicated on the drawing in the notebook entry that follows (Doc. 323).

6. Willoughby Smith served as electrician of the Gutta Percha Co., which later became the Telegraph Construction and Maintenance Co., and made a number of important contributions to cable telegraph technology. He later served as president of the Society of Telegraph Engineers (Appleyard 1939, 289). For Smith's contribution to cable telegraphy see Smith 1974 and Bright 1974.

7. Earth currents are caused by differences in the electrical potential of the earth along the length of a cable.

8. Valentia, Ireland, was the eastern terminus of the first successful Atlantic cable.

9. T. E. Weatherall was an electrical engineer with the Telegraph Construction and Maintenance Co. at Greenwich. "Supplemental List of Members to April 9, 1873," *J. Soc. Teleg. Eng.* 1 (1873): 5.

–323–

Notebook Entry: Automatic and Cable Telegraphy[1]

[London, c. June 10, 1873]

Greenwich Expts[2] ~~Metallic Circuit~~[a]

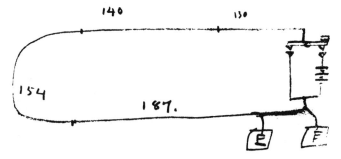

insert instrument and use perforated paper   see what speed obtainable with each section to earth direct & then with all sections on   probable result = at high speeds Cable on other side increase speed

AX, NjWOE, Lab., PN-73-00-00.1 (*TAEM* 6:834). [a]Drawing and following text written sideways on page.

1. See headnote, p. 604.
2. See Doc. 322 for a set of calculations related to this drawing.

–324–

Notebook Entry:
Automatic and Cable
Telegraphy[1]

[London, c. June 10, 1873]

Charge cable fully & kill or balance straight line by boxes Magnets[2] thus—& work quick with very small spaces[3]

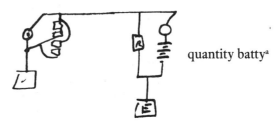

quantity batty[a]

Duplexing the static charge[a]

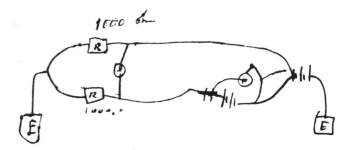

AX, NjWOE, Lab., PN-73-00-00.1 (*TAEM* 6:835). [a]Drawing sideways on page.

1. See headnote, p. 604.
2. In seeking to "balance" the line, Edison appears to be using the "reactionary discharge" magnets at the receiving end to counter static charge on the line and prevent tailing (see Doc. 317, n. 3). He proposed a similar arrangement in his U.S. Patent 147,311, whereby several magnets or induction coils were combined with a switch in order to bring any number of them into a shunt circuit. This arrangement is very similar to that found in Edison's U.S. Patent 160,405 (see patent drawing, p. 563). In another notebook from his trip to England (see Doc. 320 headnote) Edison indicates such a box in several drawings, one of which is labeled "box mag" (*TAEM* 6:880). An undated fragment describes the construction of a magnet box (Cat. 297:11[5], Lab. [*TAEM* 5:472]). Edison also apparently used such a magnet box in some of his duplex experiments (see headnote, pp. 562–63; Docs. 306 and 307; and Cat. 297:65 [*TAEM* 5:667]).
3. In two undated fragments, probably written before his trip to England, Edison indicated the use of very small holes in the perforations

to keep the line statically charged in order to overcome tailing problems. Cat. 297:12(3–4), Lab. (*TAEM* 5:479–80).

Notebook Entry:
Automatic and Cable
Telegraphy[1]

Also try this[a]

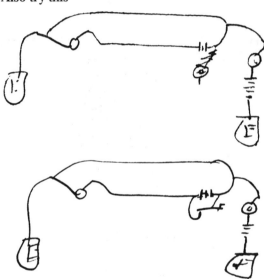

Turn quick   Keep it charged

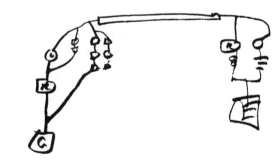

Greenwich

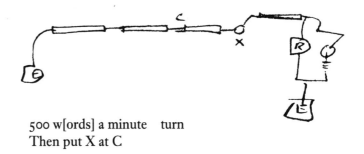

500 w[ords] a minute   turn
Then put X at C

AX, NjWOE, Lab., PN-73-00-00.1 (*TAEM* 6:838–41). ªFollowed by
"over" to indicate page turn.
    1. See headnote, p. 604.

[London, c. June 10, 1873]

–326–

Notebook Entry:
Automatic and Cable
Telegraphy[1]

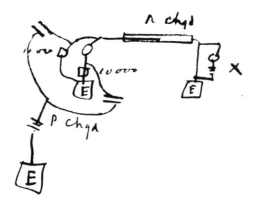

inst in bridge
Vibrator[2]
Good I think
Vary P[ositive] chg battery till it will charge cable equal to
X Trans[mitter] battery

AX, NjWOE, Lab., PN-73-00-00.1 (*TAEM* 6:842).
    1. See headnote, p. 604.
    2. See Doc. 327 for a similar vibrator drawing.

Notebook Entry:
Automatic and Cable
Telegraphy[1]

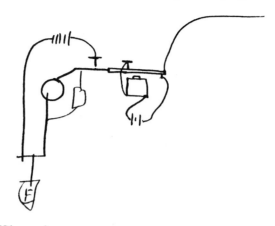

Vibrator[2]

AX, NjWOE, Lab., PN-73-00-00.1 (*TAEM* 6:854).
   1. See headnote, p. 604.
   2. Cf. Doc. 326.

–328–

[London, c. June 10, 1873]

Notebook Entry:
Automatic Telegraphy[1]

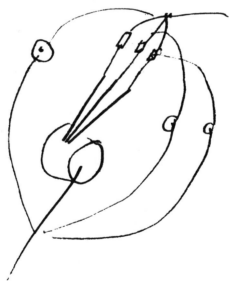

Big thing

AX, NjWOE, Lab., PN-73-00-00.1 (*TAEM* 6:855).
   1. See headnote, p. 604.

[London, c. June 10, 1873]

for a half current for a dash last half weak

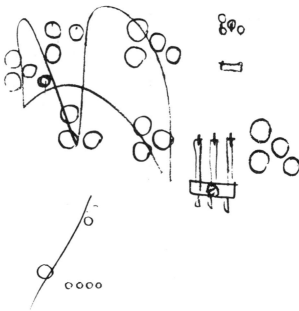

Use only one hole large for dash & make other holes small & sep pen insulated

AX, NjWOE, Lab., PN-73-00-00.1 (*TAEM* 6:859).
    1. See headnote, p. 604.

[London, c. June 10, 1873]

~~sen~~
have 3 condensers with wire coil put in at Wash[ingto]n & work condr System at NY with Ch[arle]s[to]n if good send 3 condrs ½ way bet W & Chn[2]

AX, NjWOE, Lab., PN-73-00-00.1 (*TAEM* 6:859).
    1. See headnote, p. 604.
    2. This refers to the line of the Southern and Atlantic Telegraph Co.

[London, c. June 10, 1873]

ascertain if on an artificial of say 500 or even 100 miles, there will be any difference between a battery with zinc & coke or copper or other higher metal & coke = Also on same cable see if there is any difference between or see if and record is made on at least 500 miles of artfcl cable by keeping a quantity battery of 10 cups on & make signals by putting on ten & taking off—the extra ten arranged to give extra quantity only

also ascertain if better signals can be received over an art with 50 cups, Bridge[2] put & kept on line at recg station and working[a] with 10 cups at sending station. in trying the differences take batteries out bridge but keep it on & at same resistance

Thus

Charge & discharge a large Condenser or several large Condsrs. Through a very delicate high R Engine—Revolving Armature = so as to get a perpetual revolution in the Engine =[b]

ascertain if some magnetic arrangment might not be made so as to be included within the circuit[c] to wor so that it would exactly neutralize the static charge in So many knots[3] of Cable   if these devices Could be put in the Cable & their Capacity would remain as Constant as the Capacity of the Cable = it would be valuable =

Try two insulated disks of rubber on which is a strip of Zinc & of Copper Connected together = This stands still   now another disk 100th of an inch from it revolves slowly & also[d] with immense rapidity   This disk has one Strip Copper. See if influence would generate E. & Connect to Sensitive Galvanometer =

AX, NjWOE, Lab., PN-73-00-00.1 (*TAEM* 6:863, 862, 860).
[a]Remainder of paragraph written over unrelated, canceled drawing.
[b]"perpetual ... Engine" partially written over unrelated drawing.
[c]"within the circuit" repeated at end of one page and beginning of next.
[d]Interlined above.

1. See headnote, p. 604.
2. Wheatstone bridge.
3. Nautical miles.

1 knot   8 by 8

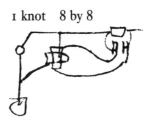

Double Condsr   high R secdy   chge C[ondenser] opposite

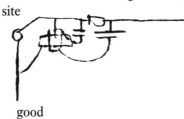

good

AX, NjWOE, Lab., PN-73-00-00.1 (*TAEM* 6:865).
1. See headnote, p. 604.

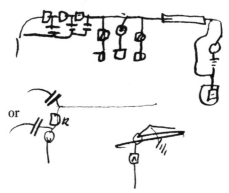

or

Try iron wires in Compensating Magnets   also in that secondary Inductn coil compensator

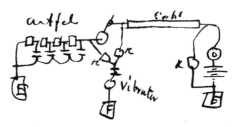

something in it

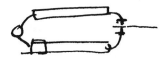

or to ground

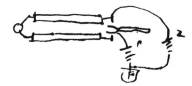

AX, NjWOE, Lab., PN-73-00-00.1 (*TAEM* 6:867).

  1. See headnote, p. 604.

–334–

Notebook Entry:
Automatic and Cable
Telegraphy[1]

[London, c. June 10, 1873]

    See if through 500,000 ohms anything can be put into the Iodinde[2] that will reduce its resistance = and obtain the mark with one cup of battery= In ltesting the sensativeness of iodized paper test with one or two cups battery through 500 000.

    Through this resistance Try and see which batty 2[a] cups Daniell[3] or grove[4] will give the mark quickest =

    It may be that Signals may be given with 20 cups grove & spaces by 20 cups Daniel

    Secondary batteries Have 6. foot square lead sheet or coke[5] secondary batters made to be used at receiving end These send back charge & act like a Condenser but have the advantage that they send[b] a long back charge & low Resistance it may be that they cannot be put in the main if so place in shunt

AX, NjWOE, Lab., PN-73-00-00.1 (*TAEM* 6:868). [a]Circled. [b]To this point, paragraph written over unrelated drawing.

  1. See headnote, p. 604.

  2. Edison used potassium iodide (KI) as a recording solution.

  3. The Daniell cell, a modification of John Daniell's constant-voltage cell, was regarded as especially well suited for closed circuit telegraph systems. Consisting of copper and zinc electrodes in dilute sulfuric acid and separated by a porous diaphragm, the 1.1-volt Daniell battery was so reliable that it was used as a standard through the 1870s. King 1962a, 241–43; Prescott 1877, 48–52; Atkinson 1910, 868.

  4. William Grove's battery used electrodes of zinc and platinum in solutions of sulfuric acid and nitric acid. It was expensive and—because the acids gave off corrosive fumes—dangerous. Its 1.9-volt output, however, was almost double that of the Daniell. King 1962a, 243; Prescott 1877, 64–66; Atkinson 1910, 869–70.

5. Secondary batteries (storage batteries) could store electricity when charged by a primary battery or generator. One type used lead sheets in a solution of dilute sulfuric acid; another used zinc and carbon electrodes. Prescott 1877, 81–82.

–335–

Notebook Entry: Automatic and Cable Telegraphy[1]

[London, c. June 10, 1873]

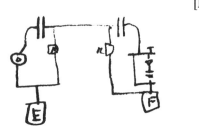 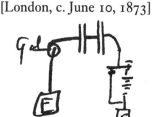

Varly Cable System[2]

Try this. obtain some Metal that will no decompose iodide & then attach it to the platina pen so as to reduce Resistance

AX, NjWOE, Lab., PN-73-00-00.1 (*TAEM* 6:869).

1. See headnote, p. 604.
2. British telegraph engineer Cromwell Varley was an expert in cable telegraphy. In 1862 he developed an artificial line system that electrically balanced cables by combining condensers and resistances. His system was widely used in cable telegraphy. *DNB*, s.v. "Varley, Cromwell Fleetwood"; Bright 1974, 639–40, 658.

–336–

Notebook Entry: Automatic and Cable Telegraphy[1]

[London, c. June 10, 1873]

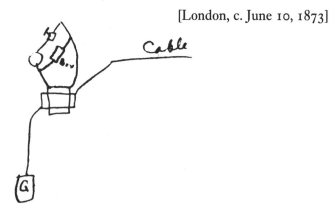

It[a] is probable that the return magnetic Charge Can be obtained freer & with less shunting of the Current by this arrangement that directly by magnets   The primary coil should have probably 5000 ohms Res & the secondary[b] about the same =[2]

It might be well to try & see if the delicacy of the paper[3] might not be increased by keeping it statically charged or by pasing a current (Local) in opposite directions through it or by the addition of several Cups in the secondary circuit & use no box on oth page—[4]

When I get home have 1000 cups glass ½ size Cable Gutta P[ercha] Cups made for testing at someones expense

AX, NjWOE, Lab., PN-73-00-00.1 (*TAEM* 6:870). [a]"Gutta P Wks Wharf Road City Road N" written at top of page. [b]Text from here to end of next paragraph written over unrelated, canceled drawing.

1. See headnote, p. 604.
2. The primary and secondary coils are the overlapping rectangles in the drawing.
3. The electrochemical sensitivity of the recording paper.
4. Here Edison refers to the drawing (which is on the facing page of his notebook). He proposes replacing his magnetic box (see Doc. 324, n. 2) with a battery of opposite polarity to the one on the main line. Instead of using the discharge from the magnets to counter the static charge, he uses an opposing current from the battery.

–337–

From Edward Johnson

New York June 11/73

My Dear "Tom,"

I am in reciept of yours this morning, and proceed to answer it Immediately as a mail goes out today.[1]

In all you say about Batteries I "Copper you & go one better"—[2] I'll explain= You remember I was using the Leclanche[3] between here & Washn. Well as soon as we got Charleston, I moved it down to the Cotton Exchange[4] & put it in operation—results not satisfactory— thought more Battery was wanted—added 100 Cups more—results no better— Whereupon it began dimly & slowly to creep into my Brain that Intensity[5] was not such a great desideratum after all—& that the Carbon Batty[6] was the best

Here I cried Halt— I had gone to the length of my rope— I had tested an Idea of some other man—& found it fruitless

Never having originated any of my Own worthy of mention—I thought my self at a Dead locke, & proceeded to put the Carbon Batty "Where it would do the most good"— That I found to be, of course with the Shunt as awe worked it to Pittsburg[7] Part plain & part magnetic—[8] And now comes the Dawn—

DeLaney[9] & I got to confabing one night as to the nature of this shunt—and we both concluded it was not a ground Shunt at all—but a Metalic Circuit— This Led D. to remark

"Well If we haven't found something new—we have found something that it will be worth while to remember." I did remember it and explained to JCR[10] next day, that my Opinion was that Instead of leaking off the surplus Current & giving us greater Intensity this Shunt arranged thus—

being simply a Metalic Short Circuit, Operated to "Increase the Consumption of Material & per Consequence, the amount of Current fed to the Line— still it had not gone through my Darkened mind yet what this Operation really Implied.

That night going over to Phila On a P[ullman]. P[arlor]. C[ar]. It struck me all amidship, & so demoralized me that sunday morning found me with Red Eyes & aᵃ Headache   I came back monday am, rushed in to J. C. & told him I was "Excited" & was Either going to Do something at last or prepare for oblivion = I explained how the Shunt Operated to Increase the "Quantity"[11] and reduce the Intensity— How in this New Departure, I found an Explanation of my failure to receive as rapidly from you at Pgh. as you did from me. (I had a Batty—Bought from a chemist—composed of very large stone Cups—Containing Double the quantity of ordinary Carbon)[12]   How—also I accounted for no end of funny, & previously inexplicable actions = (My times limited I must "Cut it" =)

In short—so satisfied him that quantity was neccessary & in it we would solve the problem of 500 a minute from Charleston = He Immedy authorized me to buy a large quantity Batty[13] & try it = It now stands in Batty room, awaiting trial— two Days have gone by but could not get wire— Your letter Comes Confirming my hopes (But also stealing my thunder)   Nevertheless—I think the Old partnership, inᵃ which Reiff bound you & I—will stand so slight a crucial test— Heres my hand Old Boy on the Extrordinary work you have accomplished among em Britishers— You certainly have had much to contend with, & I tell my Friend J. C. no other man living would have brot. success from out of such a labor-ynthine Complexity—(Big?) & that I am glad I did not go to

London— My Love to the Queen—& my regards to all the rest of the Boys Including Jack[14] & the Blonde Col.[15]

Sorry my time is so limited— I'll go you another by next train

E H J

ALS, NjWOE, DF (*TAEM* 12:1047). [a]Interlined above.

1. Edison's letter has not been found. In view of the speed of Atlantic shipping and Edison's arrival home by 25 June, this letter, as well as Docs. 338 and 339, could not have reached him until after his return.

2. In the game of faro, to "copper" a card means to bet against it. *OED*, s.v. "Copper."

3. The Leclanche battery.

4. The central cotton market in New York City. See Doc. 131, n. 1.

5. Voltage.

6. The Bunsen cell, in which the anode was a carbon rod. It could produce a considerably greater current than the Leclanche cell.

7. Nothing is known of these experiments.

8. "Plain" refers to an ordinary rheostat, a resistance with little self-induction; "magnetic" refers to a coil with an iron core and hence high self-induction.

9. Patrick Delany, a telegraph operator since 1861, was at this time probably assistant general manager of the Southern and Atlantic Telegraph Co.; he was later general manager of the Automatic Telegraph Co. He became a full-time inventor in 1880, developing inventions in telegraphy and other electrical fields. Taltavall 1893, 242–43; *NCAB*, 13:590; *New York Times*, 21 Oct. 1924, 23.

10. Josiah Reiff.

11. Current.

12. Doubling the carbon would approximately double the current.

13. A "quantity battery" would have cells arranged in parallel and would probably contain cells capable of producing a heavy current (e.g., Bunsen cells).

14. Jack Wright.

15. George Gouraud.

–338–

From Joseph Murray

Newark Thursday June 12th 1873

Sir

I Received your welcome letter after I wrote to you[1]   I stated all in that letter but since I Recived yours I wish to post you on some points   the Gen..[2] does mis you now   he wants advise in regards to Manhattan[3] and you are his best adviser on the points in question   he would like to see you very much—but stick where you are if it is profitable to you   dont feel unesay on account of shop I shall keep it square till you return   my pay roll is less than ($[a]300) per week and we are doing very well   I made (6) Regesters[4] for Bentley[5] (2550[b])

Joseph Murray was associated with Edison early in his Newark career and became his manufacturing partner in 1872.

set toy Inst[6] also (25) set of Keys Sounders & Relays   also altering Universal[7] to (15) teeth star wheel[8]   making parts lighter making them like model of Burgmans[9] Except Break which is Better and diffirent to model   also altering stock printers to 15 teeth star wheel   they work well in circuit now   I am getting along nicely dont fret about shop or your Family   all is well here I only want time to fix up all outside bills and partys which I am doing sloly but sure.. Edison sell to Fleming or Scalp Hays[10] for that (£[a]100,000) if possible   dont wait one hour let it rip   Automatic is dead here only when you are presant to give it life   business is very dull money is worse than Ever   I got ($[a]3000) note from Reiff in exchange for my note of same amount   I expect Miller[11] will get it discounted for me soon so I can use it to pay of old bills   believe me I have had hard timee of it since you left   I lost 11 lbs in one month but I shall die in the Harness if I ever do die   I often get out of small holes—   Unger draws hard on me and no mercy[12]—yet I float and mean to come out victorious   Reiff feels good over your success so does Harrington   they Blow lowd for Automatic   I met Johnson[13] he told me you wrote to him   he feels good but is not our special Friend   I know more than he thinks of   I shall tell when we meet   I keep him in check some   he condemed all our Tables and Inst—   I am doing some little for automatic and they owe us   I am nearly square with Gold & stock but Gen Give Thou & Bergman[14] Contract to alter Universal printers in opposition to us   they told Gen all new Improvements belong then and they would Get patent out for said if he did not Give them the printers to alter so you can [-]Emagime what trouble they were to me in this affair   I hope you will Kill them when you come home if I dont before—[15] I have been nearly wild with all preasure brought to bear on me   I now see clear out of the woods   Miller has done nothing since you left about Toy Inst   WU has not done anything since you left—

I will meet all notes without selling Machinery but I shall be left poore as it takes all profits to meet said demands—

I have said all about business here   I want you to stick where you are   look after stock printers   as Lord Hays is president of London stock Exchange sound him. Callahan arrived here few day ago began by telling how bad Edison printers were they were failure   E W Andrews told me there was now in circuit (100) Inst in London— find out from Hays all prospects for future orders as we are now improving them on fast speed and can make superior Inst than what is now in use   Go see all you can dont work all day & Knight and not see anything   Go outside see every place where you can Get any Ideas from—

I want you to fix Applebaugh[16] if possible   he is yours Bitter Enemy yet he is very Friendly to me   Scott[17] is working all possible way against us   Every day I have some part of Inst to take home out of order   they are all left on (Gens) desk with conplaint and he is out of sorts about Manhattan   he told me if Ever one of the altered printers give out he would not let us have anymore work   I shall fix it all right with him on those points   Wagner[18] can tell you how these fellows try to beat us by finding fault with Inst without any cause—

our stock printer altered beats all others so does Universal   they work fine   our tools is now useless for future use on Both Stock & Universal but I shall make then pay for it[19]   we shall yet flurish in all our endevours   Everything goes nicely   Batchelor does very well he is very cautious about his work— I have all confidence in his honest efforts to do well   feel perfectly easy in regards to us in Factory   my only fiars is with Private line department and scott on stock Printers   I wanted to alter Transmitter for stock Circuit but Scott said Edison Automatic Transmitter[20] would not do so Kenny[21] altered Hand Transmitter[22] to Go with our printers and the cant find fault with them now   I cut new type wheels for each Inst— Renovate all the old printers make as good as new   furnish all part for ($ª25) each   we can do well at this price—now Farewell   answer if you can   your

J. T Murray

if you want small sum say 50 I shall send it to you

wife & Baby[23] is well   fat as she can be   they dont suffer   she has had ($ª200) and did not pay one bill out of it   I dont Blame her or find fault with her   I payed intrest in Saving Bank till Jan $ª175 also 52.50ᶜ to Ross Cross &c[24] also Unger notes several other

ALS, NjWOE, DF (*TAEM* 12:1112). ªInterlined above. ᵇ"50" interlined above. ᶜDecimal point supplied.

1. Neither letter has been found. See Doc. 337, n. 1.

2. Marshall Lefferts.

3. The Manhattan Quotation Telegraph Co., organized in 1872 to compete with the Gold and Stock Telegraph Co.

4. Morse receiving instruments.

5. Henry Bentley was a pioneer in intraurban telegraphy. At this point he was building a system in Philadelphia. Reid 1879, 597–601.

6. This may have been a learner's telegraph instrument.

7. Private-line printer.

8. See Doc. 195 textnote.

9. Sigmund Bergmann; see Doc. 313.

10. John Fleming, a London merchant, was one of the men to whom Edison was trying to sell patent rights for his automatic telegraph system (Agreement between TAE, George Harrington, Josiah Reiff, and Smith, Fleming and Co., 2 Sept. 1873, DF [*TAEM* 10:1261]). Lord William Montagu Hay was chairman of the board of the Exchange Telegraph Co. in London.

11. Norman Miller.

12. William Unger held a $10,000 note on the shop (see Doc. 264).

13. Edward Johnson.

14. Henry Thau (b. 1848?) had been one of Edison's machinists. He had left Edison by early 1872 to establish an instrument-making shop in New York with a partner, David Hermann; by the time of this letter he was in business on his own. Sigmund Bergmann had made for the Gold and Stock Co. a model of the universal private-line printer which incorporated significant alterations (see Doc. 313). Wilson 1872, 539, 1193; Thau's testimony, pp. 61–63, *Wiley v. Field*; App. 1.D379.

15. Regardless of the truth of the situation or Murray's feelings about the two men, Edison maintained relations with them for many years, calling upon them for important work.

16. W. K. Applebaugh was an assistant superintendent of Gold and Stock, in charge of private lines and the bank department. Later he took an active interest in the Manhattan Quotation Telegraph Co. and the Domestic Telegraph Co. "Private Telegraphy," *Telegr.* 9 (1873): 19; Reid 1879, 622, 633.

17. George Scott.

18. G. Wagner was a Western Union employee. Reid 1879, 626.

19. Evidently the design of the printers was sufficiently altered to require new patterns and jigs. See Doc. 280, n. 16.

20. A modification of the transmitting portion of the universal private-line printer. See Doc. 211.

21. Patrick Kenny had been superintendent of Gold and Stock's manufacturing shop. G&S Minutes 1867–70, 158–59.

22. That is, a breakwheel.

23. Marion, born 18 February 1873.

24. Ross, Cross & Dickinson.

–339–

From Edward Johnson

New York June 13/73[1]

My Dear Edison—[a]

I have been cogitating and experimenting considerable of late and have come to the conclusion that I know a thing or two about the why's & the wherefores of results obtained =

1st = I started out to discover why the shunt at transmitting station <u>Improved</u> the writing, I having a notion that it was something more than a mere leak[2] with a counter acting attachment = I found that Instead of it being in the form of a leak—thus—[3]

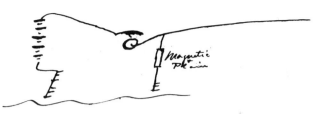

It was—ain reality no leak at all—but simply a short circuit around the Batty (as we had accidentally always made it)—thus—[4]

—Now the effect of this is—not to leak off the Battery—but simply to violently excite it into action, & force its quantity production = [5] You might say that—Putting the shunt to a seperate ground, & thus destroying the metalic connection—was in reality short circuiting the Batty & would produce the same result—but it dont—I have tried it—as in first diagram a Dozen times, and it produces no marked good results—while the Instant you put the ~~ground~~ shunt ground & the Batty ground together their is an Illumination = and wonderous results are obtained = I deduce from this that the violently exciting of the the Batty by the metalic circuit produces great quantity—without giving it an Out let—except by the wire—while the ~~g~~seperate ground—or ground proper—excites it and absorbs it at same time —so much for that. Col Reiff thinks it Impt. that you should know of this discrimination, that I have discovered existing between the Ground proper, & the metalic circuit, as you may not have thought of it yourself, and might possibly get on the Ground shunt—in some of your experiments & "Wonder why you didn't obtain the proper results"

Now for another feature = You reccollect, when the Insts. were made you proposed to put zinc to line—saying "If You would just as soon have it that way I'd prefer it, as it will save Insulation". I told you it didn't make a particle of difference to me—

Now I'm of a different Opinion—("CN"[6] calling for "Reds"[7] Hold up a min " ".--." ".--." ".--.""[8] Ha! = Inst took 20 minutes to get 16 Reds & copy them with only one man, and that on the Franklin wire from here to Washn—& the S[outhern] & A[tlantic] from there to Chastn with all the Paralel wires working & Induction fearful heavy, & not a single msg—repeated—~~a~~or correction made—or a word asked for—nothing but OK's & Ya's & a final OK Ya'—in just 25 mins from time first got him—Morse working thro. repeater at Washn—& auto working straight—Button cutting out Repeater at each Ya—& cutting in at end—Relay shunted for Indicator at Washn⁹—Hows this?— = It does make a difference By receiving ~~on~~from the Ground & using shunt at transmitting station I figure the Operation to be this =

The current from Batty is sent out of Batty at each closing—& performs the the following work—

1st Sends the working current to receiving station by the ~~ground~~ Earth—

2d Engenders in the Wire upon its return (or completion of ckt)—the static—which, passing into the earth at transmitting station shunt (or leak) flies thro the earth again on its round trip (the ckt being continuous insofar as the static current is concerned) [-] passes on to the paper just behind the working current—thus causing a tailing =

3rd Charges the magnet in the shunt violently—causing it to send a powerful countercharge on the Wire

Here are 3 operations performed by the main Batty— Now I find that to use a long shunt at transmitting station, You must "cut" it at receiving end—to get rid of the pursuing static = & this cutting—so weakens—what current has come through that the writing is very fleeting—& soon disappears— While If you use a short shunt and "cut" it like the Dickens—at transmitting station—no ~~s~~Rheo—is needed at all at Receiving end & all the current arriving is utilized—¹⁰ the effect being, to make the writing very Dark & heavy—which is the effect obtained— My theory is—that the static flows principally out of the wire at Transmtg end, as at that end—the wire being the heaviest chgd—the most static is engendered—& the operation of the magnet in the shunt is to meet & annihilate this "Major" static before he enters the earth—from the wire, & thus operates to prevent any static from passing on to the paper except what comes from the wire back into the earth at receiving end, & that of course comes underneath the paper = & is not noticeable = This looks as if I should then at all times be able to receive on Direct current by a regulating shunt at Trans end—which I now do = not using a Rheo. for either Washn or Chastn = My opinion being that by so doing you obtain these results

1st Getting Receiving station nearer centre of Circuit =

2d Annihilation of static at point where it practically originates

3rd The Consequent ability to utilize ~~all~~the whole volume of current that arrives

Am I sound—or can you pick me to pieces— I Certainly have good ground for my deductions, since I have practical demonstration of the effect that my theory Implies

Let me hear from you—I am so Infernal busy that I haven't time for a word beyond such as are actually neccessary to express my Ideas—otherwise I should bedaub your high flown sentences about 'em Britishers   My regards to Jack[11]   Yours
E H Johnson

Let me put it briefly this way— = Use your Box in a place where it does not operate to carry part of the current to the Ground—but still operates to Destroy the static current = Is'nt that quite a discovery for an amatuer—to be shure we have done it for a long time—but none of us discovered the fact &made a proper use of it—until now—I'll be a good pupil of ure's[12] yet.—this is it plainer[b]

ALS, NjWOE, DF (*TAEM* 12:1051). [a]Followed by a series of marks that may represent Morse code characters. [b]Separated from key to figure by rule.

1. See Doc. 337, n. 1.

2. A "leak" is current that is inadvertently flowing to the ground.

3. The label on the drawing is "Magnetic & Plain."

4. Compare the following sketch and description with Doc. 317. The figure label is "M & P."

5. That is, elicit a greater current.

6. Charleston, S.C.

7. The meaning of "Reds" is unknown. Perhaps short for "red hots"—that is, urgent messages.

8. ".--." was the American Morse signal for the numeral "1."

9. Johnson was receiving messages from Charleston on a chemical receiver. "All the Paralel wires working & Induction fearful heavy" would normally produce an indistinct signal, but the apparatus delivered clear messages. At the end of each message, Johnson had to acknowledge receipt by signaling Charleston with a Morse key ("OK's & Ya's"). An operator in Washington, D.C., alerted by an indicator, switched a repeater into the circuit for Johnson's acknowledgment and switched it out again for the next automatic message.

10. That is, if a high-resistance shunt is used on the transmitter, the current must also be shunted ("cut") around the receiver. However, if the transmitter shunt has little resistance and a significant portion of the current is diverted there ("'cut' it like the Dickens"), no shunt is needed at the receiver.

11. Jack Wright.

12. Andrew Ure, a British chemist and scientific writer, published a number of popular works on science and technology. *DNB*, s.v. "Ure, Andrew."

–340–

From Edward Johnson

New York, June 24 1873[a]

My Dear Edison—

I enclose a new pen—⅓rd the diameter of those in use—[1] I find by it I prolong the Dot 3 times, or thereabouts—and am thus enabled to cut as much as I please & yet get a full "Morse Key" Dot $=$[2] which is simply Increasing the size of Holes—~~with~~ minus a proportionate Increase of speed of running the paper—I also find absolute perfection of Dashes[3]— so that I can send slow & "copy by sound" without being able to detect the slightest jar in the Dashes—or any of that short jerky writing $=$

I am having this Patented in your name,[4] as also the metalic short circuit arrangement—[5] They merely go in as supplements to your patents—Try this Pen, & then give me your opinion   Hastily Yours

E H Johnson[b]

Keep[c] this to yourself as It is the only ~~D~~Size that will give maximum speed   EHJ

ALS, NjWOE, DF (*TAEM* 12:1060). Letterhead of the Automatic Telegraph Co. [a]"New York," and "187" preprinted. [b]"Hastily Yours E H Johnson" written in right margin. [c]Postscript written in upper left corner.

1. That is, a transmitting stylus for the automatic telegraph. Probably Johnson knew by this time that Edison would soon be in Newark.

2. Meaning not clear.

3. Two pens side by side were used to transmit; dashes were formed by punching overlapping holes alternately rather than by making a long hole. See Doc. 142.

4. Johnson could not legally patent an invention of his in Edison's name and did not; nor was this invention patented in Johnson's name.

5. See Docs. 337 and 339. Edison had already executed an application for this circuit (U.S. Pat. 147,313) on 23 April, before leaving for England.

Appendix 1

Edison's Autobiographical Notes

The following materials are portions of the autobiographical reminiscences that Thomas Edison wrote in 1908 and 1909. The editors plan to append these without annotation to the volumes to which the recollections pertain rather than present them as complete documents in the volumes for 1908 and 1909. In this way they will be available for reference and general study in the appropriate volumes. Only those parts referring to events of the period covered by Volume 1 (1847–1873) are reproduced here. Many of them deal with Edison's childhood and itinerant telegrapher years, a period for which there are few original documents. While independent evidence confirms and amplifies various points in these notes and flatly refutes others, many of the stories lack corroboration.[1]

Edison's anecdotes and stories derive from his memory of events that occurred as many as fifty-five years before, from recollections often told and perhaps increasingly enlarged upon, and from the needs of personal and corporate image-making. Edison prepared these autobiographical reminiscences for the official biography, *Edison: His Life and Inventions*, which Frank Lewis Dyer and Thomas Commerford Martin conceived in 1907 and published in 1910. Edison generated and reviewed these notes and encouraged publication of the subsequent two-volume biography at a time when, with considerable advertising and promotion, the Edison enterprises were being organized into a single concern, Thomas A. Edison, Inc. The Edison companies had routinely employed Edison's name, signature, and photographic image for marketing and for other public relations purposes, and the inventor himself had a long history of cultivating the press. Dyer was general counsel for the Edison laboratory, and Martin, a former Menlo Park employee, was prominent in electrical engineering circles as editor of *Electrical World* and *Electrical Engineer*, two important American technical journals.

William Meadowcroft, Edison's personal secretary, conducted interviews and collected reminiscences, including those of the inventor himself.

In the fall of 1908, Edison wrote a narrative account of his childhood and early career in a notebook; Meadowcroft then prepared a typed transcript. Later, Edison wrote accounts and outlines of early events and made notes in a second notebook. During June 1909 he wrote additional narrative and notes in response to written questions from Martin and in preparation for interviews, creating six autobiographical documents. Each of these has been designated by a letter, and every paragraph of the original documents has been numbered in sequence within the document. When all of the volumes of the book edition are published, a reader wishing to recover the original order will be able to do so. A short introduction precedes, and textnotes follow, the included portions of each document.

1. For example, Edison's note about a playmate, George Lockwood, who drowned at Milan (D92), can be both expanded in detail and carefully documented. However, his statement (A7) that the noted British engineer Robert Stephenson saw him printing his *Weekly Herald* cannot be true, since Stephenson died in 1859. In contrast to such definite cases, the editors have found nothing related to the reported incident (A24) involving objections to using "J. C." as an abbreviation. Dyer and Martin 1910, 1:18; *Huron Reflector*, 31 Aug. 1852; *DNB*, s.v. "Stephenson, Robert."

A. BOOK NO. 1

The following is a transcription of a typescript that William Meadowcroft prepared from reminiscences that were originally written by Edison in a notebook labeled "Book No. 1" and dated 11 September 1908. The contents of the notebook pertain to the period covered by this volume with the exception of the initial five paragraphs, which describe Edison's 1878 western trip, and a short paragraph (14), which relates Edison's alleged discovery of a Morse diary.

[6] After my father moved to Port Huron, he engaged in lumbering, and also had a 10 acre field of very rich land which was used for truck gardening. After the field was ploughed, I, in conjunction with a German boy of about my age, did the planting. About eight acres were planted in sweet corn, the balance in radishes, onions, parsnips, and beets, etc.; I was very ambitious about this garden and worked very hard. My father had an old horse and wagon and with this we carried the vegetables to the town which was 1½ miles distant and sold them from door to

door. One year I remember turning in to my mother 600 dollars from the farm— After a while I tired of this work as hoeing corn in a hot sun is unattractive and I did not wonder that it built up cities. Soon the Grand Trunk R.R. was extended from Toronto to Port Huron at the foot of the Lake Huron and thence to Detroit, at about the same time the war of the Rebellion broke out. By a great amount of persistence I got permission from my mother to go on the local train as a newsboy. The local train from Port Huron to Detroit, a distance of 63 miles left at 7 A.M. and arrived again at Port Huron at 9 P.M. After being on the train for several months, I started two stores in Port Huron, one for periodicals and the other for vegetables, butter and berries in the season, these were attended by two boys, who shared in the profits. The periodical store I soon closed, as the boy in charge could not be trusted. The vegetable store I kept up for nearly a year. After the railroad had been opened a short time they put on an express which left Detroit in the morning and returned in the evening. I received permission to put a newsboy on this train— connected with this train was a car, one part for baggage and the other part for U.S. mail, but for a long time it was not used. Every morning I had two large baskets of vegetables from the Detroit Market loaded in the mail car and sent to Port Huron where the German boy would take them to the store. They were much better than those grown locally and sold readily. I never was asked to pay freight and to this day cannot explain why, except that I was so small and industrious and the nerve to appropriate a U.S. mail car to do a free freight biz so monumental that it probably caused passivity. However, I kept this up for a long time and in addition bought butter from the farmers along the line and an immense amount of blackberries in the season; I bought wholesale and at a low price and permitted the wives of the engineers and trainmen to have the benefit of the rebate. After a while there was a daily immigrant train put on— this train generally had from seven to ten coaches filled always with Norwegians, all bound for Iowa and Minnesota. On these trains I employed a boy who sold bread, tobacco and stick candy. As the war progressed the daily newspaper sales became very profitable and I gave up the vegetable store, etc. Finally when the battle of Pittsburg Landing occurred (now called Shiloh) I commenced to neglect my regular business. On the day of this battle when I arrived at Detroit, the bulletin boards were surrounded with dense crowds and it was announced that there were 60 thousand killed and wounded and the result was uncertain. I knew that if the same excitement was attained at the various small towns along the road and especially at Port Huron that the sale of papers would be great. I then conceived the idea of telegraphing the news ahead, went to the operator in the depot and by giving him Harper's Weekly and some other papers for three months, he agreed to telegraph to all the stations the matter on the bulletin board. I hurriedly copied it and he sent it, requesting the agents who displayed it on the blackboard, used for stating the arrival and departure of trains, I decided that instead of the usual 100 papers that I could sell 1000, but not having sufficient money to purchase that number, I determined in my desperation to see the Editor himself and get credit. The great paper at that time was the Detroit Free Press. I walked into the office marked Editorial and told a young man that I wanted to see the Editor on important business— important to me anyway. I was taken into an office where there were two men and I stated what I had done about telegraphy and that I wanted 1000 papers, but only had money for 300 and I wanted credit. One of the men refused it, but the other told the first spokesman to let me have them. This man I afterwards learned was Wilbur F. Storey, who subsequently founded the Chicago Times and became celebrated in the newspaper world. By the aid of another boy we lugged the papers to the train and started folding them. The first station called Utica, was a small one where I generally sold two papers. I saw a crowd ahead on the platform, thought it some excursion, but the moment I landed there was a rush for me; then I realized that the telegraph was a great invention. I sold 35 papers; the next station, Mt. Clemens, now a watering place, but then a place of about 1000. I usually sold 6 to 8 papers. I decided that if I found a corresponding crowd there that the only thing to do to correct my lack of judgment in not getting more papers was to raise the price from 5 cents to 10. The crowd was there and I raised the price; at the various towns there were corresponding crowds. It had been my practice at Port Huron to jump from the train at a point about ¼ mile from the station where the train generally slackened speed. I had drawn several loads of sand at this point to jump on and had become very expert. The little German boy with the horse met me at this point; when the wagon approached the outskirts of the town I was met by a large crowd. I then yelled 25 cents apiece, gentlemen, I haven't got enough to go round. I sold all out and made what to me then was an immense sum of money. I started the next day to learn telegraphy and also printing. I started a newspaper which I printed on the train, printing it from a galley proof press, procuring the type from a junk dealer who had a lot nearly worn out. (You have a copy of the newspaper)

[7] When Stephenson who built the Victoria bridge at Montreal came over the Grand Trunk, he saw me printing an edition on the train; he bought the whole and it was afterwards mentioned in the London Times as the 1st newspaper in the world to be printed on a train in motion.

[8] I commenced to neglect my regular business until it got very low, although I managed to turn one dollar each day to my mother. The station agent at Mt. Clemens permitted me to sit in the Telegraph office and listen to the instrument; one day his little boy was playing on the track when a freight train came along—and I luckily came out just in time to pull him off the track; his mother saw the operation and fainted. This put me in the good graces of Mr. Mackenzie, the agent, and he took considerable pains to teach me, as I kept at it about 18 hours a day I soon became quite proficient. I then put up a telegraph line from the station to the village a distance of 1 mile and opened an office in a drug store, but the business was small and the operator at Port Huron knowing my proficientcy and who wanted to go into the U.S.M. Telegraph, where the pay was high, succeeded in convincing his brother-in-law (Mr. Walker) that I could fill the position all right. Mr. Walker had a jewelry store and had charge of the W.U. Tel. office. As I was to be found at the office both day and night, sleeping there, I became quite valuable to Mr. Walker. After working all day I worked at the office nights as well for the reason that press report came over one of the wires until 3 A.M., and I would cut in and copy it as well as I could, to become more rapidly proficient; the goal of the rural telegraph operator was to be able to take press. Mr. Walker tried to get my father to apprentice me at 20 dollars per month, but they could not agree. I then applied for a job on the Grand Trunk R.R. as a railway operator and was given a place nights at Stratford Junction, Canada. This night job just suited me as I could have the whole day to myself. I had the faculty of sleeping in a chair any time for a few minutes at a time. I taught the night yardman my call, so I would get ½ hour sleep now and then between trains and in case the station was called, the watchman would awaken me. One night I got an order to hold a freight train and I replied that I would. I rushed out to find the signalman, but before I could find him and get the signal set, the train run past. I ran to the Telegraph Office and reported I couldn't hold her, she had run past. The reply was "Hell". The despatcher on the strength of my message that I would hold the train, had permitted another to leave the last station in the opposite direction   There was a lower station near the Junction where the day operator slept. I started for it on foot. The night was dark and I fell in a culvert and was knocked senseless.

[9] However, the track was straight, the trains saw each other, and there was no collision. The next morning Mr. Carter, the station agent and myself were ordered to come at once to the main office in Toronto. We appeared before the General Superintendent, W. J. Spicer who started in hauling Mr. Carter over the coals for permitting such a young boy to hold such a respon-

sible position. Then he took me in hand and stated that I could be sent to Kingston States Prison, etc. Just at this point, three English swells came into the office. There was a great shaking of hands and joy all around; feeling that this was a good time to be neglected I silently made for the door; down the stairs to the lower freight station, got into the caboose going on the next freight, the conductor who I knew, and kept secluded until I landed a boy free of fear in the U.S. of America.

[11] After selling papers in Port Huron, which was not reached until about 9.30 at night, I seldom reached home before 11 or 11.30 at night, about ½ way from the station and the town and within 25 feet of the road in a dense woods was a soldiers graveyard, where 300 soldiers were buried, due to a cholera epidemic which took place at Fort Gratiot near by many years previously. At first we used to shut our eyes and run the horse past this graveyard, and if the horse stepped on a twig, my heart would be given violent movement and it is a wonder that I haven't some valvular disease of that organ, but soon this running of the horse became monotonous and after a while all fear of graveyards absolutely disappeared from my system. I was in the condition of Sam Houston, the pioneer and founder of Texas, who it was said knew no fear. Houston lived some distance from the town and generall went home late at night, having to pass through a dark cyprus swamp over a corduroy road; one night to test his alleged fearlessness, a man stationed himself behind a tree and enveloped himself in a sheet; he suddenly confronted Houston, who stopped and said—If you are a man you can't hurt me; if you are a ghost you dont want to hurt me; if you are the devil, come home with me, I married your sister.

[12] When a boy, the Prince of Wales, now King Edward, came to Canada. Great preparations were made at Sarnia, a Canadian town opposite Port Huron. About every boy, including muself, went over to see the affair. The town was draped in flags most profusely and carpets were laid on the crosswalks for the Prince to walk on; there were arches etc. A stand was built, raised above the general level, where the Prince was to be received by the Mayor. Seeing all these preparations, my idea of a prince was very high, but when he did arrive, I mistook the Duke of Cambridge for the Prince. He, (the Duke) being a large fine looking man, I soon saw that I was mistaken, that the Prince was a young stripling and did not meet expectations. Several of us started to express our belief that a prince wasn't much after all and that we were thoroughly disappointed. For this, one boy was whipped. Soon the Cunnock boys attacked the Yankee boys and we were all badly licked. I, myself, got a black eye. This has always prejudiced me against this kind of ceremonial and folly.

[13] While a newsboy on the train, one day a messenger from

the office of E. B. Ward & Company came to the train and wanted me to come quickly to the office. The firm of E. B. Ward at that time was the largest owner of steamboats on the Great Lakes. It seems that one of the captains on their largest boat had suddenly died, and they wanted me to take a message to another captain, who lived about 14 miles from Ridgeway station on the railroad. This captain had retired and had taken up some timber land and cleared part of it. Mr. Ward said he would give me 15 dollars if I would deliver the message that night. I told him I was afraid to do it alone as it was a wild country and it would be dark, but if he would pay me $25 I would get the help of another boy and do it. To this he agreed. I arrived at Ridgeway at 8:30 at night, it was raining and dark as ink. After trying 2 or 3 boys, I got one and we started out with lanterns for the place. We had the location explained, and there was only one road, if a road it could be called, all through a dense forest. We hadn't gone far before we became apprehensive of bears, the more we thought of the subject, the more stumps looked like bears. The country at that time was wild and it was a usual occurence to see deer, bear and coon skins nailed on the side of the house to dry, and I had read about bears, but couldn't remember if they were night or day prowlers. My companion proposed that we climb a tree and wait till morning. I wouldn't agree to this, as I knew that we were no safer up a tree than on the ground, as bears could climb trees and besides that message had to be delivered that night, so the captain could catch the morning train. We kept on, after a while one lantern went out, not being filled with enough oil. When we started then, within two miles of the place the other lantern went out. Then we leaned up against a tree and cried. I thought if I ever got out of this scrape alive, that I would know more about habits of animals and everything else and be prepared for all kinds of mischances when I undertook an enterprise. However, the dense darkness dilated the pupils of our eyes to make them very sensitive and we could see at times the outlines of the road and finally just as a faint gleam of daylight arrived, we entered the captain's yard and delivered the message. In my whole life, I never spent such a night of horror as this, but I got a good lesson.

[15] While working in the W.U. Telegraph office in Boston, a position obtained for me by my friend Adams, who worked in the Franklin Telegraph office, which company was competing with the W.U. Mr. Adams was laid off, and as Adams' finances had reached absolute zero centigrade, I undertook to let him sleep in my hall bedroom. I generally had hall bedrooms, because they were cheap and I needed money to buy apparatus. I also had the pleasure of his genial company at the boarding house, about a mile distant, but at the sacrifice of some apparatus. One morning as we were hastening to breakfast, we came into Tremont Row,

and saw in front of two small gents furnishing goods stores, a large crowd. We stopped to ascertain the cause of the excitement. One store put up a paper sign in the display window, which said "300 pairs of stockings received this day 5 cents a pair—no connection with the store next door". Presently the other store put up a sign, stating they had received 300 pairs, price 3 cents per pair and stated that they had no connection with the store next door. Nobody went in. The crowd kept increasing. Finally, when the prices had reached 3 pair for 1 cent, Adams says to me "I can't stand this any longer, give me a cent". I gave him a nickel and he elbowed his way in and throwing the money on the counter, the store being filled with lady clerks, said, give me 3 pairs. The crowd was breathless and the lady took down a box and drew out 3 pairs of baby socks. "Oh", said Adams, "I want mens' size". "Well sir, we do not permit one to pick size for that amount of money" and the crowd roared and this broke up the sales.

[16] Adams was one of a class of operators, who was never satisfied to work at any place for any great length of time. He had the Wanderlust. After enjoying my rather meagre hospitality on the floor of the hall bedroom, which in Boston was a paradise for an entomologist, and the boarding house run on the Banting system of flesh reduction, he came to me one day and said, good-bye Edison, I have got 60 cents and I am going to San Francisco, and he did. How, I never knew. I afterwards learned that he got a job and within a week they had a strike; then he got a big torch and sold patent medicine in the streets at night to support strikers; then went to Peru as partner of a man who had a grizzly bear, which they proposed entering against a bull in the Bull Ring in that city. The grizzly was killed in five minutes and this scheme died. Then Adams started a market report bureau in Buenos Aires. This didn't pay, then he started a restaurant in Pernambuco, Brazil. Here he did very well, but something went wrong, as it always does to a Nomad and he went to the Transvaal and ran a panorama, called "Paradise Lost" in the Kaffir Kraaks. This didn't pay and he became the Editor of a newspaper, then went to England to raise money for a railroad in Cape Colony. Next I hear of him in N.Y., having just arrived from Bogata, U.S. of Columbia with a power of attorney and $2,000 from a native of that Republic, who had applied for a patent for tightening a belt to prevent it from slipping on a pulley, a device which he thought new and a great invention, but which was in use ever since machinery was invented. I gave Adams a position as salesman for electrical apparatus. This he soon got tired of and I lost sight of him.

[17] One day a lady came to the W.U. office in Boston and stated that she had a school and would like to get one of their

operators to explain the telegraph to her scholars, illustrating the explanation with actually working apparatus. She was told to call around in the evening when I would be at work. She arranged with me to give the explanation with apparatus two weeks from that date— in a few days before I carried the the apparatus and with Adams' assistance, set it up in the school, which was in a double private house near the public library. The apparatus was set up when school was out. I was then very busy building private telegraph lines and equipping them with instruments which I had invented, and forgot all about the appointment and was only reminded of it by Adams who had been trying to find me and had at last located me on top of Jordan, Marsh & Company's store, putting up a wire. He said, we must be there in 15 minutes and I must hurry. I had on working clothes and I didn't realize that my face needed washing. However, I thought they were only children and wouldn't notice it. On arriving at the place, we were met by the lady of the house and I told her I had forgotten about the appointment and hadn't time to change my clothes. She said that didn't make the slightest difference. Adams' clothes were not of the best, because of his long estrangement from money. On opening the main parlor door, I never was so paralyzed in my life; I was speechless, there were over 40 young ladies from 17 to 22 years, from the best families. I managed to say that I would work the apparatus and Mr. Adams would make the explanantions. Adams was so embarrassed that he fell over an ottoman, the girls tittered and this increased his embarrassment, until he couldn't say a word. The situation was so desperate that for a reason I never could explain, I started in myself and talked and explained better than I ever did before or since. I can talk to two or three person, but when there are more they radiate some unknown form of influence which paralyzes my vocal cords. However, I got out of this scrape and many times afterwards, when I chanced with other operators, to meet some of the young ladies on their way home from school, they would smile and nod, to the great mystification of the operators, who were ignorant of this episode.

[18] The reason I came to go to Boston was this. I had left Louisville the second time and went home to see my parents. After stopping at home for some time, I got restless and thought I would like to work in the East and knowing that a former operator named Adams, who had worked with me in the Cincinnati Office was in Boston, I wrote him that I wanted a job there. He wrote back that if I came on immediately he could get me in the W.U. office. I had helped out the Grand Trunk R.R. Telegraph people by a new device when they lost one of the two submarine cables they had across the river, making the remaining cable act just as well for their purpose as if they had two. I thought I was

entitled to a pass, which they conceded, and I started for Boston. After leaving Toronto, a terrific blizzard came up and the train got snowed under in a cut. After staying 24 hours, the trainmen made snowshoes of fence rail splints and started out to find food which they did about ½ mile away. They found a roadside inn and by means of snow shoes, all the passengers were taken to the inn. The train reached Montreal 4 days late. A number of the passengers and myself went to the Military headquarters to testify in favor of a soldier who was on a furlough and was two days late, which was a serious matter with military people, I learned. We willingly did this, for this soldier was a great story-teller and made the time pass quickly. I met here a telegraph operator, named Stanton, who took me to his boarding house, the most cheerless that I have ever been in. Nobody got enought to eat, the bed clothes were too short and too thin, and it was 28 degrees below zero, and the wash water frozen solid; The board was cheap, being only $1.50 per week. Stanton said the usual live stock accompaniments of operators' boarding houses were absent; he thought the intense cold had caused them to hibernate. Stanton, when I was working in Cincinnati left his position and went out on the Union Pacific to work at Julesburg, which was a cattle town at that time and very tough. I remember seeing him off on the train, never expecting to see him again. Six months afterwards, while working press wire in Cincinnati about 2 A.M., there was flung into the middle of the operating room, a large tin box. It made a report like a pistol and we all jumpted up startled; in walked Stanton. Gentlemen, he said, I have just returned from a pleasure trip to the land beyond the Mississippi. All my wealth is contained in my metallic traveling case and you are welcome to it. The case contained one paper collar. He sat down and I noticed that he had a woolen conforter around his neck with his coat buttoned closely. The night was intensely warm; he then opened his coat and revealed the fact that he had nothing but the bare skin. Gentlemen, said he, you see before you an operator who has reached the limit of impecuniosity.

[19] On reaching Boston, I found my friend Adams and went over to the W.U. Office to see the manager. On entering the office, where there were perhaps 30 or 40 men, I noticed that they were dressed very neatly, which was very unusual in the offices in the West. I myself, at the time, had on a blue shirt and clothes not of the best, and somewhat wrinkled from the long journey. I was introduced to the Manager, Mr. Milliken, who I thought gave a start of surprise. Adams had told him I was an A1 man and had worked two or three years on associated press wire. After asking me to confirm Adamds' statements, he asked when I could come to work. I said "Now". Very well, come around at 5.30 and I will leave instructions. I came at 5.30 P.M. and was introduced to the

night manager, and told that I was to work N.Y. No. 1 wire. I was furnished with one of those cheap pens that the W.U. used to economize on and waited for business. I noticed considerable talk and laughter on the part of the other operators and surmised that it was at my expense, as my clothes certainly did not fit extreme Eastern civilization.

[20] After waiting ½ hour my wire was switched over on a test table in the middle of the room and I was told to take a 1500-word special for the Boston Herald. The N.Y. operator started at a rapid gate, the sending being perfect and the wire was good. After a few minutes, his gait got very rapid and I noticed he was getting up to his limit, turning my head I found nearly every operator in the office watching me from behind. I knew then they had put up a job to roast me, as they say. They had gotten Hutchinson, one of the lightning senders in the N.Y. office to do it. Now I had experimented a long time to acquire rapid penmanship. I indulged in no flourishes and each letter was separate and not connected, as I found that there was a gain in time in not connecting the letters and also that rapidity was increased by writing very small; I had been used to forcing the writing in taking press through 8 sheets of manifold paper with an agate stylus, so writing with a pen was easy. I then started writing very small and I knew I could do 4 or 5 words per minute more than he could send; after taking about ¾ of the special, Hutchinson got nervous and commenced to abbreviate. As I had to write out in full I knew that soon I would have to break, so to save the day before this took place, I opened the key and said "You seem to be tired, suppose you send a little while with your other foot" This saved me. Hutchinson quit and the special was finished by the regular man.

[21] After this, I was all right with the other operators.

[22] In the Boston office there were operators studying to enter Harvard; they were on nights. They paraded their knowledge rather freely and it was my delight to go up to the second-hand book stores on Cornhill and study up questions which I would spring on them when I got an occasion. We got our midnight lunch from an old Irishman, called the "Cake Man", who appeared regularly at 12 midnight. The office was on the ground floor and had, previous to occupation by the Telegraph Co., been a restaurant. It was literally loaded with cockroaches, who lived between the wall and the board running around the room at the floor. These were such a bother on my table that I pasted two strips of tinfoil on the wall at my desk connecting one piece to the positive pole of the big battery supplying current to the wires and the negative pole to the other strip. The cockroaches moving up on the wall would pass over the strips, and the moment the got their legs across both strips, there was a flash of light and the

cockroach went into gas. This automatic electrocution device attracted so much attention and got a ½ column description in an evening paper, that the Manager made me stop it.

[23] After being in Boston for several months, working N.Y. wire No. 1, I was requested to work the press wire. This wire was called the milk route, as there was so many towns on it taking press simultaneously. N.Y. office has reported great delays on the wire, due to operators constantly interrupting or breaking as it was called, to have words repeated which they failed to get, and N.Y. claimed Boston was one of the worst offenders. It was a rather hard position for me, for if I took the report without breaking, it would prove the Boston operator incompetent. The results made the operator have some hard feelings against me. He was put back on the wire and did much better after that. It seems that the office boy was down on this man and one night he asked me if I could tell him how to fix a key so it would not break, even if the circuit breaker was open, and also, so it could not be easily detected. I told him to jab a pen full of ink on the platinum points that there was enough sugar to make it thick enough to follow up; when the operator tried to break, the current going through the ink so he couldn't break. The next night about 1 A.M. the operator on the press wire, while I was standing near a house printer studying it, pulled out a glass insulator, then used upside down as a substitute for an ink bottle, and threw it with great violence at me, just missing my head. It would have certainly killed me if it had not missed. The cause of the trouble was that this operator was doing the best he could not to break, but being compelled to, opened his key and found he couldn't; the press matter coming right along and he couldn't stop it. The office boy had put the ink in a few minutes before, when the operator turned his head during a lull. He instinctively blamed me as the cause of the trouble. This operator and I afterwards became good friends—he took his meals at the same emaciator that I did and his main object in life seemed to be acquiring the art of throwing up wash pitchers and catching them without breaking them. About ⅓ of his salary was used up in paying for pitchers.

[24] While taking a long monotonous proceeding of a synad of some kind, which was being held at Cleveland, the word "Jesus Christ" occurred with extreme frequency, so I got to abbreviating it by using J. C. Pretty soon the assistant agent of the associated press came down stairs into the office and wanted to know what d——d operator was abbreviating Jesus Christ with J. C.; that he wouldn't stand for it. He cursed around for a while and I stopped it. It seemed very incongruous, as B.C. was regularly used.

[25] At that time the firm of Chas. Williams, Jr. was making telegraph instruments and doing experimental work for Moses G. Farmer and Gainwell, the inventors of the Fire Alarm tele-

graph system. I, as far as my money would go, was also having work done there of an experimental character. I became acquainted with a man who was having made some electrical apparatus for a sleight-of-hand performance and we did a lot of experimenting. One day I found in my copy of the Scientific American a complete description of a method of making nitroglycerin. The sleight-of hand man and myself one night after Mr. Williams had gone home started in to make some. The product came out rather brown and the article warned makers that brown nitro-glycerin was impure and dark in color, that it was due to impurities and in this condition was dangerous and might explode spontaneously. To see if the quality was O.K. we exploded a few drops and the results were so strong that we both got frightened, so we put the nitro in a pop bottle, wound waste around it, tied a cord to the end of the bottle and let it down a sewer inlet on the street at the shop. Williams, who just managed to make a living off of poor inventors, etc. afterwards became a manufacturer of telephones and died a millionaire.

[26] I established a Laboratory over the Gold room and put up a line on which I opened a stock quotation circuit with 25 subscribers, the ticker being of my own invention. I also engaged in putting up private lines upon which I used a dial instrument. This instrument was very simple and practical and any one could work it after a few minutes explanation. I had these instruments made at Mr. Hamblets who had a little shop where he was engaged in experimenting with electric clocks. Mr. Hamblet was the father and introducer in after years of the W.U. telegraph system of time distribution. My laboratory was the headquarters for the men and also of tools and supplies for these private lines. They were put up cheaply, as I used the roofs of houses as the W.U. did. It never occurred to me to ask permission from the owners; all we did was to go to the store, etc. and say we were telegraph men and wanted to go up to the wires on the roof and permission was always given.

[27] In this laboratory I had a large induction coil, which I had borrowed from Mr. Williams to make some experiments with. With this coil I had ten large cells employing nitric acid. One day I got hold of both electrodes and it clinched my hand on them so I couldn't let go. The battery was on a shelf. The only way I could get free was to back off and pull the coil, so the battery wires would pull the cells off the shelf and thus break the circuit. I shut my eyes and pulled, but the nitric acid splashed all over my face and ran down my back. I rushed to the sink which was only half big enough and got in and wiggled around for several minutes to permit the water to dilute the acid and stop the pain. My face and back were streaked with yellow, the skin thoroughly oxidized. I did not go in the street by daylight for two weeks, as the ap-

pearance of my face was dreadful. The skin, however, peeled off and new skin replaced it without any damage.

[28] On the N.Y. No. 1 wire that I worked there was an operator named Jerry Borst; he was a first class receiver and rapid sender. We made up a scheme to hold this wire so he changed one letter of the alphabet and I soon got used to it and finally we changed three letters. If any operator tried to receive from Borst he couldn't do it, so Borst and I always worked together. Borst did less talking than any operator I ever knew. Never having seen him, I went while in N.Y. to call on him. I did all the talking, he would listen and stroke his beard and said nothing. In the evening I went over to an all-night lunch house in printing house square, which was in a basement kept by a man whom they called "Snotty Oliver". Night Editors, including Horace Greeley and Raymond of the Times, took their midnight lunch there. When I went with Borst and another operator, they pointed out two or three men who were then celebrated in the newspaper world. The night was intensely hot and close. After getting our lunch and upon reaching the sidewalk, Borst opened his mouth and said: "That's a h——l of a place, a plate of cakes and a cup of coffee, and a Russian bath for 10 cents." This was about 50% of all his conversation for two days.

[29] Towards the end of my stay in Boston, I obtained a loan of money, amounting to 800 dollars to build a peculiar kind of a duplex telegraph for sending two messages over a single wire simultaneously. The apparatus was built and I left the W.U. employ and went to Rochester, N.Y. to test the apparatus on the lines of the Atlantic & Pacific Telegraph between that city and N.Y., but the assistant at the other end could not be made to understand anything, notwithstanding I had written out a very minute description of just what to do. After trying for a week, I gave it up and returned to N.Y. with but a few cents in my pocket. I slept in the Battery room of the Gold Indicator Company, owned by S. S. Laws, former college professor, who had invented an instrument for indicating the price of gold in broker's offices. I applied for a position as operator at the W.U., but had to wait a few days during which time I thoroughly studied the indicators and the complicated general sender in the office which was controlled by a keyboard manipulated by the operator on the floor of the Gold Exchange. On the third day of my arrival and while sitting in the office, the complicated general instrument for sending on all the lines and which made a very great noise, suddenly came to a stop with a crash. Within two minutes, over 300 boys (a boy from every broker in the street) rushed upstairs and crowded the long aisle and office that hardly had room for 100, all yelling that such and such broker's indicator was out of order and to fix it it at once. It was a pandemonium, and the man in charge became so

excited that he lost control of all the knowledge he ever had. I went to the indicator and having studied it thoroughly knew where the trouble ought to be and found it. One of the innumerable contact springs had broken off and had fallen down between the two gear wheels and stopped the instrument, but it was not very noticeable. As I came out to tell the man in charge what the matter was, Mr. Laws appeared on the scene, the most excited person I have seen. He demanded of the man in charge the cause of the trouble, but the man was speechless. I ventured to say that I knew what the trouble was and he said—fix it, fix it, be quick. I removed the spring and set the contact wheels at zero and the line, battery and inspecting men, all scattered through the financial district to set the instruments. In about two hours things were working again. Mr. Laws came in and asked my name and what I was doing. I told him and he requested that I should come to his private office the next day. His office was filled with stacks of books all relating to metaphysics and kindred matters. He asked me a great many questions about the instruments and his system and I showed him how he could simplify things generally. He then requested that I should call the next day. On arrival he stated at once that he had decided to put me in charge of the whole plant and that my salary would be $300 per month. This was such a violent jump from anything that I had ever received before, that it rather fazed me for a while. I thought it was too much to be lasting, but I determined to try and live up to that salary if 20 hours a day of hard work would do it. I kept this position, made many improvements, devised several stock tickers, until the Gold and Stock Telegraph Company consolidated the Gold Indicator Company.

[30] I then went into the firm of Pope and Ashley. Mr. Ashley was the editor of a telegraph journal published for telegraph operators. While with them I devised a printer to print gold quotations, instead of indicating them. The lines were started, the whole was sold out to the Stock Telegraph Company. My experimenting was all done in the small shop of one Dr. Bradley, located near the station of the P.R.R. in Jersey City. Every night I left for Elizabeth on the 1 a.m. train and then walked one-half a mile to Mr. Pope's home and up at 6 a.m. for breakfast to catch the 7 a.m. train. This continued all winter and many were the occasions when I was nearly frozen in the Elizabeth walk.

[31] Dr. Bradley was the first man to my knowledge who introduced the galvometer in telegraph and other work in this country. He was one of the old style experimenters, who would work for years on an instrument which he thought worth thousands, but which did not have any commercial value. His business sense was nil. He was rather old when I was at his shop and very irascible. On one occasion a wire connected to one of the

binding posts of a new galvanometer wouldn't come out, so he yanked it, pulled the galvanometer on the floor and then jumped on it. The hobby he had at the time I was there was the ageing of raw whiskey by passing strong electric currents through it. He had arranged 20 jars with platinum electrodes held in place by hard rubber. When all was ready he filled the cells with whisky, connected his battery, locked the door of the small room in which they were placed and gave positive orders that no one should enter. He then disappeared for three days, on the 2nd day we noticed a terrible smell in the shop, as if from some dead animal. The next day the Doctor arrived and noticing the smell asked what was dead. We all thought something had gotten into his whiskey room and died. He opened it, and was nearly overcome. The hard rubber he used was, of course, full of sulphur and this being attacked by the nascent hydrogen, had produced sulphuretted hydrogen gas in torrents, displacing all of the air in the room. Sulphuretted hydrogen gas is the gas given off from rotten eggs. As the experiment was a failure, the Doctor got very irascible, and no one dared say a word when he was around. In the sale of the Company for printing gold quotations, I was entitled to $1200, but Mr. Ashley thought that amount excessive, although his part was to be 20,000 dollars profit, for which he did absolutely nothing. Thinking that perhaps I might not get anything at all, I told General Lefferts, who was at the head of the Company making the purchase, all about my relations. He said, say nothing, do nothing, leave it to me. When the deal went through, the General handed me $1500 and said that was my share, he had saved it out when he made the payment. I was attacked savagely after that by Ashley in his paper. This was about the first time I realized that human nature had a slight tinge of selfishness connected therewith. General Lefferts, who was a very prominent man at the time, being colonel of the N.Y. Seventh Regiment, was president of the Gold & Stock Telegraph Company, which supplied tickers to Wall Street and connected with various other companies. He requested me to go to work on improving the ticker he furnishing the money for the work. I made a great many inventions, one was the special ticker used for many years outside of N.Y. in the large cities. This was made exceedingly simple as the outside cities did not have the experts we had in New York to handle anything complicated. The same ticker was used on the London Stock Exchange. After I had made a great number of inventions and obtained patents, the General seemed anxious that the matter should be closed up. One day after I had exhibited and worked a successful device, whereby if a ticker should get out of unison in a broker's office and commenced to print wild figures, it could be brought to unison from the central station and which saved the labor of many men and much trouble to the broker. He called me into his office and said, "Now, young man, I want to close up the matter of your inventions, how much do you think you should receive?" I had made up my mind that taking in consideration the time and the killing pace I was working that I should be entitled to $5,000, but could get along with $3,000, but when the psychological moment arrived, I hadn't the nerve to name such a large sum, so I said, "Well, General, Suppose you make me an offer." Then he said, "How would forty thousand dollars strike you" This caused me to come as near fainting as I ever got. I was afraid he would hear my heart beat. I managed to say that I thought it was fair. "All right, I will have a contract drawn, come around in three days and sign it, and I will give you the money." I arrived on time, but had been doing considerable thinking on the subject, the sum seemed to be very large for the amount of work, for at that time I determined the value by the time and trouble and not what the invention was worth to others. I thought there was something unreal about it. However, the contract was handed to me, I signed without reading it. The General called in the Secretary and told him to fix it up and pay the money. I was then handed a check for $40,000 on the bank of the State of New York, which was at the corner of William and Wall Streets. This was the first check I ever had. I went to the bank and noticed the window marked "Paying Teller", got in line with about a dozen men and a dozen messenger boys and slowly approached the window. When directly in front of the window passed in the check, he looked at it, turned it over and handed it back, making a few short remarks which I could not understand, being at that time as ever since, quite deaf. I passed outside to the large steps to let the cold sweat evaporate and made up my mind that this was another Wall Street game like those I had received over the press wire, that I had signed the contract whatever was in it, that the inventions were gone and I had been skinned out of the money. But when I thought of the General and knowing he had treated me well, I couldn't believe it, and I returned to the office and told the secretary what occurred. He went in and told the General and both had a good laugh. I was told to endorse the check and he would send a young man down with me to identify. We went to the bank, the young man had a short conversation with the Paying Teller, who seemed quite merry over it, I presented the check and the Teller asked me through the young man, how would I have it. I said in any way to please the bank Then he commenced to pull out bundles of notes until there certainly seemed to be one cubic foot. These were passed out and I had the greatest trouble in finding room in my overcoat and other pockets. They had put a job up on me, but knowing nothing of bank customs in those days, I did not even suspect it. I went to Newark and sat up all night with the money for fear it might be stolen. The next day I went back with it all and told the General about it, and he laughed very greatly, but said to one of his young men—Don't carry this joke on any further, go to the bank with Edison and have him open an account and explain the matter, which I did.

[32] I have too sanguine a temperment to keep money in solitary confinement, so I commenced to buy machinery, rented a shop and got some manufacturing work to do from the first shop; I moved into a large shop Nos. 10 and 12 Ward Street, Newark. I got large orders from the General to build tickers and had over 50 men, and as orders increased I put on a night shift. I was my own foreman on both shifts, one-half hour of sleep three or four times in the twenty-four hours was all I needed. Nearly all my men were on piece work and I allowed them to make good wages and never cut until their wages became absurdly high, as they got more expert. I kept no books. I had two hooks, all the bills and accounts I owed I jabbed on one hook and memorandum of all owed to myself I put on the other. When some of the bills fell due and I could not deliver tickers to get a supply of money, I gave a note/ These notes were payable at the Germania or German National Bank, I forget which. When the notes were due, a messenger came around with the note and a protest pinned to it for $1.25. Then I would go to New York, get an advance or pay it, if I had the money. This method of giving notes for my accounts and having all the notes protested I kept up over two years, yet my credit was fine; every store I traded with was always glad to furnish goods. Perhaps in amazed admiration of my system of doing business, which certainly was new. After a time I got a bookkeeper, but never could understand or believed in it, but the business got so large I had to do it. The first three months I had the bookkeeper go over the books to find out how much we made. He reported $3,000. I gave a supper to some of my men to celebrate this, only to be told two days afterwards by this alleged accountant that he had made a mistake and that we had lost $500 instead of making $3,000, and then a few days after coming to me again and said he was all mixed up and now found we had made $7,000. I discharged him and got another man, but I never counted anything thereafter as real profits, until I had paid all my debts and had the profits in the bank. Soon after starting the large shop, I rented shop room to the inventor of a new rifle. I think it was the Berdan rifle. In any event it was a rifle, which was subsequently adopted in the British Army. The inventor employed a tool-maker who was the finest and best tool-maker I had ever seen. I noticed that he worked pretty near the whole of the twenty hours. This kind of application I was looking for. He was getting $21.50 a week and was also paid for

overtime. I asked him if he could run the shop. I don't know, try me, he said. All right, I will give you $60.00 per week to run both shifts. He went at it. His executive ability was greater than any man I have yet seen. His memory was prodigious, conversation laconic and movements rapid. He doubled the production inside of three months, without materially increasing the pay-roll, but increasing the cutting speed of tools and by the use of various devices. When in need of rest he would lay down on a work bench, sleep twenty or thirty minutes and wake up fresh. As this was just what I could do I naturally conceived a great pride in having such a man in charge of my work, but almost everything has trouble connected with it. He disappeared one day and although I sent men everywhere that it was likely to be found, he was not discovered. After two weeks, he came into the factory in a terrible condition as to clothes and face. He sat down and turning to me said, "Edison, it's no use, this is the third time, I can't stand prosperity". Put my salary back and give me a job". I was very sorry to learn that it was whiskey that spoilt such a career. I gave him an inferior job and kept him for a long time.

[33] There worked at one time along the same bench, several men who in after years became very rich and prominent, One was S. Bergmann, who afterwards, when I invented the incandescent electric lighting system became my partner with E. H. Johnson in the large works, once at Avenue B and 17th Street, and who is now at the head of the great Bergmann Electric Works in Berlin, employing 10,000 men. Mr. Bergmann is many times a millionaire. The next man adjacent was John Kreuzi, who became an engineer of the Works of the General Electric Company at Schenectady, and now deceased. The next was Shuckhart, who left the bench and went back to Nuemberg to settle up his father's estate, remained and started a small electrical works, which grew into the great Shuckhart Works, the third largest in Germany, employing 7,000 men. Shuckhart died worth several millions. I gave them a good training as to working hours and hustling.

[34] I started an annex shop in Mechanic Street and also in the building occupied by the Richardson Saw Works; also one on R.R. Avenue. While running these shops I was engaged by the Automatic Telegraph Co. of N.Y., who had a line running between N.Y. and Washington to help them out of their trouble. It seems they had organized a company and built a line on the strength of some experiments by an English Inventor. The apparatus worked all right on a short wire in an office, but when put on the actual line, no results could be obtained. Connected with me was E. J. Johnson, who afterwards was associated with me in Electric Lighting and the introduction of the trolley with F. J. Sprague. After experimenting for several weeks, I devised new

apparatus and solved the problem of rapid transmission so we succeeded in transmitting and recording 1000 words per minute between Washington and N.Y., and 3500 words per minute between Phila. and N.Y. This system was put in commercial operation. These experiments, with running my four shops, made sleep a scarce article with me. Then the automatic Company wanted to spread out and have devised for them an automatic high speed telegraph, which would print the message in Roman letters instead of dots and dashes, and so they rented a large shop over the Gould factory in Newark, installed 25,000 dollars worth of machinery and gave me full charge. Here I devised and manufactured their instruments for commercial use and also started experiments on the Roman letter systems. I finished this and had a test between Phila. and N.Y., sending and receiving 3,000 words in one minute, and recording the same in large Roman letters. Mr. D. H. Craig, then the agent of the associated press became interested in the Company, of which Mr. J. C. Reiff was Vice President and Manager, and Geo. Harrington, former assistant Secretary U.S. Treasury, the President. Mr. Craig brought on from Milwaukee Mr. Sholes, who had a wooden model of a machine which was called a typewriter. Craig had some arrangement with Sholes and the model was put in my hands to perfect. This typewriter proved a difficult thing to get commercial, the alignment of the letters was awful, one letter would be $\frac{1}{16}$ of an inch above the others, and all the letters wanted to wander out of line. I worked on it till the machine gave fair results. Some were made and used in the office of the Automatic Company. Craig was very sanguine that some day all business letters would be written on a typewriter. He died before that took place, but it gradually made its way.

[35] The typewriter I got into commercial shape and is now known as the Remington typewriter. About this time I got an idea I could devise an apparatus by which four messages could be simultaneously sent over a single wire, without interfering with each other. I now had five shops and with experimenting on this new scheme I was pretty busy, at least I didn't have ennui.

TD (transcript), NjWOE, Meadowcroft. Because this transcription of Edison's manuscript is presented only as a reference text, typographical errors have not been reproduced or noted.

B. FIRST BATCH

The following is a transcription of a typescript that Edison revised. At the top of the first page is a handwritten note: "First Batch Notes dictated by Mr Edison to T. C. Martin June,

1909.— Pencil indicates Mr. Edison's revision". Only one anecdote relates to the period covered by this volume.

GEORGE LITTLE AND AUTOMATIC.

[8] In 1872 an English electrician named George Little came to this country with a system of automatic telegraphy. He got interested with him George Harrington ex-Assistant treasurer of the U.S.; Erastus Corning of Albany and General Palmer of Colorado, and others, and they formed a company to exploit the invention. When they came actually to try it one a wire, they found that while as a laboratory experiment it was successful, it would not work at all on an actual circuit. I was called in to get them out of their difficulty and I devised my automatic. It was in this automatic telegraph that the first typewriters were introduced and used. I made six of them and David H. Craig of the Associated Press who was interested with Sholes, the inventor,[a] was also interested in the automatic. This is the present Remington.[b]

TD (transcript), NjWOE, Meadowcroft. [a]", the inventor," interlined above in pencil. [b]"This . . . Remington." written in pencil.

C. SECOND BATCH

The following is a transcription of a typescript that includes Edison's revisions. At the top of the first page is a handwritten note: "Second Batch   Mr Edison's notes dictated Mr Martin   June 1909   Pencil indicates revision by Mr Edison". Only one anecdote relates to the period covered by this volume.

A CIPHER MESSAGE FOR THOMAS.

[1] When I was an operator in Cincinnati working the Louisville wire nights for a time, one night a man over on the Pittsburgh wire yelled out: "D. I. cipher" which meant that there was a cipher message from the War Department at Washington and that it was coming—and he yelled out "Louisville". I started immediately to call up that place. It was just at the change of shift in the office. I could not get Louisville and the cipher message began to come. It was taken by the operator on the other table direct from the War Department. It was for General Thomas, at Nashville. I called for about 20 minutes and notified them that I could not get Louisville. I kept at it for about 15 minutes longer and notified them that there was still no answer from Louisville. They then notified the War Department that they could not get Louisville. Then we tried to get it by all kinds of round about ways, but in no case could anybody get them at that office. Soon a message came from the War Department to send immediately

for the manager of the Cincinati office. He was brought to the office and several messages were exchanged, the contents of which, of course, I did not know, but the matter appeared to be very serious as they were afraid of General Hood of the Confederate Army who was then attempting to march on Nashville; and it was very important that this cipher of about 1200 words or so should be got through immediately to General Thomas. I kept on calling up to 12 or 1 o'clock but no Louisville. About 1 o'clock the operator at the Indianapolis office got hold of an operator on a wire which ran from Indianapolis to Louisville along the railroad, who happened to come into his office. He arranged with this operator to get ~~a relay of~~ª horses, and the message was sent through Indianapolis to this operator who had engaged horses to carry the dispatches to Louisville and find out the trouble, and get the dispatches through without delay to General Thomas. In those days the telegraph fraternity was rather demoralized, and discipline was very lax. It was found out a couple of days afterwards that there were three night operators at Lousiville. One of them had gone over ~~the~~ª toᵇ Jeffersonville and had fallen off a horse and broken his leg and was in a hospital. By a remarkable coincidence another of the men had been stabbed in a keno room and was also in hospital, while the third operator had gone to Cynthiana to see a man hung and had got left by the train!

TD (transcript), NjWOE, Meadowcroft. ªCancellation in pencil. ᵇInterlined above in pencil.

D. BOOK NO. 2

This undated notebook, labeled "Book No. 2," contains a mix of narrative passages, questions, and notes in Edison's hand. The first two pages are a memo by Meadowcroft, dated 9 January 1920, recounting the preparation and use made of this material between 1907 and 1910. The next sixty-six pages alternately present narrative passages and brief references to various anecdotes, most of which relate to the period covered by this volume. The next nine-page section is labeled "Martin's Questions." The remaining twenty-one pages contain only notes, all of which, with the exception of a few lines transcribed below, relate to events that occurred after June 1873.

[1] Notes
[2] ⟨—Dyer my wife says I am related to Perry of Battle Lake Erie fame⟩
[3] On my maternal side one of ancestors was Capt Ebenezer Elliott who fought in the Continental Army in the Revolutionary War =

[4] Cousin Commadore Perry.

[5] On my fathers side ancestors came from Holland— They were millers of grain on the Zuyder Zee in that Country— They landed at Elizabethport NJ & settled near or at Caldwell NJ. My grandfather was a Tory in the war of the revolution & had to flee to Nova Scotia losing all his property at Caldwell. He set~~ld~~tled at Digby Nova Scotia where my father was born in 1803. My great grand father was a banker on Manhattan Island & his name is signed to Continental money

[6] (I have a bill at house with his sig)

[7] My grandfather moved to Vienna Canada. My father ~~w~~settled at Milan Ohio where he became a buyer of wheat

[8] When I was 5 years of age I was taken by my Father & mother on a visit to Vienna. We left Milan & was driven by carriage to a Railroad then to a port on ~~the~~ Lake Erie & thence by a Canal boat in a tow of several to Port Burril Canada from where we drove to Vienna—

[9] I remember my grandfather very perfectly when he was 102 years of age at which age he died[a]— [We?]In the middle of the day he sat under a large tree in front of the house facing a well travelled road. his head was covered ~~with~~ completely with a large amount of very white hair ~~and~~ (he chewed tobacco incessantly.) ⟨Dont call for this⟩ & nodded to friends as they passed by—he used a very large cane & walked from his chair to the house resenting any assistance. I viewed him from a distance & never could be got ~~near~~ very close to him—

[10] ~~MI~~ remember some large pipes & especially a Molasses jug ~~which~~ a ~~truck~~trunk & several other things which came from Holland—

[11] My great grand father reached the age of 104.

[12] My father had several brothers all of which reached the age of 90 or more my father died at the age of 94—

[15] Louisville—

[16] Louisville office. Billy Lewis—

[17] Tyler & Geo D Prentice— speech 3 am— pasting up jokes from all over us for day opr   Mixer me took from main & mixer abusing him on another wire[b]

[18] Treasurer of night gang— new man, hit me   2nd time at Lvl

[19] Two oprs tramps went my[c] bed boots on   Had printing office— took A Johnson veto Dist Col bill & presidents message same day— press gave me dinner

[20] North Amer review auction shot at by police   2nd time at Lvl

[21] Room ~~of~~ over Lager beer saloon. machine shop— Boyd—[c] Shorthand man read notes Andy Johnson speech swing around circle   Full whiskey— guessed at half of it.

[22] ~~First drank Corn whisky~~

[23] Got into ~~Lvl~~ Louisville from Memphis where dischgd for putting in Repeater NY & NO together. Foley couldnt get job   I got one & after 2 days was put on press report   Arrived Linen duster & Army shoes   Ice in streets

[24] Left for NO— to go on c[h?]arter ship to Brazil—[d] arrived [----] vessell seized advised by old man not to go   back home—

[25] Memphis—

[26] Confiscated bldg— slept on floor   Drunken Opr came in one night got 4 bottles ink & poured it all over evrything

[27] Next night came in & kept throwing catridges in the fire grate.

[28] Midnight Lunch   Faro bank—

[29] Keno in church   man in pulpit

[30] Manager in confinement. signalled & friends got him out.

[31] Confederate stationary   Capitals each sheet.

[33] Sholes, ~~C~~Korty = ~~operat~~

[34] Repeater invented putting NY & NO in communication 1st after war— Manager ~~had~~ a pet   supt trying do same thing   I beat him   Noticed in Memphis appeal—got dischgd

[35] Louisville.

[36] [-----] 2nd time at Lvl in New office— Report on blind Point repeater— bad cable over Ohio River— 4 relays 4 sounders   spilt acid— eat book up ~~s~~discharged—

[37] Went to newspaper ofs   got exchanges   Cut jokes & scientific items scrap book   lost both— Newspaper man kind as I paragraphed late news to end even on takes

[39] Cincinnati

[40] ~~I~~Worked comcl wire   Di cipher   Oprs gone at Lvl— etc—

[41] Then put on press wire—

[42] Working on Duplex

[43] John Morgan guerella opr— Geo R Ellsworth—[e] Secret way sending messages   Shop in ofs building— slept there— Cooked   got run down—

[44] Presdt Lincoln message to all mayors— did know about it—

[45] Formation Union   took press got promoted— worked free to learn quick   considered crazy[f]

[46] Savings bank Lvl   Cin   NY   Boston   Copied plays for Theatre   Gilliland send   I copied—

[47] Somers & I induction coil round house. nobody could touch engines—

[48] Grave yard Ft Gratiot—

[49] Notes.

[51] When boy went Sarnia see prince Wales now King Edwards— Carpets laid Expected something great   Mistook Duke

Cambridge for prince— Indians in war Dress   Massasgua—
got likcked

[52] Published paper Pt H called Paul pry

[53] Published newspaper on Train same time had chem Lab.

[54] Had 2 stores Pt H one newspaper other vegetable—
bought most vegetables in Detroit. bot berries & butter along
RR—

[55] Walked 8 miles buy a sounder

[56] Eber B Ward 15 miles from Ridgeway in country   bears

[57] Boston. Stockings on Tremont row

[58] job got by Adams ~~snow~~   putting Repeater in at Pt H &
Sarnia   single cable— Ben T[-]ayes = got pass   snowed in
near Coburg— Soldier & funny stories   Country Inn—

[59] Met Stanton   man who threw tin box in Cincinnati ofs
2:30 am startled us all   said welcome to all   he had left from
wkg on UPRR   Boarding House Montreal. abstemous grub
frozen bedroom— went with Solider

[60] Arriving Boston small black satchel Blue shirt.

[61] Saw Milliken. [----]ing, seedy clothes   went wk NY No
1 same night   roasted on Boston specl—

[62] Hutchinson.—

[63] Coffeeman— gambling— Harvard student

[64] Ink stand Insulator   Heman Grant   viscous ink^g nearly killed me— grant spent ⅓ sal breaking pitchers   boarding
house—

[65] Cockroach killer— paper got hold it   had stop—

[66] Young ladies school— on Jordan Marsh bldg— I speechless— I never could talk to more than 4 if more present radiated
something that paralized my vocal cords—

[67] Hamblets place. Chas. Williams jr— making nitrogly

[68] Started private lines—

[69] Lab over stock Exchg   got up printer— Nitric acid ac-
cident.

[70] Synoid Meth Church JC   & agt assd press— BC OK
why not JC^h

[71] ~~Heman grant & viscous~~

[72] Jerry Borst. changed letters alphabet—

[73] Haunting Cornhill old books.

[74] Boarding Adams— Adams leaves for San Fran.

[75] Breaks— shoes

[76] ~~got~~Then got up a Duplex   went to Rochester— Expmted
bet there & ny— Chump at other end couldnt understand—
st~~u~~ruck ny Broke. ~~B~~Slept telgh ofs—

[77] SS Laws. Gold Indicator— broke down— Hired by Laws
225 month. 375 boys— Black Friday— Opr shake hands   con-
grat   we didnt have any mony   Opr a friend men in back room
of Wm Belden— Jim Fisk etc—

[78] ~~Laws sells out~~ got up printer for Laws. G&S buys him out. went into to Partnership P & Ashley— gold printer— Sold out Genl Lefferts fixed it so got something & Then in automatic. Jay Gould line to Washn— Tailing stopped, EHJ— Then ~~H~~Craig & Sholes— made 1st Remington typewriter over Goulds shops RR ave Newark— Roman letter— NY & Phila test.

[79] Kahns Museum

[80] Knew Bunnell night opr on Fre[n]ch Cable— Chicamauga junction— Capt. Van Duzer—

[81] Always saved 1 dollar for Mother—

[82] Sold bread to Norwegians— Employed extra boy— & several boys on excursions

[83] Run train from Pt H to GTR Junct Engr & fireman asleep— Mudy Eng— allowed to switch way^c frt for until got[-] into depot—

[84] Stop shoot Turkeys—

[85] One week before Christmas run off track. old cars bursted open got sick Eating

[86] Corporral of the guard No 1— Mike got in jail— I pulled Bbl over me— father got him out—

[87] 2½ years passed graveyard 12 to 1 oclock night— started horse running pass finally stopped this & at the end it got monotonous & didnt care about grave yards—

[88] Fathers Observatory— Folly that paid—

[89] Truck farming— I peddled the truck until went on train—

[90] Jumping off train at Pt H— Sand drawn by father—

[92] Milan— Lockwoods boy & I went swim he went down I waited then went home—

[93] barn on fire

[94] Set in bank. 1 suspender 40 sesions a day Employed 3 boys at Pt H nights to sell papers—

[95] Mac Walker apprentice proposition worked days & then copied press till 1 to 3 am to get expert—

[96] At St[---]ratford opr run trains head on fell culvert— string across track water between 6s— Got habit sleeping 15 min— Carter & I to Union station— W J Spicer threat send me to ~~st~~prison at Kingston Oppertune arrival 2 English swells grt joy on part Spice— Physiological moment left for Lowe Frt station caboose & USA— 1½ dolls weak board fell in culvert.

[97] Battle Pittsburgh Landing. Wilbur F Story 1000 papers—

[98] Telegh from James Clancy to mine sent news to Father &

[99] got chemical appartis made Pullman shops Detroit—

[100] ~~Alow~~ Tried acquire habit reading several words at once, so could read rapidly—

[131] G&Stock Co 40000 for sale pats   Bank wldnt pay check—

[155] 5 shops ᵰNewark

[173] After being discharged from Memphis for devising a repeater to permit NY & New Orleans to talk together the first after the war I went to Louisville— (recited before under questions) & obtained a job on the associated press wire = At that time the Telgh^c ~~office~~ operating room was in a deplorable condition. It was on the 2nd story of a delapidated building on the principal street with the battery room in the rear behind which was the office of the agt of the assoctd press—

[174] The plastering was about ⅓rd gone from the ceiling— a small stove used occasionally in the winter was connected to the chimney by a tort~a~uous stovepipe   the office was never cleaned—

[175] The switchboard for manipulating the wires was about 24 inches square the brass connections of which were black with age & the effects of Lightning which seemed to be particularly partial to Louisville   It would strike in one the wires at times with an explosion like a Cannon making that office no place for an opr with the heat disease—

[176] Arranged arround the wall were a dozen tables the ends next to the wall & about the size of those one smeets with in country hotels bedrooms for holding the bowl & pitcher for washing

[177] The copper wire connecting the instruments to the switchboard were small & apparantly were crystalized & rotten

[178] The battery room was filled with old books & messages & 100 cells of Nitric acid batteries arranged on a stand in the center of the room   This stand as well as the floor was almost eaten thru by the acid

[179] At the time I took the position there was a great shortage of telegh oprs.

[180] One night at 2 am, myself and one other opr was on duty I taking press report & the other ~~work~~ man working the NY wire   we heard a heavy tramp tramp tramp on the ricketty stairs. suddenly the door was flung open with great violence throwing it off one of its hinges   There appeared one of the best operators we had who worked days & who was of a very quiet a taciturn disposition except when intoxicated. he was a great friend of the manager of the office—

[181] His eyes were blood shot & wild. one sleeve had been torn away from his coat. Without noticing either of us he went up to the stove & kicked it over. The stovepipe fell & dislocated at every joint. it was ½ full of exceedingly fine soot which filled the room completely   this produced a momentarily respite to his labors— When the room had cleared sufficiently to see he went

around & pulled every table away from the wall & pilled them on top of the the stove in the middle of the room— then he proceeded to pull the switchboard away from the wall. It was held tightly by 4 screws. he finnally succeeded & when it gave way he fell with the board & striking on a table cut himself so that he soon became covered with blood   He then proceeded to the battery room & knocked all the batteries off on the floor—The Nitric acid ecommenced to ecombine with the plaster in the room below which was the Public receiving room for messages & bookeeping   The excess acid poured through & eat up the account books

[182] After having finished everything to his satisfaction he left—

[183] I told the other operator to ldo nothing & we would wait & leave things just as they were until the manager came

[184] In the meantime I knew all the wires coming through to the switchboard   rigged up a temporary set [-]so the NY business could be cleared up & also got the remainder of the press—

[185] At 7 oclock the day men commenced to appear   they were told to go down stairs & wait the coming of the manager   at 8 oclock he appeared   ewalked around went into battery room & then came up to me saying Edison who did this

[186] I told [-]him Billy L came in full of soda water & invented the ruin before him—

[187] He walked back & forward about a minute then coming up to my table brot his f[--]ist down & said <u>If Billy L ever does this again I will discharge</u> him—

[188] It it needless to say that there were other oprs that took advantage of this kind of discipline & I had many calls at night [--]tafter that but not with such destructive results.

[189] I remember with great satisfaction the discussions between the then celebrated journalist & poet[i] Geo D Prentice who was at that time editor of the Louisville journal & Mr Tyler of the associated press. Prentice ~~would come~~ I believe was the father of the [---]humorous paragraghs of the american newspaper. he was a poet, highly educated & a brilliant talker = ~~Tyler also but~~ he was very thin & small & I do not think he weighed over 125 lbs   Tyler who was a graduate of Harvard [----] & had a very clear ennunciation was a large man[j]   After the ~~pres~~ paper had gone to press prentice would generally come over to Tylers office & start in talking. having while in Tylers office heard them arguing on immortality of the soul etc asked permission of Mr Tyler if after going to press I might come in & listen to the conversation which I did many times after   One curious thing I never could comprehend was that Tyler had a side board with liquors & generally crackers. Prentice would pour out half a glass of what they called corn whiskey & would dip the crackers in it

& eat them   Tyler took it sans food. one teaspoonful of this stuff would put me to sleep—[k]

[190] It was the practice of the press oprs all over the country at that time tohat when a lull occurred to start in & send jokes or stories the day men had collected & these were copied & pasted up on the bulletin board— When Cleveland which was the originating off for press which it received from NY & sent it out simultaneously to Milwaukee Chicago Toledo Detriot Pittsburgh Columbus Dayton Cincinnati Indianapolis, Vincennes Terre Haute, St Louis & Louisville— Cleaveland would call first on Milwaukee if he had anything   if so he would send it[c] & Cleveland would repeat to all of us   Thus any joke originating anywhere in that area was known the next day all over— The press men would come in & copy anything that could be published which was about 3 per cent   I collected too[c] quite a large scrap book but unfortunately lost it

[191] While in Louisville the 2nd time It was the practice of the night force to go off on a picnic once a week either to Jeffersonville or New Albany   on these occasions I was made treasurer taking charge of the whole of the funds [possessed?] by each man   This was a matter of precaution as some were not able to gauge exactly their capacity for Liquor & as I did not drink I ~~could~~ [-----] was used as a sort of an Alcoholometer & refused to advance money when the limit was reached[l]   the last occasion that I acted as treasurer was the joining of the outfit by a new man from Illinois— He was told of[c] the custom & gave me his money. He became ~~drunk~~ frisky[c] with such rapidity that it surprised me & I refused to advance money whereupon I was knocked down & considerably battered before the rest of the party could intervene which they did   the ~~memenb~~ member from Illinois was rendered unconcious   I gave up this form of amusement thereafter—[f]

[192] While at Lvl the 1st time, & working on the press wire the occasion arose to take the Presidents message & at the same time Andrew Johnsons ~~vet~~ long winded veto of the District of Columbia bill— The conjunction of these two long messages was due to the fact that the Presidents message although printed & mailed to the postmasters throughout the country & to be given out on a fixed day, thru some failure of the mail ~~could~~ was not & could not reach Louisville in time for publication so it had to be sent from Cincinati west—

[193] The manager had arranged with me to come around at 10 oclock in the morning & take the veto message which made about 11 columns of [--]a newspaper   before I had finished this it became known that the Presdt Regular Message to Congress must also be received by telegh & I was asked if I thought I could do it   I said I thought I could. the last ½ column of the veto was

taken by another Opr & I started on the big message which ~~was~~ filled^c I think two pages of the newspaper. towards the end of the message I introduced a novelty which was greatly appreciated by the newspaper people. This conesisted in paragraghing the copy or ~~swriting it in sections~~ ~~so~~ ~~that~~ ~~th~~ called by the printers "takes" These sections made exactly three lines in the^c printed column, and to make ~~the~~ a good appearance there should not be to few or too many words The night editor ~~eo~~when the copy was read from the telegh ofs run his shears between these sections ~~&~~ ~~t~~ these were divided up among a dozen printers by this division of labor a column could be set up quickly & releived the Editor from doing it himself— I was 15 hours in the chair on this occasion without a moments intermission for food.

[194] For this I was given a dinner by the newspaper [-]men

[195] I never had ~~much~~ ~~of~~ ~~a~~ a high opinion of Andy Johnson after that I believed he talked too much [~~Especially~~ ~~as~~ ~~he~~ ~~had~~ ~~been~~ ~~copying~~ ~~some~~ ~~of~~ ~~his~~ ~~speeches~~ ~~from~~?] no

[196] While at Louisville the 2nd time— I attended an auction one day and bought 20 unbound^c volumes of the North American Review for 2 dollars— these I had bound & delivered to the telegh office.— One morning after ~~the~~ ~~papers~~ ~~had~~ getting through the press I took 10 volumes on my shoulder & started for home It was rather dark & while nearing home which was a room above a saloon I heard a shot & stopped. a policeman run up & grabed me by the throat. fortunately I knew him & He had yelled but I being rather^c deaf did not hear & he brot me to by the shot He supposed I had stolen the books. through all my travels I have preserved those books & have them now in my Library—

[197] While at Lvl I got for the 1st time an insight as to how speeches were reported. The associated press had a short hand man travelling with Presdt Johnson when he made his ~~el~~celebrated ~~w~~swing arround the circle in a private train delivering speeches— The man engaged me to write out from his reading the notes. He came in loaded & on the verge of incoherence— we started in. about every 2 minutes I would have to scratch out whole paragraphs & insert the same thing said in another & better way— he would frequently change words, always to the improvement of the speech. I couldnt understand this & when ~~I~~ ~~had~~ we got thru & I had copied about 5 columns I asked him that if he read from notes why these changes "Sonny" said he if these politicians ~~were~~ ~~report~~ had their speeches published as they delivered them a great many short hand men would be out of a job The best short handers ~~are~~ ~~those~~ & the holders of good positions are those who can take a lot of rambling incoherent stuff & make a rattling good speech out of it ~~This~~ ~~man~~ ~~was~~ This man was one of the most unique characters I ever saw

[198] When I left Louisville for the 1st time I with 2 other telegh oprs had saved up money & started to take position in the ~~g~~Brazillian Telegh as an advt had been inserted in some paper stating that oprs were wanted. ~~We~~ We had timed our departure from Louisville so as to catch a specially chartered steamer which was to leave N Orleans for Brazil on a certain day to convey a large number of Confederates & families who disgusted with the US were going to settle in Brazil— We arrived just at the time of the great riot where several hundred negroes were killed & the city was in the hands of a mob— The Govt had seized the steamer we intended to go in ~~for~~ to bring troops from the Yazoo river to N Oreleans to stop the rioting— myself & companions visted another shipping office to make inquiries as to the possibility of going to Brazil in other vessels   While in this office an extremely old man sat in a chair near the desk of the agent. after making inquiries & finding it impossible to find a vessell, The old man turned to me & asked why I wanted to go to Brazil   I told him whereupon he got up from his chair & shaking his boney finger in my face said that he had sailed the sea for 50 years & had been in every port in every country   that there was no country like the US   that if there was anything in a man the US was the place to bring it out & that any [~~money?~~] man that left this country to better his condition was an ignorent damned fool. I had been thinking this way myself for the last few days & the speech of the old man I considered good advice   So I ~~e~~told my companions that I was going home. ~~to~~but they were bound to go somewhere & I ~~learned~~ was told^m afterwards that both had gone to Vera Cruz & died of yellow fever.

[199] 2nd time in Louisville they had moved into a new office & the dicipline was now good. I took the press job, in fact I was a very poor sender, and therefore made the taking of press report a speciality. The newspaper men allowed me to come over after going to press at 3 am & get all the Exchanges I wanted   these I would take home & lay at the foot of the bed. I never slept more than 4 ~~t~~or 5 hours so I would [~~work?~~] awake at 9 or 10 & read those papers till dinner time   As I thus kept posted & knew from their activity every member of Congress & what committees they were on & all about the topical doings as well as the prices of ~~dif~~ breadstufs in all the primary markets, I was in a much better position than most operators to call on my imagination to supply missing words & sentences which frequent in those days with old rotten wires badly insulated especially on stormy nights— on these nights I had to supply in some cases ⅕ of the whole matter—pure guessing ~~but~~ but I seldom got caught except once   There had been some kind of [~~an meetin?~~] a convention in Virginia in which John Minor Botts was the leading figure. there was great excitement about it and there had been

two votes taken on the 2 days of the convention. There was not doubt but the vote the next day would go ~~the~~ ~~same~~ a certain way ~~My~~ A very bad storm came up about 10 oclock at night I my wire worked very bad. then there was a cessation of all signals then I made out the words Minor Botts The next was a N York item— I filled in a paragraph about the convention & ~~that~~ how[c] the vote went as I was sure it would but the next day I learned that instead of there being a vote the convention had adjourned one day[f]

[200] ~~while~~ One night I went into the new battery room to get some sulphuric acid for a battery I had made. The carboy slipped & I couldnt lift it back so about 4 gallons of acid started in to leak through to the private office of the manager below. In the morning I was notified to appear before the manager He said that he couldnt afford to keep me any longer & I left for Cincinnati where I for the 2nd time got a job ~~on~~& in few days was put on[n] the press wire

[201] 1st job—

[202] After leaving Stratford junction, I got a position as operator on the Lakeshore & Michigan Southern at Adrian Mich [~~working?~~--- ----] in the [---] Division Supts office. as usual I took [---]the night job which most oprs disliked but which I preferred as it gave me more leisure to Experiment. I had obtained from the station agent a small room & had established a little shop of my own. one day the day opr wanted to get off & I was on duty. about 9 oclock the Supt handed me a dispatch which he said was very important & I must get it off at once the wire at the time was very busy & I asked if I should break in & got orders to do so[o] ~~but~~ acting under the orders of the supt I broke in & tried to send the dispatch but the other opr would not permit it & the struggle continued for 10 minutes. I finally got possession of the wire & sent the message in about 20 minutes the Supt of Telegh who then lived in Adrian ~~but~~ & went to his office in Toledo every day happened this day to be in the WU office up town & it was the Supt who I was struggling with. He was livid with rage when he arrived[p] & discharged me on the spot. I told him the Genl Supt told me to break in & send the dispatch he then turned to the Supt who had witnessed this burst of anger[q] & said Mr H did you tell this young man to break in & send your dispatch he repudiated the whole thing— Their families were socially close & I ~~became~~ ~~a~~ ~~wanderer~~ was sacrificed[r] My faith in human nature ~~got~~ [------------- ------- ----- ---- ------------ ----] got a slight jar—

[203] I then went to ~~Toldeo~~ Toledo & got a job at Ft Wayne Ind on the Pittsburgh Ft Wayne & Chicago RR now leased to the PRR— This was a day job & I did not like it I then got a place in the ~~WU~~ ~~office~~ ~~at~~ ~~Ind~~ W Union office in Indianapolis—work-

ing a way wire [-] but I was very ambitious to be able to take press report. while the position was a day position I ~~cou~~ taking no interest in anything except the telegh came around every night & on an adjoining table ~~copied~~ to the regular press opr would copy press until about 1 am & then go home ~~I was~~ but it came faster than I could write it down legibily— at this time I conceived the idea of taking two old Morse Registers which recorded the dots & dashes by indenting a continuous strip of paper   the indenting point being worked by a lever & magnet   [-]I arranged these 2 instruments so I could receive the regular press signals at ~~high spee~~ their regular rate & record the same on the strip of paper   of course I could have read & copied[s] the signals from the paper but taking by sound was the ambition of all oprs the old Registers being obsolete. ~~but~~ but I arranged the second register [-]so that the strip passing through it the indentations were made to actuate a delicate double lever causing the ~~cir~~ Local circuit of a sounder or receiving instrum to be opened & closed corresponding exactly to the original signals   This it did with great perfection   When press was coming over the wire the primary register recorded them at the rate of 40 words per minute. The paper strip passing into the 2nd Register repeated these signals audibly on the sounder but at the rate of 25 or 30 words per minute according to the speed of the clockwork, which could be varied at pleasure. [-]by the aid of another day[c] opr who was ambitious I got permission from the press man[t] to put this in circuit & together we took press for several nights my companion keeping the apparatus in adjustment, & I copying. The reg Press opr would go to the theatre or take a sleep— only finishing the report after 1 am— Soon ~~th~~one of the newspapers complained of bad copy etc towards the end of the report ie[u] from 1 am till 3 & requested that the opr taking the report up to 1 am which was ourselves[v] take it all ~~(ourselves)~~ as the copy was unobjectionable. This led to an investigation by the manager & the scheme was forbidden   Of course having more time I could make better copy than the regular opr

[204] This instrument many years afterwards was applied by me to teleghy for transferring messages from one wire to any other wire simultaneously or after any interval of time. It consisted of a disk of paper the indentations being formed in a volute spiral exactly as the disk phonghs of today & it was this instrument which gave me the idea of the phonograph while working on the Telephone—[f]

[205] Not liking Indianapolis I obtained a situation in ~~C~~the WU ofs in Cincinnati on a way ~~wa~~ire as a plug opr. Operators were designated 1st class oprs & plug oprs, the latter being inefficient & there was very little association socially between the two classes.

[206] ⊥I worked a wire which ran to Portsmouth Ohio. I kept up the practice of coming around nights to copy press & would willingly act as a substitute for any opr who wanted to get off for a few hours   The few hours in most cases meant all night. however I didnt care, requiring little sleep but I was bound to become proficient in the very shortest time. The salary I received was 80 dollars per month   I made some extra by copying plays for the theatre, by using the telegh.

[207] Whil One night I came around and was working a local wire when the little Dutch boy who carried press came up in the office & said Lincoln was shot   Nobody believed him but he stoutly maintained that it was on the bulletin board at the Inquirer office & that there was a big crowd in the street. We found that it was true & that one of the oprs had taken a short special from Washington without sensing it, probably thinking of his girl or something else which is not unusual, some oprs become so expert that they work unconciously   I knew of an instance where a press opr fell asleep & still continued to write it down correctly, the manifold boy guiding his hand when it his agate stylus^w got over the edge of the [-]sheet & he^c was working writing on the table. This peculiar state of the brain doing intellectual work unconciously should be investigated

[208] This same night about 10 oclock we received & sent from Washington & long message which was sent to every mayor in Ohio notifying them of the death of the president.

[209] Sometime after this the oprs at Cleveland started in to form a Union of all the oprs in the US, to be called the Telegraphers Union of the US.

[210] A committee of 3 were to come to Cinncinati to init form & initiate the [-]oprs here   At that time there were 8 oprs working nights. the formation of the Union resulted in a jollification   The imbibing of large quantities of Brewery Serum although most of the men was immune to that anesthetic & only 2 men turned up for work. among the missing was the press opr— When Cleveland called up the different cities & Cincinnati signal came for press^x I made up my mind that I would to try my hand at it as some report was better than none at all, and that I couldnt see how I would be discharged for the attempt & to prevent delaying the report by interrupting for repetitions I determined I would get what I could & not interrupt.

[211] An agate stylus was used and 5 copies were taken simultaneously by the use of oiled tissue & black paper   One copy was an office copy to settle disputes.

[212] I stuck to the wire till 3 am—the ocopy looked fine if viewed as a whole as I could write a perfectly straight line across the wide sheet which was not ruled, and there were no flourishes but the individual words would not bear close inspection. When

I missed a understanding a word there was no time to think what it was so I made the an illegible one to fill in trusting to the printers to sense it   I knew they could read anything because Mr Bloss an Editor of the Inquirer made such bad copy that one of his editorials in manuscript was pined pasted up on the notice board in the office with an offer to of 1 dollar to any man who could read twenty consecutive words. nobody ever did it. When I got through [-]I was too nervous to go home so waited the balance of the night for the day manager Mr Stevens to arrive to see what was to be the outcome of this Union formation & my efforts   He was an austere man & I was afraid of him—

[213] I read the morning papers which were [-]came out about 4 am & the press report read perfectly which greatly surprised me [---- --- -----]   I went to work and th on my regular wire. there was considerable excitement but nothing was said to me neither did Mr Stevens examine the copy on the office hook which I was watching with great interest. However about 3 pm he went up to the hook grabbed the whole, looked at it as a whole without examining it in detail for which I was thankfulʸ & jabbed it back on the hook   I knew then I was all right. He then walked over to me & said young man I want you to work the Louisville wire nights your salary will be 125. Thus I got from the plug classification to that of a 1st class man.

[214] While working at Stratford Junction Canada, I was told by one of the freight conductors that in the freight depot at Goodrich there was a heap several boxes old broken up battery   I went there & found over 80 cells of Groves nitric acid battery one of the electrodes of each cell being made of sheet platinum— The operator there who was also agent when asked if I could have the tin part of these batteries readily gave his permission thinking they were tin   I removed them all amounting to several oz— [---] platinum was even in those days very expensive & I only owned 3 strips. &I was overjoyed at this acquisition & these strips & the reworked scrap are used to this day in my laboratory over 40 years

[215] After working at Cincinnati (1st time) for several months a friend of mine at Memphis teleghd me that he could get me a job at Memphis Tenn. as I wanted to see the country I accepted it & went to work nights on the N York side. The telegh was still under Military Control not having been turned over to the original owners the Southern Telegh Co but [----- -- --- -- -------]   In addition to the regular force there was an extra force of 2 or 3 operators & some stranded ones which was a burden to us as board was high— One of the stranded oprs was a great source of worry to me   he would come in at all hours & either throw ink around, or make a lot of noise. one night he built a fire in the grate & started to throw pistol catridges in   these

would explode & I was twice hit by the bullets which left a black & blue mark. another night he came in & got from some part of the building a lot of stationary with Confederate states printed on head   He was a fine operator & wrote a beautiful hand. He would take ~~all~~ a sheet write ~~A then~~ Capital A then another sheet with the A made another way, & so on through the alphabet, each time crumpling the paper up in his hand & throwing it on the floor— He would keep this up until the room was filled nearly flush with the tables then he would quit.

[216] Everything at that period was wide open   demoralization reigned supreme. there was no head to anything. At night myself & companion would go over to a gorgeously furnished Faro bank & get our midnight lunch. ~~Th~~[--] Everything was free

[217] there was over twenty keno rooms running   one of them that I visited was in a [~~method~~?] Baptist Church   The man with the revolving wheel being in the pulpit & the gamblers in the pews. I was rather pleased than otherwise when I was discharged for the invention of the repeater

[218] While there the manager was arrested for somtething I never understood & incarcerated in a military prison about ½ mile from the office. the building was ~~in th~~ in plain site of the office & 4 stories high^z He was strictly incommincado. One day thinking he might be confined in a^c room facing the office I put my arm out of the window & kept signalling dots & da~~t~~shes by the movement of the arm. I tried this ~~a~~several times for two days finally he noticed it & putting his arm out through the bars of the window^{aa} we established communication   he sent several messages to his friends & was afterwards set free

[219] When working in Cincinnati the 2nd time the office had been moved & the discipline was very much better— I was put on press nights which just suited me   I rented a room in the top floor of an office building bought a cot & an oil stove   bought a foot lathe & some tools & cultivated the acquaintance of Mr Sommers supt of telegh of the Cinncinnati & Indianapolis RR who gave me permission to take such scrap apparatus as I desired & which was of no use to the Co— Sommers was a very witty man & fond of experimenting himself & we worked on a self adjusting Telegh Relay which would have been very valuable if we could have got it. I soon became the possessor of a second hand Rhumkoff induction Coil which although it would only give a small spark would twist the arms & clinch the hands of a man so he couldnt let go of the electrodes. both Sommers & I were delighted with this apparatus   one day we went down to the round house of the RR & ~~got behind~~ connected up the spring^c wash tank in the room with the coil one electrode going to Earth— above this [---]~~at~~wash room was [----] a flat roof. we bored a hole through the roof & could see the men as they came in— the 1st man that came in dipped in the water   the floor

being wet formed a circuit & up went his hands   he tried it the second time with the same result. he then stood against the wall & had a puzzled Expression. we surmised that he was waiting to for somebody else to come in which was shortly after, with the same result. they then went out & soon the place was crowded & there was considerable excitement various theories were broached to explain this curious phenomenon   We enjoyed the sport immensely

[220] Not long after this the came to work at Cincinnati office a man by the name of Geo R Ellsworth. This man was the telegh opr of Morgan the Confederate Guerella General who gave so much trouble to the Union by raiding & capturing or destroying[bb] stores   El Ellsworth tapped wires, read messages & sent false ones, and did an immense amount of mischief generally by his superior ability as an Opr   [-]It is well known that one opr can recognize another by the way he makes his signals & [-]Ellsworth possessed the art of imitating these peculiarities & therefore he easily deceived the Union Operators   We soon became acquainted & he wanted tome to[c] invent a secret method of sending dispatches so an intermediate operator could not tap the wire & understand it   He said that if it could be accomplished he could sell it to the government & get a large amount of money. this suited me & I started in & succeeded in making such an instrument which had the germ of the Quadruplex in it aft Thafterwards invented by myself & now generally used throughout the world   This apparatus Quadruplex[c] permitted the sending of 4 messages over one wire simultaneously. By the time I had succeeded in getting the apparatus to work Ellsworth suddenly disappeared & it was only years afterwards that I heard that the Tameness of a telegh office was obnoxious (& perhaps other reasons[cc] & that he had become a gun man in the panhandle of Texas & had been killed. from his appearance I never would have thought such a thing possible   Many years afterwards I used this little device again for the same purposes   At Menlo Park NJ I had my Laboratory. there were several WU wires cut into the office [----] Lab which was used by me in experimenting with nights   one day I sat near an instrument which I had left connected during the night. The wire I soon found was a private wire between NY or Phila & I heard among a lot of stuff a message which surprised me. a week after that I had occasion to go to NYork & visited the NY office of the lessee. I told him asked him if he hadnt sent such a message   the expression that came over his face was a sight   he asked me how I knew of any message   I then told him the circumstances & suggested that he better cipher such communications or put on a secret sender. the finality of the interview was that I installed my old Cincinnati apparatus &which was used thereafter for many[c] years—

[221] Martins Questions

[222] While working in NY for WU you are said to have stayed & slept in its shop or laboraty all times. have u any recolctn—ͤ

[223] Its said you started Lab of ur own on getting Laws position— where was it   was it not fm tr u moved to Newark shop—ͤ

[226] Do u recal exact condition when u helped out the Laws gold system & got berth for so doing   R W Pope gives brief interesting account of it

[227] Date of that occurance seems to been Black Friday Sept 4 1869— Had u bn long in NY before that day & how did you happen to be in Laws headqutrs

[228] Do u recall anything abt ur 1st residence or workshop in NY

[229] Mention is made of a trip to Rochester NY in 1869 to try your duplex on Lines of Pacific & Atlantic   see Dicksons book p 56–7)   This seems all wrong   F L Pope is named as assisting—ͤ

[233] Any recollections of John Kruezi   The family appears to have been few if any records of his. He wasnt kind of man to keep diary but always ready to talk about those early days—

[236] Are any details u recollect abt Murray. he seems to have been a good associate & stand byͤ

[237] Did WU put up money at this time or did u have outside customers—ͤ

[239] The period 1871–7 began ur active tkg out patents abt a fortnight ever since   Wld like remarks on patent & patent system its utility or futility—could we get along without it   How could it be improved   Has it not been of some real benefit to you   If you could collect all the royalties justly due what would it amount to & how much have you actually got

[244] In early days of telegh didnt lots opr knock around country as you did   it seems to have been one of the habits   had it anything to do with the war   Did the Civil war impress you much.

[245] Geo L Anders in Wms shop

[246] Adams says— going along sts Boston I wld say Milt do you see anything I cld apply elec toͤ

[247] Liar—ͤ

[250] Is it true you worked upon the typewriter at any early stage of its development—ͤ

[252] Who was Geo Harrington   ﬀ are there any details or annecdotes abt him   Delany comes later   What about him— What was his actual share of the work   were they forerunners of Batchelor—

[266] Geo Little got people into auto & failed— I brot in to save the day. Expmts. E H J. Delany Charleston— pie— typewriter Craig asscd press—

[269] Dean at Newark & at Goerck—

[272] Carrol D Wright took out 1st pat—

[280] Black friday   Wm Belden— opr back end 60 Bdway

[317] 1873 visit English PO Test. Telgh st. Think I have this—

[379] Supplied funds to Thau & Herman   took no paper He repudiated whole thing

AD (photographic transcript), NjWOE, Archives Office; "mme", possibly in another hand, written in bottom margin of several pages. a"at . . . died" interlined above. bThis and paragraphs through number 102 (except 22, 25, 33, 35, 39, 49, and 55) overwritten with large "X"; paragraphs 269, 280, and 317 also thus overwritten. cInterlined above. d"to . . . Brazil—" interlined above. e"Geo R Ellsworth—" interlined above. fFollowed by horizontal line. g"viscous ink" interlined above. h"BC OK why not JC" interlined above. i"& poet" interlined above. j"was a large man" interlined below. k"would . . . sleep—" written over centered horizontal lines. l"& . . . reached" interlined above. m"was told" interlined above. n"in few days was put on" interlined above. o"& . . . so" interlined above. p"when he arrived" interlined above. q"who . . . anger" interlined above. r"was sacrificed" interlined above. s"& copied" interlined above. t"from . . . man" interlined above. uCircled. v"which was ourselves" interlined above. w"his agate stylus" interlined above. x"for press" interlined above. y"for . . . thankful" interlined above. z"& . . . high" interlined above. aa"through . . . window" interlined above. bb"or destroying" interlined above. cc"(& . . . reasons" interlined above.

E. NOTES

Taken from a notebook that has five pages in Edison's hand, these "Notes" are numbered consecutively from 1 to 33.

[13] 13   Geo Little, got Geo H ex asst Tresr US & ~~others~~ Genl Palmer & others by Lab expts which were failures when ~~t~~oried on actual lines— I was called in— pie. Typewriter. DH Craig Assd Press

[19] 19 = Carrol D Wright, took out 1st pat Vote Recorder #

AD, NjWOE, Lab., N-09-06-27. All notes except those numbered 26 and 27 overwritten with a check mark.

F. NOTES

This notebook includes sixteen pages in an unlabeled section in Edison's hand relating to the Dyer and Martin biography. They are preceded by a memo to Edison from William Meadowcroft dated "June 28/09" stating that the notes on the following page had been copied. Only three items concern events during the period covered by this volume. There is a typed version of the notes in the William H. Meadowcroft Collection at the Edison National Historic Site. The last fifteen pages are a biographical sketch of Edison's former employee Sigmund Bergmann.

[1] What was matter with episode about War dept cypher for Genl Thomas when Hood was raiding Tenn & Lousiville officer couldnt be reached by wire   this is historical & interesting—

[3] My version of the stocking in Boston[a] Episode it appears to me is better put than you have it & its actually true—

[4] The J. C. Synod episode left out, also Kahn museum—

AD, NjWOE, Lab., N-09-06-28. [a]"in Boston" interlined above.

Appendix 2

*Bibliographic Essay: Edison's Boyhood
and Itinerant Years*

The references here should assist researchers seeking to explore further Edison's early years, but they are not designed to cover the field exhaustively.

No official records or contemporary announcements exist for Edison's birth. Family records and later accounts agree on the date and place. Much of this material can be found in the Edison Biographical Collection (EBC), primarily in the "Birth" and "Family—General" folders. Included are photocopies of the "Family Record" pages of Nancy Edison's Bible (ENHS catalog no. E-355, now missing); W. A. Galpin to H. Lanahan, 23 June 1926 and n.d.; and Report of J. McCoy, 2 July 1926. See also M. Wadsworth to TAE, 24 Mar. 1912, GF. Records for the births and deaths of the other Edison children are found in photocopies from various family Bibles in EBC and in Milan and Port Huron burial records.

The most authoritative source on the Edison family background and on Samuel and Nancy prior to their move to Milan is Simonds 1940. On William Mackenzie's rebellion and Samuel's role in it, see J. Bartlet Brebner, *Canada: A Modern History,* ed. Donald C. Masters (Ann Arbor: University of Michigan Press, 1970), 215–40; Charles Lindsey, *The Life and Times of Wm. Lyon Mackenzie* (Toronto: P. R. Randall, 1862), 399; Edgar McInnis, *Canada: A Political and Social History,* rev. and enl. ed. (New York: Rinehart and Co., 1959), chap. 10; and Willis F. Dunbar, *Michigan: A History of the Wolverine State,* ed. George S. May (Grand Rapids, Mich.: William B. Eerdman, 1980), 282–85. On the sentiments of Milan's residents toward the rebellion see E. A. Theller, *Canada in 1837–38* (Philadelphia: Henry F. Anners, 1841), 2:291; and Francis P. Weisenburger, *The Passing of the Frontier, 1825–1850* (Columbus: Ohio State Archaeological and Historical Society, 1941), 356–62. On Samuel's brief stays in Detroit, Mich., and Peru, Ohio, see Erie County Common

Pleas Journal 3:595–98, OSECo; Huron County Probate Court Record of Settlement, 2:218, 224, ONHCo; and an obituary of Samuel Edison excerpted from the *Norwalk (Ohio) Reflector,* in "Samuel Edison—Funeral," EBC.

For the Edison family in Milan see U.S. Bureau of the Census 1967a, roll 392; idem 1964, roll 676; Erie County Record of Deeds 3:96, 6:201–2, 14:335–36; Erie County Common Pleas Court Execution 3:456, 554, OSECo; Erie County Common Pleas Record 3:231–32, 597–98, OSECo; Erie County Probate Court Marriage Book 1:309, OSECo; Church Records 2:42, First Presbyterian Church, Milan, Ohio; Grace Goulder, *Ohio Scenes and Citizens* (Cleveland, Ohio: World, 1964), 81–91, 235; and *Centennial Celebration* (Milan, Ohio: First Presbyterian Church, 1918). Regarding Samuel's shingle-making see *Erie County Reporter,* 29 Sept. 1887; and Henry Howe, *Historical Collections of Ohio,* 2 vols., Centennial ed. (Columbus, Ohio: Henry Howe and Son, 1889–91), 1:580. Beyond these, the principal sources for Milan's history and the Edison family in it are Frohman 1976 and James A. Ryan, *The Town of Milan* (1928; reprint, Columbus: Ohio Historical Society, 1974).

The precise date of the Edison family's move to Port Huron is not documented. According to Caroline Farrand Ballentine, "The True Story of Edison's Childhood and Boyhood," *Michigan History Magazine* 4 (1920): 178, the family arrived in the spring of 1854 on the river steamer *Ruby.* Pitt Edison is known to have been briefly active as a partner in a livery stable in April 1854 (*Port Huron Commercial,* 13 Apr. 1854, 2). The *Port Huron Commercial* for 6 April 1854 (p. 3) lists "Edison, Thos." as having a letter waiting for him at the post office. This may refer to young Tom, his uncle Thomas, or someone else. However, no one else of that name is known to have been in that area at that time.

Samuel Edison derived income from several sources during Edison's boyhood in Port Huron. Evidence that he was in the grocery business is in Mich. 69:61, RGD. Samuel's land dealings can be partly traced through the indexes to Register of Deeds and Register of Mortgages, MiPHStCo. For his ventures in the timber business see Register of Deeds 1:266, MiPHStCo. The repercussions of the Panic of 1857 apparently had a severe impact on Samuel's timber and land operations but did not impoverish the family (O. M. Carter to Thomas Martin, 14 Dec. 1907 and c. 1907, EBC; *History of St. Clair County* 1883, 443–45, 550, 630). Circuit Court Minutes 2:299–300, 308, and 5:4, 22, MiHC, show that Samuel was forced to forfeit property to pay debts. In one case the judgment against him was recorded as uncollectable because the sheriff could find "no property" to seize and sell. At the same time, Nancy is recorded as the purchaser of 200 acres, 120 of which brought $1,000 in a sale three years later (Register of Deeds 5:605 and 11:173, MiPHStCo). By 1860 Samuel was again listed as a substantial landowner in U.S. Bureau of the Census 1967b, roll 559.

Contrary to the claims of Conot, Samuel Edison was never appointed keeper of the lighthouse near Fort Gratiot. He did build an observation tower next to the edge of the pine grove near the Edison house. Built before 1859, the tower, according to local tradition, made money for the family for a brief period. The location of the tower is shown on the U.S. Army Corps of Engineers map, "Part of the West Shore of Lake Huron including the Mouth of the St. Clair River," surveyed and drawn in 1859 under the direction of George G. Meade, Lake Huron No. 1, *Survey of the Northern and Northwestern Lakes*. The tower is discussed in Bancroft 1888, 258; Caroline Farrand Ballentine, "The True Story of Edison's Childhood and Boyhood," *Michigan History Magazine* 4 (1920): 181–82 (with illus.); Dorothy Marie Mitts, *That Noble Country: The Romance of the St. Clair River Region* (Philadelphia: Dorrance and Co., 1968), 206–8. See also *Port Huron Commercial*, 30 July 1853, 2; and *Port Huron Press*, 17 Aug. 1864, 3.

For evidence of Edison's schooling by his mother see TAE to Pupils of the Grammar Schools of New Jersey, 30 Apr. 1912, GF. Evidence of more formal schooling is in Doc. 1, nn. 1, 7; evidence of attendance in Sunday School is in Mary Hoyt to TAE, 9 Feb. 1921, and TAE to Mary Hoyt, 16 Feb. 1921, GF. It is said that Edison's mother read to him a good deal and his father rewarded him for each book mastered. People who knew him stated that Thomas's reading included Hume's *History of England* and Gibbon's *Decline and Fall of the Roman Empire*. Edison indicated that he first read Paine at the age of thirteen. See Thomas Edison, "Introduction," *The Life and Works of Thomas Paine*, ed. William M. Van der Weyde (New Rochelle, N.Y.: Thomas Paine National Historical Assoc., 1925), vii; and *The Diary and Sundry Observations of Thomas Alva Edison*, ed. Dagobert D. Runes (New York: Philosophical Library, 1948), 151–58. For evidence of Edison's other readings as a youth see O. M. Carter to Thomas Martin, 14 Dec. 1907, and "Sunday 1 Mr. Edison's Interview," galley proof, EBC; and "Making Good Friends with Lady Luck," *Literary Digest* 100 (16 Feb. 1929): 54–55.

For Edison's attendance at the Rev. Mr. Engle's school see George Engle to TAE, 13 Aug. 1885, DF (*TAEM* 77:147); Mary Engle to TAE, 4 Oct. 1886, DF (*TAEM* 79:152); William Brewster to TAE, 10 May 1883, DF (*TAEM* 64:180); Brewster to TAE, 22 Dec. 1891, GF; and T. H. Lester to TAE, 6 Apr. 1892, GF. Brewster's sketch of the school mentioned in his 1891 letter was published in Conot 1979, 9; the original can no longer be found.

For Edison's friends and activities while growing up in Port Huron see George Mann to TAE, 25 Feb. 1910; John Miller to TAE, 16 Feb. 1903; H. S. Palmer to TAE, 11 Mar. 1898; Ambrose Robinson to TAE, 13 Apr. 1923; George Twiss to TAE, 4 Oct. 1894 and 27 May 1910; and J. L. Wilson to TAE, 14 Feb. 1922; all in GF. See also J. D. Cormack, "Edison," *Good Words* 42 (1901): 157; and "The Genesis of a Genius," *Magazine of Michigan* 1 (1929): 20. For further aspects of the history of Port Huron and the Edisons' lives there see Jenks 1912; Endlich 1981; Mary C. Burnell and Amy Marcaccio, *Blue Water Reflections* (Norfolk, Va.: Donning Co., 1983); Register of Communicants, 25–26, First Congregational Church, Port Huron, Mich.; and *First Congregational Church, Port Huron, Michigan* (Port Huron, Mich.: First Congregational Church, 1965).

The precise date Edison began working on the Grand Trunk Railway is uncertain. The railway opened on 21 November 1859 (*Detroit Daily Advertiser,* 23 Nov. 1859, 2; *Sarnia Observer and Lambton Advertiser,* 25 Nov. 1859, 2), and there is some evidence that suggests Edison was on the first train running between Port Huron and Detroit (G. W. Carleton to TAE, 21 June 1879, DF [*TAEM* 49:271]; John Donahugh to TAE, 3 Jan. 1905, GF). For Edison's venture in selling produce see App. 1.A6; and N. W. King to TAE, 9 Nov. 1914, and John Talbot to TAE, 26 Nov. 1920, GF. For Edison's account of his hearing loss see TAE to Arthur Brisbane, 13 Feb. 1906, GF; Dyer and Martin 1910, 37; and *The Diary and Sundry Observations of Thomas Alva Edison,* ed. Dagobert D. Runes (New York: Philosophical Library, 1948), 44, 48–56.

While working on the train, Edison had a small chemical laboratory and he may also have had a small lab or workshop at home. References to these are found in Alexander Stevenson to TAE, 2 Mar. 1881, DF (*TAEM* 57:76); James Clancy to TAE, 3 Mar. 1889, and Ambrose Robinson to TAE, 13 Apr. 1923, GF; O. M. Carter to Thomas Martin, 14 Dec. 1907, EBC; and Stamps and Wright 1982.

References to Edison's early interest in telegraphy are found in John Raper to TAE, 5 Apr. 1878, and John Thomas to TAE, 26 July 1882 and 17 Aug. 1884, DF (*TAEM* 15:467, 64:559, 71:660); and Fred Betts to TAE, 12 Feb. 1931, Louis Huber to TAE, 16 Feb. 1920, James McCrae to TAE, 24 Jan. 1931, George Sent to TAE, 14 and 18 June 1907, and John Thomas to TAE, 16 Jan. 1898, GF. References to his study of telegraphy with James MacKenzie include App. 1.A8; TAE to N. Eisenlord, 24 Dec. 1889, LB-035:273, Lbk.; O. M. Carter to Thomas Martin, 14 Dec. 1907 and 2 Apr. 1908, and M. Walker to O. M. Carter, 31 Mar. 1908, EBC; and Henry D. Benner to TAE, 20 Oct. 1909, N. Eisenlord to TAE, 11 Dec. 1889, and Mrs. R.

Brooks to TAE, 4 Jan. 1912, GF. References to Micah Walker's store include O. M. Carter to Thomas Martin, 14 Dec. 1907 and 2 Apr. 1908, and Micah Walker to O. M. Carter, 31 Mar. 1908, EBC; and J. L. Wilson to TAE, 14 Feb. 1922, GF.

Edison's prowess as a telegraph receiver and his failure as a sender of Morse code are referred to in App. 1.A8, 20, D193, 199, 210–13; "Tom Edison's Operating Days," *Operator* 9 (1 Apr. 1878): 4; "More Edisonian Reminiscences," ibid. 11 (15 Aug. 1880): 7; and Charles Mixer to TAE, 22 July 1920, GF. In Edison's marginalia on Mixer's letter he specifically states that as a sender he was "a complete failure." For references to Edison's attempts to improve his telegraphic skills see Dyer and Martin 1910, 79–80; and TAE to C. H. Ames, 16 Nov. 1899 (in the possession of Jessie Kent of Lakeland, Fla.). The Edison National Historic Site possesses a copy of Benn Pitman, *The Manual of Phonography* (Cincinnati: Phonographic Institute, 1860), inscribed in Edison's hand as follows: "A. Edison, Louisville, Ky, April 4, 1867."

Eighteen sixty-three is probably the year during which Edison took the position at Stratford. Edison's handwritten application for membership in the Old Time Telegraphers' and Historical Association, dated 5 December 1904, states that he began telegraphing at Stratford in 1863. However, William Simonds (1940, 55) reproduces an item from what he claims is a May 1864 newspaper stating that Edison had just left his job at "Thomas" Walker's store to work as a telegrapher at Stratford. This newspaper item cannot be found. The inaccuracy of Walker's first name and the imprecision of the date suggest that Simond's record of the date also may be incorrect.

Simonds (1940, 56) states that Edison boarded at 19 Grange Street, Stratford. Several newspaper clippings in the Edison file at the Stratford-Perth Archives refer to Edison's staying at that address. But acccording to James Anderson, archivist at the Stratford-Perth Archives, the assessment records for Stratford indicate that this building was not erected until 1866. Conot (1979, 16) claims that Edison went to Washington, D.C., in 1863 to work on military telegraphs. However, an examination of the letter on which he bases that claim (John Schultz to TAE, 19 Jan. 1905, GF) and Edison's reply, written on that letter, reveals no evidence that Edison made such a trip.

For references to Edison's rigging an alarm to wake him so that he could send half-hourly signals while a telegraph operator for the Grand Trunk Railroad, see App. 1.A8; Robert Wagner to TAE, 24 July 1928, GF; and "More Edisonian Reminiscences," *Operator* (15 Aug. 1880), 7, in which William Stanton is quoted as claiming that Edison's inattention to duty resulted in a train collision in which lives were lost. Conot (1979, 17, 497) cites the

Stratford Herald [*sic*], 18 December 1863, as a source. The *Stratford Beacon* (18 Dec. 1863, 2) reported that a train ran off the rails because a switch was incorrectly set. In contrast, Edison states (App. 1.A9) that there were two trains and no derailment. According to James Anderson, the *Perth Herald* and the *Stratford Beacon* carried three items relating to train accidents in the Stratford area during the time when Edison might have worked there. The dates of these accidents were 18 December 1863, 24 February 1864, and 16 March 1864. In no case was a near collision or loss of life reported. However, the newspapers might not have noted a near collision. Regarding Edison's work as a telegraph operator for other railroads, see App. 1.D202–3; and TAE marginalia, Samuel Insull to TAE, 7 Dec. 1909, GF.

For references to Edison's obtaining telegraph registers from John Wallick—superintendent of Western Union Telegraph Co. at Indianapolis—in order to slow down the incoming signal and allow him to take the message, see App. 1.D203; and E. L. Parmelee to TAE, 14 Mar. 1891, GF. For references to Indianapolis and its railways see William Robeson Holloway, *Indianapolis: A Historical and Statistical Sketch of the Railroad City* (Indianapolis: Indianapolis Journal Print, 1870); Jeannette Covert Nolan, *Hoosier City: The Story of Indianapolis* (New York: Julian Messner, 1943); *Indianapolis Daily Sentinel*, 16 Dec. 1864, 3, and 14 Jan. 1865, 2–3; and *Indianapolis Daily Journal*, 2 Dec. 1864, 2.

Regarding Edison's friends in Cincinnati, see App. 1.D219–20; Jot Spencer to TAE, 8 Apr. 1878, DF (*TAEM* 15:478); Nat Hyams to TAE, 11 Feb. 1912, GF; "Thomas Alvey Edison," *Cincinnati Commercial*, 18 Mar. 1878, 5; and *Cincinnati Daily Enquirer*, 6 Sept. 1865, 3. For general references to Cincinnati see Margaret Walsh, *The Rise of the Midwestern Meat Packing Industry* (Lexington: University Press of Kentucky, 1982); and Charles Cist, *Cincinnati in 1841: Its Early Annals and Future Prospects* (Cincinnati: Privately published, 1841).

For a general description of Memphis and for Edison's opinion of the city see Shields McIlwaine, *Memphis Down in Dixie* (New York: E. P. Dutton and Co., 1948); App. 1.D215–18; and TAE draft reply, with Mr. Dealy to TAE, 12 Jan. 1909, GF. Regarding Edison's experiments while in Memphis, see App. 1.D173; Quad. 72.16 (*TAEM* 9:259–69); Derrick Dyer to TAE, 9 July 1875, DF (*TAEM* 13:407); O. J. McGuire to TAE, 12 Feb. 1912, GF; "Tom Edison's Operating Days," *Operator* 9 (1 Apr. 1878): 4; TAE marginalia, Clyde Grissam to TAE, 7 Apr. 1919, GF; and *Telegr.* 2 (1865–66): 27, which stated that the *Memphis Bulletin* of 26 November 1865 had reported the first direct connection of New York and New Orleans. Edison remembered being fired for using the repeater between New York and New Orleans; but if he did this in November 1865, he was not fired for it, because he was still in Memphis at the end of March 1866.

Edison wrote (App. 1.D23) that there was "Ice in [the] streets" when he arrived in Louisville. Weather reports in the *Louisville Daily Courier* for March and April indicate that winter's last frost occurred on 9 April, suggesting that Edison arrived in Louisville sometime before mid-April. Edison had to find his own way to Louisville via Decatur, Ala., and Nashville, Tenn., because his Memphis boss would not give him a railway pass. See "Thomas Alva Edison," *Operator* 9 (15 May 1878): 4; and Lathrop 1890, 431. For general references to Louisville see George R. Leighton, *Five Cities: The Story of Their Youth and Old Age* (New York: Harper and Bros., 1939); Isabel McLennan McMeekin, *Louisville: The Gateway City* (New York: Julian Messner, 1946); and Allen J. Share, *Cities in the Commonwealth: Two Centuries of Urban Life in Kentucky* (Lexington: University Press of Kentucky, 1982). Regarding Edison's trip to New Orleans and his return to Louisville, see App. 1.D198; Vandal 1983; *New York Times*, 5 Aug. 1866, 5; ibid., 3 Sept. 1866, 4; and "Thomas Alva Edison," *Operator* 9 (15 May 1878): 4. Appendix 1.D197 refers to Edison's being in Louisville during President Johnson's visit to that city, which took place on 9 September 1866.

The date of Edison's return to Cincinnati is uncertain. The *Cincinnati Commercial* periodically published a list of undelivered letters in the possession of the Post Office. The list for 4 August 1867 shows the name "Edison A." The list for 11 August 1867 does not contain his name, suggesting that he had already arrived in Cincinnati and picked up his mail.

Edison's boyhood home had been commandeered for use by the Fort Gratiot Military Reservation in 1864 and was bought the following year by Henry Hartstuff, agent in charge of the reservation. Regarding the fate of both the house and the tower, which were gone by 1870, see Stamps and Wright 1982, 10–14. Regarding Samuel's occupation as a justice of the peace at the time of Edison's 1867 return, see Election Records 37:63, 75, MiHC. Samuel's land operations are partly reflected in Register of Deeds, MiPHStCo. Samuel was apparently also a contractor; see PN-69-08-08, Lab. (*TAEM* 6:757–93). Mrs. M. J. Powers said she boarded with the Edisons at Fort Gratiot and that during one winter Edison was home sick (Powers to TAE, 18 Feb. 1880, DF [*TAEM* 53:88]). Regarding the Port Huron and Gratiot Street Railway, see Jenks n.d.

Appendix 3

The American Patent System

America's patent laws and practices were a crucial feature of the world within which Edison the inventor matured.[1] The U.S. Patent Office administered these laws and collected fees from inventors in order to meet its operating expenses. These fees, quite modest in comparison with their British counterparts, made U.S. patents accessible to many inventors.[2] American patents provided inventors with exclusive ownership of their creations for seventeen years. In return, inventors had to disclose to the public the details of their invention. The patent system, established under the U.S. Constitution, was intended to provide economic incentive for the inventor, who could assign all or partial rights in a patent to other individuals or companies. Inventors frequently sold their patent rights for a flat sum or in exchange for royalty payments. Some also obtained financial support for past or ongoing inventive activity or for help in establishing businesses to exploit their inventions.

To obtain a patent, the inventor submitted to the U.S. Patent Office an application consisting of a proper specification, carefully measured drawings, and, in most cases, a physical model of the invention.[3] The application required careful preparation so that it clearly stated its claims and described the invention and how it worked. After the inventor submitted the application, an official examiner scrutinized it and determined whether the description and claims were satisfactory. If the examiner concluded that the application was acceptable and that the invention was new, useful, and unknown prior to the time of submission, a patent could be issued. If an application was rejected, the inventor could amend the written description or the claims but not the drawings or model. Such amendments were allowed in order to make the application clearer or to make the claims more precise or narrow. The cycle of rejection and amendment could involve many rounds and take many years before the patent was issued or the application was abandoned.

If two or more persons submitted applications for substantially the same invention, the Patent Office issued the patent to the inventor with the best claim to priority. This often made inventive activity much like a race, particularly when others were known to be working on the same problem. It also placed a premium on careful record-keeping and secrecy. To establish priority, inventors at this time often submitted caveats to the Patent Office. Caveats were preliminary applications in which the inventor made claims to one or more potential inventions without presenting the detail required in a formal application. The Patent Office noted the subject matter of the caveat and placed it in a confidential archive. If within one year another inventor filed an application on a similar process or device, the Patent Office notified the holder of the caveat, who then had three months to submit a formal application.[4]

When two or more inventors submitted similar patent applications or when one inventor submitted an application for an invention that had already been patented by another, these applications and patents were declared in interference with each other. The Patent Office then held a quasi-judicial hearing before the examiner of interferences to determine which inventor had priority. The proceedings often included depositions, testimony, exhibits, and arguments. The inventor could appeal the decision of the examiner first to the commissioner of patents and then to a federal court in the District of Columbia. The Patent Office could not rescind an issued patent, but it could issue a patent for the same invention to a second inventor who established priority. The issuance of a patent did not, therefore, provide the inventor with a definitive claim. Economically significant patents were often challenged in court on grounds of priority, novelty, and ownership.

Many inventors turned to professional patent solicitors and agencies to ensure the cogency of their claims in applications and to defend their patents in interference proceedings. Because any "person of intelligence or good moral character" could appear as a patent attorney, many mid-nineteenth-century patent solicitors had no formal legal training. Commonly they had some technical training and had made a study of the patent laws and procedures. For example, Franklin Pope, a telegraph engineer and inventor who was associated briefly with Edison, served as a patent expert for the Western Union Telegraph Company and then established his own practice as a patent attorney. Only attorneys admitted to the bar, however, could bring suit in the federal courts regarding infringement or ownership of a patent, and a number of such attorneys specialized in patent law.

General and specialized technical journals of the day publicized and promoted the patent system. Patent solicitors also helped educate inventors about the system. Munn & Company, the nation's largest patent agency, published the *Scientific American*. This popular mechanics' magazine featured news of the Patent Office and promoted inventions. Likewise, many other technical journals presented articles about the Patent Office and patent law, editorialized about needed reforms, and published information about the latest patents issued. Munn & Company also published several editions of its handbook on patent law with instructions for obtaining patents. Other patent solicitors also published such guides and some acted as propagandists for the patent system through speeches and articles.[5]

When describing the patent system's contribution to American progress, nineteenth-century advocates usually pointed to major inventions such as Samuel Morse's telegraph, Cyrus McCormick's reaper, and Elias Howe's sewing machine. Most patents and inventions, however, represented improvements or variations of existing technologies. Some of these later became as essential to an industry as the patent that gave it birth. Improvements often allowed a technology to develop in new ways or made its use more profitable. An important improvement provided competitive advantages to the firm that controlled the patent and, in a few cases, even prevented competition by denying use of an essential invention. While a patented variation might not improve the operation of a technology, the firm already possessing key patents for a technology might use it to prevent someone else from patenting or using an alternative method not embodied in those patents. Such noncompetitive use of patents received little public attention until late in the century, when large firms gained increasing power in American society. As firms competed for larger market shares, control not only of inventions but of the inventors themselves often became an important strategy. As a telegraph inventor, Edison became a central figure in telegraph competition involving Western Union.

1. This account of U.S. patent law and practice is derived from Munn & Co. 1878, which includes copies of the patent laws from 1836 to 1870; and U.S. Patent Office 1875.

2. The 1861 Patent Act set fees that remained in force for several decades. These included $15 for filing an application, $10 for filing a caveat, and $20 for issuing a patent. Additional fees were collected during interference proceedings, when recording an assignment, or for design patents. Prior to 1852 the fee for a patent for the entire United Kingdom cost a minimum of £310. In that year, the fee was reduced to £25, but renewal fees of £50 after three years and £100 after seven years were required to keep the patent in force for the full term allowed. Munn & Co. 1876, 76; Davenport 1979, 57–58.

3. When a model could illustrate an invention, the Patent Office required the inventor to submit one. The growing number of electrical and chemical patents precipitated the dropping of this requirement about 1880, however. Ferguson and Baer 1979, 11; U.S. Patent Office 1880, 13.

4. A caveat could be renewed annually. The caveat procedure fell into disuse during the late nineteenth century, and Congress abolished it in 1910. Curtis 1873, 332–33; Deller 1937, 977.

5. On Munn & Co. see Borut 1977. An early guide was Phillips 1837. Henry Howson, a Philadelphia solicitor, served on the U.S. Patent Association's Committee on Publication and published several pamphlets on the patent system. See, for example, Howson 1871, 1874, and 1878.

Appendix 4

Edison's U.S. Patents, 1868–1873

The following list contains all patents for which Edison executed an application prior to July 1873. It is arranged in chronological order by execution date, which is the date on which Edison signed the application and the date in the patenting process that comes closest to the time of actual inventive activity. The application date is the date on which the U.S. Patent Office received and recorded the application.

	Exec. Date	*Appl. Date*	*Issue Date*	*Pat. No.*	*Title*
1.	10/13/68	10/28/68	06/01/69	90,646	Electrographic Vote-Recorder
2.	01/25/69	02/17/69	06/22/69	91,527	Printing-Telegraphs
3.	08/17/69	08/27/69	11/09/69	96,567	Printing-Telegraph Apparatus
4.	08/27/69	09/04/69	11/09/69	96,681	Automatic Electrical Switch for Telegraph Apparatus
5.	09/16/69	10/27/69	04/26/70	102,320	Printing-Telegraph Apparatus (with Franklin L. Pope)
6.	02/05/70	04/11/70	05/17/70	103,035	Electro-Motor Escapements
7.	04/12/70	04/14/70	06/07/70	103,924	Printing-Telegraph Instruments (with Franklin L. Pope)
8.	05/24/70	05/27/70	07/02/72	128,608	Printing-Telegraph Instruments (with Franklin L. Pope)
9.	06/22/70	06/28/70	05/09/71	114,656	Telegraphic Transmitting Instruments
10.	06/22/70	06/28/70	05/09/71	114,658	Electro-Magnets for Telegraph Instruments
11.	06/29/70	07/06/70	01/24/71	111,112	Governors for Electro-Motors
12.	09/06/70	09/21/70	05/09/71	114,657	Relay-Magnets for Telegraph Instruments
13.	11/17/70	11/22/70	03/28/71	113,033	Printing-Telegraph Apparatus
14.	01/10/71	01/14/71	03/28/71	113,034	Printing-Telegraph Apparatus

Exec. Date	Appl. Date	Issue Date	Pat. No.	Title
15. 07/26/71	08/04/71	01/23/72	123,005	Telegraph Apparatus
16. 07/26/71	08/04/71	01/23/72	123,006	Printing-Telegraphs
17. 07/26/71	08/04/71	02/27/72	123,984	Telegraph Apparatus
18. 08/12/71	08/18/71	03/19/72	124,800	Telegraphic Recording Instruments
19. 08/16/71	08/18/71	12/05/71	121,601	Machinery for Perforating Paper for Telegraph Purposes
20. 11/13/71	11/18/71	05/07/72	126,535	Printing-Telegraphs
21. 11/13/71	11/18/71	12/10/72	133,841	Type-Writing Machines
22. 01/03/72	01/12/72	05/07/72	126,532	Printing-Telegraphs
23. 01/17/72	01/24/72	05/07/72	126,531	Printing-Telegraphs
24. 01/17/72	01/24/72	05/07/72	126,534	Printing-Telegraphs
25. 01/23/72	01/30/72	05/07/72	126,528	Type-Wheels for Printing-Telegraphs
26. 01/23/72	01/30/72	05/07/72	126,529	Type-Wheels for Printing-Telegraphs
27. 02/14/72	02/19/72	05/07/72	126,530	Printing-Telegraphs
28. 02/14/72	02/19/72	05/07/72	126,533	Printing-Telegraphs
29. 03/15/72	03/22/72	10/22/72	132,456	Apparatus for Perforating Paper for Telegraphic Use
30. 04/10/72	04/16/72	10/22/72	132,455	Paper for Chemical Telegraphs etc.
31. 04/18/72	04/25/72	11/12/72	133,019	Electrical Printing-Machines
32. 04/26/72	05/07/72	06/18/72	128,131	Printing-Telegraphs
33. 04/26/72	05/07/72	07/02/72	128,604	Printing-Telegraphs
34. 04/26/72	05/07/72	07/02/72	128,605	Printing-Telegraphs
35. 04/26/72	05/07/72	07/02/72	128,606	Printing-Telegraphs
36. 04/26/72	05/07/72	07/02/72	128,607	Printing-Telegraphs
37. 05/06/72	06/06/72	09/17/72	131,334	Rheotomes or Circuit-Directors
38. 05/08/72	06/06/72	01/14/73	134,867	Automatic Telegraph Instruments
39. 05/08/72	06/06/72	01/14/73	134,868	Electro-Magnetic Adjusters
40. 05/09/72	06/06/72	08/27/72	130,795	Electro-Magnets
41. 05/09/72	06/06/72	09/17/72	131,342	Printing-Telegraph Instruments
42. 05/28/72	06/06/72	09/17/72	131,341	Printing-Telegraph Instruments
43. 06/10/72	07/09/72	09/17/72	131,337	Printing-Telegraphs
44. 06/10/72	07/09/72	09/17/72	131,340	Printing-Telegraphs
45. 06/10/72	07/09/72	09/17/72	131,343	Transmitters and Circuits for Printing-Telegraphs
46. 06/15/72	07/09/72	09/17/72	131,335	Printing-Telegraphs
47. 06/15/72	07/09/72	09/17/72	131,336	Printing-Telegraphs
48. 06/29/72	07/09/72	09/17/72	131,338	Printing-Telegraphs
49. 06/29/72	07/09/72	09/17/72	131,339	Printing-Telegraphs
50. 06/29/72	07/09/72	09/17/72	131,344	Unison-Stops for Printing-Telegraphs
51. 10/16/72	10/22/72	01/14/73	134,866	Printing-Telegraph Instruments
52. 10/16/72	10/22/72	05/13/73	138,869	Printing-Telegraphs

Exec. Date	Appl. Date	Issue Date	Pat. No.	Title
53. 10/31/72	11/05/72	09/23/73	142,999	Galvanic Batteries
54. 11/05/72	11/09/72	08/12/73	141,772	Circuits for Automatic or Chemical Telegraphs
55. 11/09/72	11/11/72	02/04/73	135,531	Circuits for Chemical Telegraphs
56. 11/26/72	12/03/72	01/27/74	146,812	Telegraph-Signal Boxes
57. 12/12/72	01/15/73	08/12/73	141,773	Circuits for Automatic Telegraphs
58. 12/12/72	01/15/73	08/12/73	141,776	Circuits for Automatic Telegraphs
59. 12/12/72	01/15/73	05/12/74	150,848	Chemical or Automatic Telegraphs
60. 01/21/73	02/18/73	05/20/73	139,128	Printing-Telegraphs
61. 02/13/73	02/18/73	05/20/73	139,129	Printing-Telegraphs
62. 02/13/73	02/18/73	07/01/73	140,487	Printing-Telegraphs
63. 02/13/73	02/18/73	07/01/73	140,489	Circuits for Printing-Telegraphs
64. 03/07/73	03/13/73	05/13/73	138,870	Printing-Telegraphs
65. 03/07/73	03/13/73	08/12/73	141,774	Chemical Telegraphs
66. 03/07/73	03/13/73	08/12/73	141,775	Perforators for Automatic Telegraphs
67. 03/07/73	03/13/73	08/12/73	141,777	Relay-Magnets
68. 03/07/73	03/13/73	09/09/73	142,688	Electrical Regulators for Transmitting-Instruments
69. 03/07/73	03/13/73	11/17/74	156,843	Duplex Chemical Telegraphs
70. 03/24/73	07/29/73	02/10/74	147,312	Perforators for Automatic Telegraphy
71. 03/24/73	07/29/73	02/10/74	147,314	Circuits for Chemical Telegraphs
72. 03/24/73	07/29/73	05/12/74	150,847	Receiving Instruments for Chemical Telegraphs
73. 04/22/73	04/26/73	04/27/75	162,633	Duplex Telegraphs
74. 04/23/73	05/16/73	07/01/73	140,488	Printing-Telegraphs
75. 04/23/73	06/27/73	02/24/74	147,917	Duplex Telegraphs
76. 04/23/73	06/27/73	05/12/74	150,846	Telegraph-Relays
77. 04/23/73	07/29/73	02/10/74	147,311	Electric Telegraphs
78. 04/23/73	07/29/73	02/10/74	147,313	Chemical Telegraphs
79. 04/23/73	07/29/73	03/02/75	160,405	Adjustable Electro-Magnets for Relays, etc.

Bibliography

Abernathy, J. P. 1887. *The Modern Service of Commercial and Railway Telegraphy, in Theory and Practice.* 6th ed. Cleveland: n.p.

American Compound Telegraph Wire Co. 1873. *Trade Catalogue.* New York: n.p.

Appleyard, Rollo. 1939. *The History of the Institution of Electrical Engineers (1871–1931).* London: Institution of Electrical Engineers.

Armstrong, James. 1873. *Armstrong's Directory of the Inventions, Manufactures, and Manufacturers of Newark, New Jersey.* Newark, N.J.: n.p.

Atkinson, E., trans. 1910. *Elementary Treatise on Physics, Experimental and Applied.* Ed. A. W. Reinold. From *Ganot's Eléments de Physique.* New York: William Wood and Co.

Atkinson, Joseph. 1878. *The History of Newark, New Jersey.* Newark: William B. Guild.

Ayrton, William E. 1891. *Practical Electricity: A Laboratory and Lecture Course.* 6th ed. London: Cassell & Co.

Baker, Edward C. 1976. *Sir William Preece, F.R.S.* London: Hutchinson and Co.

Bancroft, William L. 1888. "A History of the Military Reservation at Fort Gratiot, with Reminiscences of Some of the Officers Stationed There." *Michigan Pioneer and Historical Society Historical Collections, 1887* 11:249–61.

Barney, William C. 1872. "Automatic Telegraphy." *Telegr.* 8:185.

Bishop, Joseph Bucklin. 1918. *A Chronicle of One Hundred and Fifty Years: The Chamber of Commerce of the State of New York, 1768–1918.* New York: Charles Scribner's Sons.

Borut, Michael. 1977. "The *Scientific American* in Nineteenth Century America." Ph.D. diss., New York University.

Boston Directory. (Printed annually.) Boston: Sampson, Davenport and Co.

Boyd, Andrew, and W. Harry Boyd, comps. 1868. *Boyd's Directory, of Elizabeth, Rahway, and Plainfield.* Elizabeth, N.J.: n.p.

Bright, Charles. 1974 [1898]. *Submarine Telegraphs: Their History, Construction, and Working.* New York: Arno Press.

Brittain, James. 1970. "The Introduction of the Loading Coil: George A. Campbell and Michael I. Pupin." *Technology and Culture* 11:36–57.

Brock, Gerald W. 1981. *The Telecommunications Industry: The Dynamics of Market Structure.* Cambridge, Mass.: Harvard University Press.

Brooks, David. 1876. "Report on Telegraphs and Apparatus." In *Reports of the Commissioners of the United States to the International Exhibition, Held at Vienna, 1873,* ed. Robert H. Thurston. Washington, D.C.: GPO.

Brown, Henry T. 1896. *Five Hundred and Seven Mechanical Movements.* 18th ed. New York: Brown & Seward.

Bruce, Robert V. 1973. *Bell: Alexander Graham Bell and the Conquest of Solitude.* Boston: Little, Brown & Co.

Butrica, Andrew J. 1987. "Women in Telegraphy: Transatlantic Contrasts and Parallels." *Proceedings of the One Hundred First Annual Meeting of the American Historical Association.* Ann Arbor, Mich.: University Microfilms.

Calahan, E. A. 1901a. "The Evolution of the Stock Ticker." *Elec. W.* 37:236–38.

———. 1901b. "The Forerunner of the Telephone." *Elec. W.* 37:312–13.

———. 1901c. "The District Telegraph." *Elec. W.* 37:438–39.

Carosso, Vincent P. 1979. *More Than a Century of Investment Banking: The Kidder, Peabody and Co. Story.* New York: McGraw-Hill.

Catlin, George B. 1923. *The Story of Detroit.* Detroit: Detroit News.

Conot, Robert. 1979. *A Streak of Luck.* New York: Seaview Books.

Craig, Daniel H. 1870. Letter to the Editor. *Telegr.* 7:26.

———. 1871. "Automatic Telegraphy." *Sci. Am.* (n.s.) 24:4.

———. 1872. Letter to the Editor. *Telegr.* 8:354.

———. 1888. *Machine Telegraphy.* New York: n.p.

Crampton, Emeline Jenks. 1921. *History of the St. Clair River.* St. Clair, Mich.: St. Clair Republican.

Culley, Richard S. 1863. *A Handbook of Practical Telegraphy.* London: Longman, Green, Longman, Roberts & Green.

———. 1871. *A Handbook of Practical Telegraphy.* London: Longmans, Green, Reader, and Dyer.

Current, Richard N. 1954. *The Typewriter and the Men Who Made It.* Urbana: University of Illinois Press.

Curtis, George T. 1873. *A Treatise on the Law of Patents.* 4th ed. Boston: Little, Brown and Co.

Dain, Floyd R. 1968. *Education in the Wilderness.* Lansing: Michigan Historical Commission.

Davenport, Neil. 1979. *The United Kingdom Patent System: A Brief History.* Portsmouth, England: Kenneth Mason.

Davis, Charles H., and Frank B. Rae. 1877. *Handbook of Electrical Diagrams and Connections.* 2d ed. New York: D. Van Nostrand.

Deller, William. 1937. *Walker on Patents: Deller's Edition*. New York: Baker, Voorhis and Co.

Drummond, James O. 1979. "Transportation and the Shaping of the Physical Environment in an Urban Place: Newark, 1820–1900." Ph.D. diss., New York University.

Dub, Julius. 1863. *Die Anwendung des Elektromagnetismus, mit besonderer Berücksichtigung der Telegraphie*. Berlin: J. Springer.

DuBoff, Richard B. 1983. "The Telegraph and the Structure of Markets in the United States, 1845–1890." In *Research in Economic History*, ed. Paul Uselding, vol. 8. Greenwich, Conn.: JAI Press.

Dun & Bradstreet. 1974. *Dun & Bradstreet: A Chronology of Progress*. New York: Dun & Bradstreet.

Dyer, Frank, and T. C. Martin. 1910. *Edison: His Life and Inventions*. 2 vols. New York: Harper and Bros.

Elizabeth Directory. (Printed annually.) Elizabeth, N.J.: Fitzgerald and Dillon.

Endlich, Helen. 1981. *A Story of Port Huron*. Port Huron, Mich.: Privately published.

Farmer, John S., and W. E. Henley, comps. and eds. 1970 [1890–1904]. *Slang and Its Analogues*. New York: Arno Press.

Farmer, Moses. 1870. Letter to the Editor. *Sci. Am.* (n.s.) 23:388.

Ferguson, Eugene S., and Christopher Baer. 1979. *Little Machines: Patent Models in the Nineteenth Century*. Greenville, Del.: Hagley Museum.

Fischer, Henry C., and William H. Preece. 1877. "Joint Report upon the American Telegraph System." UKLPO.

Ford, William F. 1874. *The Industrial Interest of Newark, N.J.* New York: Van Arsdale & Co.

Frohman, Charles E. 1976. *Milan and the Milan Canal*. Sandusky, Ohio: Privately published.

Gabler, Edwin. 1986. "Kid-Gloved Laborers: Gilded Age Telegraphers and the Great Strike of 1883." Ph.D. diss., University of Massachusetts.

Garratt, G. R. M., David C. Goodman, and Colin A. Russell. 1973. *Science and Engineering*. Milton Keynes, England: Open University Press.

Gibson, F. M., and M. H. A. Lindsay. 1962. "Electric Protection Services: A Study of the Development of the Services Rendered by the American District Telegraph Company." DSI-NMAH(E).

Gold and Stock Telegraph Co. 1872. *Directions for Putting Up and Using the Two Wire Universal Stock Printer*. New York: Gold and Stock Telegraph Co.

———. 1873. *"Universal Stock" Printer Instructions No. 2*. Washington, D.C.: Gold and Stock Telegraph Co.

———. 1876. *Instructions to Agents Using the Universal Stock Printer*. New York: Gold and Stock Telegraph Co.

———. n.d. *Edison's "Universal Printer" for Private Lines No. 3*. Instructions. N.p.

Goulding's Business Directory of New York. 1873. New York: L. G. Goulding.

Gray, Elisha. 1868. The Induction Relay. *J. Tel.* 1:1.

Hall, Thomas. 1874. *Hall's Illustrated Catalogue of Telegraphic Instruments and Materials.* Boston: n.p.

Healy, Clarence L. 1905. "Stock Ticker." *Teleg. Age* 23:121.

Higgins, Frederick. 1877. "A Description of the Automatic Step-by-Step Type-Printing Telegraphic Apparatus Used by the Exchange Telegraph Company." *J. Soc. Teleg. Eng.* 6:122–43.

Hindle, Brook. 1981. *Emulation and Invention.* New York: New York University Press.

Hirsch, Susan E. 1978. *Roots of the American Working Class: The Industrialization of Crafts in Newark, 1800–1860.* Philadelphia: University of Pennsylvania Press.

History of St. Clair County, Michigan. 1883. Chicago: A. T. Andreas and Co.

[Holbrook, A. Stephen.] 1882 [1872]. *Newark Industrial Exhibition: Report and Catalogue.* Newark, N.J.: n.p.

Holbrook's Newark City Directory. (Printed annually.) Newark, N.J.: A. Stephen Holbrook.

Hopkins, G. M. 1873. *Combined Atlas of the State of New Jersey and the City of Newark.* Philadelphia: G. M. Hopkins and Co.

Hotchkiss, Horace L. 1969 [1905]. "The Stock Ticker." In *The New York Stock Exchange,* ed. Edmund C. Stedman. New York: Greenwood Press.

Hounshell, David. 1984. *From the American System to Mass Production, 1800–1932.* Baltimore: Johns Hopkins University Press.

Howson, Henry. 1871. *A Brief Inquiry into the Principles, Effects, and Present State of the American Patent System.* Philadelphia: n.p.

———. 1874. *What We Owe to Patents.* Philadelphia: U.S. Patent Association.

———. 1878. *Patents and the Useful Arts.* Philadelphia: n.p.

Hunt, Robert, ed. 1864. *A Supplement to Ure's Dictionary of Arts, Manufactures, and Mines.* New York: D. Appleton and Co.

Huntington, F. M. 1875. *The Telegrapher's Souvenir: A Work Comprising Compilations and Original Articles . . . Intended to Be Instructive, Interesting, and Amusing. Not Only to the Fraternity, But to Strangers as Well.* Paterson, N.J.: Lyon and Halsted.

Indianapolis City Directory. 1865. Indianapolis: Hawes & Co.

Israel, Paul B. 1984. "Changing Patterns of Support for Innovation: Edison and the Telegraph Industry, 1868–1878." Thomas A. Edison Papers, Rutgers University. Typescript.

———. 1986. "Invention and Corporate Strategies: Western Union and Competition." Paper presented at the annual meeting of the Society for the History of Technology.

Jehl, Francis. 1937–41. *Menlo Park Reminiscences.* 3 vols. Dearborn, Mich.: Edison Institute.

Jenkins, Reese V., and Paul B. Israel. 1984. "Thomas Edison: Flamboyant Inventor." *IEEE Spectrum* 21:74–79.

Jenks, William Lee. 1912. *St. Clair County, Michigan: Its History and Its People.* 2 vols. Chicago: Lewis Publishing Co.

———. n.d. *History of Port Huron Street Railways*. William Lee Jenks Collection, MiD-B.

John. 1870. "Telegraphic Quiet.—The New Printing Telegraph Instrument." *Telegr.* 7:146.

Johnson, J. Stewart. 1966. "Silver in Newark: A Newark Three Hundredth Anniversary Study." *Museum* (n.s.) 18.

Johnston, William John. 1880. *Telegraphic Tales and Telegraphic History*. New York: W. J. Johnston.

Jones, Francis Arthur. 1924. *Thomas Alva Edison: An Intimate Record*. Rev. ed. New York: Thomas Y. Crowell Co.

Josephson, Matthew. 1959. *Edison: A Biography*. New York: McGraw-Hill.

Kieve, Jeffrey L. 1973. *The Electric Telegraph: A Social and Economic History*. Newton Abbot, England: David & Charles.

King, W. James. 1962a. "The Development of Electrical Technology in the Nineteenth Century: 1. The Electrochemical Cell and the Electromagnet." *United States Museum Bulletin 228*. Washington, D.C.: Smithsonian Institution.

———. 1962b. "The Development of Electrical Technology in the Nineteenth Century: 2. The Telegraph and the Telephone." *United States Museum Bulletin 228*. Washington, D.C.: Smithsonian Institution.

———. 1962c. "The Development of Electrical Technology in the Nineteenth Century: 3. The Early Arc Light and Generator." *United States Museum Bulletin 228*. Washington, D.C.: Smithsonian Institution.

Klein, Wolfgang. 1979. "Aus der Entwicklung der elektromagnetischen Telegrafenapparate." *Archiv für Deutsche Postgeschichte* 2:147–65.

Knight, David. 1967. *Atoms and Elements: A Study of Theories of Matter in England in the Nineteenth Century*. London: Hutchinson and Co.

Knight, Edward H. 1876–77. *Knight's American Mechanical Dictionary*. 3 vols. New York: Hurd and Houghton.

Lamb, Martha J. 1876. Newark. *Harper's New Monthly Magazine* 2:1–20.

Lathrop, George Parsons. 1890. "Talks with Edison." *Harper's Monthly Magazine* 80:431.

———. 1894. Edison's Father. *Once A Week* 5:7.

Laussedat, A. 1865. *Notice biographique sur Gustave Froment*. Paris: J. Hetzel.

Lief, Alfred. 1958. *"It Floats": The Story of Procter & Gamble*. New York: Rinehart & Co.

Lindley, Lester G. 1975 [1971]. *The Constitution Faces Technology: The Relationship of the National Government to the Telegraph, 1866–1884*. New York: Arno Press.

Little, George. 1868. "New Telegraph Recording." *Telegr.* 5:77.

———. 1874a. Letter to the Editor. *Telegr.* 10:182–83.

———. 1874b. Letter to the Editor. *Telegr.* 10:189.

Lovette, Frank H. 1944–45. "Western Electric's First Seventy-five: A Chronology." *Bell Telephone Magazine* 23:271–78.

McKelvey, Blake. 1973. *American Urbanization: A Comparative History*. Glenview, Ill.: Scott, Foresman & Co.

Marsh, J. O., ed. 1980. *The Letters and Papers of David Edward Hughes (1831–1900), Part I*. Microfilm Archive for Victorian Technology, no. 1. Manchester, England: Northwestern Museum of Science and Industry and Department of History of Science and Technology, UMIST.

Martin, T. C., and J. Wetzler. 1887. *The Electric Motor and Its Application*. New York: W. J. Johnston.

Massachusetts, Middlesex County. 1912. *Index to Probate Records of the County of Middlesex, Massachusetts, 2nd ser., 1870–1890*. Cambridge, Mass.: n.p.

Maurice, Klaus, and Otto Mayr, eds. 1980. *The Clockwork Universe: German Clocks and Automata, 1550–1650*. Washington, D.C.: Smithsonian Institution.

Maver, William, Jr. 1892. *American Telegraphy: Systems, Apparatus, Operation*. New York: J. H. Bunnell & Co.

Mayhew, Ira. 1858. *Reports of the Superintendent of Public Instruction of the State of Michigan for the Years 1855, '56, and '57: With Accompanying Documents*. Lansing: Hosmer & Kerr.

Memphis City Directory. 1866. Memphis: Bingham, Williams & Co.

Moise, Robert, and Maurice Daumas. 1978. "La periode d'approche." In *Histoire générale des techniques*, ed. M. Daumas. 5 vols. Paris: Presses Universitaires de France.

Munn & Co. 1867. *The United States Patent Law: Instructions How to Obtain Letters Patent for New Inventions*. New York: Munn & Co.

———. 1878. *The United States Patent Law: Instructions How to Obtain Letters Patent for New Inventions*. New York: Munn & Co.

National Telegraph Co. 1869. *Report of the Executive Committee of the National Telegraph Company to Subscribers of Its Capital Stock on Little's Automatic System of Fast Telegraphy*. New York: National Telegraph Co.

Nier, Keith A. 1975. "Physics in Nineteenth Century Britain as a Socially Organized Category of Knowledge: Preliminary Studies." Ph.D. diss., Harvard University.

Noad, Henry M. 1859. *A Manual of Electricity*. 4th ed. London: Lockwood and Co.

Orton, William. 1870. "Executive Order No. 110." *J. Teleg.* 3:279.

Phillips, Walter Polk. 1876. *Oakum Pickings: A Collection of Stories, Sketches, and Paragraphs Contributed from Time to Time to the Telegraphic and General Press*. New York: W. J. Johnston.

———. 1897. *Sketches Old and New*. New York: George Munro's Sons.

Phillips, Willard. 1837. *The Inventor's Guide: Comprising the Rules, Forms, and Proceedings for Securing Patent Rights*. New York: Collins, Keese and Co.

Plainfield Merchant's City Directory. 1875. Elizabeth, N.J.: Freie Presse.

Plum, William R. 1882. *The Military Telegraph during the Civil War in the United States.* 2 vols. Chicago: Jansen, McClurg & Co.

Pope, Franklin. 1867. "A Dissertation upon Construction." *Telegr.* 3:131–32.

———. 1868a. "The Double Transmission System of Telegraphy" (Parts 1, 2). *Telegr.* 4:290, 297–98.

———. 1868b. "The Invention of the Inverted Cup." *Telegr.* 5:117.

———. 1869. *Modern Practice of the Electric Telegraph.* New York: Russell Bros.

———. 1871a. *The Telegraphic Instructor: A Hand-book for Students of Telegraphy.* New York: F. L. Pope.

———. 1871b. "The Western Union Telegh Co. Manufactory." *Telegr.* 8:105.

———. 1874. Letter to the Editor. *Telegr.* 10:15.

———. 1889. *Evolution of the Electric Incandescent Lamp.* Elizabeth, N.J.: Henry Cook.

Popper, Samuel H. 1951. "Newark, N.J., 1870–1910: Chapters in the Evolution of an American Metropolis." Ph.D. diss., New York University.

Post, Robert. 1976. *Physics, Patents, and Politics: A Biography of Charles Grafton Page.* New York: Science History Publications.

Preece, William. 1884. "The Progress of the Telegraph." In *The Practical Applications of Electricity.* London: Institution of Civil Engineers.

Prescott, George B. 1863. *History, Theory, and Practice of the Electric Telegraph.* Boston: Ticknor and Fields.

———. 1866. *History, Theory, and Practice of the Electric Telegraph.* Boston: Ticknor and Fields.

———. 1870. "Automatic Telegraphy." *Sci. Am.* (n.s.) 23:292.

———. 1877. *Electricity and the Electric Telegraph.* New York: D. Appleton and Co.

Rainwater, Dorothy T. 1975. *Encyclopedia of American Silver Manufacturers.* New York: Crown Publishers.

Reid, James D. 1879. *The Telegraph in America.* New York: Derby Bros.

Ricord, F. W., ed. 1897. *History of Union County, New Jersey, Illustrated.* Newark: East Jersey History Co.

Rocke, Alan J. 1984. *Chemical Atomism in the Nineteenth Century: From Dalton to Cannizzaro.* Columbus: Ohio State University Press.

Rosenberg, Robert, and Paul B. Israel. 1986. "Intraurban Telegraphy: The Nerve of Some Cities." Paper presented at the annual meeting of the American Historical Association.

Sabine, Robert. 1867. *The Electric Telegraph.* London: Virtue Bros. & Co.

———. 1872. *The History and Progress of the Electric Telegraph.* 3d ed. London: Lockwood and Co.

Saint-Edme, Ernest. 1869. "Physique industrielle." *Annales industrielles* 1:502–4.

Schallenberg, Richard H. 1978. "Batteries Used for Power Generation during the Nineteenth Century." In *Selected Topics in the History of Electrochemistry*, ed. George Dubpernell and J. H. Westbrook. Princeton: Electrochemical Society.

Schellen, Heinrich. 1888. *Der Elektromagnetische Telegraph.* 6th ed. Brunswick: Vieweg & Sohn.

Scott, J. M. 1972. *Extel One Hundred: The Centenary History of the Exchange Telegraph Company.* London: Ernest Benn.

Shaffner, Taliaferro P. 1859. *The Telegraph Manual.* New York: Pudney & Russell.

Shaw, G. M. 1878. Sketch of Edison. *Popular Science Monthly.* 13:487–91.

Shiers, George. 1971. "The Induction Coil." *Sci. Am.* (n.s.) 224:83.

Siemens, Werner von. 1891. "Der Induktions-Schriebtelegraph von Siemens & Halske, Beschreibung für den Gebrauch desselben." In *Wissenschaftliche und Technische Arbeiten von Werner Siemens*, vol. 2, 2d ed. Berlin: J. Springer.

———. 1968. *Inventor and Entrepreneur: Recollections of Werner von Siemens.* Trans. W. C. Coupland. New York: Augustus M. Kelley.

Sikes, William Wirt. 1867a. "About Newark." *Northern Monthly Magazine* 1:385–402.

———. 1867b. "More About Newark." *Northern Monthly Magazine* 2:1–20.

Simonds, William Adams. 1940 [1934]. *Edison: His Life, His Work, His Genius.* New York: Blue Ribbon Books.

Smith, J. E. 1871. *Manual of Telegraphy Designed for Beginners.* 4th ed. New York: L. G. Tillotson & Co.

Smith, Willoughby. 1974 [1891]. *The Rise and Extension of Submarine Telegraphy.* New York: Arno Press.

Stamps, Richard B., and Nancy E. Wright. 1982. *Preliminary Report of the Archaeological Excavations at Thomas A. Edison's Boyhood Home, Port Huron, Michigan.* Working Papers in Archaeology. Rochester, Mich.: Oakland University.

Stanton, William. 1880. "More Edisonian Reminiscences." *Operator,* 15 Aug., p. 7.

Starring, Charles R., and James O. Krauss. 1968. *The Michigan Search for Educational Standards.* Lansing: Michigan Historical Commission.

Sterling, Adaline W. 1922. *The Book of Englewood.* Englewood, N.J.: Committee on the History of Englewood.

Taltavall, John B. 1893. *Telegraphers of Today.* New York: John B. Taltavall.

Ternant, A. L. 1881. *Les Télégraphes.* Paris: Librarie Hachette.

Thompson, Robert Luther. 1947. *Wiring a Continent: The History of the Telegraph Industry in the United States, 1832–1866.* Princeton: Princeton University Press.

Ulriksson, Vidkunn. 1953. *The Telegraphers: Their Craft and Their Unions.* Washington, D.C.: Public Affairs Press.

Urquhart, Frank J. 1913. *A History of the City of Newark, New Jersey.* 2 vols. New York: Lewis Historical Publishing Co.

U.S. Bureau of the Census. 1964. *Population Schedules of the Seventh Census of the United States, 1850*. National Archives Microfilm Publications Microcopy No. 432. Washington, D.C.: National Archives.

———. 1967a. *Population Schedules of the Sixth Census of the United States, 1840*. National Archives Microfilm Publications Microcopy No. 704. Washington, D.C.: National Archives.

———. 1967b. *Population Schedules of the Eighth Census of the United States, 1860*. National Archives Microfilm Publications Microcopy No. 653. Washington, D.C.: National Archives.

———. 1967c. *Population Schedules of the Eighth Census of the United States, 1870*. National Archives Microfilm Publications Microcopy No. 593. Washington, D.C.: National Archives.

———. 1975. *Historical Statistics of the United States: Colonial Times to 1970*. Washington, D.C.: GPO.

U.S. Patent Office. 1872. *Decisions of the Commissioner of Patents for the Year 1871*. Washington, D.C.: GPO.

———. 1875. *Rules of Practice in the United States Patent Office*. Washington, D.C.: GPO.

———. 1880. *Rules of Practice in the United States Patent Office*. Washington, D.C.: GPO.

Van Arsdale & Co. [1873?]. *The Successful Businessmen of Newark, N.J.* Syracuse, N.Y.: Van Arsdale & Co.

Vandal, Gilles. 1983. *The New Orleans Riot of 1866: Anatomy of a Tragedy*. Lafayette: University of Southwestern Louisiana.

Walsh, Justin E. 1968. *To Print the News and Raise Hell! A Biography of Wilbur F. Storey*. Chapel Hill: University of North Carolina Press.

Watson, Thomas A. 1926. *Exploring Life: The Autobiography of Thomas Watson*. New York: D. Appleton and Co.

Weiher, Sigfrid von, and Herbert Goetzeler. 1977. *The Siemens Company: Its Historical Role in the Progress of Electrical Engineering*. Trans. G. N. J. Beck. Berlin: Siemens Aktiengesellschaft.

Welch, Walter L. 1972. *Charles Batchelor: Edison's Chief Partner*. Syracuse, N.Y.: Syracuse University.

Williams, Charles, Jr. 1876. *Price List for 1876*. Boston: Spooner.

Williams' Cincinnati Directory. 1865. Cincinnati: Williams Directory Co.

Wilson, H., comp. (Printed annually.) *Trow's Business Directory of New York City*. New York: John F. Trow.

Credits

Reproduced with permission of the American Philosophical Society: Docs. 42, 46–47, 49. Reproduced with permission of the AT&T Corporate Archive: illustration on p. 68 (neg. 2751). Reproduced with permission of the Thomas A. Edison Papers, Rare Book and Manuscript Library, Columbia University: Docs. 130, 164, 273. Reproduced with permission of Dun & Bradstreet and the Baker Library, Harvard University: Docs. 252, 254, 260, 265–66. From the collections of the Henry Ford Museum and Greenfield Village: Docs. 3–5, 25, 43 (negs. B99238, B99239), 51–53, 56, 60, 76 (negs. B99252, B99253), 82, 87, 95, 97 (neg. 99172), 99–100, 129, 158 (negs. B99248, B99249, B99250), 178, 195 (negs. B99244, B99245, B99247), 249 (negs. B99232, B99233), 255 (negs. B99230, B99231), 289 (negs. B99234, B99235), 290 (neg. B99241); frontispiece (neg. B37330); illustrations on pp. 6 *bottom* (neg. B99173), 55 (neg. ECP-95), 85 (neg. B99259), 133 (neg. B99258), 154 *bottom* (neg. B101407), 161 (neg. B99170), 171 *right* (neg. B99243), 232 (neg. B99246), 346 (neg. B99251), 372 *bottom* (neg. B99256), 385 (neg. 13318), 463 *top* (neg. B99171), 584 (neg. B99236). From the collection of Charles Hummel: Doc. 199. Courtesy of Dr. Charles F. Kempf: illustration on p. 7. Courtesy of Newark Public Library: illustration on p. 363. Courtesy of the New-York Historical Society: Docs. 144, 148, 168, 185, 204, 261. Reproduced courtesy of Post Office Archives, London: Docs. 146, 318–19; illustration on p. 186 (ATF). Reproduced with permission of Dorothy C. Serrell: illustration on p. 196.

Courtesy of Edison National Historic Site (all designations are to *TAEM* reel:frame unless otherwise indicated): illustrations on pp. 5, 6 *top*, 27, 76 (12:14), 103, 126 *bottom* (Cat. 551:3), 130 (Cat. 551:5), 154 *top* (Cat. 551:2, 47), 209 (Cat. 551:8), 210 (Cat. 551:23), 211 (Cat. 551:50), 217 (27:169), 219 (27:201), 272 (27:169), 274 (5:399), 303, 347 *top* (96:572), 347 *bottom* (27:136), 369 (5:456, 591), 372 *top* (5:164), 373 (Cat. 551:19, 41), 379 (27:136), 402 (Cat. 551:15), 436 (27:168, 174), 437 (27:168), 462 (5:345), 473, 488 (96:570), 514 (27:136), 563 *left* (5:809), 592 (6:906), 600 (6:899).

Index

Boldface page numbers signify primary references or identifications; italic numbers indicate illustrations. Page numbers refer to headnote or document text unless the reference appears only in a footnote.

Accounts. *See* Finances

Adams, Joseph, **71**, 75; as witness, 72, 74, 75

Adams, Milton, 19, 51–52, **63**, *65*, 122, 128, 633–36, 651, 664; agreement with Welch, 89; article by, 63–65; printing telegraph with TAE, 89

Adrian, Mich., 15, 16, 658

Agreements, TAE's: with Anders, 118–19; with Andrews, 147, 151–53, 155–56, 210, 296; with Ashley, 163–69, 173n.2; with Automatic Telegraph Co., 209n.1; with Craig, 180–82, 197n.2, 473; with Craig (draft), 184–85; with Field, 147, 151–53, 155–56; with Gold and Stock Telegraph Co., 151–53, 155–56, 163–69, 173n.2, 208–9, 283–91, 314, 643, 653; with Harrington, 190–96, 209n.1, 226, 269–71, 515, 542n.4; with Hills, 105–6, 173; with Lefferts, 210, 296, 483–84; with Orton, 522n.1; with Plummer, 105–6, 173; with Pope, 163–69, 173n.2; with Reiff, 506–7, 515, 542n.4; with Ropes, 105–6, 211; with Unger, 494–95; with Welch, 118–19

Albert E. Kent & Co., 296–97

Allen, William, 176, **216n.2**, 217n.5

American compound telegraph wire, 66–67, **86–89**; for Automatic Telegraph Co., 248; for

Financial and Commercial Telegraph Co., 142; for National Telegraph Co., 184n.3; price of, 248

American Compound Telegraph Wire Co., 66–67, **86**, 121

American District Telegraph Co., 379, **411n.2**, 519n.2; Andrews in, 153n.2; and TAE's district telegraph system, 427n.1

American Institute fair, 216n.2; TAE's ticket to, *217*

American printing telegraph, **186–88**, *188*, 292; at American Institute fair, 216n.2; and cotton instrument, 233; polarized relay in, 204n.2; Pope's relay switch in, 171n.11

American Printing Telegraph Co., 162n.5, 170n.4, 187, 210n.5, **212**; agreement with Pope, Edison & Co., 176–77, 212–15; dissolution of, 225–26; TAE and, 177, 219; founding of, 176; Gold and Stock Telegraph Co. buys, 187, 225; Lefferts in, 170n.6; stock assignments, 215

American Telegraph Co., 12, 89n.3; Calahan in, 127n.4; combination printer and, 62n.1; Grace in, 254n.1; Lefferts in, 170n.6; Phelps in, 135n.2

American Telegraph Works, 156n.1, 178, **190–95**, 224; advertising card, *272*; and Automatic Telegraph Co., 379; Batchelor in, 495n.9; book-

keeping, 193–94, 196, 308; Clark at, 504n.1; closes, 515n.5; Craig and, 195n.5; Craig mentions, 246; and Edison and Murray, 515; and Edison and Unger, 473n.7, 497n.6; Hyde at, 257n.2; insurance, 220; inventory, 278–79; leasing shop for, 189; location of, *148*; ownership of, 195n.5, 227, 277–78; payroll, 308; staff, 212, 220; work for Stearns and Mendell, 493n.5

—manufacturing, 218–19; cotton instrument, 218n.2, 220, 233n.2, 272n.2; large perforator, 265n.2, 301–2, 340n.2; paper-wetter, 266n.6; three-key perforator, 265n.3

Ammonium nitrate, 264n.2

Anders, George, **104**, 122, 128, 132n.1, 664; agreement with TAE and Welch, 118–19; and Boston instrument, 109; and Financial and Commercial instrument, 133n.1; and magnetograph, 82n.2, 104; and polarized relay, 110n.9

Anderson, Frank, 253

Andrews, Elisha, **151**, 460; in American District Telegraph Co., 379; British patents of, 177, 225, 273–74, 296n.1; Calahan mentions, 490; and TAE's district telegraph system, 427n.1; letter to TAE, 299; Murray mentions, 620; and

60; Lefferts and, 175n.3, 223nn.2,5; motors, 218, 220, 268; and National Telegraph Co., 177; paper feed, 303; perforator, 178n.9, 268, 358, 360, 362n.1; receiver, 360; speed of, 150n.8, 185–86, 358; tablet, 268; tested by Hicks, 258n.9; transmitter, 360; and Western Union, 186n.2

—Wheatstone's system, 591–92; perforator, 178n.9, 250, 266n.4, 281; speed of, 150n.8, 250, 592n.3, 594

Automatic writer. *See* Automatic telegraphy, TAE's system, receivers (printing)

"Baby" printing telegraph, 463–64, *464*

Bain, Alexander, 303; automatic system, **63**, *65*, 170n.6, 223n.5, 592, 595, 597–98; facsimile telegraph, 83, **90**; Serrell and, 196n.2

Baker, H. N., 127n.6

Bakewell, F. C., facsimile telegraph, 83, **90**

Baldwin, A. J., 247

Bankers' and Brokers' Telegraph Co., **103**, 107, 121, 254n.1

Bank printer, 210n.3

Barnitz, David, **504**; as witness, 538, 545

Bartlett, 499

Barton, Enos, 402n.5

Batchelor, Charles, 293n.1, **495**, 621, 664; account with Edison and Murray, 500, 517; as witness, 495

Batteries, 10, 67; Bunsen, **599n.3**, 619n.6; carbon, 617; copper sulfate in, 113n.3; cups, **113n.5**, 117n.3; current of, TAE's query on, 139; Daniell, 10, *11*, 117n.3, 238, **615**; TAE's, 198–99, 481; in TAE's automatic system, 596, 617–18, 623; in TAE's duplex, 508, 510–11, 579–81; electrodes, 314n.2, 612; Farmer's thermo-electric, 67; Grove, 10, *11*, 596, **615**; gutta-percha, 606, 617; inductive, 613; intensity, **91**, 617–18; Leclanché, **98n.5**, 617; Leclanché, process for manufacturing, 198; manganese, 97;

Menotti, 607n.4; quantity, **96n.6**, 608, 613, 618; secondary, 615; for sewing machine, 238; trough, 199n.3

—local: in TAE's automatic system, 412–14, 417, 426–27; in TAE's duplex, 569n.2; with Financial and Commercial instrument, 133; with gold printer, 161; with polarized relay, 461; with Pope's relay switch, 171n.11; with printing telegraphs, 157–58; with universal private-line printer, 446, 489n.5

Belden, William, 651, 665

Bell, Alexander Graham, 67n.7, 253n.2

Bennett, James Gordon, 255

Bentley, Henry, 619

Bergmann, Sigmund, **578**, 620, 645, 666

Bertram, Harry, 504

Birmingham gauge, 599nn.11,12

Black Friday, 651, 664, 665

Blavier, E. E., *Nouveau traité de télégraphie électrique*, 540

Blockley, Frank, as witness, 573

Blowhard, 25n.4

Blue vitriol, 113n.3

Bombaugh, Charles, 34

Bonelli, Gaetano, facsimile telegraph, 83, **90**, *95*

Book, TAE's, 514, 546; topic list, 539–40

Bookkeeping, 644; at American Telegraph Works, 193–94, 196, 308

Boorum, P. J., 364n.2

Borst, Jerry, 640, 651

Boston, Mass., 19, 51–53, 56, 102–4, 595n.5, 633–40, 651, 666; TAE honeymoons in, 385n.2; electrical instrument manufacture in, 77–81; stock-quotation service, 103, 107, 111–12, 639

Boston Commercial Telegraph, 67

Boston instrument, 70n.2, 102–3, **108–10**, *110*; agreement concerning, 105–6; and American printing telegraph, 187; Anders and, 105n.10; and cotton instrument, 232; and Financial and Commercial instrument, 132; and gold printer, 161; patent application, 107–8; patent assignment, 105–7, 146, 158–

60; payment for, 139–40; polarized relay in, 187n.4; transmitters, 115–16. *See also* Patents (U.S.), TAE's, No. 91,527

Botts, John Minor, 657–58

Boyden, G., 190n.4

Bradford, H., 498

Bradley, Capt. Alva, 19n.3

Bradley, Leverett, **141**, 641–42; account with Edison and Murray, 500; perforator, 182n.3

Brazil, 17, 29, 657

Break, 10; in duplex, 101n.7

Breakwheels, **133**, *133*; in TAE's district telegraph system, 405–6, 408, 410, 415–21, 424–25; with Financial and Commercial instrument, 161; insulation on, 453; motor-driven, 368n.2; shortcomings of, 368n.2; Unger's, 404n.1, 443; universal private-line printer, 342, 368, 452, 453, 466–67, 469–70; with universal stock printer, *371. See also* Universal private-line printer, pulsator

Bright, Edward, 33

Britain: automatic telegraphy in, 250; TAE in, 515, 591–92; nationalization of telegraph system in, 34n.2, 252n.1, 514; stock-quotation service in (*see* Exchange Telegraph Co.). *See also* British Post Office

British Post Office, 591–92; central office address of, 599n.9; Culley in, 34n.2; and TAE's automatic system, 513–15, 591–98; Scudamore in, 252n.1; and Wheatstone automatic system, 178n.9, 591–92

Bronk, Toney, 28

Brooks, David, 142; insulators, **142**, 248

Brown, James, **508**, 510–12

Brown, James (Mrs. Craig's agent), 486

Brown and Sharp gauge, 67n.3

Bryan, 247

Buchanan, Albert, 27

Buchanan, Alexander, 27n.4

Buchanan, Carrie, **6**, 27n.4, 212

Buchanan, Fred, 27

Buell, Madison, 139

Bulkley, Charles, 42n.1

Bunnell, Jesse, 652

Burlingame, Anson, 83, 84n.2

Burns, Patrick, 54n.8, 75n.2
Burton, Thomas, *Anatomy of Melancholy*, 7

Cable telegraphy: artificial line, 616n.2; batteries, 607n.4; condensers on, 549n.1; to Cuba, 121; joints, 607; polarized relays on, 38n.1; recording apparatus, 334; transatlantic, 182n.1, 334; transmission speed of, 604; Varley's system, 616
—TAE's system, 549–50, **586–90**; experiments, 591–92, 604–17; transmission speed of, 604
Calahan, Edward, **126**; in American District Telegraph Co., 379; bank printer, 210n.3; district telegraph system, 379, 411n.2; TAE, relationship with, 490n.1; and TAE's printing telegraphs, 299; in Gold and Stock Telegraph Co., 126, 135, 297n.7; Griffin mentions, 297, 298; lawyer's instrument, 210n.3; letter to unknown person, 490; patent interference vs. TAE, 519; on universal stock printer, 620
—stock printer, 104n.4, 108, 125–26, **153n.4**, *154*, 157, 232, 299n.3; in TAE–Gold and Stock Telegraph Co. agreement, 151; TAE's unisons for, 388–90, 470; and financial instrument, 225; patent reissue, 130
Canada. *See* Montreal, Que.; Sarnia, Ont.; Stratford Junction, Ont.; Toronto, Ont.; Vienna, Ont.
Candee, A. B., as witness, 271
Capacitance, in tailing, 149, 505n.6. *See also* Static discharge
Capacitors. *See* Condensers
Carbon, electrode polarity of, 314n.2. *See also* Graphite
Carter, Mr., 631
Carter, Robert, 198
Caselli, Giovanni, facsimile telegraph, 82n.5, 83, **90**, *95*
Caveats, **675**; application form, 74n.1
—TAE's, 177, 224; district telegraphy, 379, 417–27; electromagnet, 210; fire alarm, 71–74;

in Gold and Stock agreement, 288–91; multiple telegraphy, 552–53; perforators, 263n.2; printing telegraphy, 221–22, 224, 288–91; for railroad, 498–99; typewriters, 355; universal stock printer lock, 381–83
Cells. *See* Batteries
Center screw, 452n.2
Chamberlain, Henry, 82n.9
Chamberlain, Walter, 82n.9
Channing, William, 68n.8
Charles T. and John N. Chester, 122n.9
Charleston, S.C., 264, 492, 503–4, 617, 624–25
Charles Williams, Jr., 51, **67**, *68*, 81, 83, 123, 638–39, 651; bills from, 117; and Boston instrument, 109; magnetographs, 127; receipts from, 113, 114; Redding in, 128n.1
Chauvassaignes, Paul, automatic system, **63–65**, 90
Chemical printer. *See* Typewriter, chemical
Chemical telegraphy. *See* Automatic telegraphy
Chemistry: laboratory, 8, 25n.3, 670; text, 20n.16
Chester, Charles, 122n.9; dial telegraph, 121
Chester, Steven, 122n.10
Chester, Partrick & Co., **121**, 137n.3
Chicago, Ill.: private lines in, 178n.2; stock-quotation service, 211n.3
Chicago instrument, 178n.5, 211n.3; and American printing telegraph, 187; and cotton instrument, 233; in Gold and Stock agreement, 287; polarized relay in, 161n.4, 204n.2; relay in, 288
Chile, universal private-line printer to, 488
Chinese, transmission of, 83
Cincinnati, Ohio, 16, 18, 43n.1, 48n.2, 58n.3, 59n.1, 65n.1, 97n.1, 297n.1, 635–36, 650–51, 658–60, 662–63, 672, 673
Cincinnati Free Library, 29, 33
Cincinnati Industrial Exposition, 488
Civil War, 10–12, 31
Clancy, James, 652

Clark, Horace, 255
Clark, J. Latimer, 34n.5
Clark, James, 45n.1, 46n.1
Clark, S. M., 503; letter from Johnson, 503–4
Clockwork, in facsimile telegraphs, 90
Coburn, A. D., as witness, 271
Cockroach-killer, 637–38, 651
Code, telegraph, 10, 260n.4
Coding machine, 457
Collins, J. B., 122n.3
Combination printer. *See* Phelps, George, combination printer
Combination repeater, 61–62
Commercial News Department (Western Union), 147, 170n.6, 211n.3, 298n.12
Commissioner of Patents (U.S.), letter from Serrell, 497–98
Commodity exchanges, 217n.1
Compound wire. *See* American compound telegraph wire
Compressed air: in TAE's automatic system, 345, 350–51; in TAE's perforator, 265
Condensers, **520**; adjustable, 550n.3; in TAE's automatic system, 587, 612, 614; in TAE's cable system, 549; in electric motors, 613; in Stearns's duplex, 101n.3, 507, *521*
Conductivity, electrical, TAE's table of, 35
Continental Sugar Refinery, **120n.2**; instruments for, 128; testimonial from, 120
Contracts. *See* Agreements; Patent assignments
Cook, Joseph, 216n.2
Cooper, Gage, 306n.4
Cooper, Mrs. M., 306
Copper sulfate, 113n.3, 481
Copying printer. *See* Typewriters
Corning, Erastus, **240**, 241, 647
Corning, Erastus, Jr., 243n.2, 280n.7
Cotton instrument, 177, **232–34**, *234*, 383n.7; abandonment of, 301; alterations in, 299n.1; British patent, 225; and Calahan's printer, 153n.4; manufacture of, 218, 218n.2, 220, 233n.2, 254–55, 272; new switches for, 299; order for, 217; parts of, 348n.6; as screw-

thread-unison patent model, 276n.2; shifting mechanism of, *232, 273*

Craig, Daniel, 177, **180**, *182*, 646, 647, 652, 664, 665; and American Telegraph Works, 195n.5, 280n.6; on arrangement of perforator holes, 243–44; and Associated Press, 182n.1, 249n.5, 257; in Automatic Telegraph Co., 242n.1, 591; and automatic telegraphy in Britain, 514, 591; TAE mentions, 220; and TAE's district telegraph system, 246, 356–57, 359, 427n.1; and TAE's letter of introduction to Farmer, 183–84; Harrington and, 591; Hoadley and, 249n.6; on ink recorder, 302; interest in TAE's inventions, 280n.6; interest in Little's system, 223n.2; Jones mentions, 283; on labor for automatic system, 178n.9, 239, 247, 251–52, 256–57, 360–61; Lefferts mentions, 222; on Little and TAE, 303, 359–60; patent assignments to, 181, 184–85; relationship with TAE, 226–27; and typewriters, 182–83, 227, 244; and Western Union, 252, 255
—agreements: with TAE, 180–82, 197n.2, 473; with TAE (draft), 184–85; with Little, 240, 356n.1; with National Telegraph Co., 240, 356n.1
—letters: from TAE, 218–19; to TAE, 182–83, 185–86, 235, 239–40, 243–44, 245–46, 249–50, 253, 259, 268–69, 281–82, 295–96, 356–60; to Lefferts, 247–49, 254–57, 326–28, 360–62, 485–87; to Scudamore, 250–52

Craig, Mrs. Daniel, 248, 254, 486

Crawford, Alexander, 24

Credit reports. *See* R. G. Dun & Co., credit reports

Crosby, Halstead, and Gould, 6on.1

Culley, Richard, **34**, 591; *Handbook of Practical Telegraphy*, 34; letter from Lumsden, 595–98; perforator, 266n.4

Cummings, D. D., 123

Curran, J. C., 29

Darrow, W. E., 295

Dashes, long and short, 260n.4

Davenport, William, 293n.3

Davidson, Robert, 293n.3

Davis, Daniel, 83n.15

Dead motion, 401

Dean, Charles, **468**, 517, 665

Delany, Patrick, **617**, 664

Detroit, Mich., 7

Detroit Young Men's Society, 7; TAE's membership card, *6*

Dial telegraphs, **78–79**; Chester's, 121; TAE's, **179**; Hamblet's, 78–79, 119n.2; Kramer's, 200; Siemens and Halske's, 158n.3; Wheatstone's, 82nn.3,4. *See also* Magnetograph

Diplex telegraphy, **32n.6**, 533n.6; with duplex, 531n.18; Kramer's, 539. *See also* Double transmission; Duplex telegraphy; Multiple telegraphy
—TAE's, 513, 519, 550, 557n.1, 577; circuit designs, 143–45; conception, 584n.2; in 1869 multiple telegraph, 101n.6

Distortion. *See* Tailing

District telegraphy, **411n.2**; Calahan's, 379, 411n.2, 519; Craig and, 246
—TAE's, 415–17; Craig and, 356–57, 359, 427n.1; Morten and, 280n.7; patent interference vs. Calahan, 519; transmitters, 404–10

D. N. Skillings & Co., **120**; instruments for, 127–29, 135, 150; letter to TAE, 120

Domestic Telegraph Co., 411n.2, 622n.16

Domestic telegraphy. *See* District telegraphy

Double transmission, **31**. *See also* Diplex telegraphy; Duplex telegraphy; Multiple telegraphy
—TAE's, 52, 53, **56–58**, 102, 104, 135; advertisement, *98*; Laws and, 132; on New York–Boston line, 143; patent assignment to Welch, 116, 137, 140; in *Telegrapher*, 50n.3, 98, 118; test of, 116, 121, 640, 651, 664; Welch and, 71n.1; Welch's interest in, 70–71, 100n.2, 104, 116

Downer Kerosene Oil Co., 120n.1, **128**; instruments for, 128

Dub, Julius, *Die Anwendung de Elektromagnetismus*, 200, 540

Dujardin, Pierre, 233n.5

Du Moncel, Theodose, *Traité theorique et practique de télégraphie électrique*, 540

Duncan, Sherman & Co., 490

Dunnells, 297

Duplex telegraphy, 31, **32n.6**, 522; break in, 101n.7; bridge, 33n.9, *530*, 530n.17, 569n.2; compensation, 48n.1; differential, 32, 48n.1; in TAE's book, 539–40; Frischen, Siemens, and Halske's, 32n.7, 101n.3, 539, 578n.4. *See also* Diplex telegraphy; Double transmission; Multiple telegraphy
—TAE's, 493, 540–41; artificial line, 554; caveat, 417–27; circuit continuity, 573–76; circuit designs, 545–47; condensers in, 520; differential, 573n.2; early designs for, 48–49; equipment for, 545; experiments, 17, 508–12, 522n.1, 522–29, 531–32, 534, 538, 548, 555, 556; induction coils, 511, 526, 528, 546–48, 564–67, 570–73, 576–77; inductive compensation, 520–21, 525–28, 564–67, 570–76; magnet boxes, 608n.2; "no balance," 519; patent applications, 514, 522n.1, 529n.1, 554, 556–62, 564–76, 579–84; polarized relays, 507–10, 526, 528, 571–72, 582–84; relays, 37n.1; reverse currents, 507–11, 528, 541, 550, 582–84; rheostats, 562–64, 568–69, 571, 574–75, 579–80; in secret signaling system, 99n.2; shunt circuit and equating battery, 579–81; static discharge, 564–67, 570–76; and Western Union, 522n.1, 534, 555, 556
—Stearns's, 32n.7, 48n.2, **99**, *101, 521*, 539; vs. TAE's, 99–100; test of, 104; and Western Union, 493, 507

Dyer, Frank, 627

Dyer, Joseph, 121

Eagan, James, 293n.1

Eames, J. W., 595; letter to Fischer, 593–95

Little, George (*continued*)
356n.1; with Harrington, 354–
56; with National Telegraph
Co., 240, 356n.1
Little distributor, 272
Livermore, Isaac, 83
Livingston, Montgomery, **157**,
196n.2; letter from Pope, 157–
58
Local circuit, 30. *See also* Batteries, local
Lockwood, George, 628n.1
London Stock Exchange, 296n.1,
642. *See also* Exchange Telegraph Co.
Louisville, Ky., 17, 649–50, 653–
58, 673
Ludwig, Edwin, 297
Lumsden, David, 598; letter to
Culley, 595–98

Machine shop, 189; equipment,
515–16; expenses, 111; power,
273, 518n.2
Machine Telegraph Co., 254n.2
MacKenzie, James, 8, 631
McManus, John, 277; agreement
with Harrington et al., 277–79
Magnet boxes, **608**, 616, 625
Magnetic saturation, 76–77
Magnetoelectric generator. *See*
Magnetos
Magnetograph, 82n.3, 102–4,
114n.8, **119n.2**, 120n.1, 123,
128, 639; advertisement, *119*;
agreement concerning, 118–19;
demand for, 121–22; lines for,
124n.1; patent assignment to
Anders and Welch, 104, 118;
price of, 122; on private lines,
103, 176; Welch and, 71n.1
Magnetos, 119n.2, **336**; in Edmands and Hamblet's dial telegraph, 78
Manhattan Quotation Telegraph
Co., 619, 621, 622n.16
Manufacturing: automatic repeater, 267n.8; cotton instrument,
218, 218n.2, 220, 232, 272; experimental duplex instruments,
519; Financial and Commercial
instrument, 141; ink recorder,
267n.9; Laws's gold indicator,
122n.9; Leclanché battery, 198;
Little's system, 268–69; magnetograph, 119n.2; process,
518n.16; typewheels, 197–98,

228; typewriter, 218; universal
private-line printer, 209, 218,
379, 492, 514; universal stock
printer, 233n.3, 272n.2, 301,
346, 347n.2, 348n.6, 379, 492,
514. *See also* American Telegraph Works; Edison and Murray; Edison and Unger; Murray
and Co.; National Telegraph
Works
—perforators, 218–20, 260, 269;
large, 265n.2, 294–95, 301–2;
three-key, 265n.3
Marsh, J. J., 280n.7
Martin, Thomas, 627–28; questions for TAE, 663–65
Mechanical printer. *See* Typewriter
Medill, Joseph, 256
Mellen, William, 242n.1, **277**;
agreement with Harrington et
al., 277–79
Memphis, Tenn., 16–17, 650,
653, 661–62, 672
Mendell, W. H., 493n.5
Menlo Park, N.J., 97n.1, 297n.1,
663
Mercantile Agency. *See* R. G.
Dun & Co.
Mercury, in governors, 415–16,
423–24, 460
Metals, relative cost of, 154n.5
Microphone, 109n.5
Milan, Ohio, 3–5, *4*, 667–68;
Samuel Edison in, 649; telegraph in, 9
Miller, John, **306**, 307
Miller, Norman, **513**, 514; TAE's
power of attorney, 585; Murray
mentions, 620; pays Edison and
Murray's bill, 499n.1
—letters: from TAE, 519, 534,
538, 545, 555; from Orton, 521
Milliken, George, 51, **86**, 636,
651; co-inventor of compound
telegraph wire, 68n.8, 86; hires
TAE, 89n.3
Montreal, Que., 636
Morris, Francis, 486
Morse, Samuel, 9, 252; Baltimore-Washington telegraph
line, 83n.15
Morse code, 10, 260n.4
Morse register. *See* Registers
Morse system, compared to automatic system, 252, 256–57,
326–27, 361

Morten, Alex, 280n.7
Mosher, T. B., as witness, 560,
562, 580, 584
Motors, gas, 360
—electric: early applications of,
293n.3; TAE's, 479–81, 613; in
TAE's automatic system,
268n.12; in TAE's perforators,
260–62; in TAE's printing telegraphy, 368n.2, 371, 374,
431–33, 471–72; in TAE's
roman-letter automatic system,
318; in TAE's typewriter, 434–
38; Froment's, 82n.5; Page's,
91, *96*; Paine's, 229; in printing
telegraphy, 293n.3; in universal
private-line printer, 292–93,
401, 403, 451, 453, 455–56,
468n.6, 470. *See also* Sewing
machine
Mt. Clemens, Mich., 8, 630–31
Multiple telegraphy, 31–32. *See
also* Diplex telegraphy; Double
transmission; Duplex telegraphy; Quadruplex telegraphy
—TAE's: caveat, 552–53; circuit
designs, 143–45, 507–12, 545–
47, 550–51, 576–77; vs.
Stearns's, 99–100
Munn & Co., 529n.1, **554**, 556,
585, 676; TAE mentions, 555;
letter from TAE, 554
Murray, Joseph, **282**, 302, 493,
664; account with Edison and
Murray, 517; letter to TAE,
619–21; partnership with TAE,
378; as witness, 313, 316, 326,
332, 337, 339, 342, 345, 352,
367–70, 374–75, 383, 385,
403, 415, 429, 431, 433, 436,
442–43, 445, 447–48
Murray and Co., **380n.1**, 493;
credit reports on, 469, 484–85,
496; location of, *148*

Nationalization: of American telegraphy, 237n.1, 358; of British telegraphy, 34n.2, 252n.1,
514
National Telegraph Co., 182n.2,
249n.3; address of, 184n.4;
Craig and, 177; and TAE-Craig
agreement, 181; Grace in,
254n.1; and Little's automatic
system, 186n.2, 223n.2; New
York–Washington line, 184n.3
—agreements: with Automatic

speed of, 182n.3, 360; Wheatstone's, 178n.9, 250, 266n.4, 281
—TAE's, 189, 226–27, 228, 282, 477; agreement with Craig concerning, 180–82; arrangement of holes in, 235, 243–44, 259, *379*, 474, *474*, 612, 626; British patent for, 537–38; caveats for, 263n.2; clockwork in, 263; compressed air, 265; cost of, 251; Craig's ideas for, 239; doubled paper, 259; electric, 260–62, 534–35; European patents for, 250; folded paper, 264n.9; hand, 265; keys, 441; large, 258, 265, 269n.4, 281, 294, 301–2, 339–40, 379, 431, 534; levers, 390–94; linkages, 394; manufacture of, 218–20, 260, 269; with movable die, 263, 263n.4; operators for, 251; paper feed, 319–20, 322–24, 339–40, 392–94, 441, 476–78; punches, 197, 230–31, 329–30, 393, 449, 478; punching mechanisms, 260–63, 440–41, 449–50; revolving knife, 230–31, 262; rock shaft, 339–40, 441–42; roman-letter, 317, 328–32; sample tape from, *379*; six-key, 253n.5, 294, 379, 473–75; small, 295; speed of, 220, 251, 256, 327; test of, 359–60; three-key, 253n.5, 258, 259, 265n.3, 281, 474n.3; used remotely, 263n.4; and Wheatstone's, 281
Perpetual motion, 613
Perry, Oliver, 648–49
Phelps, George, **135**, *135*, 225, 255; combination printer, **61**, *63*, 108, 109n.5, 157, 293n.3; financial instrument, **209n.2**, *209*, 211n.3, 225, 292, 342n.1, 401; patent infringement vs. TAE, 187n.7; printing telegraphs, 275n.5, 293n.3, 402n.4; and screw-thread unison, 275n.5. *See also* Patents (U.S.), other inventors', Phelps
Philadelphia, Pa., stock-quotation service, 103
Philadelphia Centennial Exposition, TAE's automatic telegraph at, 359n.1
Philosophical Magazine, 81

Phonograph, 659; and practice instrument, 36n.1
Pinckney, George, as witness, 175, 222, 427, 590
Pistons, as governors, 290, 405–6, 415–16, 419–20, 460
Pittsburgh, Fort Wayne, and Chicago Railroad, 15, 658
Platinum, 661; in automatic telegraph pens, 208, 264n.2, 616
Plummer, William, 103, **105**, 146; agreement with TAE, Ropes, and Hills, 105–6; draft agreement with Ashley and Hills, 159–60; release from agreement with TAE, 173
—letters: from TAE, 107, 139–40, 158–59; from Hills, 107
Polarized relays, 30, **38n.1**; in Edmands and Hamblet's dial telegraph, 78; with induction coil, 59; in Pope's U.S. Pat. 103,077, 162n.5; weaknesses of, 108–9
—TAE's: in American printing telegraph, 187, 204n.2; in automatic repeater, 185n.2; in automatic telegraphy, 265; in "baby" printer, 463; in Boston instrument, 108–9, 187n.4; in Chicago instrument, 161n.4, 204n.2; in combination repeater, 61–62; in cotton instrument, 232, 299n.1; double-tongued, 202–3; in double transmitter, 100; in duplex, 507–10, 526, 528, 540–41, 571–72, 582–84; early, 38, 40; in facsimile telegraphs, 91, 94; in Financial and Commercial instrument, 132; in gold printer, 161; with local battery, 461; in multiple telegraphy, 551; in printing telegraphy circuits, 69, 143–45, 204, 221–22, 287–89; residual magnetism in, 459; in universal stock printer lock, 382
Polishing machine, 218–19
Pope, Electra, 172; financial statement from, 172
Pope, Franklin, **114**, *129*, 210n.5, 641; affidavit to Hanaford, 140; agreement with TAE, Ashley, and Gold and Stock Telegraph Co., 163–69; American printing telegraph, 186–88; and American Printing Telegraph Co., 176, 215, 225–26; and double

transmitter, 104, 121, 664; as electrical engineer, 138; expertise of, 137n.3; Financial and Commercial instrument, 132–34; financial statement from, 172; gold printer, 160–62; in Gold and Stock Reporting Telegraph Co., 129; inventions with TAE (*see* American printing telegraph; Financial and Commercial instrument; Gold printer); and Laws's gold indicator, 125; Lefferts and, 170n.6; letter to Livingston, 157–58; loans to TAE, 172n.2; *Modern Practice of the Electric Telegraph*, 114–15, 137n.3; Nonpareil telegraph apparatus, 228n.9; patent assignments to Gold and Stock Telegraph Co., 163–69; in Pope, Edison & Co., 137n.2, 212–15; relationship with TAE, 125–26, 146, 226; on repeaters, 32n.4; as *Telegrapher* editor, 50n.3
Pope, Henry, 172n.1
Pope, Ralph, 172n.1, 664
Pope, Edison & Co., 126, **136–37**, 146–47, 641–42, 652; dissolution of, 177; as electrical engineering firm, 138; and Financial and Commercial Telegraph Co., 142; founding of, 129n.1
—agreements: with American Printing Telegraph Co., 176–77, 212–15; with Gold and Stock Telegraph Co., 159n.2, 163–69, 210n.5, 213–14
Pope and Edison printer. *See* American printing telegraph
Port Huron, Mich., 7, 15, 17, 18–19, 628–32, 650–51, 652, 669–70; Edison family in, 5–6, 18, 668–69, 673; Samuel Edison in, 308–9, 628
Port Huron and Gratiot Street Railway Co., 304–5, 307, 497n.3, 673
Port Huron and Lake Michigan Railroad, 307
Port Huron Union School, 23–24
Potassium ferrocyanide, 264n.2
Potassium iodide, 264n.2, 589
Powers of attorney, TAE's: to Harrington, 270–71, 537–38; to Lefferts, 483–84; to Miller, 514, 585